Contemporary Biotechnology and Bioengineering

He Xiaoxian
He Po
Ding Yong

 SCIENCE PRESS
Beijing

 Alpha Science International Ltd.
Oxford, U.K.

Contemporary Biotechnology and Bioengineering
392 pgs. | 86 figs. | 13 tbls.

Copyright © 2014, Science Press and Alpha Science International Ltd.

Authors
He Xiaoxian
He Po
Ding Yong

Responsible Editors
Xi Hui
Yang Fang

Co-Published by:

Science Press
16 Donghuangchenggen North Street
Beijing 100717, China

and

Alpha Science International Ltd.
7200 The Quorum, Oxford Business Park North
Garsington Road, Oxford OX4 2JZ, U.K.

www.alphasci.com

ISBN 978-7-03-037043-3 (Science Press, Beijing)

All rights reserved. No part of this publication may be reproduced, stored in a retrieval system, or transmitted in any form or by any means, electronic, mechanical, photocopying, recording or otherwise, without prior written permission of the publisher.

Printed in China

Translation Committee

Directors:

He Xiaoxian
Shaanxi University of Science and Technology
He Po
Shaanxi University of Science and Technology

Associate Director:

Ding Yong
Shaanxi University of Science and Technology

Members:

Gong Guoli
Shaanxi University of Science and Technology

Gao Xiuzhi
Beijing University of Agriculture

Gong pin
Shaanxi University of Science and Technology

Liu Huan
Shaanxi University of Science and Technology

Liu Changmeng
Shaanxi University of Science and Technology

Chang Xiangna
Shaanxi University of Science and Technology

Preface

Bioengineering technologies are the core of the 21st century high-tech revolution, and the biotechnology industry is a pillar industry in the 21st century. Currently, bioengineering technologies play a strong role in solving the major problems of food, health, environment, resources, population and energy, which human beings now have to be faced with. Biotechnologies are widely used in medicine and health, agriculture, light industry, chemical and energy, and so on, and have promoted modification of the traditional industry technologies and formation of the emerging industries. Bioengineering technologies have permeated almost all disciplines and are closely related with daily life of people, economic and social development. There will be more outstanding talents of various disciplines to participate in research and development in the edge fields which cross with life sciences in the 21st century.

In order to strengthen and improve students' consciousness of high-tech, train high-quality talents, making college students in higher level consciousness of bioengineering technologies for their future job and the meaning of life, understand the research situation in bioengineering technologies and the development trend of life sciences in the world, the authors of this book extensively collected and borrowed domestic and foreign advantages of similar materials and technological literature in recent years, and finally this book, "Contemporary Biotechnology and Bioengineering", as a teaching textbook for non-bioengineering undergraduates in higher education institutions, was written and now has completed. By reading and learning from this book, most students have a basic understanding of trends and hot spots on life sciences in the 21st century and the knowledge of the major disciplinary development direction of bioengineering technologies and the areas to be covered in the future, so that during their major study students can find the integration point between bioengineering technologies and their own specialty, and become proficient in application of bioengineering technologies to their own specialty, so as to inspire their thoughts to pioneer and innovate, and lay the foundation for their future professional development. We hope that you will find this book a useful revision aid and a stimulus for further study.

This book was translated from Chinese version "Contemporary Biotechnology and Bioengineering" published in 2005 by Science Press, Beijing, China. The contents of the book include 13 chapters, covering not only basic knowledge and basic theories of both contemporary bioengineering technologies, but also their application in various fields and the impact on human society. In the case of constrained class number, teaching content should be concise rather than exhaustive, and should give students the space for studying themselves. For this book, Chapter 1: Introduction; Chapter 2: Science basis for contemporary bioengineering; Chapter 3: Gene and genome; Chapter 4: Gene engineering; Chapter 5: Cells and cell engineering; Chapter 6: Enzyme and enzyme engineering; Chapter 7: Microorganisms and fermentation engineering; Chapter 8: Contemporary bioengineering technologies and agriculture and light chemical industry; Chapter 9: Contemporary bioengineering technologies and biomaterials; Chapter 10: Contemporary bioengineering and biological medicine; Chapter 11: Infection and immunity; Chapter 12: Biodiversity and environmental management; Chapter 13: Safety and social ethics of contemporary bioengineering technologies.

Because of fast development of contemporary bioengineering technologies, and involvement in a wider range of knowledge, and owing to limitation of the author's knowledge level and writing skill, the errors in this book are inevitable, I would like to welcome valuable advice from all readers, and all the compiling staff will be greatly appreciated.

HE Xiaoxian
In Xi'an
June 2013

Acknowledgments

We would like to thank our families, colleagues and friends for their support and help we were writing this book. Thanks are also due to the staff at Science Press, Beijing, China for their encouragement and assistance during this time. Finally, we would like to thank Alpha Science International Ltd. Oxford, U.K. for their support and help English version, "Contemporary Biotechnology and Bioengineering" were successfully published.

We would like to dedicate this book to our parents.

Contents

Preface
Chapter 1　Introduction..1
　1.1　Life Sciences: the Leading Discipline in the 21st Century...1
　1.2　The Contents and Characteristics of Contemporary Bioengineering.........................9
　1.3　Development and Social Impact of Contemporary Biotechnologies.......................13
　1.4　Contemporary Biotechnologies and Sustainable Development..............................16
Chapter 2　Science Basis for Contemporary Bioengineering...21
　2.1　Structure and Function of Proteins and Enzymes..21
　2.2　Structure and Function of Nucleic Acids...27
　2.3　Structure and Function of Polysaccharides..36
Chapter 3　Gene and Genome..39
　3.1　Concept of Gene and Its Essence...39
　3.2　Gene Isolation and Synthesis...44
　3.3　Gene Expression and Regulation...46
　3.4　Gene Mutation..58
　3.5　Heredity and Human Diseases...64
　3.6　Genome...69
Chapter 4　Gene Engineering...78
　4.1　Tool Enzymes for Gene Engineering...81
　4.2　Gene Engineering Vectors..88
　4.3　Target Genes...95
　4.4　Ligation of a Target Gene and Vector DNA..101
　4.5　Introduction of Recombinant DNA into Receptor Cells.......................................103
　4.6　Screening of Recombinants...108
　4.7　Achievements and Application of Contemporary Gene Engineering..................112
Chapter 5　Cells and Cell Engineering...116
　5.1　Basic Concept and Technology of Cell...116
　5.2　Culture Characteristics and Nutritional Requirements of Animal and Plant Cells.............122
　5.3　Plant Cell Engineering...126
　5.4　Animal Cell Engineering...132
　5.5　Chromosome Engineering...136
　5.6　Research of Stem Cells..138
　5.7　Embryo Engineering and Animal Cloning Technology..142
　5.8　Research Progress and Future Prospects of Cell Engineering..............................150

Chapter 6 Enzymes and Enzyme Engineering 153
- 6.1 Overview of Enzymes and Enzyme Engineering 153
- 6.2 Fermentation Production of Enzymes 162
- 6.3 Production of Enzymes with Immobilized Cell Fermentation 171
- 6.4 Application and Molecular Modification of Enzymes 175
- 6.5 Research Progress in Enzyme Engineering 178
- 6.6 Protein Engineering 181

Chapter 7 Microorganism and Fermentation Engineering 187
- 7.1 Introduction to Fermentation and Fermentation Engineering 187
- 7.2 Elementary Knowledge on Microorganisms 195
- 7.3 Fermentation Process Control 204
- 7.4 Fermentation Equipments 212
- 7.5 Downstream Processing Technology 218

Chapter 8 Contemporary Bioengineering Technologies, Agriculture and Light Chemical Industry 232
- 8.1 Contemporary Bioengineering Technologies and Chemical Industry 234
- 8.2 Contemporary Bioengineering Technologies and Agriculture 240
- 8.3 Contemporary Biotechnologies and Food Industry 244
- 8.4 Contemporary Bioengineering Technologies and Paper Industry 253
- 8.5 Contemporary Biotechnologies and Leather Textile Industry 257

Chapter 9 Contemporary Bioengineering Technologies and Biomaterials 259
- 9.1 Introduction 259
- 9.2 Natural Biomaterials 261
- 9.3 Medicinal Biomaterials 262
- 9.4 Tissue Engineering Materials 269
- 9.5 Intelligent Biomimetic Materials 277
- 9.6 Nanoscale Biomaterials 283

Chapter 10 Contemporary Bioengineering and Biological Medicine 286
- 10.1 Application of Contemporary Biotechnology in Antibiotic Industry 290
- 10.2 Gene Diagnosis and Gene Therapy 293
- 10.3 Genetic Engineering Vaccine 298
- 10.4 Application of Contemporary Biotechnology in New Drug Development 305

Chapter 11 Infection and Immunity 311
- 11.1 Infection of Pathogenic Microorganisms 311
- 11.2 Immune Response in Human Body 314
- 11.3 Antigens and Antibodies 319

Chapter 12 Biodiversity and Environmental Management 324
- 12.1 Loss of Biodiversity and Its Reasons 324
- 12.2 Significance of Biodiversity Protection 327

12.3	Biomonitoring and Evaluation of Environment	330
12.4	Bioremediation of Polluted Environmental	335
12.5	Treatment of Water Pollution	339
12.6	Atmospheric Purification	348
12.7	Biotreatment of Solid Wastes	352

Chapter 13 Safety and Social Ethics of Contemporary Bioengineering Technologies 355

13.1	Ethical Problems with Human Cloning	355
13.2	Bioweapons and Biowarfare Agents	361
13.3	Safety of Genetically Modified food	371
13.4	Research and Application of Human Genes and Its Impact	376

References 380

Chapter 1

Introduction

Life sciences are the key discipline which promotes natural science and makes social progress, and also one of the backbone disciplines in the 21st century. Research subjects of life science relate to activity phenomena and their nature of organisms, as well as to interrelationship between lives and environment. Reviewing the history of development of science, it is easy to see that it mainly concentrated on physics, wherein the theory of relativity proposed by Einstein pulled contemporary physics upward to a glorious peak in the first half of the 20th century, and in the next half of the 20th century, its core tended to life sciences obviously. Among all the Nobel Prize winners from the 1950s to the recent years, we can see that the highest science award has been more and more in the field of life sciences, even some of the so called chemistry awards were won as the object of study and the achieved breakthrough fruit related with life activities.

In the 20th century, life sciences made the most brilliant progress. Protein, sugar and deoxyribonucleic acid (DNA) molecular structures have been clarified in succession. Test-tube baby, transgenic plants and transgenic animals appeared one by one. The human genome project was started up; and the first cloned sheep "Dolly" appeared in the world. All of these rapid progressive events have shown that the 21st century is the century of life sciences to more and more people. This includes the meaning of two aspects. Firstly, life sciences will become the leading discipline, and provide new solutions and methods for research and development of other disciplines; and secondly, bioengineering industry will become a mainstay industry in the 21st century, and bioengineering agriculture and bioengineering pharmaceutical industry are the two important parts of the bioengineering industry. Bioengineering will finally solve the vital problems influencing human survival, including world's population, grain, environmental pollution, health, energy sources and so on.

1.1 Life Sciences: the Leading Discipline in the 21st Century

The 20th century is a stage of rapid development of life sciences, especially in the last 20 years of the 20th century, people's attention is more attracted to the speed of development of life sciences. Genetically modified food (GMF) has been laid out on the table of ordinary people; genetic therapy technology has begun to save lives of patients with genetic diseases; and the significant breakthrough in cloning technology has already made it become possible to make copies of animals. The dreams of mankind for several thousand years will come true one by one, with life science century replacing the physics century, and the world will appear more miraculous with development of life sciences. It can be expected in the 21st century from both development and crisis, that life sciences will become the leading discipline of natural sciences.

1.1.1 The Origin and Development of Life Sciences

As a discipline with a long history, Life Sciences was originated in ancient times, formed in the modern times, and developed fast in the contemporary times.

In the ancient times, people came into contact with all kinds of animals and plants in production practice, observed reproduction and death of organisms, and changes of the celestial bodies, and so formed the life view with simple materialism thought that all things were changing. But the people at that time couldn't explain unpredictable life phenomena, which formed the original religious idea that everything had its own soul. About six to seven thousand years ago, Chinese agriculture reached a relatively high level; five to six thousand years ago, original animal husbandry developed; three thousand years ago; people began to breed silkworm indoors; and Chinese people applied vaccine to prevent disease earlier than western people by over 800 years. The Book of Songs, a famous ancient Chinese literary work written in the period of Spring and Autumn and the Warring States, recorded over 260 kinds of animals and 350 kinds of plants. Qi Min Yao Shu, a monograph on agricultural sciences written by Jia Sixie in the Northern Wei Dynasty, systematically summarized the agricultural technical achievements before the 6th century AD. It was an agricultural encyclopedia in ancient China. Meng Xi Bi Tan, written in the Song and Yuan Dynasties, contained a lot of data about anatomy, physiology and Traditional Chinese Medicine (TCM), in addition to much biological knowledge on taxology, morphology and geographical distribution, domestication of animals, and so can be regarded as a milestone in the Chinese science history. Inner Canon of Yellow Emperor, written in the Qin and Han Dynasties, systematically discussed structure and physiology of the human body, and explained about disease-related knowledge. Shen Nong's Herbal Classic, written in the Han Dynasty, recorded 365 kinds of herbal drugs. In the Ming Dynasty the famous scholar Li Shizhen's Compendium of Materia Medica explained about 1892 Chinese medicines. Li's taxological and evolutionary thought made him become a biologist giant prior to Linnaeus and Darwin. All the research achievements before the 16th century mainly stressed on the aspects of medicine (including pharmacy) and agricultural sciences, they had characteristics of empirical property and applicability, adopting the methods including visual description, classification and anatomy to make knowledge on morphological classification and anatomy enrich rapidly.

After the 16th century, with rapid development of the European Renaissance Movement and capitalism, the real experimental natural sciences with a new style emerged in the world, and life sciences also made new progress. In the 17th century, Harvey discovered the blood circulation, and found that the blood vessels formed a closed system through which the blood circulated rapidly around the body and was pumped by the heart. He employed mechanical and chemical quantitative experiments in his research work, which laid a basis for physiology. Robert Hooke, one of the founders of the early scientific revolution, for the first time observed and described cells with a homemade microscope, and opened the door to the microscopic world. Linnaeus established the first scientific biological classification system, so as to establish the order in the kingdom of living things, and laid the basis for systematics as an independent discipline. Wolff proposed the theory of gradual formation of the chicken embryo by studying on development of the chicken embryo with experiments, and established scientific embryology. Overall, from the 16th to the 18th century, the outstanding achievements in the area of life sciences are successive establishments and developments of some branch disciplines of life sciences, such as physiology, taxology, embryology, etc.

The 19th century was the century of sciences. Life sciences also made all-round progress, wherein the significant progress was establishment of the cell theory, the evolutionary theory and genetics. Schleiden and Schwann established the cell theory, that the cell is the basic unit of structure and function in all of organisms, which explained consistency of the basic structures of organisms at the cellular level. The 20th century, especially after the 1950s, with widening and deepening of new theories and new methods in modern physics, chemistry, mathematics and computer science, life sciences had great changes and development, from a static, qualitative descriptive discipline to the dynamic, precise and quantitative one, and so experimental biology progressed into a new stage of comprehensive development.

In 1900, Mendel's principles of genetics were discovered once again and proved, which opened a prelude to modern genetics. Morgan put forward the gene theory in 1926, marking formal establishment of modern genetics. Morgan Genetics bridged the gap between embryology and the evolutionary theory, directly promoted the cytology, and laid the foundation for new big integration of biology, facilitated biological research from the cellular level to the molecular level. In the second half of the 20th century, development of biochemistry laid foundation for molecular biology and molecular genetics. In 1941, Beadle and Tatum proposed the doctrine of "one gene one enzyme", which combined gene with protein function. Avery's bacterial transformation experiment in 1944 and the Hirsch phage infection experiment in 1952 proved that DNA was the carrier of genetic information. In 1953 Watson and Crick established the DNA double helix structure model, and laid the foundation of molecular biology, with this as a breakthrough creating a new era to clarify the essence of life activities at the molecular level. Since then, the central dogma in molecular biology was put forward, and so the unified genetic codes in the living kingdom were deciphered, which demonstrated consistent development and relationship in the living kingdom at the molecular level, making life sciences enter a new era of molecular biology.

As a growing point of the contemporary life sciences, molecular biology has infiltrated into various branches of the life sciences, and meanwhile a series of emerging disciplines, such as molecular cell biology, molecular neurobiology, molecular structural biology, molecular taxology, molecular developmental biology, molecular virology, etc., were derived giving organic relation between life activities at various levels to explore laws of life activities in essence from a comprehensive multi-disciplinary viewpoint, so as to open up a new situation of modern life sciences. On the other hand, recombinant DNA succeeded in 1973, and genetic engineering has established since then. After 1980s, contemporary biotechnologies with genetic engineering as a backbone emerged worldwide as a high-tech industry. Conversion of biotechnologies to powerful productivity has shown its broad application prospects. The most ambitious research project "Human Genome Project"(HGP) in modern life sciences began in 1990, it was successfully completed with drawing of the working draft in June 2000, and then its follow-up work went well and was very exciting. In terms of macro-biology, contemporary ecology has developed into a multi-level comprehensive discipline with human research as the main topic, and is playing an increasingly important role in solving the global issues affecting human development.

Life activities are the most advanced, most complex phenomena in nature. So, life sciences have unlimited driving force, and activate more and more scientists such as physicists, chemists, mathematicians and, computer scientists to participate in the life science research, which greatly facilitate development of life sciences in recent decades. For today's life sciences, quantitative analysis under very strict conditions, such as physical, chemical and mathematical methods, can be adopted.

In short, contemporary life sciences in the 20th century have made enormous and remarkable progress. Contemporary life sciences have entered the development stage of great sciences, and are developing rapidly at every depth and every width in a state of leading natural sciences.

1.1.2 The Meaning of Life Sciences

"Life Sciences" has meanings at both broad and narrow sense. In a broad sense, "life sciences" are a conventional view and concept. Their definition about systems, methods and, means is uncertain, and internal logic systems in them are not strict. They generally refer to all the scientific activities for research and the phenomena of life. In a narrow sense, "life sciences" are a scientific system with a strict internal logical structure, based on traditional biology, containing two big parts of basic disciplines and applied ones, by means of biotechnologies, after undergoing formation and development of the discipline, and cross and fusion of the discipline and other ones. "Life sciences" mentioned herein are narrowly defined, and their definition is determined from a scientific division point of view, according to the developmental trend of contemporary disciplines and perspectiveness of discipline division in the 21st century. They are different from conventional life sciences in a broad sense.Life sciences relate to studies on substance composition, structures, functions, biodiversity, various life phenomena and laws of life activities of human and other organisms, and relation between the mankind and the nature and between organisms and environment, at the molecular, genetic, cellular, individual, group and ecological levels. Activities of lives are the highest form of motion of matters; and whether structures, functions, metabolism, reproduction, inheritance, development, evolution, or reactions and activities helpful for individuals and races are detailed embodiment of various matter motions in lives. With biology as a basic and backbone discipline, life sciences include various fields of biology, such as microbiology, botany, zoology and, molecular biology which are core disciplines, and at the same time also includes some applied disciplines such as agriculture sciences and engineering and medicine which are based on microbiology, botany, zoology and molecular biology. Life sciences have their own frontier of the development of discipline in various fields and at various levels, and researches in all the fields have their application prospects. The general trend of life sciences is interdisciplinary and comprehensive development by means of mutual infiltration with mathematics, physics, chemistry, astronomy, geology, economy, philosophy and other disciplines, with support from multidisciplinary research fruits such as research on trace elements in the human body.

In recent years, the HGP has made fast progress and great achievements, and such research enters a new era of post-genomic, proteomics, bioinformatics. Research fruits of bioinformatics will expand new areas of life sciences and accelerate the process of their research.

1.1.3 The Relation Between Life Sciences and Contemporary Bioengineering Technologies

1.1.3.1 Relation between life sciences and contemporary bioengineering technologies

Life sciences and technologies relating to substances of lives, may be divided into traditional life science and technology and contemporary life science and technology. The former includes traditional technologies such as brewing, making sauce, breeding, etc., employed by people for thousands of years. And the latter mainly includes genetic engineering, cell engineering, fermentation engineering and enzyme engineering, which currently, has become the frontier field of rapid development.

So, what is bioengineering? Bioengineering, also known as biotechnology, is a multi-disciplinary applied discipline, integrating the latest research fruits in contemporary life sciences (including biochemistry, microbiology, molecular genetics, cellular biology and immunology) with advanced engineering technologies. Bioengineering relates to a wide range of heterogeneous content, and scholars in different parts of the world give it different names. Americans called it"biotechnology and bioengineering"; Europeans "molecular biological engineering"; Frenchmen "génie biologique", i.e.

biological engineering; and the English use the term "biotechnology" to show the technical content of the bioengineering, also to represent the project content, but different from "bioengineering". The word, "Engineering", means the procedure of completing production of goods or providing labour for the society to get benefits, which is characterized as a large-scale, strong complexity and need for coordination between multidisciplinary knowledge and technology; but "technology" refers to use of a specific skill during implementation, and hence the two terms have notable difference. Bioengineering is industrialization of biotechnology. In bio-engineering field, the "technology" process and the "engineering" process cannot separate from each other, but they have their individual independence relatively. The definition of bioengineering accepted widely which was given by International Economic Cooperation and Development (OECD) in 1982, bioengineering refers to those technologies processing raw materials in order to provide products or service for mankind, by utilizing principles of natural sciences and engineering science, and relying on biocatalysts. Biocatalysts mentioned herein can be natural organisms, including microorganisms, animal and plant cells obtained with tissue culture technology, and new types of organisms with specific characters constructed with recombinant DNA technology and cell fusion technology, as well as enzymes or other ingredients produced by these new types of organisms.

People have to know that the life sciences and bioengineering have a clear difference. Life sciences refer to acquirement of biological knowledge, but bioengineering is application of biological knowledge. In most cases, bioengineering processes conduct at a low temperature, low consumption, and generally employ cheap raw materials as substrate.

In fact, bioengineering technologies are organic combination between life sciences and engineering technologies. Now, biotechnology has become closely related to a multi-disciplinary comprehensive interdisciplinary subject, which has close association with microbiology, biochemistry and chemical engineering. It is comprehensive technology based on the life sciences, and utilizing features and functions of organisms to design and construct new substances or new strains with expected performances, and combine with engineering principle to manufacture products or provide service.

Contemporary biotechnology provides various conveniences for human life, mainly including: ①more accurate diagnosis, prevention or cure of infectious diseases and genetic disorders; ②increasing crop yields effectively, get plants with many excellent characters, including resistance to insects, antifungal, antivirus, and adversity resistance; ③developing microbes which can produce chemical drugs, biological polymers, amino acids, enzymes and all kinds of food additives; ④creating livestock and other animals with more desirable characters; ⑤simplifying procedures for removing pollutants and wastes from environment.

At present, life sciences and technology have already gone into common people's houses, and have become an important methods for enhancing whole people's quality of daily life. For example, in some states of America the state-scale DNA databases have been already set up, and are used for cracking criminal cases and identifying criminals. DNA databases can be used for systematic examination of fetus chromosomes in order to determine whether a fetus carries a defective gene, and to improve population quality (although this practice still has dispute). Genetically modified plants and transgenic animals step by step "go" on the table of people, providing for people nutritional food with a more reasonable diet and of higher-quality. Genetic engineering drugs such as insulin, interferon and interleukin have been volume-produced, and put on the market. Polymerase chain reaction technology (PCR) and Southern hybridization method are also gradually widely used diagnosis of disease in molecular. In short, contemporary biotechnology has begun to penetrate into people's life in various ways.

1.1.3.2 Relation of bioengineering and other disciplines

Contemporary bioengineering is a comprehensive interdisciplinary subject relating to biology and engineering, and has a feature of knowledge-intensive, covering the widest range of disciplines in natural sciences. Now that the discipline needs to utilize principles of life activities, the knowledge about organism structure, function, metabolic activity and growth rhythm must be grasped. It is based on all the sub-disciplines of life sciences, such as cell biology, microbiology, physiology, biochemistry and molecular genetics, and combines with some other frontier basic disciplines beside biology, such as chemistry, physics, chemical engineering, mathematics, microelectronics, computer technology and informatics, to form a multi-disciplinary mutual infiltration of comprehensive discipline, wherein major theoretical and technological breakthroughs are fundamental. In addition, in all the areas of biotechnology, a lot of modern sophisticated instruments such as electron microscopy, High Performance Liquid Chromatography (HPLC), DNA synthesizer, biological mass spectrometry, DNA sequence analyzer, capillary electrophoresis instrument, are applied. These instruments are fully automated devices, and are all controlled by computer.

In recent years, cell engineering, tissue engineering and animal cloning have become an indispensable part of biotechnology, and represent the most promising trend in future. In fact there are no obvious gaps between biotechnology and pharmaceutical industry and pharmacy, and they have very close association. Agricultural biotechnology includes variety improvement, molecular breeding, bioreactor, etc.; environmental and marine biotechnology includes ecological restoration, biodiversity conservation, sustainable utilization of biological resources; the biological information service system has already been an important part of biotechnology. The importance of pharmaceutical products which are an eternal need for human, pharmaceutical industry has been never declined, and their development needs new ideas, new approaches. For traditional drug screening process, at first multi-index animal tests should be carried out, then human trials are needed, which has some defects, such as long process, less combination and low efficiency. In the new century, mankind has begun to understand the behaviour of genetic development at the molecular level, and knowledge about the laws of life activities has undergone a qualitative leap. Gene isolation, gene amplification, gene recombination and somatic cell cloning technologies have all turned into reality; structures and functions of some important proteins have been proved; the partial mechanism about signalling between the intracellular and the extracellular membrane and transmission of nerve impulses, and the photosynthesis mechanism about microbial and plant have been understood somewhat; and for the first time, the genetic recombination attempts for some purposes have been carried out, and based on this the strategy of cloning has been put forward. At present, sheep, cows and mice were successfully cloned by mankind. Using the cloning technology, the nucleus of human skin cells is transplanted into unfertilized eggs of the cattle, to cultivate a kind of universal cells and embryonic stem cells, which can develop into all kinds of cells in the human body. This technology has opened a new area for human organ transplantation.

A new revolution in agricultural sciences and technology is triggered by bioengineering. Genome work framework map of China's rice and database have been completed. Two-line hybridized rice technology is established. The genetically modified gene engineering strain constructed by using the new *Bacillus thuringiensis* insecticidal gene has a strong insecticidal effect and can substitute for chemical pesticides. Using molecular marker assisted breeding technology produced a series of fine varieties. With transgenic technology, the fluorescent genes are transferred into the silkworm to obtain a fluorescent genetically modified silk successfully, which lays a good foundation for further improvement of silk quality.

Genetic engineered bacteria can efficiently produce high quality L-methionine, and its all quality indices have attained the international standards and the standards of Chinese Pharmacopoeia (2000 edition) according to the tests performed by the state statutory drug test institution. The expression level of α-acetolactate decarboxylase, produced by genetically engineered bacteria, reaches more than 900U/mL in fermented broth, and as high as 4-5 times the world's existing highest expression level. This technology has been now widely used in many breweries. Phytase, which is produced by engineered yeast in China, is a novel animal feed additive, and this technology reaches the international advanced level and is commercialized in Jiangxi province.

In modernization of traditional Chinese medicine, analysis and research on genes and enzymes of plants and their biochemistry and structure-activity are carried out; and effective ingredients of plants are extracted by biotechnologies. Plant cell reactors are used to produce paclitaxel, ginkgo lactone, artemisinin, shikonin, ephedrine, etc.; and animal cell reactors can produce monoclonal antibodies, interferon, growth hormones, growth factors, enzymes, etc. Organ transplantation is also making important progress by utilizing transgenic technology in animals. In addition, gene therapy has extended to treatment of cancer, cardiovascular disease and Acquired Immune Deficiency Syndrome (AIDS) apart from genetic diseases. A batch of gene therapy solutions and genetic engineering drugs have progressed into the clinical trial stage and gene therapy has made great progress. Therapeutic vaccines are beginning development of industrialization, and it is expected to progress into clinical trials soon. Completion of the HGP and discovery of functional genes of a certain disease will make bioengineering technologies play an increasing role in treatment and diagnosis.

Energy and the environment are the major themes of sustainable development. Serious pollution and increasing shortage of fossil fuels require for a new clean alternative energy. Among renewable clean energy classes, hydrogen energy made from biomass has broad prospects. Various alcohols produced with starch as raw material can be replaced novel alternative energy material, such as cellulose. Genetically modified oil crops can produce the important chemical raw materials for alternative energy, such as degeneration fatty acids. Polylactic acid may be used to produce biodegradable plastic and genetically modified crops can be utilized to manufacture PHA (polyhydroxyalkanoates, one kind of biodegradable plastic). All these products may replace petrochemical products and are better than it. In addition, pollutant biodegradation technology based on recombinant microbes will become the backbone of the environmental protection industry. For example, it has been discovered that many microbes have biodegradation action on a variety of contaminants in soil, water and air. The crude oil desulfurization engineered bacteria with high efficiency and specificity, developed by USA Bioenergy Corporation, have far more advantages over traditional desulphurization methods. Through transformation of microorganisms and their degradation action, exciting research fruits have been achieved in both pollution control and resource application of organic agricultural wastes.

Future development of biotechnology depends on width and height of the technology platform, and now many platforms, including recombinant DNA technology, cell culture and DNA chips, have been established. Through them many fruits are achieved or some technologies are industrialized. Gene therapy, genetic engineering drugs, genetically modified plants, cloned animals and diagnostic reagents all attribute to these technology platforms. Some new platforms have been formed in the 21st century. The first is the gene-level platform, in which all the whole genome sequences from dozens of microorganisms and four model organisms (yeast, *Caenorhabditis elegans*, *Drosophila*, and *Arabidopsis thaliana*) have so far entered the database, and the draft of human genome sequence also has just been completed, which means that hundreds of thousands of genes and proteins encoded by these genes can be used in manipulation of genetic engineering and protein engineering, so as to expand the scope of biotechnology industry very much. The second platform is biological chip, it is the intersection and integration of a variety of high-techs in the molecular biology, chemistry and physics field. Both DNA chip extended and other silicon chips containing all kinds of biological molecules will combine with nanotechnology to make the chips manipulated *in vitro* develop into the components that can perform some function *in vivo*. The third platform is stem cell biology, which is the basis of animal cloning and tissue and organ cloning. The key for the developing technologies is control over differentiation and development of totipotent or pluripotent stem cells. For example, neural stem cells can develop into various types of cells in the nervous system. Improvement of the platform will bring organ transplant in medicine and excellent livestock breeding in agriculture with revolutionary progress. The fourth platform is bioinformatics which has been widely used in genome and proteome research. With discovery of functions of most genes and proteins, bioinformatics will have a new development prospects, i.e., on a computer simulating biochemical metabolic processes in cells, even simulating the course of evolution, which will make biology really go into the new theoretical period. The high technology how to use computers to design new types of organisms will also become a reality. The fifth platform is neuroscience. Nowadays, a big science program about neurobiology is carrying out internationally. Higher nervous activities in the human body, such as feeling, cognition and thinking will eventually be analyzed at the molecular level and cellular level. In the near future, new biotechnologies will appear on this platform. On one hand, they will bring Gospel for human mental illness; and on the other hand, based on this, highly intelligent computers and robots will be developed.

1.1.4 Life Sciences: the Leading Discipline in the 21st Century

Contemporary biotechnology has been paid attention by many people in various fields in the development history of nearly 20 years. 21st century is called the life science century by many experts, and contemporary biotechnology industry can be called the rising sun industry in the 21st century. Due to the rapid development, contemporary biotechnology is applied widely. Also, contemporary biotechnology has advantages over any other technology, i.e. it helps in sustainable development. Faced with a series of serious problems which directly related to the whole human survival, such as population expansion, resource exhaustion, environment pollution and so on, more and more people deeply realized the necessity and urgency of sustainable development of new technologies and new industries. Since biotechnology is based on organisms (animals, plants, cultured microorganisms cells, etc.) to manufacture products, the raw materials are renewable. And at the same time, few pollutants are produced in the production of products, and hence destruction of environment is little. What's more, recombinant microorganisms can even eliminate pollutants in environment. In view of the above characteristics of the biotechnology industry, clean and economical biotechnology will gain inevitably greater development in the 21st century.

At the turn of the century, biotechnology has shown us the grand plan in the new century, from the pharmaceutical revolution to the green revolution, from the new energy to the sustainable ecological environment, which indicates that biotechnology progresses into a rapid and steady development period. Publication of the draft of the recombinant human genome became the headline news in 2001; in December of the same year, the gene sequencing task of the 20th pair of chromosome of human was completed, which marked completion of a new chapter in the bible of human lives.

Hou Yunde, vice dean of Chinese Academy of Engineering, pointed out that information technology and biotechnology in our country were the key technologies relating to economic development and national destination in the new century, and will become an economic growth point for the innovation industries. Biotechnology will solve a series of problems, such as disease prevention, population expansion, food shortages, energy shortage and environmental pollution, etc., for the human being. In recent 20 years, the share of biopharmaceuticals in the whole drugs and biological products across the globe has reached 13%. So far, over 100 classes of biotechnology drugs in the world have been formally approved to be launched on the market, and the annual output value has reached 100 billion dollars. And over 1000 classes of biopharmaceuticals have been studied in the clinical trial phase. About half of people around the world have used the biotechnology products. Due to broad prospects of the biotechnology industry, in 2004 biopharmaceutical enterprises in the United States invested more than 30 billion dollars in research and development of new drugs, which was greater than that in the preceding year by 19%. Therefore, the biotechnology industry should be positioned as a key industry.

1.2 The Contents and Characteristics of Contemporary Bioengineering

Contemporary bioengineering is an interdisciplinary subject which integrates multidisciplinary theories, technology and engineering principles. According to the operation object and purpose, they can be divided into four main parts, genetic engineering, cellular engineering, enzyme engineering and fermentation engineering. In a broad sense, bioengineering also includes biomedical engineering (e.g. detection and analysis of the human body information, recovery medicine, etc.), and even bionics.

1.2.1 The Contents of Contemporary Bioengineering

1.2.1.1 Genetic engineering

Genes are a particular fragment in DNA molecules. Different genes carry different genetic information which based on the corresponding DNA fragment sequence. Accurate replication of genes ensures transmission of the genetic information from generation to another generation. A gene transcribes to messenger ribonucleic acid (mRNA), so as to instruct synthesis of the specific protein. And so genes are expressed to complete certain life activities. So genes are the foundation of heredity and the "blueprint" of life activities.

Microorganisms may be used to produce antibiotics, amino acids, nucleic acid and enzyme preparations for mankind, because they have the genes to synthesize these substances. The silkworm, containing silk fibroin genes, can spit out the silk. Legume plants can fix nitrogen in the atmosphere, and so can grow well in the infertile soil, because the symbiotic microorganisms of the root of legume plants have nitrogen fixation genes, these organisms can get what they want, each has its own features. But microorganisms can't produce some precious drugs in medicine, such as insulin, interferon, etc., because these microorganisms do not have the related genes, but human organs have. For centuries, humans have bred many new species with excellent characters from parents through artificial selection and cross-breeding methods. However, these conventional breeding methods cannot alter the genetic characteristics of organisms directionally according to people's intentions, and especially, there are still insurmountable gaps in cross-breeding between different species. People have been looking forward to a method for directional breeding for new varieties.

In 1973, for the first time American scientist Cohen recombined two different DNA molecules *in vitro*, and made the recombinant DNA express in *Escherichia coli* (*E.coli*), and finally found the technology and engineering for directional modification of organisms, i.e. genetic engineering. This year was called "first year of genetic engineering", and Cohen became the first founder in the history of genetic engineering. However, birth of genetic engineering is not an accidental event, and is associated with a series of important achievements, such as, discovery of the genetic material DNA (or genes) in the 1940s, elucidation of molecular mechanism and molecular structure (double helix model) of the genetic material DNA in the early 1950s, determination of the transmission pattern of genetic information, and research results of "tool enzymes" and molecular vectors used for genetic engineering. These achievements have laid a solid foundation for the birth of genetic engineering. On the other hand, the birth of genetic engineering resulted in revolutionary change of life science, marked that the human has entered a new era for modification and construction of new lives, which showed a very promising future.

Genetic engineering is a new technology rising in the 1970s. Its main principle is that the genetic material, DNA, is isolated from an organism with artificial methods to conduct cleavage and splicing and recombination *in vitro*, and then the recombinant DNA is introduced into host cells or an individual so as to alter their genetic characters. Sometimes, in order to get gene products (peptides or proteins), the gene is a large number of expressed in the new host cell or an individual at a high level. The technology for creating new organism and endowing them with special functions through DNA recombination *in vitro* is called genetic engineering, also called recombinant DNA technology.

Since genetic engineering has emerged over 30 years, the related research work has a leap in its progress. For example, "Engineered bacteria" can produce drugs for mankind, such as insulin, interferon, hepatitis B vaccine, etc. Genetically modified animal can produce pharmaceuticals and nutrient products of top quality. Genetically modified crops can resist various plant pests and diseases. HGP has been completed, and this is a miracle for mankind. The human body contains approximately 30,000 genes, and the gene map will be mapped out so as to understand their functions. Undoubtedly, the genome map will become a guide to diagnosis and treatment of genetic diseases, or gene therapy, and exploration for drugs, cancer and AIDS.

With development of the HGP, draft sequences of the human genome have been completed, and then the post-genomic era comes. Studies on gene structures and functions will provide new ways not only for disease diagnosis and treatment, but also for new drugs, new vaccines, new diagnostic reagents in R & D, and bring new changes for these. The HGP has a far-reaching significance and will be able to benefit mankind, and its goal is entirely achievable.

1.2.1.2 Cell engineering

So-called cell engineering refers to a procedure, in which cells as basic units are cultured and propagated *in vitro*, or some biological properties of cells are altered according to people's intention, so as to improve breeds and create new ones, or to accelerate breeding of plants or animals, or to obtain one useful substance. Therefore, cell engineering includes plant cell, zooblast culture technology *in vitro*, cell fusion technology, organelle transplant technology, cloning technology and stem cell technology, immobilized cells, and large-scale culture of animal and plant cells. But some people also believe that immobilized cells should be divided into enzyme engineering, and large-scale culture of animal and plant cells into fermentation engineering.

In appropriate environmental conditions, two kinds of animal cells are put together, and after adding a fusion-promoting agent, the two kinds of cells may be fused with each other to form hybrid cells. The fusion genes from cells of the two parents can be expressed in the process of cultivation. This technology has the potential to break barriers of distant hybridization, so as to modify species and to create new ones. In 1975 British Milstein invented monoclonal antibody technology by utilizing cell fusion technology. Simply speaking, so-called clone is an asexual reproduction mode induced artificially. An asexual plant is propagated by means of cuttage to form new plants, and the propagated plants are propagated in the same way to obtain more plants. All these plants originate from one cutting and form an asexual reproductive line. If a cell is selected to culture, the cell can be divided endlessly to form many cells so as to constitute an asexual cell line. If a DNA fragment is isolated from one genome to connect with a vector, and then is transformed into a bacterium to amplify endlessly so as to make copy number of the gene on the DNA fragment increase continuously. All these copies of the gene come from the same recombinant vector, and so can be regarded as an asexual reproductive line, i.e. a clone. The word "clone" can act as both a verb and a noun, whether cloning a gene, or get a gene clone means the same.

1.2.1.3 Enzyme engineering

Enzyme engineering refers to a procedure to produce some products or to reach a purpose by utilizing the function of specific catalysis of enzymes or cells, and organelles, with the aid of production processes and bioreactors. Enzymes are the essential biological catalysts for metabolism, synthesis, degradation and conversion of substances in organisms. They have their own catalytic characteristics of high specificity, and mild action conditions such as ambient temperature and pressure. Fermentation with organism cells virtually are processes of biochemical reactions catalyzed by all kinds of intracellular enzymes. Enzymes are macromolecule proteins, and can be isolated and purified from fermented broth, using the method for purifying protein. Microbial fermentation is a procedure to convert substrates to the required products through many steps of reactions by using the multi-enzyme system in microorganisms. For bioconversion with enzymes, at first enzyme is extracted and purified for requiring, and then is mixed with the necessary substrate. The conversion rate of enzymes is far higher than cells. If pure enzymes are immobilized on the carrier to prepare immobilized enzymes, their conversion rate will increase greatly and product purification and recovery process will be simplified much. For example, asparaginase is prepared as an enzymatic bioreactor; and with fumaric acid supplied continuously as substrate in industry, the rate of conversion of aspartic acid from fumaric acid can reach more than 95%, and after reaction the product is almost aspartic acid.

If chemical signals of specific reaction between immobilized enzymes or enzymes produced by immobilized cells and substrates are converted to electrical ones, one certain chemical change will be measured quantitatively through changes in electric potential, current and conductivity. If you want to monitor glutamate content in some fermentation tank, glutamate decarboxylase or the *E.coli* strain containing the enzyme will be immobilized first, and coupled with the CO_2-sensitive electrode. Such device can be called a biological sensor (biosensor) or an enzyme electrode. If we put this device in a fermentation tank, the following reaction will happen in the glutamate fermentation tank:

$$\text{Glutamate} \xrightarrow{\text{glutamate decarboxylase}} \text{Carbon dioxide}$$

Released carbon dioxide reacts with the sensitive electrode. When the glutamate content is in between 100-300mg/L, its mass concentration has linear correlation with the electric potential changes. So the potential changes can be directly measured as the glutamate content. In the past, in order to monitor fermentation status in-tank, samples were taken from the fermenter periodically in different batches. Therefore, unlike the complicated procedure in the past, these biosensors can be placed directly in a fermenter to conduct continuous and dynamic monitoring, which is called online measurement and control. The biosensors have many advantages, such as rapid detection, high sensitivity, continuous monitoring, precision Now they are widely used to monitor CO_2, NH_3 and O_2 in a fermenter, or urea, uric acid, protein, cholesterol and triglycerides in the blood and urine. It has already been an important monitoring method in environmental protection, fermentation industry, and clinical application.

1.2.1.4 Fermentation engineering

So-called fermentation engineering refers to a biotechnology system for producing particular useful products with the aid of contemporary engineering technologies (mainly automation, high efficiency, function diversity and scaling of fermenter or bioreactors), by utilizing some microorganisms including "engineered microorganisms", or for using directly microorganisms in some industrialized production. Since fermentation has close association with microorganisms, fermentation engineering is also called microbial engineering or microbial fermentation engineering. Production of any fermented product, not only "engineered microorganism", but also conventional microorganism can be achieved in a way of fermentation. Fermentation engineering can be divided into two parts: one is the fermentation part, including seed (strain) breeding system, fermentation system, culture medium sterilization system and air sterilization system. The other part is the extraction part (post-treatment procedure), including a series of unit operations, such as separation, ion exchange, electro-dialysis, gel filtering, extraction, evaporation, distillation, crystallizing, drying, and packing. The former is the key to the whole production process, and also is a prerequisite for fermentation production. The latter is very important for product quality in fermentation engineering and plays a significant role in industrial production. Whole production both parts must be an organic combination for the sake of optimizing. Therefore, the whole fermentation process from raw material to the final products is a complete production system. In order to ensure that the achievements in laboratory and intermediate trials expand to industrial production as soon as possible and maintain the normal operation, so as to attain to the optimal conditions and high quality of fermentation products, a series of problems, including processes and equipments in the whole fermentation process and a quality of technician, also must be studied and resolved. Combination of fermentation with chemical synthesis is an extremely important aspect in optimization of product production. For example, firstly, after production of penicillin with fermentation method, the available synthesis method is adopted to further make a variety of antimicrobial drugs with higher efficacy and without side effect. Secondly, in order increase production capacity, the organic combination of fermentation engineering with cell immobilization technique or cell co-immobilization technique is a development trend in fermentation production. Thirdly, the organic combination of fermentation engineering with enzyme technology is helpful for fermentation production. For example, application of enzyme immobilization technique or co-immobilized technique, and cross linking enzyme (coupling enzyme) technology in industrial fermentation will play a more effective role of catalysis. As for cross-linked enzymes, generally, their crystal has the characteristics of enzymes, such as high catalytic activity and selectivity, and mild catalytic conditions, as well as operating stability and simple recycling process of non-homogeneous chemical catalysts; and they will manifest their superiority in fermentation industry.

These several parts of bioengineering technologies can form their own complete independent systems, but they have close relation and interaction in most cases. Without mutual infiltration and dependence on each other, perhaps bioengineering technologies would not form such deep and broad impact and momentum. The basis and starting point of bioengineering technologies is various physiological functions of organisms, which refer to capacity to conduct substance synthesis, degradation and conversion during growth, development and reproduction of organisms. All kinds of biochemical reactions in all types of organism cells are catalyzed by various intracellular enzymes, and specific structures and functions of various enzymes are controlled by certain genetic genes. So, from this sense, genetic engineering and cell engineering can be deemed to the core basis of biotechnology. These "engineered strains" or "engineered cell lines" may be prepared by utilizing genetic engineering and cell engineering, and can be used to produce more and better products, and gain greater economic benefit. But enzyme engineering and fermentation engineering are often the key factors for large-scale production and biotechnology industrialization. Thus, it is very important to understand several major parts of bioengineering technologies as a unified whole.

1.2.2 The Characteristics of Contemporary Bioengineering Technologies

Contemporary bioengineering includes genetic engineering, cell engineering, enzyme engineering and fermentation engineering, which have close association with each other and mutual infiltration. Genetic engineering is the core content of contemporary bioengineering, and can promote development of other bioengineering technologies. "Engineered bacteria" or "engineered cells" are obtained by modifying gene of bacteria or cells through genetic engineering technology. For example, enzyme-producing microorganisms modified through genetic engineering may produce enzymes in higher yield. The most significant characteristic of genetic engineering is construction of new organisms according to people's wishes through recombinant DNA technology *in vitro*. These organisms was also called "engineered organisms", or "genetically modified organisms", which play an important role in medicine, agriculture, industry, environmental protection, energy, and people's real life. Besides, genetic engineering technology can break the species barrier for genetic recombination and gene transfer, and can complete the long course of organism evolution under laboratory conditions quickly, which means construction of species can be completed faster than that in the nature. This just shows where the power of genetic engineering technology is. Just due to progressiveness, superiority, irreplaceable and promising market prospect, as well as cross penetration and integration with other high-techs. Genetic engineering technology makes contemporary biotechnology industry become a new growing point. This shows that contemporary bioengineering has the following characteristics: ①namely high investment, high risk, high yield (or high return); ②high integration between knowledge and technology;③research institutions mainly engaged in technological research and development, or small companies will become the main force of technology innovation; ④financial community will be keenly involved in the biotechnology industry, and make a large investment. Based on these, contemporary biotechnology industry will certainly develop gradually in almost all fields including medicine, agriculture, environmental protection and others, which is a trend in today's world. So we should grasp the opportunity or make steady progress.

1.3 Development and Social Impact of Contemporary Biotechnologies

1.3.1 Development of Contemporary Bioengineering Technologies

Contemporary biotechnology products attribute to the contemporary biotechnology, DNA recombination technology and protoplast fusion technology. In 1953, the American scientist James Watson and the British scientist Francis Crick published a paper in the famous journal "Nature", whose title is "Molecular structure of nucleic acid", and this paper illustrated the double-helix structure of DNA. In 1973, the S.Cohen research group pioneered *in vitro* recombinant DNA and succeeded in transformation of *E.coli*. Because of discovery of the double helix structure of DNA and fulfillment of gene transfer in laboratory, it is possible to design new living organisms according to the people's desire.

With genetic engineering, target products may be obtained in a way of "pregnancy in another belly", by culturing these recombinants. In 1975, the two British scientists Kohler and Milstein, invented hybridoma technology, and they used lymphocytes (from spleen, can produce antibodies) and myeloma cells (*in vitro* reproduce indefinitely) to get the hybrid cells, specially called hybridoma through protoplast fusion technique, which not only can produce monoclonal antibodies but also reproduce indefinitely *in vitro* culture. This monoclonal antibody can be used for clinical diagnosis reagent, or biochemical therapeutic agents.

In 1969, the first immobilized enzyme was used in studies on optical resolution of the DL-amino acids in Japan. At present, immobilized isomerase is often used to produce fructose-glucose syrup and immobilized acylase to produce 6-aminopenicillanic acid. Immobilized enzymes have certain uses in clinical diagnosis and treatment, and can also be used as biosensor to determine substrate concentration for enzyme.

In 1977, Boyer first obtained the gene clone of growth hormone inhibitory factor, using gene manipulation. Then, in 1978, Gilbert got the gene clone of rat insulin. In 1982, the first gene engineering product, recombinant human insulin produced with a recombinant microorganism finally came into being.

Contemporary bioengineering endows biological reaction process with new vitality to improve product fermentation yield, but it is hard to extract and purify target products from media. This is because target products have very low concentration, and sometimes they are still included in cells. In addition, in order to get recombinant cells, high-density culture technology is always adopted. However, in practice, high-density cells are obtained with the culture conditions for high-density, but high-concentration products cannot be obtained, because recombinant cells have instability, and so the introduced recombinant plasmid containing the exogenous gene easily lose from the host cell so the exogenous gene can't express. Therefore, in DNA recombinant process, stability of the recombinant plasmid should be enhanced, and the culture process conditions for enhancement of stability also should be investigated.

Large scale culture of plant cells started earlier than animal cells. Plant cell culture technology can be applied to production of some precious plant secondary metabolites, such as alkaloid, steroids and so on, which also belong to contemporary biotechnology products.

Traditional bioengineering has provided many important products and a number of important drugs for humans. They play a significant role in enrichment of human life, improvement of human health, and promotion of social progress. Since its birth, contemporary bioengineering has shown great vitality in the pharmaceutical industry, and has provided many precious drugs which cannot be obtained with traditional technologies, such as insulin, human growth hormones, human urokinase, human brain hormone, human erythropoietin, bone formation protein, human calcitonin, hepatitis B vaccine, monoclonal antibodies, superoxide dismutase (SOD), etc. These shows that pharmaceutical industry is the first beneficiary during development of contemporary bioengineering.

Contemporary bioengineering will promote industrial and agricultural development greatly. In industrial production, in addition to many precious drugs, many kinds of amino acids, organic acids, vitamins and single cell protein (SCP), are produced through contemporary bioengineering. In agriculture and husbandry production, contemporary bioengineering has already provided plant growth hormone, alkaloids, pesticides, bio-fertilizer, pesticide residue remover, aftosa vaccine and so on. These products ensure and promote development of agriculture and animal husbandry. In addition, through genetic engineering and cell fusion technology, human may obtain genetically modified animals and genetically modified plants or hybrid crops.

Traditional bioengineering has made a great contribution for human society prosperity, we can say that there are not civilization human contemporary if without it. But in today's world, all kinds of problems, such as population explosion, rapid expansion of human production activities, too much energy consumption, resource exhaustibility, rapid development of oil and chemical industry, abuse of chemical pesticide, severe environmental pollution, and destruction of ecological balance, make social prosperity difficult to maintain, and human fate also suffers from a serious threat. Traditional bioengineering technologies have been powerless, while contemporary bioengineering exhibits strong vitality in all these aspects.

1.3.2 Social Impact of Contemporary Bioengineering Technologies

Rapid development of contemporary bioengineering technologies has huge impact on the society. In recent years from the movie, especially American movies, such as "Jurassic Park," "Spider-Man ", "Renewable People", "Danger Zone," and so on, we can also feel social impact of development of life science. In these movies, design of many plots of the story and making of special effects are derived from important discoveries and breakthrough in contemporary life sciences. For example, dinosaur regeneration in "Jurassic park" is based on the principles of gene clone and gene expression. The founder of the park is a very great biological scientist, he and his assistants excavate dinosaur fossils, extract all the genes of dinosaurs from a prehistoric mosquito which once soaked the dinosaur blood, and then make the dinosaur regenerate in the appropriate environment through bioengineering technologies, so as to deduce a tortuous and thrilling tale. Sometime ago, the globally popular film, "Spider-Man", tells about a legend that a student occasionally acquires biological functions of a transgenic spider. With this unique ability, the student made a series of chivalric events, so he is known as "Spider-Man". Sophisticated production and innovative idea of the film attract a number of audiences. Its premiered box-office income exceeds 70 million dollars. From the above examples we can see that more and more people, including those without any association with life sciences such as film producers, pay attention to development of life sciences, beside biological sciences workers and learners in the field, because they also know how to make full use of the latest achievements of contemporary life sciences, to obtain the huge box office income. This is just the charm of contemporary bioengineering development. The DNA double helix structure is one of the creative scientific discoveries in the 20th century, and has changed all aspects of life sciences. With further research on gene functions and human genome, its significance is more and more obvious, and it is worthy of one of the three greater discoveries in the 20th century. The most influential three big science programs in the 20th century are as follows: Manhattan Project, which has led to emergence of the atomic bomb; Apollo Project, which provides human being with the chance to go to the moon to explore, and has opened a new chapter in human space sciences; and HGP, deciphering all the human genetic codes, which is closely related with knowledge about human themselves, and has immeasurable influence on development of biomedicine in the future.

Although achievements in the area of contemporary life sciences and technology have brought enormous social and economic benefits for human beings, undeniably, the enormous power of biotechnology has brought many unexpected shocks to the human society, and unexpected consequences may occur. As contemporary nuclear physics can be used to benefit mankind, and also can be used to make an atomic bomb to destroy human beings thousands of times, contemporary biotechnology is also a double-edged sword. It can benefit mankind, and may be used to produce deadly biological weapons. Some lunatics may use it to clone human beings and supermen to destroy social peace for the entire human race. Therefore, whether contemporary biotechnology would lose lose control and become the murderer of destruction of human beings, what studies should be supported and encouraged, or what should be banned, has become a common concern of people.

Basically, regarding the effects of contemporary biotechnology on human and natural environment, people concerned about the following problems: ①whether genetically modified organisms cause harm to other organisms or environment; ②whether use and development of genetic engineering organisms will reduce genetic diversity of the nature; ③whether a person can become the object of genetic engineering operation; ④whether gene diagnosis procedures would infringe upon personal privacy; ⑤whether those genetically engineered modified organisms can become privately owned property; ⑥whether economic aid of contemporary biotechnology will affect or restrict development of other important technologies; ⑦whether emphasis on business success means that contemporary biotechnology will only benefit the rich but the poor cannot afford it; ⑧whether agricultural biotechnology will completely change traditional farming methods; ⑨whether contemporary biotechnology drug therapy will suppress traditional treatment methods which have the same therapeutic efficiency; ⑩whether application of contemporary biotechnology patents will prevent the free idea communication among scientists.

From a sociological perspective, the human is different from other organisms; from a biological perspective, the human is a member of the living kingdom. Understanding of life phenomena is applicable for the human, the technology to modify organisms can be applied to the human, and thus development of life sciences will bring about some ethical issues. Development of reproductive biology and invention of contraceptive pills have been important for population control and also have great impact on sexual relations and family relations. In the 21st century, with development of bioengineering technologies, there are more social ethical issues. Cloning technology can perform human asexual reproduction, i.e. human being oneself can be reproduced on a large scale. If genetic engineering technology is applied to human in the same way as to transformed animals and plants, some people will be the objects of modification activities, while others are the subjects, and some people can be created according to other people's purpose. This inequality is far beyond difference in wealth and political status. Such problems make development of life science and technology become a concern to the whole society.

1.4 Contemporary Biotechnologies and Sustainable Development

1.4.1 Challenges in the 21st Century

In the 21st century, mankind is faced with many challenges, the first is population problem. China's population is gradually growing by 13-14 million every year, that figure is projected to increase to 1.6 billion by 2030. According to the data provided by the United Nations, some researchers estimate that global population minimum can amount to 7.3 billion by 2050. Now, there are 6 billion people on the earth, by then, after increasing so many people, what would happen to the crowded earth? The second is supply of food. There is only one earth for humans. So many people want to eat, while the arable land is limited, then, what can we do? According to China's present situation, if the population reaches 1.6 billion by 2030, food productivity must increase by 30% from now to 2030 so as to meet the needs of people. With development of economy, the arable land decreases every year, and so food supply will be come a serious problem. The third is health problem. With increase of the population, improvement of people's living standard and extension of human lifespan, the requirements for human health are higher than in the past relatively. But industrial development and environmental deterioration result in increase in morbidity of some diseases, including cardiovascular diseases, tumors, and senile diseases such as Alzheimer's disease, Parkinson's disease and other diseases nervous system. From the related data it is known that it would take about 20 years in China for the population proportion of the old at the age of more than 65 to increase from 10% to 20%, but in Japan 23 years, the United States 57 years, Germany 61 years, and Sweden 64 years. It will take the time in China less than the Unites States by 30 years to reach the same aging population, but our economic conditions are far worse than the United States and other developed countries, so it may bring a lot of social problems. Across the world, many dangerous diseases appear constantly. AIDS still shows a growth trend in many countries and regions, including the African continent, and Southeast Asia, South Asia, Russia and some areas of China. Mad Cow Disease (bovine spongiform encephalopathy or BSE) makes a lot of people talk about cows as a terror. In addition to these diseases which were not heard of in the past, taking an example, tuberculosis, which had been thought to be under control across the world, in these years, has relapsed worldwide. The reason is that some of tuberculosis bacteria acquire resistance on antibiotics, so that common antibiotics are hard to treat this disease. The fourth is hereditary diseases. In China, especially because of cultural and traditional concepts, and some mountainous and remote areas in isolated state, the phenomenon of selection of a spouse still occurs only in a small geographical area so that even marriage between close relatives still exists, which causes much higher incidence of genetic diseases than in developed countries. At present even in some coastal areas, the proportion of genetic disorders is also increasing. The fifth is environment issue. Due to rapid development of science and technology in the 20th century, various aspects of human life have been greatly changed. With their great wisdom, power and creativity, human beings improve the environment to make it more suitable for their own survival. However, social activities of humans damage the whole environment spontaneously. Therefore, while people's life standards are improved greatly, the natural environment for human survival is also worsening, Global population situation, resources strain, environment and other problems more and more seriously threaten survival and development of human society. The last is resource and energy issue. While reveling in their own victory, human beings have already got stuck deeply in predicament of the global problems which they have never run into in their own development history. At present the resource and energy issue is very serious, which arose from too much consumption of natural resources and damage of environment for a long time, Nowadays, all kinds of crises, such as cultivating land crisis, forest crisis, freshwater crisis, energy crisis and so on, are threatening human survival and development directly.

1.4.2 Contemporary Bioengineering Technologies and Sustainable Development

In recent 20 years, rapid development of bioengineering and biotechnology has opened up a broad prospect for world agriculture, medical industry, pharmaceutical industry and environmental protection. As "the most important technology which may change the future industrial and economic structure for the whole society", life sciences and bioengineering technologies, are increasingly attracting attention of many countries all over the world. As a country with a population of 1.3 billion, China is faced with huge challenges in population, health, agriculture, energy, environmental protection, and many other fields. And hence, it is very important to develop life sciences and bioengineering technologies for fulfillment of sustainable social and economic development.

1.4.2.1 Bioengineering and agriculture

Zhu Lilan, the former minister of Chinese Ministry of Science and Technology, discussing population, food, survival and health, resources and ecological environment protection problems

China and the world are faced with, pointed out that life sciences had a special meaning to the world especially China, because China was a nation with the heaviest population in the world. There's an old Chinese saying, "hunger breeds discontentment." The first thing we are faced with is to solve the people's food problems. Now, engineering agriculture, based on protein engineering, cell engineering and enzyme engineering, and constructed through comprehensive use of genetic engineering, will promote transmission from traditional agriculture of the "two-dimensional structure" which consists of animals and plants, to the one of the "three-dimensional structure" which consists of animals, plants and microorganism; and from utilization of land organisms resources to that of ocean ones. This will create the "big future agriculture", which means development of diverse resources including both land and ocean resources, and artificial energy without limitation of climate, and implementation of industrialized production which saves both soil and water, and does not pollute environment, so as to attain to the strategic goal of sustainable development within the 21st century, and face up to a new agricultural century.

Bioengineering is the basis for agricultural sciences, and development of basic agricultural science is source of development of the whole agricultural sciences. Agriculture's first task is to solve the problem of food shortage. Both the world population and the food demand increase, but the arable land area decreases. And therefore, unit area yield of a crop must be increased greatly by breeding excellent plant varieties. These excellent varieties may be obtained by utilizing genetic engineering technology to perform gene transfer. They may have many useful characters including resistance to drought, low temperature, salt, pests and diseases, and stress, in order to meet the demand for food. People require both enough food and top quality food, and so food must also meet the nutritional requirements. Application of gene engineering can improve quality of food and livestock products, such as increasing the protein content in grains, and making protein composition of livestock and poultries closer to that of human. To realize agricultural sustainable development, we must overcome disastrous effects of chemical pesticides, restrict use of chemical fertilizers and pesticides, and vigorously carry out research on genetic engineering of nitrogen fixation, in order to breed new varieties of pest-resistant crop, and implement biological control to reduce dependence on chemical pesticides.

1.4.2.2 Bioengineering and human health and longevity

People know that genes are the main factors determining all the human characters; and during the HGP which is carrying out actively almost every week some phenomena of gene deficiency resulting in human diseases can be found. We believe that the corresponding relation between genes and diseases can eventually be eradicated. For instance, the genes for control over formation of the eyes, the relevant genes of Alzheimer's disease, schizophrenia and AIDS, and a gene associated with bone marrow cancer have been discovered. Some people such as possessing impulsiveness and extremism also have their corresponding genes; and even suicide phenomenon is related with genes. The writer Ernest Hemingway and his father and brother all committed suicide, and this probably results from hereditary factors of the family. Obesity also has the corresponding gene, namely the hunger control genes. One of the goals of HGP is to clarify what genes are related with health and diseases, and how to control these so called "disease genes" at the genetic level, in order to maintain human health.

Life science theories have laid a foundation for medicine, each new result in life sciences will give a new look to diagnosis, treatment and prevention of diseases and health care. With development of life sciences and technologies more new technology drugs can be produced, and more efficient extraction technology can be applied. Genetically modified microorganisms can improve efficiency of fermentation or can produce new fermentation products. It is predicted that drugs will be produced mainly with fermentation of genetically engineered microorganisms by the middle of the 21st century, and this will improve the level of medical and healthcare service to a great extent.

The human genome contains about 30,000-50,000 genes, and positioning and deciphering all of genetic information are the goal of HGP. After the huge research project is completed, it is possible to understand fundamentally outbreak mechanisms and prevention solutions of various genetic diseases, cancers and cerebral and cardiovascular diseases. Modification of composition of the human gene will become an important approach to strengthen physique and prevent diseases.

1.4.2.3 Bioengineering and energy

Now we have to change the energy structure, explore diverse energy sources, and focus on the fourth generation energy, natural gas. Resources are an important basis for human survival and development. Because of population growth, coal and oil, resources will be exhausted certainly some day, and nuclear fuel materials will be depleted. In order to solve the energy problem thoroughly, we must utilize solar energy and renewable resources. People have high hopes for biotechnology to solve the energy problem. Agricultural side products fermentation by microorganism produced alcohol to substitute for gasoline as a fuel and has been used in Brazil. The research on breeding plant varieties with high content of fuel oil has made progress. Based on the mechanism of decomposition of water molecules into hydrogen and oxygen in plant photosynthesis and the structure of catalytic enzymes, artificial photosynthesis is simulated, and thus solar energy is utilized to decompose water so as to obtain hydrogen fuel, and such renewable energy source can be supplied endlessly. The study is more attractive for mankind.

1.4.2.4 Bioengineering and ecological balance

During long-term evolution, adapation and interaction with environment, the ecological system in the earth has been formed, and human beings have been evolved in such ecological system. Because of greed and selfishness, humans develop natural resources in a predatory way to create the present "material civilization". At present, harmful substances produced in various activities of humans are emitted into the atmosphere, water and soil, which results in resource depletion, water pollution, environmental degradation and frequent disasters. The environment problem gets increasingly serious. People must realize that serious damage of the earth's environment is actually giving up an approach to survival.

From scientific point of view, biodiversity, including genetic diversity, species diversity and ecosystem diversity, is the basis for human survival. Protection of environment and biodiversity has been an increasing concern in many countries of the world. According to estimation of scholars, there are approximately 5-50 million biological species on the globe, but now natural species are reducing at the rate 1000 times that of natural extinction. Some scholars estimate that at the present rate, within 20 to 30 years 1/4 of organism species will disappear, and more than half of them by the end of the 21st century. Biodiversity in China has a great significance to the globe. China has abundant animal and plant resources, and extensive experience in the resource utilization, but here species disappear faster than in other countries of the globe. For example, China has about 1 million terrestrial species, which account for about 1/10 of the total number across the globe. Endangered plants account for 10% across the globe, while in China it is 15%-20%, far higher than in Europe and the United States. Nature is a large ecosystem in which any problems will affect the whole system balance. Therefore, protection of environment and biodiversity is responsibility of every citizen, and every person should have a good sense of environmental protection.

Hu Jintao, the former general secretary of the Chinese Communist Party, put forward new scientific concept of development at the third session of the party's sixteenth plenary conference. One of the ideas of scientific concept of development is: "to adhere to concept of sustainable development that economic development should adapt to population, resources and environment, striving to create a development road to sustainable economic development, overall social progress, sustainable resources use, recyclable resource utilization, constant environment improvement and good ecological cycle fulfilling harmonious development of the economic society." Its core is "sustainable development", which first is a strategy for tackling the more and more serious crisis emerging during development of the economic society, including ecological destruction, environment pollution, less energy and resource scarcity. Protect earth environment, maintain the balance of the earth's ecological system, the slogans such as "protecting environment in the earth", and "maintaining balance of the ecological earth system", are now speaking loudly. Bioengineering will play an important role in solving the problems through joint efforts of human beings in the 21st century.

To make human get out the difficult position of the current global crisis, implement the strategy of sustainable development, and fulfill harmony of economic development with population, resources and environment. Besides restriction of certain policies, laws and regulations, we should abide by the laws of ecology, reasonable development and utilization of resources. At the same time, we should make full use of contemporary bioengineering technologies to produce alternative energy, dispose general wastes, municipal wastes and industrial sewage. We should modify species and create new plant and animal varieties of high quality and high yield to solve food crisis and protect biodiversity. Thus, contemporary life sciences and contemporary bioengineering technologies play very important roles in solving resources and ecological environmental problems across the globe.

Chapter 2

Science Basis for Contemporary Bioengineering

Organic macromolecules with complex structure in organisms, mainly including proteins, nucleic acids, polysaccharides and lipids, are polymers that are composed of monomers such as amino acids, nucleotides, monosaccharide and fatty acids, respectively. Nucleic acids and proteins are the most important biological macromolecules, and the most important basic substances for life. Amino acid classes and linkage bonds between amino acids, and nucleotide classes and linkage bonds between nucleotides are all identical across the whole living kingdom. Hereditary traits of organisms are manifested by various proteins synthesized by organisms themselves and interactions between proteins and nucleic acids.

2.1 Structure and Function of Proteins and Enzymes

2.1.1 Classes and Element Composition of Proteins

Proteins generally exist in all organisms, and are biological macromolecules that are composed of natural amino acids and are formed by linking amino acids through peptide bonds. Proteins have many different classes, mainly including the structural proteins, contractile proteins, storage proteins, defensive proteins, transport proteins, hormone proteins, etc. Even those prokaryotes such as *E.coli* with simple structure, contain more than 3000 kinds of proteins. The more complicated structures and functions of organism are, the more varieties proteins have, and for example, the human body contains about 100,000 kinds of proteins. Various proteins have certain relative molecular weight, complicated molecular structure and particular biological functions; and they are the main substance to express biological traits. Proteins are not only the composition to constitute structure of cells and organisms, but also participate in almost all of the life activities. For example, enzymes may catalyze various biochemical reactions in organisms, and many important hormones can regulate metabolism.

Some proteins participate in gene expression and regulation, or transmission of all kinds of signals, or immune defense function; and some proteins have the function of storing and transporting substances. Enzymes are one special class of proteins; which can catalyze biological chemical reactions in cells and can change the rate of biochemical reactions as a catalyst, but themselves do not change at all. All reactions in the cell are based on the action of enzyme. However, in spite of their different proteins structures and functions, proteins are almost all composed of common 20 kinds of amino acids (Table 2.1), which link together through peptide bonds in protein molecule, and are generally called protein amino acids in the nature, because they are necessary for synthesis of proteins.

Table 2.1   The 20 kinds of amino acids constituting proteins[categorized based on the polarity of R groups (pH7)]

Side chain(R groups) non-polarity	Side chain (R groups) uncharged polarity	Side chain(R groups) positively charged polarity	Side chain(R groups) negatively charged polarity
Ala Leu	Gly Cys	Lys	Asp
Pro Trp	Ser Asn	Arg	Glu
Val Ile	Thr Gln	His	
Phe Met	Tyr		

These amino acids have common characteristics: ①they are L-amino acid (except glycine); ②they are α-amino acid (except the proline), that is, their amino group and carboxyl group are linked on α-carbon atoms, their structures differ only in groups of their R side chain. The 20 kinds of amino acids are encoded by the known genetic codons in the genetic code dictionary, and there are codons only for the 20 amino acids in the dictionary.

In fact, amino acids of proteins in the nature are more than 20 kinds, and those beyond the 20 kinds of amino acids are derived from them by furthering and modifying them after synthesis of proteins. Other over 200 kinds of amino acids are not the components of proteins in nature, and are generally called non-protein amino acids.

The main elements of proteins are carbon, hydrogen, oxygen, nitrogen and a small amount of sulfur. Some proteins also contain mineral elements such as phosphorus, copper, iron, iodine, zinc, molybdenum. One important feature of proteins is the nitrogen content which is always 15%-17%, with the average content of 16%, in spite of sources of protein samples. It means that the sample contains 6.25g protein when 1g nitrogen element exists in it.

2.1.2   Spatial Structures and Functions of Proteins and Enzymes

2.1.2.1   Primary structure and functions

International Union of Pure and Applied Chemistry (IUPAC) defined in 1969 that primary structure of proteins is the sequence of amino acids in the peptide chain. The knowledge points that are needed to clarify the primary structure of proteins are: ①the number of polypeptide chains in protein molecules; ②the terminal amino acid residue class of each peptide chain; ③amino acid sequence of each peptide chain; ④configuration of disulfide bonds within chains and between chains.

Insulin is the first protein whose chemical structure was discovered. In 1955, Frederick Sanger and others revealed all the chemical structure of bovine insulin successfully with enzymatic and chemical methods. Insulin is a protein hormone secreted by pancreatic cells in mammalian. It not only can regulate glucose metabolism, promote the utilization of glucose and glycogen synthesis, but also can regulate the metabolism of lipid and protein. Bovine insulin molecule (monomer) is composed of A and B chains, and in total has 51 kinds of amino acid residues. The A chain is composed of 21 kinds of amino acid residues, and has disulfide bonds in chain; and the B chain, has 30 kinds of amino acid residues. There are two disulfide bonds between A chain and B chain, which link these two chains and constitute the complete primary structure. The structure of the bovine insulin is shown in Fig.2.1.

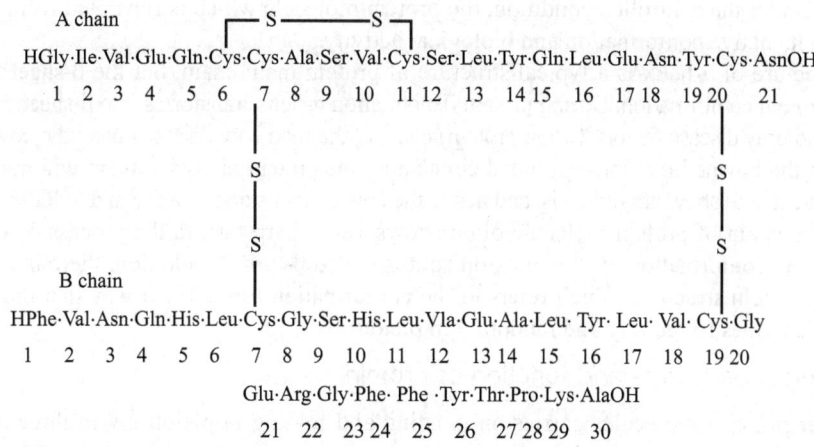

Fig.2.1 The structure of the bovine insulin

From the bovine insulin chemical structure, we can see that A chain and B chain are composed of the certain amino acid according to the specific sequence. The amino acid sequence of protein is usually called the primary protein structure. Which involve the relations between amino acid residues at different sites in multi-peptide chains and the biological function. Research results indicate that the amino acid may be changed, or be cleaved, or be replaced with other residues at some sites, but this does not affect the biological activity; however, some sites must be excised in order to activate some proteins; and at some sites the amino acid residues are conservative by nature, cannot be deleted, Or be replaced, otherwise the protein will lose activity. For example, one disease called sickle cell anemia is epidemic in Africa, and one abnormal hemoglobin (Hb-S) is synthesized in the blood of patients with this disease. The difference between abnormal hemoglobin (Hb-S) and normal hemoglobin (Hb-A) only lies in the change of the 6th residue in the N-terminus of the β-chain's N end:

Hb-A	N-terminus	Val-His-Leu-Thr-Pro-Glu	C-terminus
Hb-S	N-terminus	Val-His-Leu-Thr-Pro-Val	C-terminus

The hemoglobin has four peptide chains. If only one residue is replaced respectively in the two β-chains, the physiological function alters. Under the physiological pH condition, R group of Glu is negative charged, but the R group of Val is neutral, which make the surface charge of Hb-S to reduce and the isoelectric point increase, so that the molecules will agglomerate abnormally and the solubility will decrease, Finally, the red blood cells shrink into the sickle shape, which have reduced capacity of the oxygen carrying to become very fragile, and are very easy to have hematolysis.

2.1.2.2 Secondary structure and function of protein

Secondary structure mainly refers to the mode of chain's own twist, such as α-helix, β-pleated sheet, that is, the conformation of main chain. Alpha-helix refers to the polypeptide chain twist like screw, but β-sheet refers to the peptide chain which has folded indention conformation that is formed by stretching fully two or more chains, accumulating laterally and juxtaposing parallelly in the direction of peptide chain's major axis, then forming folded conformation, such as silkworm and spider silk fibroin. The natural protein molecule has one kind or several kinds of particular conformation which is associated with its biological activity. This kind of natural conformation is quite stable. Under the controlled condition, the protein molecule which is separated from the cell can still maintain its native conformation and biological activity.

The structure of α-helix is a typical structure of protein main chain, but the β-sheet is a general structure in protein conformation. During protein denaturation α-helix transforms into β-sheet. For example, when "the mad cow disease factors"(prion protein) causing the mad cow disease enters the cow's stomach, and arrives in the bovine brain through blood circulation; the protein in bovine brain will transform from α-helix to β-sheet, and show amyloidosis, and hence the cow cannot stand steadily and will die.

The main chain of protein molecule often shows 180° sharp turn in the process of twisting and folding, and this conformation of the inflexion spot is called β-turn. In addition, there is irregular curl in secondary protein structure, which refers to the conformation formed in a way that the main chain of polypeptide twines irregularly and randomly in pleiotropy.

2.1.2.3 Tertiary structure and function of protein

Globular proteins molecules carries on twisting and folding in pleiotropy in three-dimensional spaces on the basis of the primary and secondary structure, then forms the certain conformation that is roughly spherical and called tertiary structure of protein.

The tertiary structure includes relations of spatial distribution between all atoms and atomic groups in the main chain and the side chain of protein molecule. But it does not refer to relations of distribution between subunits or molecules. The characteristics of tertiary structure are: ①roughly spherical shape; ②forming so-called "the hydrophilic surface, hydrophobic core"; ③conformation stability mainly depending on hydrophobic interaction; ④containing the biological activity sites after forming the tertiary structure. There is a very deep cleft frequently seen on the spherical molecule surface, and the active site locates in it. The so-called active site of globular proteins molecules, also called the active center is a small spatial region in the information of tertiary structure which is composed of minority of essential groups and is responsible for playing biological function of the protein molecule. If the tertiary structure of protein is destructed, the conformation of the active center does not exist and the biological function will be lost. Therefore, globular proteins must have tertiary structure to keep the biological activity of molecules.

2.1.2.4 Quarternary structure and function of protein

Quarternary structure refers to the spatial arrangement and relations between subunits in a protein molecule, and the stability mainly depends on hydrophobic interaction among subunits. The number and the types of subunits also belong to the concerning content of quarternary structure, but the structure does not include conformation of subunits themselves. Some globular protein molecules are associated with two or more tertiary structure subunits, and so they usually are called oligoprotein. One tertiary structure unit of oligoprotein is called a subunit.

2.1.2.5 Relation between all the structures of protein

Primary structure determines spatial structure which will form automatically as soon as the primary structure of certain protein is synthesized. When forecasting the spatial structure of a new protein, we firstly should search whether there are any proteins homologous with the new protein exist in the known protein spatial structure database. When we find the homologous protein, the spatial structure of the new protein may be similar to that of the found protein. In the procedure of forecasting, we should choose the method to avoid more blindness. Different proteins may have the same spatial structure unit. Many databases of protein spatial structure units have already been established. And it is predicted that the patterns of protein's spatial structure probably has 600 kinds in the nature. After these pattern protein databases are established completely, forecasting work will be merely to search for the same structure pattern in the database. Amino acid residues in protein are independent from each other. The entire spatial structure is the result of interaction among all amino acid residues. This mechanism is fundamental to understanding the mechanism of protein folding. Like English grammar, the grammar of 20 letters of the amino acids is the law of interaction between amino acid residues. The protein folding code is considered as the second biological code after discovery of the genetic code, which substance does the code exist on? It may be on protein, or may be on transfer ribonucleic acid (tRNA), and it has not come to a conclusion yet. The relation between the protein structures is shown in Fig.2.2.

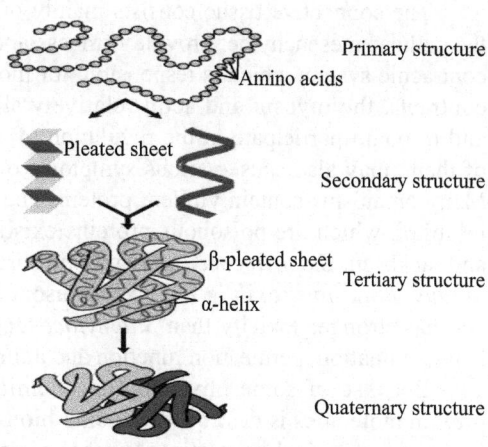

Fig.2.2   Relation between protein structures

The protein has the morphological functions. Proteins participate in constitution of all of tissue structures in organisms, and are main parts of dry material of structures such as animal's horn, wool, hoof, meat, etc.

Proteins have physiological activity. Metabolic activities and every physiological phenomenon in organisms are procedures in which various proteins are involved. For example, enzymes have function of catalysis, hormones regulation, and immunoglobulin anti-infection. Enzymes are a kind of proteins with function of catalysis, and also called biocatalyst. All the reactions are performed with catalysis of enzymes. Quantity, classes, activity and catalyzing ability of enzymes all may be regulated and controlled by genes. Enzymes have most strong ability of catalysis, in which required the mild physiological condition. Among regulators in animals and plants produced spontaneously due to exogenous and endogenous stimulation, most of them are proteins or short-chain peptides, such as growth hormones, interferon, leucocyte growth factor (LGF) and so on.

Antibodies are a kind of globulins with specific immune function, produced by cells of the lymphatic system with stimulation of antigens. The human body can recognize the foreign exogenous antigen invaded into the body, such as protein molecules, virus and bacteria, and the immune system will produce the corresponding antibody protein. Once antibodies are produced, they can recognize and bind exogenous antigens, and form the antigen-antibody complex. This complex can activate complements, further form the membrane-attacking complex, which can kill the cells carrying antigens effectively and eliminate interference of antigen.

Proteins can act as one nutrient (exogenous protein). Exogenous proteins will decompose and generate amino acids through digesting, and be absorbed by the human body to synthesize their own protein. Among reasonable nutrition requirement 10% of energy is obtained through oxygenolysis of protein. Proteins are the essential nutrients in animal and the human body. The human body cannot synthesize 8 kinds of essential amino acids (threonine, valine, methionine, tryptophan, phenylalanine, leucine, isoleucine, lysine) and two semi-essential amino acids (arginine, histidine), and these amino acids must be supplied by eating food every day.

The connective tissue consists mainly of structure proteins, which exist between tissues and in the cellular mesenchyme, provide firmness and protection for the cells and the tissues. The muscular contractile system which is responsible for movement is also composed of proteins, when the muscle contracts, the myosin and actin relatively slide, and complete organism movement, tropomyosin and troponin participate in the regulation of muscle contraction. Heterologous proteins, even a few of them, may also cause various symptoms of poison and even death once invading into organisms. Many organisms contain virulent proteins and peptides, such as agglutinin, trypsase and the amylase inhibitor, which are poisonous proteins existing in the encarpium of castor oil plant, soybean, pea and jackbean. *Staphylococcus* enterotoxins are poisonous proteins which come from *Staphylococcus aureus*. *Botulinus* toxin is a kind of poisonous protein which is produced by *Bacillus botulismus*, and has stronger toxicity than *Staphylococcus* enterotoxins. Proteins also play a significant role in hemoglutination, permeation function and animal memory, etc.

Because of some physical and chemical factors, the spatial structure in order of natural protein molecules is destroyed, causing biological activity loss that accompanies abnormal change in physicochemical properties, but the primary structure cannot be destroyed. After denaturation, physicochemical properties change. For example optical rotation changes, viscosity increases, and agglutination occurs because of reduced solubility. Biochemistry properties also change, and denatured proteins are easier to have hydrolysis by protease compared to natural proteins. Additionally, biological activity changes, for example, enzymes will lose catalytic activity once denaturated, hemoglobin will lose the ability of transporting oxygen, denatured antibody protein will lose immune activity. Protein denaturation also has advantages in some aspects such as food processing, disinfection and sterilization, silkworm cocoon's wiredrawing, termination of enzyme reactions. Disadvantages mainly manifest in biological separation and purification, because denaturation influences extraction of protein. The human body will be aging and the skin will be rough and dry because proteins denaturation gradually and their hydrophilicity reduces relatively. Cataract caused by ultraviolet radiation, is mainly a result of denaturation and coagulation of proteins in the lens of eyes.

2.1.3 Research Progress of Polypeptide and Protein Drug

With rapid development of molecular biology, biochemistry technology, the research of polypeptides has made a startling and epoch-making leap. Thousands of peptides existing in organisms have been found. It is also found that all of the cells can synthesize peptides. At the same time, almost all of the cells are regulated through peptides, which involves in hormone, nerve, cell growth and reproduction, and other fields.

Studies on peptides aim at applying them to medication and health care, detection and other fields, which will benefit all mankind. In fact, development and application of peptide drugs have become a reality, and they play a unique efficacy. For example, neurotensin (NT) can reduce blood pressure, and promote contraction of the intestine and the uterus; Derivatives of endorphins and encephalin have a strong analgesic action. Thyroid-releasing hormone (TRH) is a kind of peptide that can promote milk secretion in puerpera. A ring-shaped 14-amino-acid peptide can cure diabetes, stomach ulcer and pancreatitis. Oxytocin used clinically is a polypeptide. Peptides used widely albumin peptide, thymic peptide, serum thymus factor (FTS) all can cause differentiation of the immune T cells. Recently, China and Japan have started to use sugar peptides as the auxiliary treatment of cancer, and the mechanism is activating the lymphatic system, and so on. The medical protein chip (peptide chip) developed with biological technology is only in the size of fingernail, and the antigen molecules related with nephritis, ulcer and gastric cancer are placed on it. Just through reading instrument of chip, we can detect the functional status and variations of the related disease. Its function has been equivalent to a large or medium-sized laboratory, and it has higher efficiency than traditional medical tests by hundreds to thousands of times and the client nearly has no pain. Wide application of peptide chips has initiated a technology revolution in medical clinical tests.

In the 21st century, the biggest disease in the mankind will be the diseases caused by ways of life. Peptides will play out its great power in improving and adjusting the way of people's life. Maybe one day, in the morning we will take peptide adding vitamins, providing energy for one-day work; at noon we will take peptides of comprehensive nutrition for simple lunch; facing dinner in high content of lipid and high quantity of heat, we will take lipid-decreasing peptide, gastrointestinal required peptide, meeting the desire for delicious, but not making us forget dietary control; and before sleeping, we will take sleep-inducing peptide so as to have good dreams repeatedly. Besides, peptide to eliminate fatigue quickly, chocolate peptide candy to resist on strain, peptide nourishing cream and anti-aging peptide to keeping youth beauty, peptide ginseng tablet to let the old man enjoy sunset good time, and polypeptide gum to resist on radiation of the screen display, all of these of functional peptides are not scientist's fantasy imagination, but you are coming into our peptide world.

Protein medications are defined as the configuration protein administered *in vitro* for treating human diseases, and the feedstock of protein medications mainly contains natural biological material, including the human body, animals, plants and microorganisms and so on. Their production method can be divided into general biological method, genetic engineering method and the chemical synthesis method. In the above three methods, the protein medications, that are manufactured by general biological method, account for the overwhelming majority. However, it is estimated that protein drug from the human body or animals, or plants will decrease in the future, so as to reduced possibility that medications undertake pollution of bacteria, while with the method of biological technologies for drug production, protein medications and vaccines will become more effective and safer.

The sales figure of protein preparations on the global market is approximately 30 billion dollars, and it is estimated that the figure will amount to 43 billion dollars by 2004. On the average, the price of protein medications manufactured through genetic engineering technologies, is higher than that of those drug products manufactured by general biological method. For example, for the recombinant cytokine drugs, the average price per gram drug substance reaches as high as 440 thousand dollars, and the average unit price of protein vaccines is about 370 thousand dollars. On the contrary, for some enzyme drug products in large productivity (average unit price, 12 dollars/g), blood preparations (average unit price, 8 dollars/g) and protein antibiotics (average unit price, 5 dollars/g), their prices are relatively lower. The drug products with high price like erythropoietin and insulin, the sales figure of the two medications accounts for 40% in the protein drug market; for others such as interleukin (IL), blood factor (BF), and cell colony stimulation factor (CSF), their individual market share surpasses 10%. These genetic engineering drugs with small quantity and high quality are the main product on the entire protein medication market.

2.2 Structure and Function of Nucleic Acids

Nucleic acids are one kind linear polynucleotide, and their basic composition unit is nucleotide. In nucleic acid molecules, the most common basic group is purine base, such as adenine(A), guanine(G), and pyrimidine base, such as cytosine(C), uracil (U), thymine(T). Mononucleotide is composed of a pentose molecule, a basic group and a phosphoric acid molecule. The phosphoric acid and the pentose in all of nucleotide molecules is the same in structure, but the basic group is not. In ribonucleotide molecules, the nucleoside is formed through N-C bond condensation reaction between the first carbon atom of pentose (deoxyribose or ribose) and the first nitrogen atom of pyrimidine or the ninth nitrogen atom of purine, and then the pentose hydroxyl undergoes phosphoric acid esterification to form nucleotide. Nucleotides are divided into ribose nucleotide (RNA) and deoxyribose nucleotide (DNA), and different nucleotides are named according to the first basic group (Table 2.2).

Table 2.2 Composition and structure of nucleotides

Nucleic acids	Nucleotides (abbr.)	Nucleosides	Basic groups
RNA	Adenylic acid (A,AMP)	Adenosine	Adenine
	Guanylic acid (G,GMP)	Guanosine	Guanine
	Cytidylic acid (C,CMP)	Cytidine	Cytosine
	Uridylic acid (U,UMP)	Uridine	Uracil
DNA	Deoxyadenylic acid (dA,dAMP)	Deoxyadenosine	Adenine
	Deoxyguanylic acid (dG,dGMP)	Deoxyguanosine	Guanine
	Deoxycytidylic acid (dC,dCMP)	Deoxycytidine	Cytosine
	Deoxythymidylic acid (dT,dTMP)	Deoxythymidine	Thymine

Notes: Nucleotides have already become a taxon including ribose nucleotide and deoxyribose nucleotide. In nucleic acid molecules, they are indicated by A, G, T, C, respectively; under free state, deoxyribose nucleotides, dA, dG, dT and dC.

It has have discovered that hereditary characteristic is determined by genes through long-term studies, but the genes linearly distribute in chromosome, and the chromosome is composed of nucleic acids and proteins. In 1944, Avery's *pneumococci* transformation experiment *in vitro* had proven that the genetic material was DNA, but not protein. Hershey and Chase's transfection experiment of T_2 phage in *E.coil* in 1952 helped verify that DNA was the genetic material. In 1953 James Watson and Francis Crick came up with the double helix structure model of DNA. DNA is made up of two nucleotide chains, which are double helix structure with complementary base pairs. Such structure model enables many heredity phenomena be explained thoroughly and reasonably at molecular level, and has become a milestone in development history of molecular biology.

2.2.1 Structure and Function of RNA

RNA is unbranched linear ribonucleotide polymer, and it contains 4 kinds of basic groups, which are A, G, C and U. The composition of RNA basic group does not look like DNA that has the strict rule of A-T, G-C. Only partial constitution is double helix in the structure of RNA. In animals, plants and microorganism cells, they all include 3 kinds of main RNA, namely ribosome ribonucleic acid (rRNA), transfer ribonucleic acid (tRNA) and messenger ribonucleic acid (mRNA), and additionally eukaryotic cells also contain a few small nuclear RNAs (snRNA).

2.2.1.1 Structure characteristics of tRNA

All tRNAs, whether from animals, plants or microorganisms, have many common grounds in structure. It is the RNA for transferring amino acids.

(1) The relative molecular mass of tRNA is approximately 25, and is composed of 70-90 nucleotides, the sedimentation coefficient is about 4S.

(2) There are many rare basic groups in the basic group composition, such as clouble hydrogen

uracil (DHU), pseud ouridine (Ψ) and methylation purine (mG, mA).

(3) The 3'- terminus is… CpCpAOH, and it is used for accepting one activated amino acid, and so called the accepting terminus. The 5'-terminus mostly is pG…, or pC.

(4) The secondary structure of tRNA exhibits clover-shape, and it is shown as Fig.2.3. There are five integral parts in this clover structure, and they are amino acid arm, dihydro-uracil loop, anticodon loop, extra loop and TψC.

(5) Secondary structure of tRNA must conduct further folding and twisting in the cell, so as to make free energy inside of molecules minimize. The clover-shape structure of tRNA again curves, and then forms the structure of inverted L-shape in a whole, which is called tertiary structure of tRNA, shown as in Fig.2.4.

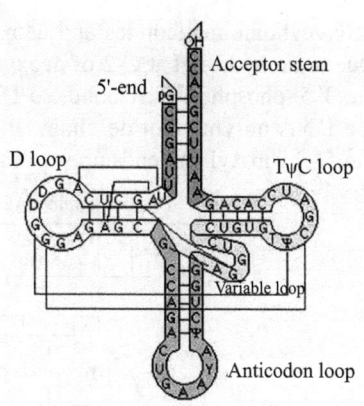

Fig.2.3　The secondary structure of tRNA

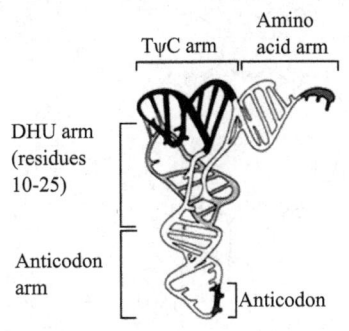

Fig.2.4　The tertiary structure of tRNA

Among the three kinds of RNA, tRNA has the smallest molecular weight and constant types. Its complete spatial structure has been researched thoroughly and can be synthesized artificially. The first nucleic acid synthesized artificially by our Chinese scientists in the world is tRNA.

2.2.1.2　Structure characteristics of mRNA and rRNA

In relationship between of DNA and protein, mRNA plays the role of media. The functions of mRNA lie in transcribing base sequence on DNA template strand into the base sequence of RNA molecules (mRNA), and then translating base sequence on mRNA into amino acid sequence of protein molecule. In most of eukaryotic cells, there is polyadenine in a length of about 200 base pairs at the 3'-terminus of mRNA, but it is not in prokaryotic mRNA. The structure of the polyadenine is related with mRNA transportation from cell nucleus to cytoplasm, and also with a half-life of mRNA. The newly synthesized polyadenine of mRNA is longer than ageing. There is a special structure (also known as cap-like structure) in the 5'-terminus of eukaryotic cells, and this special 5'-terminus has function of anti-nuclease hydrolysis. Some mRNA in the virus has the similar 5'-terminus. This special structure may be related to translation initiation in the process of protein biosynthesis.

Ribosomal RNA, rRNA, is the site for synthesis of various functional proteins. Ribosome RNA is divided into three kinds. Prokaryotic ribosome RNA is composed of 5S rRNA, 16S rRNA and 23S rRNA, which respectively contains 120, 1541 and 2904 nucleotides residues, and rRNA is a stable molecule, its precursor is larger than the mature rRNA molecules.

2.2.2　Structure and function of DNA

2.2.2.1　Primary structure of DNA

The primary structure of DNA is a the linear or ring polymers, which is made up of a large number of four kinds of deoxyribonucleotides (deoxyadenine nucleotides, deoxyguanine nucleotides, deoxycytidine nucleotides and deoxythymidine nucleotides), through the 3', 5'-phosphodiester bond. Due to no hydroxyl at C-2 of deoxyribose molecule, C-1 connect with the base and only formation of the 3',5'-phosphodiester bond, so DNA does not have side chain. Fig.2.5 shows a small fragment of the DNA polynucleotide chain. Polynucleotide chain begins with phosphoric acid connected to the 5'- hydroxyl and ends up with the 3'- hydroxyl of deoxyribose.

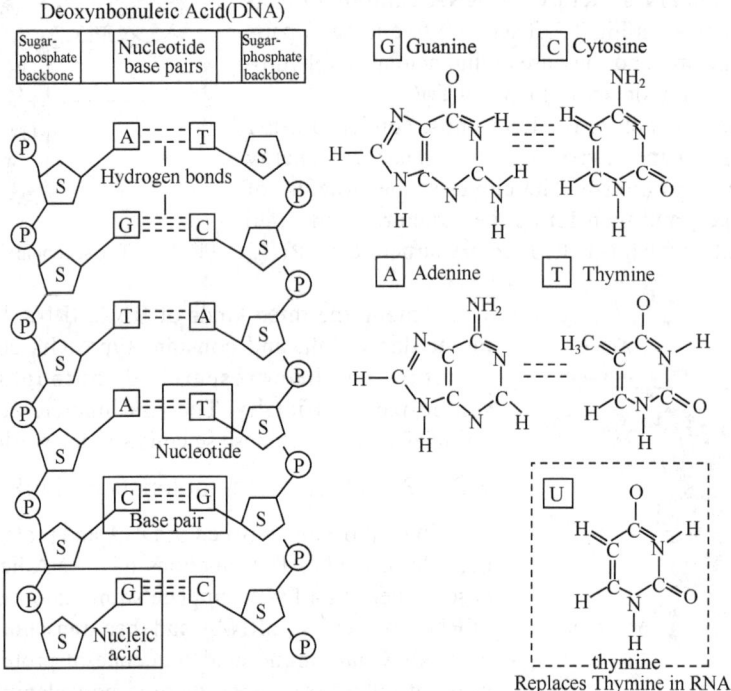

Fig.2.5 A small fragment of DNA polynucleotide chain

2.2.2.2 Double helix structure of DNA

DNA can form high compression chromosomes which are the carrier of genetic information. According to the results of quantitative determination of DNA components and X-ray diffraction data, James Watson and Francis Crick proposed a model for the structure of DNA in 1953. According to this model, it was predicted that DNA would exist as a helix of two complementary antiparallel strands, which would surround each other in a rightward direction and stabilized by hydrogen bond between bases in adjacent strand. A lot of research results prove that this model of the double helix structure is almost correct. The model points are as follows:

(1)DNA molecules are the double helix structure, which consists of two antiparallel polynucleotide chains around a central axis. The direction of the polynucleotide chain depends on the trend of the phosphodiester bond between nucleic acids. Usually the 3'→5' is positive direction, and the two chains are right-handed helix. From any fixed position in the helix, one strand is oriented in the 5'→3' direction and the other in the 3'→5' direction. On its exterior surface, the double helix of DNA contains two deep grooves between the ribose-phosphate chains. The two grooves are of unequal size and termed as the major and minor grooves. One is deep, and the other is shallow.

(2) The bases in the two chains stack inside of the spiral, and the base plane and the vertical axis are perpendicular. Phosphoric acid connects to ribose by 3′, 5′-phosphodiester bond and forms the main chain framework of DNA, with the main chain on the outside of the spiral and the bases on the inside of the helix. The distance between the two chains is 2nm, and the angle of between the two ribonucleotide chains is about 36°. So there are 10 nucleotides along the central axis of each rotation of a circle. The height (thread pitch) of each rotation is 3.4 nm. The relation between chromosomes and DNA and the structure of DNA are shown in Fig.2.6.

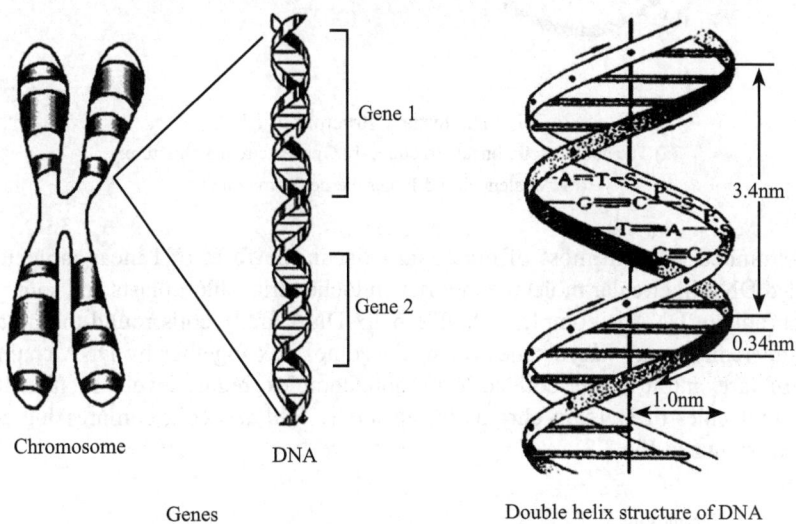

Fig.2.6 The relation between chromosomes and the structure of DNA

(3) Two nucleotide chains link together by forming hydrogen bonds between bases. Purine bases form hydrogen bonds with pyrimidine in the crucial phenomenon of base-pairing. In any given molecule of DNA, generally, the concentration of A is equal to that of T and the concentration of C is equal to that of G. This means that A will only pair with T, and C with G. According to this pattern, known as Watson-Crick basepairing, the basepairs composed of G and C contain three hydrogenbonds, whereas those of A and T contain two This makes G-C basepairs more stable than A-T ones.

Because of three kinds of working forces, double helix structure of DNA is stable. The first force is hydrogen bonds between complementary base pairs; the second one is base stacking force caused by interaction between π electrons of the aromatic bases; and the third one is ionic bond formed between negative charges of phosphate residues and cation in medium, and in fact there are a large number of DNA-binding regions such as Na^+, K^+, Mg^{2+}, Mn^{2+} in the cell.

Described above is the type B structure of DNA, and in addition, there are Type A and Type C structures. we need not to introduce one by one.

2.2.2.3 Tertiary structure and function of DNA

Most of double-stranded DNA is linear, but DNA, in certain viruses, mitochondria, chloroplasts and some bacteria, is double-stranded circular. These circular DNA can further twist and fold into superhelix, i.e. tertiary structure in the cell, see Fig.2.7.

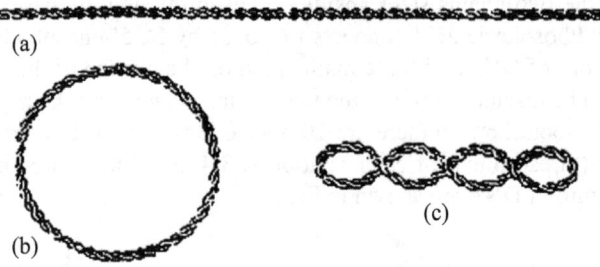

Fig.2.7 The tertiary structure of DNA
(a) The double helix linear structure; (b) Opened circular structure;
(c) Covalent closed-loop superhelix structure

Double-stranded DNA in most of organism exist in the form of Linear molecule, but some double-stranded DNA in circular molecule in virus, mitochondria, chloroplasts and some bacteria. The structure of chromatin DNA is complex. Double helix DNA firstly coils round the proteins forming karyosome (supercoil), many karyosomes (or nucleosome) link together by DNA chain constituting a bead-like structure, and it further coils to form more complex, higher level structure. It is estimated that the giant molecules of DNA in chromatin repeatedly fold and coil, compressing about 8000 to 10,000 times, as shown in Fig.2.8.

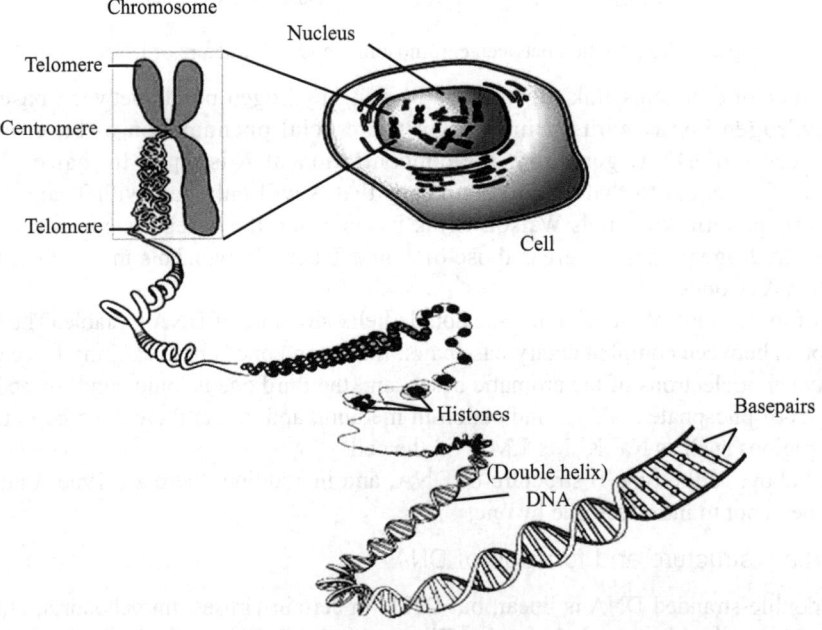

Fig.2.8 The relation of DNA and chromatin and cell

2.2.2.4 Properties of DNA

Properties of nucleic acids are closely related with its composition and structure. Nucleic acids consist of purine base, pyrimidine bases, phosphoric acid, ribose and deoxyribose. Structure of nucleic acids is characterized by great molecular, conjugated double bond, hydrogen bond, glycosidic bond and the phosphodiester bond, enol alkyl, free amino radical, phosphate group, and these structures are basis of characteristics of nucleic acids.

1) DNA denaturation

When temperature increases, the three-dimensional structure of nucleic acids will be destroyed, so that the regular helical double-stranded structure of nucleic acids changes into a irregular single-stranded "coils". Therefore, the process of transition from the natural state of DNA molecules to the denatured state is called degeneration or melting. Heating the double-stranded DNA or natural DNA, the binding force between the two chains are damaged and the two chains are separated, and so denatured DNA is single stranded. When pH value is greater than 11.3, DNA is very stable to the alkaline hydrolysis, so that the method is often used for DNA denaturation in genetic engineering.

Although DNA is relatively stable, but it is a dynamic structure, and the double-stranded areas are often opened into single-stranded bubble. This phenomenon, that a part of the double-stranded areas of the double-stranded DNA are opened, is known as respiration. Respiration phenomenon can make specific protein interact with DNA molecule and "read" its encoded information. Because there are three pairs of hydrogen bonds between G-C base pairs, only two pairs between A-T base pairs, so respiration phenomenon is prone to occur in the A-T rich region of DNA molecules than the G-C rich one.

2) Renaturation and hybridization of DNA

After DNA denaturation, the double-helical chains separate. If the solution is rapidly cooled, two single chains remain separate, but if slowly cooled, the two chains may specifically recombine and restore to the double-helix. This phenomenon that denatured DNA restores its original structure and nature is called renaturation or annealing. The re-formation of DNA is known as renaturation of DNA. Renaturation is of great significance in the genetic engineering. It can be used to display the genetic correlation between different organisms and detect specific DNA, so as to examine whether some of the sequences in DNA of specific organisms appear more than once, and determine the position of the specific base sequences in DNA.

The renaturation occurrence must have two conditions. Firstly, saline concentration in the system must be enough to eliminate the electrostatic repulsion of the phosphate groups in the two chains, usually using 0.15-0.5mol/L NaCl; secondly, temperature must be high enough to destroy the random intrachain hydrogen bonds, but it may not be too high, otherwise a stable pairing cannot be formed and maintained.

When DNA of different sources forms renaturation DNA molecules, such renaturation is called hybridization. The hybridization technology is an important means to study nucleic acid function. Nitrocellulose membranes can combine with single-stranded DNA firmly, but it cannot combine with the double-stranded DNA. The important purpose of this technology is detecting sequence homology between single-stranded DNA and RNA molecules, and can be called DNA-RNA hybridization. The detailed procedure to copy RNA molecules from the specific DNA molecules is as follows: put filterable membrane binding single-stranded DNA in a solution containing radioactive RNA, wash the filterable membrane after renaturation, and according to radioactive RNA on the membrane detect whether hybridization occurs. Changes of denaturation and renaturation of RNA are similar with

DNA, but the extent of change is smaller than DNA, because there is only part helix structure in the RNA. Nucleic acid denaturation are not involved in breaking of covalent bond between nucleotides like protein denaturation, thus does not cause reduction of the relative molecular mass.

DNA denaturation, renaturation and hybridization in experiments on genetic engineering are often used to develop a variety of new research methods for genetic engineering, which can be applied to various fields of life science research.

2.2.3 The Central Dogma of Molecular Biology

Understanding of life phenomena fundamentally changed after the double helix of DNA was put forward in 1953. James Watson and Francis Crick believed that the heredity, variation, development and growth would follow a rule: the central dogma of molecular biology. The central dogma of molecular biology refers to the biological information transferring from DNA to RNA and then to the protein, which constitutes the basis of the entire molecular biology. The central dogma of molecular biology is shown in Fig.2.9.

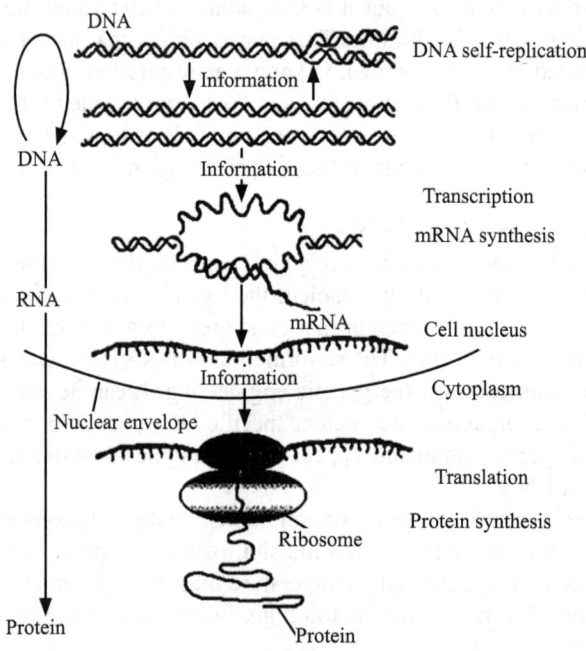

Fig.2.9 The central dogma of molecular biology

2.2.3.1 Semi-conservative replication of DNA

In the process of life-sustaining, only when the genetic materials are transmitted accurately from the parental generation to the filial generation, it can maintain stability of the species. Modern biology has fully proved, except a small number of RNA viruses, the genetic materials of all other organisms connecting the parental generation with filial generation are DNA. The genetic information of organisms encoded on the DNA molecules in the form of codons, is manifested as a particular nucleotide order, and passes from the parental generation to filial generation by DNA replication. In the process of growth and development of filial generation, the filial generation shows hereditary character similar to the parental generation through expression of genetic information. In the living kingdom, molecular mechanisms of transmission and expression of genetic information in various organisms are the same.

Genetic information is stored in the nucleotide order, so the genetic information determining its structure specificity can only come from itself in synthesis of DNA. Commonly so called replication of DNA, also called semi-conservative replication, refers to the process of using the original DNA as a template to synthesize DNA, and is strictly proceeding in the form of the semi-conservative replication. So-called semi-conservative replication refers to the procedure of DNA replication, that hydrogen bonds between the DNA molecular bases firstly break, the two chains open, the double helix is unwound and separate, and then respectively with each fundamental chain as a template, in accordance with the principle of complementary of base pairing, under the action of DNA polymerase, the new sub-strands complementary with the original template are formed respectively. One DNA molecule will form two identical DNA molecules of filial generation through DNA replication. DNA replication begins with the initial point (ori), which mostly is bilateral and symmetrical. In this process, one strand is always from the parental DNA, and the other strand is newly synthesized in the filial DNA molecules. The result of replication is forming of two progeny DNA molecules with identical nucleotides sequence, and the new DNA molecule contains an old strand and a new one.

Conditions of DNA replication are as follows: ①DNA duplicase include DNA polymerase, DNA ligase; ②template is the direction of $3'\rightarrow 5'$ single-stranded DNA; ③substrate is four kinds of deoxyribonucleoside triphosphate; ④primers are a fragment of nucleotide; ⑤the $3'\rightarrow 5'$ of exonuclease, its role is correcting mispairing, and magnesium ion plays a role of activator.

2.2.3.2 Transcription of DNA

On the DNA molecules within the cell, only the fragment of gene encoding can transcribe and synthesize the corresponding RNA (mRNA, tRNA and rRNA). The process of RNA synthesis using DNA as template is called transcription of DNA, while synthesis of DNA using RNA as a template is called reverse transcription. The process of RNA synthesis using original RNA as a template is called RNA replication.

There are three regions at least in a transcribing gene (Fig.2.10), and the sequence for transcribing RNA site in the middle of gene, it is called a transcription unit. There are some conserved sequences in the up-stream of the transcription unit, which are regions where RNA polymerase recognizes and binds to, and is known as the transcriptional promoter, after RNA polymerase binds to the promoter and then moves on to the transcription unit, transcription begins. In the down-stream of transcription unit is the transcription terminator, which offers the signal for transcription termination and withdrawal of synthesized RNA.

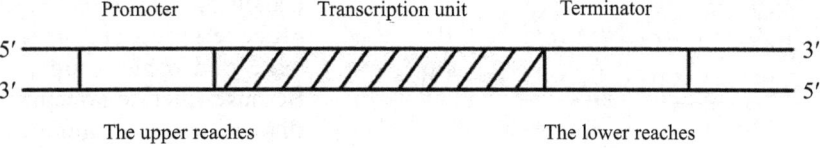

Fig.2.10 Gene structure schematic diagrams

Transcription RNA in eukaryotic cells will need to be modified, adding a cap sequence in the 5′-end of the mRNA, which plays the role of identifying translation, and make it protected from destruction of nuclease. Poly A in the 3′-end of the mRNA is added. In most eukaryotic cells, gene transcription units contain one to several non-coding intergenic regions of various lengths, known as introns making coding regions separated. A process that cuts intron by enzyme, and makes exons link together to become mature mRNA, is known as mRNA splicing.

2.2.3.3 Translation of RNA

Translation is a process carrying corresponding amino acid to the ribosome-mRNA complex by the tRNA according to the genetic information on the mRNA assembling a specific peptide chain. Large and small subunits of the ribosome are free in the cytoplasm. In protein synthesis, they have to bind with mRNA to form complete ribosome, in order to play the function of translation.

2.3 Structure and Function of Polysaccharides

Polysaccharides are macromolecular compounds through polymerization of monosaccharides or monosaccharide derivatives, including starch, cellulose, hemicellulose, chitin, pectin and so on. Starch is the carbon source used by the most microorganisms, but cellulose, hemicellulose, chitin, pectin, by only some special microorganisms. All of polysaccharides have certain biological functions.

2.3.1 Structure and Function of Starch

Most of microorganisms can use starch as the carbon source to growth. Starch is a kind of macromolecular substance formed by linking glucose through glycosidic bonds. Starch is divided into amylose and amylopectin. The former is straight-chain molecule formed through α-1,4-glycosidic bonds. Because the latter contains many side chains, which are linked by α-1,6-glycosidic bonds, and the straight-chain is linked by α-1,4-glycosidic linkages. In natural starch, generally amylose is about 10%-20%, and amylopectin is about 80%-90%. Starch structure is shown in Fig.2.11.

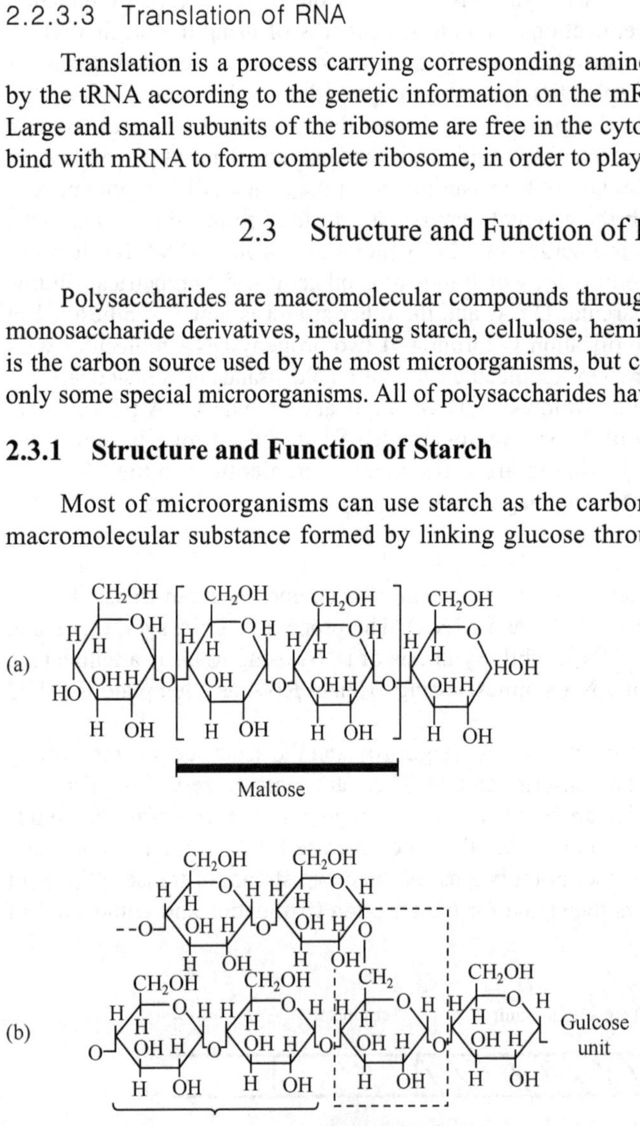

With enzyme method or acid method, starch can be hydrolyzed into a mixture of glucose, maltose, maltotriose and so on, which is called glucose syrup and used in candy, cake and other food production. Because fructose sweetness is higher than sucrose, in general, glucose obtained with starch hydrolysis may be converted to fructose by the glucose

Fig.2.11 The structure of starch
(a) Amylose; (b) Amylopectin

isomerase in the industry, which is known as fructose-glucose syrup with high sweetness and low dosage. Diabetes patients should not take up glucose and fructose, and may commonly use sorbitol as the sweetener. Considering weight control, people increasingly require for those sweeteners with low non-energy source, high nutrition and safety, while biotechnology can solve this problem. Dipeptide sweetener is an important food additive produced with biotechnological process. Aspartame with high sweetness and nutrition is composed of aspartic acid and phenylalanine.

2.3.2 Structure and Function of Cellulose

Cellulose is the main component of plant cell wall, and is a straight chain macromolecular compounds with β-1,4-glycosidic bond linking glucoses. Human and animals cannot digest cellulose, and some microorganisms have the ability to break down cellulose, which is relatively stable in the environment, not soluble in water and it also can be broken down into monosaccharide by cellulase Cellulase is a generic term for a hydrolysis enzyme of cellulose, or the cellulase complex. The structure of cellulose is shown in Fig.2.12.

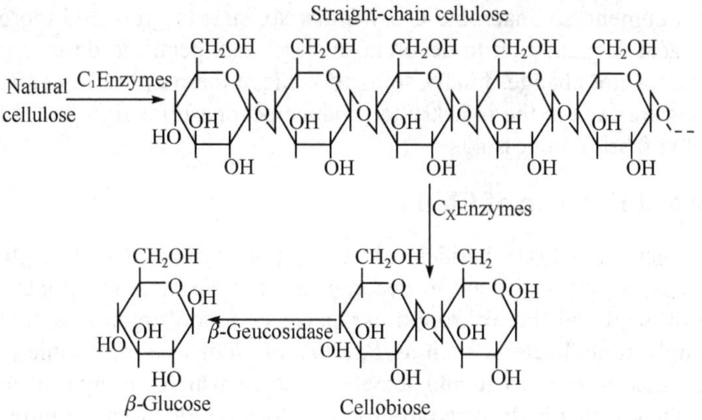

Fig.2.12 The structure of cellulose

Cellulose material is rich in nature. Plant stem and leaves, straw and bran husk all contain a great amount of cellulose. In addition to paper-making raw materials and crude animal feedstuff, in the past most of cellulose was used as a fuel or waste for disposal. If the ability of microbial decomposition of cellulose can be effectively utilized, a large number of farm and side products and wastes can be turned into glucose, products of single-cell protein (SCP), so as to make cellulose transform into fermented foods, fermented feedstuff, and raw materials of fermentation industry, and turn wastes into wealth, benefiting mankind.

Cellulose is the main raw material of the construction industry, paper industry and textile industry. Cellulose yield is 10^{11} t/a synthesized through photosynthesis in plant in the earth. If it is burned as a wood directly, energy utilization efficiency is only less than 10%. Therefore, application of bio-tech to energy production from cellulose is also one way to solve energy crisis faced by mankind.

2.3.3 Structure and Function of Pectin

Pectin, which plays the role of "glue" in plant tissue, exists widely in higher plants, especially in the tissue of fruits and vegetables. It is one important component of intercellular substance and primary cell wall in higher plants. Pectin is high polymer of straight chain, and its component units are linked by α-1,4-glycosidic bonds between D-galacturonic acid. The majority of carboxyl groups on the D-galacturonic acid can be esterified to form methyl ester by methanol. Pectin without methyl ester is called as pectic acid. The pectin substance includes pectic acid and pectin resin acid.

The natural pectin often refers to protopectin (insoluble pectin). It is converted into water-soluble pectin by the protopectinase, and is further decomposed to remove the methyl ester groups by pectin esterase, to produce pectic acid, and at last is hydrolyzed to form galacturonic acid by polysemigalacturonatase (PG), by cutting α-1, 4-glycosidic bond. The latter is broken down in the glycometabolism pathway to release energy.

Traditional hemp soaking and degumming procedure used by many Chinese people is an embodiment of utilization of pectin decomposing bacteria. Generally, hemp stalks are soaked in water to make them ferment naturally, plant tissues absorb water to expand by soaking, and exude some soluble substances such as sugar and others which are useful for growth of aerobic bacteria, lactic acid bacteria and yeast. These microorganisms may consume fermentable substances so as to result in an anaerobic environment, so anaerobic *Clostridium* sp. start to grow and reproduce. *Clostridium* sp. conduct butyric acid fermentation to decompose starch and pectin to destroy plant tissues, while the tough cellulose does not change, finally so as to achieve the purpose to degum the hemp stalks. Now microorganisms are used in the industry to produce pectinase for fruit juice clarification, orange excystation and cotton fabrics processing.

2.3.4 Structure and Function of Chitin

Chitin is a nitrogenous polysaccharides formed by polymerizing N-acetyl glucosamine through β-1,4-glycosidic bonds, which is difficult to decompose. It is the main component of the fungal cell wall, the insect body wall and the arthropod carapace; common organisms cannot break down it and utilize it; and only some bacteria such as *Bacillus chitinovorus*, and some actinomycetes such as *Streptomyces*, are able to synthesize and secrete chitinase which can hydrolyze chitin to produce chitobiose which is then further hydrolyzed to produce N-acetyl glucosamine through chitobiase. The latter is further decomposed to produce glucose and ammonia. Chitin can be processed to get chitosan through chemical method. Both chitin and chitosan have a wide scope of uses, and can be used to process and absorb heavy metal ions in polluted water in environmental protection, to make artificial skin for treatment of burn patients, and to prepare surgical suture lines. Due to hypolipemic efficacy and stimulation of immune responses, recently chitin has been widely used in health foods.

Some polysaccharides derived from plant and fungus, such as lycium barbarum polysaccharide, angelica sinensis polysaccharide, lentinan, pachymaran and so on, have strong function of activation of the immune system, and so have been used for accessory treatment of cancer patients. Some polysaccharides derived from marine organisms have anti-HIV effects. In short, development of the polysaccharide products has great promise.

Chapter 3

Gene and Genome

Gene is a fragment of base sequence with specific functions in DNA or RNA molecules, and DNA molecules are the carrier of genes. The transformation experiment in bacteria with indisputable facts has proved that conversion factor making hereditary characters transformed is DNA instead of protein or RNA. However, it must be stressed that all of genes are not made up of DNA in the living kingdom. Some animal viruses, plant viruses and phages have a genetic system based on RNA but not on DNA. For example, in studies on tobacco mosaic virus (TMV), A. Gierer and G. Schramm at first found that RNA molecules could transmit genetic information, and meanwhile they still demonstrated that the RNA component of TMV could induce synthesis of new viral particles in infected leaves.

3.1 Concept of Gene and Its Essence

3.1.1 Contemporary Concept of Genes

With continual progress in molecular biology and molecular genetics, we can research on structure and functions of genes at the molecular level and have discovered many different kinds of genes, so as to deepen our understanding about essence of genes and enrich theoretical basis of genetic engineering. And then concept of genes expands gradually to become one phrase with the most rapid change in modern life science.

1) Super genes

Super genes refer to several closely linked genes controlling one character or a series of related characters. Super genes address close link between members in their arrangement.

2) Gene cluster

Several members in one gene family arrange closely in one chromosome to form a gene cluster which is also called super genes. For example, there are two gene clusters, α-gene cluster and β-gene cluster, in the globulin family in human beings.

3) Gene family

Gene family refers to a series of genes with identical origins, similar structures and related functions. It is evolved from the same progenitor by means of reproduction and variation. Each member in one gene family may gather into a cluster arranged on one chromosome, and also may disperse on different chromosomes.

4) Super gene family

Super gene family, also called gene super-family, is integration of several gene families with approximately identical structures but with no absolutely same functions, formed from the same progenitor gene by means of various modes of variation.

5) Pseudogenes

In some multi-gene families, some members do not produce functional proteins, but have similarity with the others in structure and DNA sequence, and so are called pseudogenes. Pseudogenes have certain homology with functional genes, but some reasons such as gene mutation lead to loss of their functions. As for origin, it is now considered that pseudogenes may be formed through gene transposition mediated by reverse transcriptase.

6) Split genes

In the past researchers agreed with that genetic coded of one gene were juxtaposed together continuously to form an entire gene entity without any gap. However, later in studies on gene structure in eukaryotic organisms it was found that there was one fragment of non-encoding sequence between two encoding sequences, which might make one gene be split into various discontinuous regions. So those disconnected genes with discontinuous encoding sequences are called split genes. The encoding sequences are called exon, and the non-encoding sequences intron.

7) Movable genes

Genes are mostly fixed on one site of the chromosome, but some genes have movability on the chromosome, and such genes are called movable genes, also called transposable elements or transposable factors. The concept of movable genes was proposed at first by American geneticist Dr. Barbara Mc Clintock in 1951. She thought controller factor was integrated on one gene locus that might lead to new mutation of the gene; but after the controller factor was excised from the chromosome, the phenotype of the gene locus would restore to be normal. These elements can move on the chromosome, and some elements have no gene products but locate on the gene locus controlled by them and act as receptors of other regulatory factors. Regulatory factors may move actively and dominate their receptors to move.

8) Overlapping genes

In traditional opinion, code of a gene is arranged on the DNA chain, and overlap between genes cannot occur because each gene is read in turn. DNA sequence of one gene cannot overlap with that of another gene. However, with development of DNA sequencing technique, it has been found that sometimes nucleic acid sequences are shared among different genes, i.e., the nucleic acid sequences in some phage and viruses are overlapped. Such overlap between genes may occur both in a head-tail way and in a nesting way. Such two kinds of genes are called overlapping genes or nested genes.

9) Extra-chromosomal genes

Chromosome is the carrier of genes, and genes in organisms mainly locate on the chromosome. In prokaryotes, there are no obvious demarcation line between cytoplasm and cellular nucleus in which chromosomes are generally naked circular DNA molecules, i.e., no protein molecules are bound on DNA molecules. But chromosomes in eukaryotes locate in the cellular nucleus, and are generally linear molecules bound to protein molecules. In either prokaryotes or eukaryotes, some genes exist out of chromosomes, and such genes are called extra-chromosomal genes. Transmission of these genes do not conform to Mendel's laws of genetic separation and free combination, and is so called non-Mendel inheritance.

3.1.2 Features of Genes

Gene has many characteristics, such as discontinuity, repeatability, mobility and overlaps. There are gaps within a gene, and generally, those sequences which bring about gene gaps are called introns in which sequences may be automatic excised in RNA synthesis. Discontinuousness of genes is found only in eukaryotic organisms. Based on comparison of the genetic structure of eukaryotes and prokaryotes, we found that genomic DNA in almost all eukaryotes (except the single-cell yeast) have repeated sequences, and repetitive sequence of copy number can be as high as millions of copies of the above in some cases. In human DNA, repetitive sequence (at least 20 copies) of DNA account for around 30% of the total DNA. Mobility of gene refers to that some genes may drift at different sites on the same DNA molecule, but overlapping feature of gene refers to that the same base at the same site may encode for more than two genes, i.e., one base may be used over twice, which is cubic distribution of genes. This may lead to the sum of various genes sequences greater than the actual sequence length of nucleic acid in number.

3.1.3 Representative Symbols of Genes

Initial gene symbols were represented with the first capital letter of English names indicating some one associated character. At present the following naming rules are adopted. ①Each gene is represented with three italic small letters which should be taken from the previous three letters of one or one group of English words indicating feature of this gene. ②Different genes controlling the same phenotype should be represented with different italic capital letters after the three letters. For example, *trp* represents the tryptophan gene and various tryptophan genes are represented with *trp*A and *trp*B respectively. ③Those mutated genes should be represented by adding the symbol "−" at the top right corner. For example, leucine-deficiency phenotype is presented with *leu*⁻. Drug-resistant genes should be represented by adding "r" at the top right corner and adding "s" represents drug-sensitive genes. For example, the streptomycin-resistant gene is represented as str^r. ④The phenotype of some mutated gene should be represented with the three corresponding letters, but the first letter should be capitalized. For example, the gene symbol of lactose fermentation deficiency phenotype is *lacZ*⁻, and then its phenotype symbol is LacZ⁻. ⑤When gene deficiency exists on chromosomes, the deficiency gene symbol should be put in the blank after the symbol"Δ". For instance, Δ (*lac*, *pro*) represents that the chromosomal segment from lactose fermentation gene to proline gene is deficiency.

3.1.4 Relation between DNA Base Sequences and Amino Acid Sequences

DNA molecules are carrier of genes, and then, is every fragment of DNA a gene? According to the traditional concept of genes, genes on chromosomes or DNA molecules are linked one by one like beads through those non-inherent material. Gene exchange occurs only between genes but not within a gene. In other words, genes are not only functional units of inheritance, but also gene exchange units and gene mutation units. However, later many studies, especially those with T4 phage as research material, have indicated that it is not the fact.

The experiment on T4 phage infection of *E.coli* has shown that a fragment of nucleotide acid sequence is just one cistron which is equivalent to one DNA or RNA unit of genes and can encode one entire peptide chain. Such peptide chain may be a protein with biological activity, and also may bind to another peptide chain to form a multi-functional protein. One cistron is a functional unit, and is composed of many mutable sites between which gene exchange may occur. In modern literals on genetics, the two terms, cistron and gene, are interchangeable. Generally, one cistron is just one gene, it consists of about 1500 nucleotide acid pairs, and is linear structure composed of a group of mutation units and recombination units. Therefore, concept of cistron indicates that gene is not the smallest unit, and is dividable; and not all the DNA sequences are genes, while only some specific nucleotide acid fragments are the encoding regions of genes.

Now it is known that genes are "blueprint" of all RNA and protein molecules in a cell. Final products of some genes are RNA molecules, such as rRNA genes, tRNA genes and other small RNA genes. Most of genes have final products of peptides or proteins. These protein molecules are synthesized with mRNA mediated. The relationship of nucleotide acid sequence and amion acid sequence is shown in Fig.3.1.

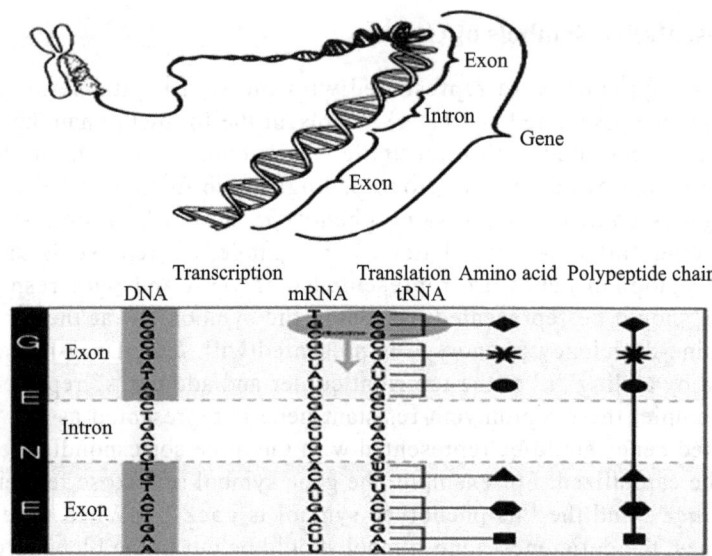

Fig.3.1 The relationship of nucleotide acid sequence and amino acid sequence

Early in 1912, when studying on human alcaptonuria, A.Garrod put forward that such disease resulted from lack of some enzyme-catalyzed reaction. However they were the scholars, G.W. Beadle and E. L. Tatum who clearly came up with the one gene-one enzyme hypothesis for the first time. They used X-ray to treat *Neurospora crassa* for mutagenesis and obtained a large number of auxotrophic mutants. Further studies have shown that every mutation results from deficiency of one single gene, and so it is thought that every metabolic reaction occurring in organisms is controlled by one specific enzyme. And this enzyme is the synthesized product of some specific gene. Once a gene mutates, the corresponding protein controlled by this gene varies subsequently, and it even loses gene activity. Just as for structure of a gene, mutation is only a random event and probably destroys gene functions, and so a great amount of mutation may bring about a non-functional gene.

The hypothesis of one gene-one enzyme once played an active role in promotion of research on genetics, but at that time people did not understand molecular essence of a gene at all. Furthermore, this hypothesis neither explained how a gene controlled over synthesis of an enzyme, nor related to the concept that a gene guided amino acids to assemble peptide chains of protein.

In studies on structure of protein, it has been found many proteins are composed of multiple sub-units. This kind of proteins are called multimeric proteins. Those proteins whose all subunits are the same belong to homomultimeric proteins, and are encoded by a kind gene. Those proteins whose all subunits are different belong to heteromultimeric protein, and are encoded by multiple genes. For example, hemoglobin is a heteromultimeric protein which is composed of two α-subunits and two β ones. Each type of the subunits is a different peptide chain, and a product of a different gene. Therefore, mutation of any one gene encoding α subunit or β subunit may reduce function of hemoglobin. In order to make it applicable for any heteromultimeric protein, the wording of "one gene-one enzyme" was revised as "one gene-one polypeptide chain", so as to reflect essence of genes more exactly.

A gene is a DNA fragment with genetic characteristics, and genetic information is encoded on nucleic acid molecules and primarily located on DNA molecules. Genetic information on DNA molecules may guide synthesis of mRNA, rRNA and tRNA. Every gene has its own genetic code (Fig.3.2). Although a gene is just only a small fragment on DNA molecule, it has most important functions. Nucleotide acid sequences on DNA molecules determine those on mRNA, while those nucleotide acid sequences on mRNA encode amino acid sequences of protein. Therefore genes determine synthesis of protein, protein determines metabolism, and metabolism determines various characters.

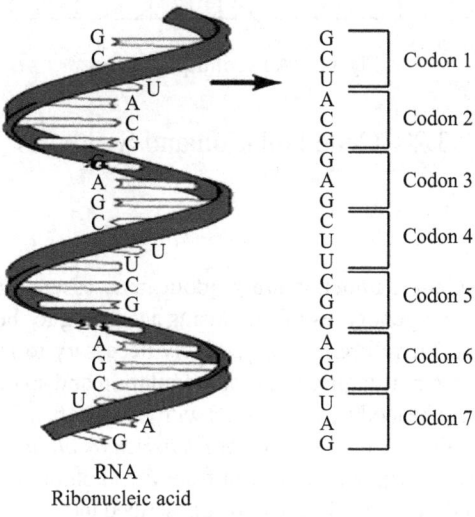

Fig.3.2 Triplet codons on a gene

DNA transmits genetic information to mRNA in a form of codes. During synthesis of protein, ribosomes move from 5'-end to 3'-end of mRNA, reading one codon every time. Every amino acid matches with the corresponding codon on mRNA through its own reverse code on tRNA, and then links into one polypeptide chain.

Since RNA consists of 4 bases, 64 triplet codons (Fig.3.3) are constituted. Among them, 3 codons are special for termination of synthesis of polypeptide chains, i.e. terminators. The remaining 61 condons correspond to 20 kinds of amino acids, and hence most amino acids have more than one corresponding codon. Therefore, these codons are degeneracy.

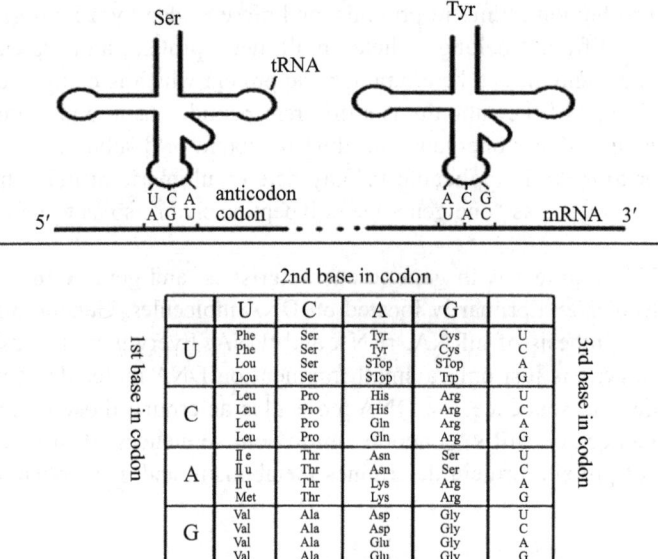

Fig.3.3   A list of triple codons

3.2   Gene Isolation and Synthesis

3.2.1   Gene Isolation

With development of science, understanding about gene essence is more and more profound. The indirect method for studying genotypes of organisms according to their phenotypes can't meet the need for science development. Therefore, it is objectively necessary to isolate the relevant genes, and conduct a series of assays on their structure, function, regulation and so on in a test tube directly.

In 1969, the research group headed by Dr. R. Beckwith from Harvard University in the United States initially isolated a special gene, β-galactosidase gene of *Escherichia coli lac* operon. They used two specific transduction phage, λ and Φ80, to capture *lac* operon from *E.coli* chromosome in opposite orientation. H strand is the sense strand (the DNA chain which can be transcribed into mRNA) of phage λ with *lac* operon gene, and L strand is the sense strand of phage Φ80. Therefore, H strand of phage Φ80 must contain the complementary sequence of the sense *lac* operon on H strand of phage λ. If these two H strands are mixed together, the only complementary nucleotide sequence is the part equivalent to *lac* operon as they are from different phages. Hence, double-stranded DNA structure is formed in this small section, and other sections still maintain in single-stranded state because they can't be complemented. The single-stranded

DNA can be hydrolyzed by deoxyribonuclease S1 which acts specifically upon single-stranded DNA. As a result, the pure β-galactosidase gene of *lac* operon can be acquired. But the theoretical significance of this experiment is far beyond its practical application value. Because it created a successful precedent of single-gene isolation, and aroused people's enthusiasm for isolation of genes from different aspects and in different ways, thus promoted progress of genetic research work. Now there are many methods for isolating specific genes, including biochemical and genetic methods, nucleic acid hybridization, restriction endonuclease cleavage, and so on. But it is generally acknowledged that the most effective method is gene cloning technology and polymerase chain reaction (PCR) technique which can isolate the individual target gene from a genome directly. To achieve efficient isolation, it needs to firstly build an experiment system that can infinitely amplify the target gene for isolation, and to secondly develop a sensitive molecular probe for target gene detection. Such probes can be RNA or DNA molecules with high specific activity which have radioactive labels and can pair with the specific sequences of target genes. After hybridizing with target genes, they can be detected through autoradiography, so as to achieve the purpose of gene isolation. Application of probes with radioactive labels significantly improves sensitivity of gene isolation, and it is the most common and effective method for gene isolation.

During gene isolation, to acquire the probe with high purity and specificity is the crucial step. It can be obtained in the way that the specific mRNA synthesizes cDNA via reverse transcription. Also it can be chemically synthesized according to the nucleotide sequence of encoding region of a target gene deduced from the amino acid sequence of the corresponding protein. Besides, the specific probe can be prepared based on protein antigen-antibody reaction as well. Applying the methods above, it is possible to isolate any gene of desired clones only if protein product of the gene is known and can be separated. At present, scientists have preliminary capacity to clone and isolate the target gene. Then with development of improved safe host strains and cloning vectors, gene cloning strategy has changed consequently, i.e., cDNA should be firstly cloned and then the target gene is isolated. Since the 1980s, gene cloning and DNA sequencing have become more simple and effective. Along with progress in synthesis of oligonucleotide, mammalian cell culture and transformation, and enzymes used for the DNA and RNA manipulation, scientists have further improved their ability of cDNA cloning. Based on these methods, scientists can obtain full-length cDNA more easily, and it is possible to clone cDNA from mRNA in low abundance. Especially due to development of PCR technology in recent years, it provides a powerful means for scientists to isolate and clone genes. That is to say, almost all cDNA of desired genes can be isolated and cloned. Even for genes encoded unknown product, they can also be isolated through those methods, such as mRNA difference display, positional clone technology, and so on.

3.2.2 Gene Synthesis

Today, scientists cannot only isolate natural genes but also synthesize related genes in the laboratory with chemical methods. The group headed by H. G. Khorana, Massachusetts Institute Technology, USA, performed early in this study. In 1970, yeast alanine tRNA gene (77bp) was completely synthesized by them for the first time. In 1976, complete synthesis of tyrosine tRNA gene of *E.coli* was accomplished by them. This gene is small, and only has 200bp, and its sequences have been determined in 1967. With diverse nucleotides comprising this gene as raw materials, Khorana and others. synthesized short oligonucleotide fragments of 10-15 nucleotides. After acquiring about 40 such fragments, they linked and sewed up these fragments with ligase to form double-stranded DNA molecules. Its sequence is exactly the same as the natural tyrosine tRNA gene in *E.coli*. The synthetic gene contains both the DNA fragments encoding structural genes (1-126bp), and two regulatory fragments, including the promoter fragment (52bp) and the terminator fragment (21bp). Khorana claimed that this synthetic gene indeed showed unique biological activity and took effect as a gene after it was transferred into *E.coli* cells. However, this *E.coli* tyrosine tRNA gene obtained via chemical synthesis can't generate peptide products, though it can exhibit gene activity *in vivo*.

In October 1977, the research group led by Doctor H. W. Boyer linked the chemically synthesized human brain hormone gene, i.e., the somatostatin gene, to the *lac* operon and introduced it into *E.coli*. This is the first time for human beings to transfer a eukaryotic gene successfully into a prokaryote cell, and the gene can be transcribed and translated to produce protein with bioactivity by DNA recombination technology. Theoretically, genes at any size can be synthesized with chemical method. But if the gene is relatively large, it will be very difficult to synthesize with chemical method. Therefore, in terms of the genes with high molecular weight, it is more appropriate to isolate natural genes directly from organisms. So it generally tends to conduct combination between chemical synthesis and natural isolation.

The artificial synthetic gene can be a gene that has already existed in organisms, and also be redesigned in accordance with people's desires and special needs. Therefore, it provides powerful approaches for humans to manipulate genetic information, correct genetic diseases and create excellent new organism types. It is a major productive leap in gene research.

3.3 Gene Expression and Regulation

Protein synthesis, also called gene expression (Fig.3.4), is a process in which genes are constantly transcribed and translated to produce various proteins. Each cell owns a complete set of gene regulation system to make all kinds of proteins synthesized only when needed, reduce unnecessary wastes, avoid harmful consequences, and maintain normal metabolisms.

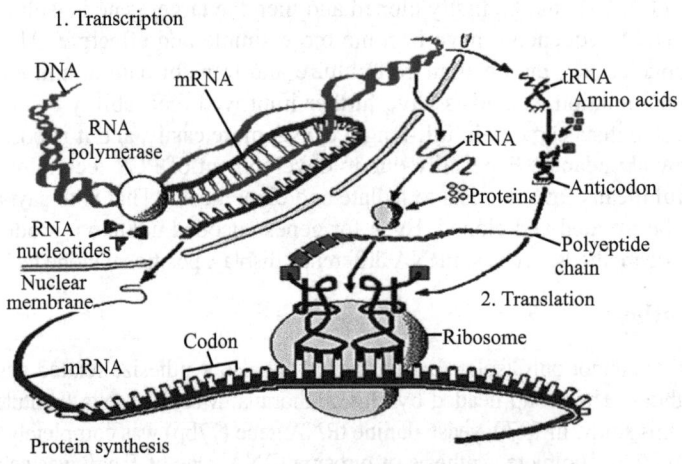

Fig.3.4　Gene expression

3.3.1　Gene Expression and Regulation in Prokaryotes

Gene expression depends upon gene regulation. It involves the following four kinds of genes, including promoter gene, regulatory gene, operator gene, and structural gene. Usually a structural gene is the gene encoding basic housekeeping proteins in a cell, such as metabolic enzymes, transport protein, cytoskeleton and so on. These structural genes can produce certain enzyme systems and structural proteins in a cell after transcription and translation. Thus they are directly related with development and performance of biological traits. A regulator gene simply describes a gene encoding a protein (or RNA) that controls expression of other genes. Regulator genes can be transcribed to mRNA and thus produce protein, i.e. repressor. While the operator locates adjacent to the structural gene, it can't be transcribed into RNA, but can control transcription speed of the structural gene so as to restrain structural gene activity, so the operator gene is also called control element. Normal living cells can demonstrate harmonious life functions, adapt to the changing conditions, and show different characteristics in different environment only through close cooperation between these genes. A promoter, unable to be transcribed into RNA, usually locates close to the operator gene and controls initiation of mRNA synthesis.

In 1961, French molecular biologists F. Jacob and J. Monod proposed the operon model to illustrate regulation of enzyme synthesis in *E.coli* (Fig.3.5). According to their definition, operon is a complete bacterial gene expression unit which is composed of several structural genes, one or several regulator genes, and the control element. The control element includes an operator and the promoter sequence. For example, *lac* operon of *E.coli* is made up of three structural genes and an operator gene. In general, the regulator gene *lac I* is transcribed into mRNA and then translated into the repressor protein, which is a repressor of the operator gene. Once they bind together, the structural gene *lacZ*, *lacY*, and *lacA* will be closed. Consequently, synthesis of the three corresponding enzymes encoded by these genes (β-galactosidase, permease, and acetylase) stops. However, as the metabolite (lactose) accumulates in the cell and needs for more enzyme, it will bind with the repressor to form a complex. As a result, the repressor can't bind to the operator and the structural gene will restore its transcription activity and produce the three enzymes which participate in lactose metabolism continuously. When the metabolite is removed by these enzymes, the repressor can freely move to the binding sites of the operator to cease expression of the structural gene. Obviously, lactose plays an inducer role. From the example of *lac* operon, it's known that the inducer is a small molecule and controls transcription activity of the structural gene through binding or dissociation of the regulator gene.

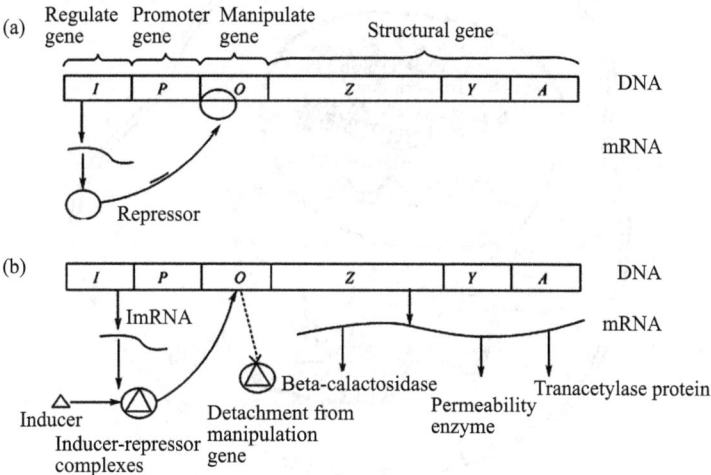

Fig.3.5 Structure and gene regulation of lac operon in *E.coli*
(a) Repression state; (b) Induction state

E.coli can produce a set of enzymes utilizing lactose and grows with lactose as unique carbon source. When *E.coli* is cultured in medium with only lactose but no glucose, lactose will bind with the repressor to change their configuration and make them inactivate. As a result, the repressor is unable to bind with the operator, and RNA polymerase can pass the operator and reach the structural gene so that the structural gene is transcribed and translated to produce the three enzymes to use lactose. As both glucose and lactose are present in the medium, bacteria just utilize glucose instead of lactose, because RNA polymerase can't bind with the promoter and the structural gene can't be transcribed and translated in this case.

In 1969, R. J. Britten and E. H. Davidson put forward a hypothesis about gene regulation model in eukaryotic cells. According to this model, a DNA fragment linked with the regulator is termed as a sensing gene to which cells can send signals as needed. This signal may be transmitted via chemical molecules in cell. Once the signal arrives at the sensing gene, the regulator begins to work to produce activator RNA which gives an instruction to the receptor gene adjacent to the structural gene. Then, the structural gene is transcribed into mRNA and translated into the corresponding protein.

It's demonstrated from the model above that the molecule controlling gene expression on lactose utilization in *E.coli* is protein inhibitor, but activator RNA in eukaryotic cells. During the process of lactose utilization, once the repressor appears, the structural gene will inactivate. This control mode is called negative regulation. On the contrary, the structural gene begins to work once activator RNA binds to the receptor gene in eukaryotic cells. This regulation is termed as positive regulation.

3.3.2 Gene Expression and Regulation in Eukaryotes

Regulation of gene expression in eukaryotic cells is much more complicated than that in prokaryotic cells and expression procedure is often closely correlated with the whole physiological processes.

Eukaryotes have nucleus coated with the nuclear membrane. Transcription of genes occurs in the nucleus and translation in the cytoplasm (Fig.3.6).

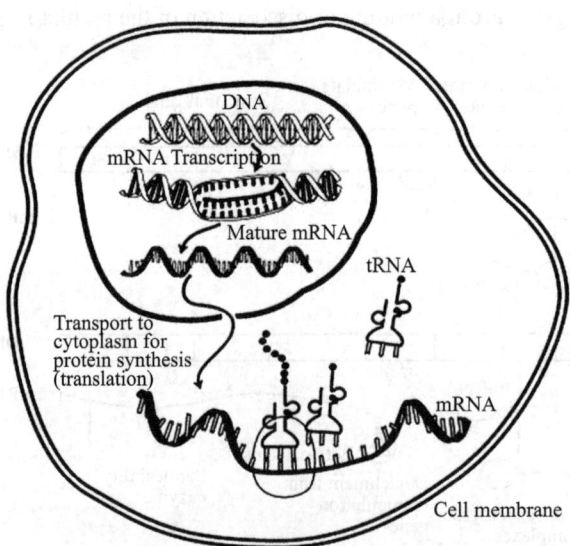

Fig.3.6 Transcription and translation in eukaryotes

In eukaryotic cells, there are introns that can't be expressed and many other non-encoding sequences for gene regulation in most genes. Only if transcribed mRNA precursor is processed to become mature mRNA it enters into the translation stage (Fig.3.7).

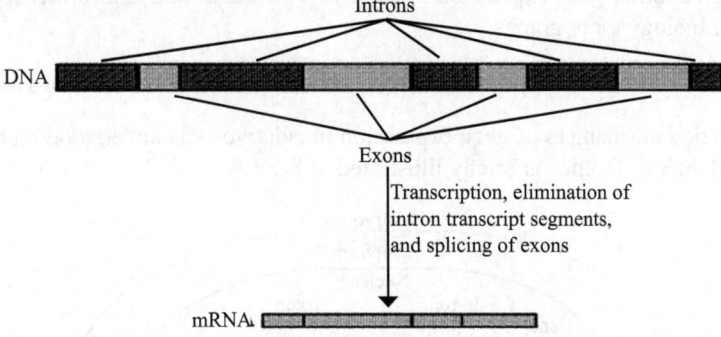

Fig.3.7 Excision of introns in eukaryotic genes

Eukaryotic chromatin is formed by binding between DNA and histones. DNA and histone fold and twine to form super nucleosome, and then the nucleosome and chromatin further fold and condense to form the super-structured metaphase chromosome. The structure of chromatin plays a role of an overall controller in gene expression (Fig.3.8).

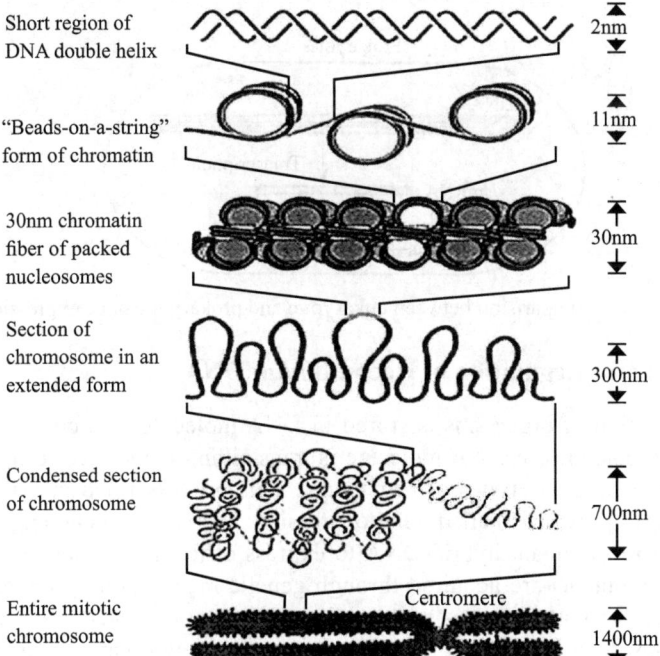

Fig.3.8 Structure of eukaryotic chromatin

Some hormones have inductive regulation on gene expression and chemical modification of DNA in genome can also alter gene expression. In addition, high differentiation mechanism which demands for selective control of gene expression is present during eukaryotic cell development process. Knowledge on regulation of gene expression in eukaryotes is quite rich, and further learning can refer to some molecular biology monographs.

3.3.3 Comparison of Gene Expression Between in Prokaryotes and in Eukaryotes

So far, regulation mechanism of gene expression in eukaryotes is understood much less than that in prokaryotes and their difference is briefly illustrated in Fig.3.9.

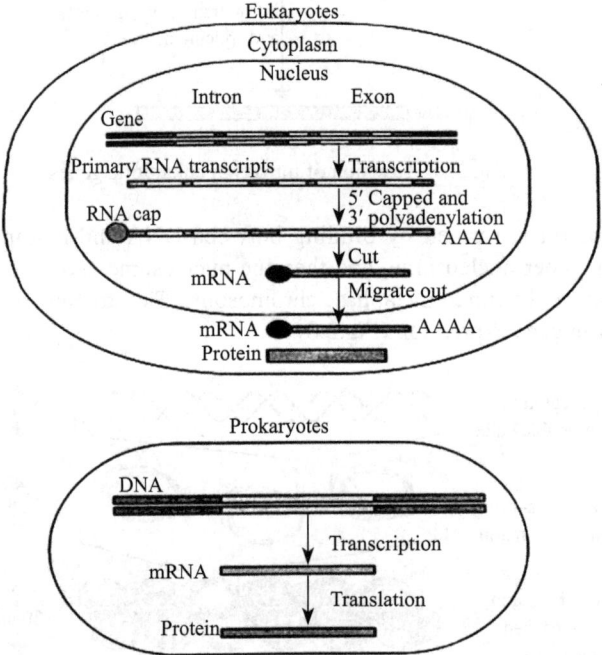

Fig.3.9 Comparision between eukaryotic and prokaryotic gene expression

3.3.4 Expression and Regulation of Recombinant DNA

Genetic information of organisms is stored in DNA molecule of a cell in the form of genes, and one of the basic functions of DNA molecule is transmitting the carried information into protein to determine the genetic phenotypes of organisms. This process for transcribing DNA to RNA and then translating to protein is called gene expression. The aim of gene engineering is to make exogenous genes of recombinant hybrid DNA to express efficiently in host cells and get purified products. After recombinants are acquired through genetic manipulation, expression efficiency of target genes becomes the most important problem. Regulation of each step in gene expression, whose process includes transcription, translation, and post-processing, will have influence on efficiency of recombinant gene expression. The expression systems in the common bacteria, yeast, insects, and mammals have individual distinguishing features and thus different expression efficiency.

At present, the core of gene engineering research is expression efficiency of target genes which is also an important problem for industrialization of engineered bacteria. In order to improve gene expression efficiency, we should comprehensively take into account expression way, biological activity and expression level of target products, and so on. How to raise expression efficiency of target genes is an interdisciplinary research subject which should be guided by the following requirements. ①Target protein should have biological activity without any treatment. ②If the translated protein needs structural modification, the correlated genes should be present. The target product should be consistent with natural protein as far as possible, and have the strongest biological activity. ③The target protein can be secreted and especially the extracellular secretary system should exist. In a word, the product expression should be in high level and its isolation be simple. ④The proportion of cells without the recombinant plasmid should be reduced as far as possible by means of plasmid design, culture optimization and others to maintain stability of the plasmid, so that the target gene can exist stably and expressed in the host cell. ⑤Host cells for high-density culture are selected and the cell density should be increased as far as possible with proper culture methods to promote product expression. ⑥Easier product isolation and purification should be considered in process of the host cell selection, plasmid construction and medium design.

There are various advantages to ensure priority of *E.coli* in gene engineering. It is one of the most common systems of gene expression and extensive studies have been conducted on genetics and molecular biology of *E.coli*. But with regard to a specific gene, whether it can be expressed efficiently in *E.coli* will depend on the gene structural traits, the host, vector construction, cell culture and so on. The chief drawback of *E.coli* system is that it can't carry out various post-translational modifications. As a result, it affects biological activity of eukaryotic proteins and the expressed proteins usually form an insoluble inclusion body. In addition, *Bacillus subtilis* is frequently-utilized system of prokaryotic expression which is easy to conduct various genetic manipulations, suitable for secretary expression of industrial enzymes at high level. However, engineering strains constructed from this strain are not stable enough. Yeast is a widely-used eukaryotic expression system, which can expresses proteins with complicated structure and conduct post-translational glycosylation and secretary expression. Even though yeast and insect cells can conduct the post-translational modification (glycosylation, etc.) of the target protein, it is still inferior to a mammalian cell system in both glycosylation degree and glycosyl variety. Hence, more and more attentions have been paid to mammalian cell systems recently and diverse approaches are being adopted to improve the animal cell culture technology and expression yield. The features of target protein in these different expression systems are shown in Table 3.1.

Analyzed from the two aspects of construction of gene expression system and expression process of target genes, expression efficiency depends on not only characteristics of host cell and construction of expression vector, but also on culture engineering of recombinant strain. From expression systems, it is mainly demonstrated at the two levels of transcription and translation.

The major factor influencing transcription of exogenous DNA is strength of promoters. A promoter can bind with RNA polymerase of a host cell exclusively and start transcription and synthesis of mRNA. Most of exogenous promoters, especially eukaryotic ones, can't be recognized by RNA polymerase of *E.coli*. Thus exogenous genes must be inserted under control of *E.coli* promoters. However, the promoters to start exogenous gene expression with extremely strong activity may seriously damage the normal growth metabolism of recombinant strain, thus the appropriate one should be chosen. Moreover, the terminal signals affect the transcription, and so man-made genes must be fit for suitable terminators in order to reduce energy consumption and ensure accurate transcription. Strong promoters often go with strong terminators.

Table 3.1 Comparison of features among different target gene expression systems

Feature	Cell				
	E.coli	*B. subtilis*	*S. cerevisiae*	Mycete	Insect*
High growth rate	E	E	VG	G	P
System applicability	E	G	G	F	R
Expressing level	E	VG	VG	VG	G
Feasibility of cheap medium culture	E	R	E	E	P
Protein folding	F	F	F	F	E
Simple glycosylation	No	No	Yes	Yes	Yes
Complex glycosylation	No	No	No	No	Yes
Low protease activity	F	P	G	G	VG
Product secretion	VG	E	VG	E	E
Safety	VG	VG	E	VG	E

*Glycosylation form is different between insect cells and mammalian cells.
Notes: E-Excellent; VG-Very good; G-Good; F-Fine; P-Poor.

The important factor at translation level influencing exogenous gene expression is the translation initiation region. Translation takes place in ribosome, so mRNA must contain a ribosome binding site (called SD sequence). As for artificially synthetic genes, optimization of codon is also very important. The favorite codons of a host cell should be adopted to maintain energy balance of purine and pyrimidine base pairing reaction. Post-translational modifications, including excision of N-formyl methionine of new peptide bonds, disulfide bond formation, glycosylation, and post-processing of peptide bond, and etc, will also have effect on gene expression.

Gene expression is a very complicated system. Besides the two major factors mentioned above, the vector stability and physiological state of host cells will influence expression level of target genes.

3.3.4.1 Efficient expression of target genes

During research and development of the first generation gene engineering products, people discovered that expressed yield of somatostatin and insulin in *E.coli* was quite low. The reason was that most expressed proteins were degraded by intracellular proteases. But when a target product and β-galactosidase were fusion expression, fusion protein can accumulate in high level in the cell, resulting in efficient fusion expression strategy of target genes. Fusion protein often forms an insoluble inclusion body without activity which can obtain activity only after dissolution and renaturation. With the help of high-density culture, and engineering bacteria growth and induced phase expression separation technique, the inclusion body yield can reach a high level. What's more, with development of protein renaturation technology in recent years, activity recovery of target proteins has been greatly improved. Therefore, with regard to protein without post-translational modifications, utilizing *E.coli* as a host cell which grows fast and needs simple media, and adopting expressing strategy of insoluble fusion protein is still a good choice for improving the gene expression efficiency.

The fusion protein, obtained by adopting the new strategy to fuse a target protein with the bacteria protein with a purification tag, cannot only resist on degradation of proteases, but also isolate and purify the target protein through the affinity reaction between the protein with a purification-tag and its corresponding antibody.

When *E.coli* was used a gene engineering bacterium to express target proteins at the primary stage, it was found that concentration of soluble target proteins in the cell was very low, and high-concentration expression could lead to formation of insoluble inclusion bodies. Recent studies have shown that if a target protein can resist on protease attack or protease-deficient host bacteria are used, the target gene may express the soluble target protein at high level in the cell. Many target proteins have been expressed in a soluble form by reducing promoter strength and culture temperature. For example, using weaker promoter and culture at 25 ℃, concentration of soluble human interferon α2b be expressed in recombinant *E.coli* to reach above 1.0g/L. Soluble target proteins often have biological activity, without complicated separation process. Therefore, it is a very prospective new strategy for improving expression of target genes.

3.3.4.2 Metabolic regulation in host cells

Over-expression of exogenous genes will add load to host cells, influence its metabolism and growth and, consequently effect synthesis of target proteins. In order to maintain the normal growth rate of a host cell, stability of the recombinants and efficient expression of target gene, it is necessary that adopt appropriate method to regulate cell metabolism. At present, there are many regulation methods, such as nutrient concentration control, product-induced expression, induced vector replication, and product secretion promotion.

Under the stable conditions of recombinants, in order to make target genes express efficiently during culture of host cells, firstly cell biomasses as high as possible (i.e. high-cell-density culture) should be obtained by controlling the nutrient concentration. In gene engineering, mutant strains of vitamin or amino acid auxotrophy are usually employed as hosts. A small amount of vitamins are required and they can be added in essential amount at the beginning of culture. But if at the beginning of culture the essential amount of amino acid is added, it may be possible that amino acid may inhibit cell growth due to excessive amount. So in the event of constant pH, continuous feed of amino acid mixture and glucose is adopted during the whole culture period to maintain their invariable concentration, so that the bacterial cell can grow at a constant rate to raise expression efficiency. Taking culture of *E.coli* C600 as an example, yield of the dry strain can reach the level of 60g/mL.

Gene product can be induced expression. Usually, as a host cell, when it is in growth stage, product synthesis rate is very low. Only when the host cell enters production period, genes will start to express protein products. This is a growth rule that organism cells themselves reduce existing loads. Usually cell does not need drug induction, but if λcI_{857} temperature-sensitive repressor is used to regulate gene expression, the repressor protein expressed by *cI* gene of the recombinant DNA will inhibit expression in the downstream genes of the λ_{PL} promoter at 32 ℃. During this period, the cells grow fast. Once turning into the production stage, the culture temperature is raised to 42 ℃, and then the *cI* gene inactivates and doesn't express the repressor protein. Consequently, the P_L promoter is out of repression and the exogenous gene expresses, resulting in increase of expression efficiency.

Expression vector can be induced replication. The host cell usually inhibits replication of the recombinant plasmid in its growth period and induces it in its production period to increase copy number of the plasmid and improve expression efficiency of the target gene. Plasmid pCZ_{101} is the best example of temperature-controlled induction of DNA replication. When the recombinant pCZ_{101} which carried the bovine growth hormone gene is introduced into the host cell which is cultured at 25 ℃, the copy number of the plasmid contained in each cell is only 10 and the host grows rapidly. Then the culture temperature raises to 37 ℃, and the cell growth rate decreases. Yet the recombinant plasmid replicates in so large number of 1000 copies/cell that the exogenous gene expresses efficiently.

Secretion of expressed protein can be regulation. Whether the expressed protein products of exogenous genes in host cells can be secreted outside the cells, it is crucial for expression efficiency. On one hand, the products accumulated in the cells are easy to be damaged by intracellular proteases; on the other hand, the increased load will bring about low expression efficiency. So, if the products can be easily secreted outside the cells, the above problems will be avoided and expression efficiency can be improved. Nowadays, many technologies are adopted to promote protein secretion. Generally, proteins rely on its short N-terminal signal peptide, which has a highly hydrophobic core, to secrete outside smoothly. Once protein synthesis is completed, the signal peptide will guide the protein to move outside through the cell membrane. Then the signal peptide is hydrolyzed by signal peptide enzyme and the functional protein releases.

For *E.coli* as an expression system, if a signal peptide gene is added at the plasmid design stage, it is possible to fulfill secretary expression of a target protein. There are two types of secretary expression of target gene, target protein secret to the periplasm directly and secret to outside of cell by periplasm. The process of secretion to the periplasm has the following advantages: ①the type and quantity of proteases in periplasm are far less than that in the intracellular protoplasm, resulting in reduced protease attack; ②highly oxidized environment of the cell periplasm is more conducive for the correct protein folding and solubility; ③with the aid of peptide enzymes the ligated signal peptide that will be excised as the expressed protein is secreted to the cell periplasm, so as to produce a mature target protein; ④the target protein in the periplasm can be secreted to the medium through simple penetration. Therefore, during isolation protein interferences resulting from cell disruption can be avoided. As for the expression system which can secrete the target protein outside, in addition to the advantages mentioned above, it can further simplify isolation process of the target protein, and more importantly it can reduce concentration of the intracellular product, which can greatly improve expression level, especially for those product-inhibition-presented expression systems. So construction of secretary expression vectors (especially extracellular secretary ones) is one important development trend for expression of target genes at high efficiency.

It is possible to fulfill secretary expression of target protein in bacteria by fusing the target protein with the mammalian signal peptide, but it will be more effective to adopt the signal peptides from *E.coli* itself. The common signal peptides of *E.coli* are PhoA, Lamb, OmpA, ST II, and so on. In fusion with these signal peptides, many proteins, such as human growth hormone, interferon, human epidermal growth factor, cattle growth factor, and etc., have been expressed in a secretary way. The first commercialized protein expressed in *E.coli* is human growth hormone which was formally on the market in 1993. Wong and others have made remarkable achievements in gene fusion. Thay fused human epidermal growth factor gene with the signal peptide OmpA gene to fulfill its extracellular secretion. Besides, with combination of the plasmid optimization and bacterial culture engineering, the extracellular hEGF yield is as high as 380mg/L, and it is the highest extracellular expression level of small peptides in the current world and almost the same to the intracellular level, indicating the huge potential of secretary expression to improve expression efficiency of target genes. However, some studies have shown that fusion of target protein with a signal peptide can't guarantee that the products can be secreted outside the cell, and the target protein secreted to the periplasm is not always soluble, which can be improved only by adding non-metabolic sugar and reducing expression rate. Thus, the expression system using *E.coli* as a host cell still has many problems that remain to be solved for successful secretary expression.

Another type of secretary expression system exhibits a solution to damage the cell wall structure. For example, the target gene and the gene encoding cell wall lyase are transformed into a host cell simultaneity, and their expression is induced after the bacteria growth. On one hand, the target protein begins to express, on the other hand, the expressed cell wall lyase will damage the structure of cell wall so as to release the expressed target protein to extracellular. This method has been successfully applied in polyhydroxy methane acid production with genetic engineering bacteria. Also, some people transferred an expression vector into the mutated leaky host cell and realized extracellular secretion of the target protein. As cellular physiology of the host bacteria mentioned above is in abnormal conditions, gene engineering bacteria are hard to achieve efficient expression in the practical culturing process.

In gene engineering, secretion of the target gene product can be realized with the help of natural signal peptide, since the expressed fusion protein is easy to be secreted to outside the cell. In addition, it can also promote secretion of the recombinant protein through culture of immobilized cells. For instance, immobilized genetic engineered *Bacillus subtilis* can secrete 50% of the expressed insulin outside the cell.

3.3.4.3 Measures to improve expression level of gene engineering bacteria

After a recombinant bacterium has been constructed, its physiological metabolism and culture conditions are the two important factors influencing expression efficiency of the target gene. It is mainly shown in three aspects. ①Different with the traditional cell, the recombinant bacteria are apt to lose the plasmid, and the specific growth rate of host bacteria without plasmid is faster than that of host bacteria with plasmid. While prolonging culture process, the host bacteria without plasmid will account for a larger proportion, which will seriously impact on the expression efficiency of the target gene. ②The recombinant bacteria should not only maintain the normal growth but also express exogenous genes. So the energy distribution phenomenon presented in the recombinant bacteria may be confined their high-density culture. ③The target protein expressed in the recombinant bacteria is a foreign substance of the cells and usually toxic to the host bacteria to a certain extent. Moreover, some organic acids, like acetic acid and etc, will accumulate in cell culture process and seriously inhibit cell growth and expression of the target gene.

E.coli has many merits for efficient expression of the exogenous protein and meets many basic functional requirements as a gene engineering host bacterium. However, during culture of engineered bacteria (especially high-density culture), the inhibition by-product acetic acid often accumulates in large quantity and seriously inhibits bacterial growth and target gene expression. Many scientists take many measures to reduce synthesis of acetic acid. Consequently, the accumulated acetic acid level is greatly decreased in culture process of *E.coli* and expression level of exogenous genes is obviously improved.

1) To reduce formation of acetic-acid-like inhibitory by-products

There are main inhibitory by-products that are produced in the culture procedure of engineered *E.coli, Bacillus subtilis*, and mammalian cells, respectively, such as acetic acid, propionic acid, and lactic acid. Their accumulation not only influences cell growth but also restrains product expression. Consequently, how to reduce production of organic acids (acetic acid, etc.) in gene engineering fermentation, it is an important problem. Appropriate upstream technologies in gene engineering can reduce organic acid synthesis, and the right strategy adopted in cell culture can achieve good results. Engineered *E.coli* is taken as an example for illustration as follows.

The higher the bacterial growth rate, the higher the specific formation rate of acetic acid. In the synthetic medium, once the growth rate of recombinant bacteria exceeds a critical value, it will bring about acetic acid accumulation. Riesenberg, and others. found that acetic acid can be detected as the dilution rate was over $0.2h^{-1}$. When the specific growth rate was controlled within $0.11h^{-1}$, the acetic acid production rate could reduce significantly and the bacteria density reached 110 g (DCW)/L. Although low specific growth rate of recombinant bacteria reduces acetic acid yield, at the same time it is unfavorable for product expression. So an appropriate specific growth rate should be chosen to realize fermentation in high density and high-level expression.

As the culture temperature of engineered *E.coli* fall within 26-30℃ from 37℃, the nutrient absorption rate can reduce by bacterium and thus formation of organic acid decrease. In the fermentation of recombinant *E.coli* KS467 which can produce Proapo A-I through induction, when the culture temperature decrease from 37℃ to 30℃, the concentration of acetic acid drops from 10g/L to 5g/L.

In addition, acetic acid can be removed from fermentation broth by fermentation and isolation coupling method, so as to realize high-density fermentation and high-level expression in the recombinants. Organic acids accumulation can also be reduced and efficiency of gene expression can be improved in the recombinants by restricting amount of fed glucose which can eliminate the "glucose effect".

In the culture process of genetic engineering bacteria, dissolved oxygen is an important factor influencing growth of engineered bacteria and exogenous gene expression. Usually, as the growth density of recombinant bacteria is up to 30-50g/L, dissolved oxygen will become the growth restriction factor. Modification of *E.coli* make it can grow under lean oxygen environment, that is a new fundamental strategy to remove dissolved oxygen restriction. Now, a hemoglobin gene associated with transportation of oxygen has been found in a bacterium. If it is integrated into the chromosome of *E.coli*, the host cell can survive under the lean oxygen condition, its growth density and exogenous protein expression rate will be improved. Furthermore, when the hemoglobin gene is introduced into other host strain, such as *B.subtilis* and *streptomycete*, it also increases bacteria density and improves expression level.

2) To improve plasmid stability in engineered bacteria

Plasmid stability in engineering bacteria is an important aspect to improve expression level and it is necessary to study on improvement in both plasmid construction and culture method. In plasmid construction, antibiotic resistance genes are usually introduced and this not only provides convenience for screening of recombinant bacteria, but also creates possibility for increasing proportion of cells containing plasmid in the culture process. Only if a certain amount of a proper antibiotic is added in the medium, it will inhibit growth of cells without plasmid. In addition, the *par* and *cer* should be added to plasmid when constructing a vector. The *par* sites can make plasmid distribution more uniform during cell division and *cer* sites are able to prevent polymer plasmid formation. Thus they can improve plasmid stability in terms of the resources.

During culture process of engineering bacteria, appropriate decrease of the bacterial growth rate is helpful to increase the vector copy numbers in the recombinant cells and prevent from generating cells without plasmid in cell division, so as to increase the expression rate of target protein. Multi-stage culture strategy is to separate cell growth period from induced expression period. At the early stage of batch culturing technique, the inducer isn't added and the target protein isn't expressed. So cells can utilize all of carbon resources and energy for high-level growth and can avoid the plasmid instability caused by induced expression. When the cell density reaches a higher level, the inducer is added into the medium and the target protein can be expressed at higher efficiency. Some researches put forward a cycle control strategy of culture condition, it can reduce growth advantage of plasmid-free host bacteria, increase the proportion of bacteria containing plasmid, and therefore promote the expression yield. Moreover, it is helpful to improve plasmid stability by immobilized cell cultivation.

3) High-density culture of engineered bacteria

If high-density culture of engineered bacteria can be realized, it cannot only improve target product yield, but also reduce culture volume, benefit separation and extraction of the products and decrease the production cost. To achieve this aim, it depends on not only construction of upstream recombinant expression systems, but also strategy of culture engineering of recombinant bacteria.

Firstly, high-density culture of recombinant bacteria can be realized through fed-batch fermentation technology. During batch culture, increasing the initial nutrient concentration can't increase the cell density but results in substrate inhibition and "glucose effect" which causes accumulation of organic acids and influences cell growth and product expression eventually. Fed-batch fermentation is the key technology to realize high-density culture of recombinant bacteria and the feeding way of nutrients should be designed reasonably according to growth characteristics of engineered strain and expression type of the product. In fed-batch culture, glucose as the restrictive matrix should be fed at a constant speed. For microbial cells in a fermentation tank, concentration of nutrients gradually decreases, specific growth rate of cells drops slowly, while the total biomass basically keeps linear increase. Pan and others. adopted constant fed-batch strategy to produce human growth hormone and the terminal cells concentration reached $120OD_{525}$. Jung and others. used this technology to produce interferon, the obtained cells concentration reached 46g (DCW)/L, and the specific yield of interferon was 17mg/g strain. Variable flow speed or gradient increase of the flow speed is a kind of controling strategy which can promote cell growth by adding more nutrients at higher cells density. Exponential fed-batch culture strategy is a simple and effective feed supplement method, which can make matrix concentration remain low in a reactor. Such approach cannot only reduce production of harmful metabolites like acetic acid and so on, but also keep the cells increase exponentially at a specific growth rate. Moreover, the feeding rate can be varied to control cells growth rate for sustainable and to contribute expression of exogenous gene. Exponential fed-batch culture technology has already been widely used in high-density culture of recombinant bacteria in order to achieve a high level of exogenous gene expression.

Secondly, choose an appropriate host cell to realize high-density culture of recombinant bacteria. Different species and subspecies of different or even the same host bacteria have great effect not only on exogenous gene expression but also on high-density culture of corresponding recombinant bacteria. Culture densities and expression levels of different subspecies of *E.coli* can vary 2-5 times in the same conditions, so optimization of high-density culture should take the factors of host bacteria into account so as to choose the most appropriate host strain. Expression modes and induction methods of recombinant bacteria will also affect cell density and product expression level. After protein products of recombinant cells are expressed, the target proteins often suffer from various modifications, such as protein oxidation, deamination, degradation and so on. The nature of modified proteins is so close to that of the target proteins that they are difficult to be separated from the target proteins, thus this influences authenticity and quality of target protein expression. If used for disease treatment, these protein modifications will cause many side-effects. From the point of upstream of gene engineering, host cell selection is a very important factor. The selected host cell should not or less produce enzyme system that can cause protein denaturation or degradation as far as possible. As for culture methods, it can reduce exposure time of the target protein by detaching the growth stage and protein expression stage, resulting in less opportunity for modification of target proteins. In addition, low culture temperature can decrease activity of modification enzymes, and is good for improving quality of target proteins.

Thirdly, optimization of medium components is very important in high-density culture of recombinant bacteria. The culture media for *E.coli* can be divided into synthetic medium, half synthetic medium, and composite medium. Half synthetic medium is often used in high-density culture. The concentration and the proportion of each component of the medium should be appropriate, otherwise the growth of the bacteria will be offected. For specific culture system, the medium components usually need optimization. What's more, using modern online or offline assay methods, fermentation dynamic model of engineering bacteria is established according to feedback information of cell metabolism, and advanced fed-batch technology should be developed. Actual bacteria growth in a fermentation tank should be elucidated further to reduce harmful metabolites and determine the conditions for high-density culture in order to realize the high-density fermentation prolluction.

3.4 Gene Mutation

Genes locating in the chromosome determine the biological character through transcription and translation. Differences between organisms in fact lie in genes. As the saying goes "we reap what we sow", this concise and plain sentence really reflects the universal law of species transmission from generation to generation in the living kingdom. It has long been found that the characters of organisms (including plants, animals, and even humans) can be transmitted to the offspring, which is called heredity. This is also why children's skin colour, features height and etc. always take after their parents. Mysterious genetic phenomena existing widespread in the colourful biosphere, it is just nucleic acid in the cell nuclei. Nucleic acid and protein together make up the chromosomes. The nucleus of each mature cell of the human body contains 23 pairs (46) of chromosomes and ten thousands of human genes just are contained in the invisible tiny chromosomes. In the 1760s, nucleic acid regarded as the genetic material was found in the pyocyte nucleus. In 1840s, scientists found that the nucleic acid extracted from one capsular and pathogenic *pneumococcus* can change hereditary characteristics of another non-capsular and non-pathogenic *pneumococcus* to make it become pathogenic. The mystery is apparently that nucleic acid transferred the genetic information for capsule and pathogenicity to another *pneumococcus*. Therefore, nucleic acid is the carrier of genetic information and also the material basis of heredity.

Heredity is usually determined by genes. When a cell divides, intranuclear chromosomes may be replicated accurately to generate a new set of chromosomes whose deoxyribonucleotide sequence and structure are exactly the same to the parent cell. So the parents' genetic information can be completely and correctly transmitted to the offspring. Among 23 pairs of the chromosome in human fertilized eggs, the 23 ones of chromosomes from father sperm cells inherit all of the father's genetic information and the other 23 ones from mother ova also faithfully retained the mother's genetic information. Therefore, developed characters of the offspring almost are the parent's "copies" and this is just the science behind the common saying that "we reap what we sow". DNA is the genetic material, so gene is the DNA fragment locating in chromosomes. Different genes determine different characters of organisms, and this also means a specific gene holds genetic information of a specific character. Hence, a gene is actually the basic heredity unit.

Genes determine human hereditary traits and various individual traits are presented through various proteins. It can be well imagined that genetic information must control protein synthesis so that the information carried in genes can be delivered to proteins to maintain specific genetic traits. Therefore, DNA is the genetic information carrier and protein is just embodiment of genetic traits. Protein molecules are formed by ligating numerous basic units (amino acid). There are 20 kinds of amino acids in human body. Different arrangements of amino acids in protein molecules form various proteins which perform specific functions respectively and represent specific biological traits. Surprisingly, all kinds of simple and advanced organisms in the wonderful living kingdom just use a set of simplified genetic codes and perform the similar heredity law that genetic information is transmitted from DNA to messenger RNA and then to protein. However, gene alteration can also bring misfortunes pains to organisms and even human beings, for every coin has its two sides.

During the long-time processes in which human adapts themselves to external environmental changes for survival, genes in the nucleus are unconsciously going through extremely slow development, evolution, and improvement. Therefore, genetic stability and gene evolution are the driving force for species evolution in the living kingdom. For example, congenital hereditary diseases caused by some gene variations in the body usually accompany patients for life, such as Hemophilia, in which patients have lack of a clotting factor, once the patients' organs or tissues are slightly injured, they will bleed endlessly. Due to lack of the blood coagulation factor, the blood is difficult to coagulate into clot in the wounds. For some hereditary diseases, loss or change of just one deoxyribonucleotide in one gene will bring about malfunction of the whole gene and cause a lifelong disease. In addition to congenital genetic diseases, those acquired diseases can also be caused by gene mutations. When chemical carcinogens enter the human body, they will bind with human DNA molecules so as to interfere with transmission of normal genetic information. This results in malignant damage of cells and causes human cancers.

Gene mutation, also called point mutation, refers to gene structure changes caused by DNA base replacement, addition, or loss. The mutation occurring in the natural conditions is termed as spontaneous mutation and the mutation caused by artificial physical factors or chemical agents is known as induced mutation. Gene mutation is the major cause of genetic variation and the main factor of organism evolution. Artificial mutation is an important method for creating novel species of organisms.

3.4.1 Features of Gene Mutation

Gene mutation, as an important reason of biomutation, is marked by the following main features:

Firstly, gene mutation is spread widely in the biosphere and can occur in lower organisms, animals, plants, and human beings. For example, the short branch of cotton, the short rod and glutinousness of paddy, the white eye and vestigial wing of fruit fly, the pinkish-gray feather of pigeon, and achromatopsia, diabetes, albinism and other genetic diseases in humans, are all mutation traits.

Secondly, gene mutation occurs in random. It can occur in any period of organism ontogenesis and any cell of an organism. Generally speaking, the later gene mutation happens, the less mutation part show in the ontogenetic process. Gene mutation can occur not only in somatic cells but also in reproductive cells. The mutation occurring in the reproductive cells can be directly transmitted to the offspring through the fertilization effect, namely transmissibility. The mutations occurring in somatic cells generally can't be passed on to their progeny.

Thirdly, frequency of gene mutation is very low for an organism in the natural state. It is estimated that in higher organisms, in one hundred thousand to one hundred million reproductive cells, only one cell will go through gene mutation, and the mutation rate is 10^{-5}-10^{-8}. Gene mutation rates in different organisms are different. For example, the mutation rate in bacteria and phage is lower than that in plants and animals. As for different genes from the same organism, its mutation rate is even different.

Fourthly, most of gene mutations are harmful to the organisms, because every creature is the product of long-term evolution and is in perfect coordination with environmental conditions. Gene mutation may destroy the harmony and thus is often harmful to organism survival. For example, the vast majority of human genetic diseases which are caused by gene mutations pose severe threat on human health. In another example, albino seedlings in plants are also caused by gene mutation. These seedlings can't carry out photosynthesis to produce organic substances due to lack of chlorophyll and ultimately go to death. However, there are a few beneficial gene mutations. For example, the disease-resistant and drought-resistant mutations in plants and the drug-resistant mutation in microorganisms are favorable to survival of organisms.

Fifthly, gene mutation is non-directional. A gene can take mutation in different directions and produces more than one allele. But every gene does not mutate without any restrictions. For example, gene mutation of mice hair colour is only limited in pigment and cannot be beyond this range.

3.4.2 Types of Gene Mutation

Gene mutation can be classified according to mutagenesis modes, phenotypic changes, genetic material changes and so on.

According to characteristics of mutant phenotype, mutation is divided into the following five types.

1) Morphological mutation

It is the mutation causing cell or colonial morphology changes, like mutants losing capacity to produce spores, capsule and flagellum.

2) Biochemical mutation

It refers to the mutation without morphology effect, such as auxotrophic mutant which can grow only by adding a substance to the medium due to metabolic disorder. For example, the adenine mutant strain, represented as Ade^-, can grow only in the medium containing adenine. In contrast, strains that can also grow in the medium without adenine are called wild type strain, represented as Ade^+, this strain can grow both in basic culture medium and complete medium, while the auxotrophic mutant can grow only in complete medium. Thus they are easy to distinguish. In addition, the mutation enhancing microbial antibiotic resistance belongs to biochemical mutation, too.

3) Lethal mutation

It is the mutation leading to individual death or vitality decrease.

4) Conditional lethal mutation

It refers to the mutation making the strain survive under certain conditions but lethal in other conditions. The most typical example is temperature-sensitive mutation. The temperature-sensitive mutant can grow at a certain temperature but not at another temperature. Usually, this kind of mutation is caused by protein structure change because of amino acid variation. Such protein (or enzyme) can maintain its spatial structure and normal biological activity only at the permissible temperature. It will denature and be dysfunctional as the temperature reaches the limited value. For example, the temperature-sensitive mutant of phage T4 can grow and reproduce normally at 25℃ in *E.coli* cell and form plaque, but can't grow at 42℃.

5) Resistance mutation

The mutant strain with some resistance on a certain drug is called resistant mutant. For example, a streptomycin resistant mutant strain can grow in the culture medium added with 1000U/mL of streptomycin, while the wild type strain can't grow.

According to the way of gene structure change, gene mutation can be divided into base substitution mutation and frameshift mutation.

1) Base substitution mutation

The mutation in which a mistake base pair replaces the correct one is called base substitution mutation. For example, in the DNA molecules, the GC base pair is replaced by CG, AT, or TA, and the AT base pair is replaced by TA, GC, or CG. Base substitution affects only the codon in which it occurs. That is to say, a single base alteration just changes the involved codon but not others.

There are two reasons and ways to cause base substitution mutation. One is the base analogs incorporation. For example, during culture of *E.coli*, after 5-bromouracil (BU) is added to the medium, a part of thymine of DNA will be replaced by BU, resulting in alteration of AT into GC, or GC into AT. The other one is certain chemical substance, such as nitrite, nitrosoguanidine, diethyl sulfate, nitrogen mustard, and so on. Besides, the ultraviolet irradiation can also cause base substitution mutation.

2) Frameshift mutation

The insertion of one or more bases into gene, or it is deleted from the gene that can alter the DNA reading framework and the whole sequence distal to the insertion or deletion site to produce an abnormal polypeptide chain consequently. Reason of frameshift mutation is insertion of some dye molecules like acridine into the DNA, it makes DNA replication disorder and results in frameshift mutation.

According to genetic information alteration mode, gene mutation can be divided into three types: synonymous mutations, missense mutation, and nonsense mutation (Fig.3.10).

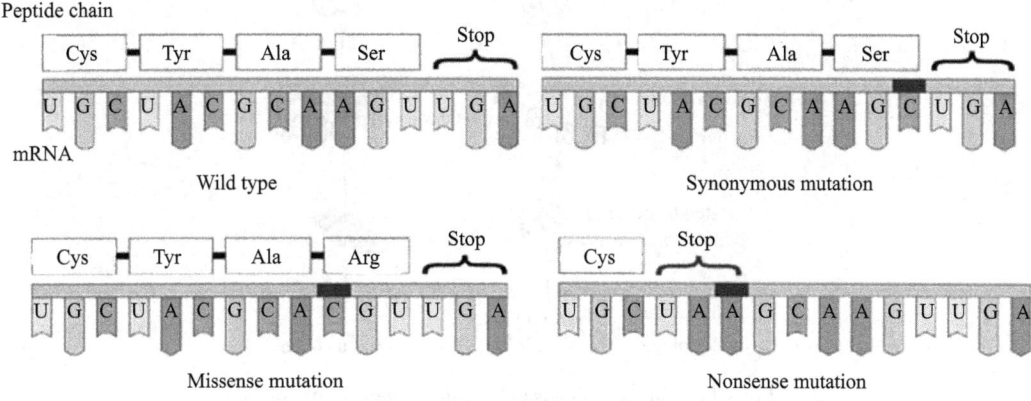

Fig.3.10 Types of mutations

1) Synonymous mutation

Sometimes a single DNA base pair alteration will not affect the amino acid sequence of its coding protein, because the altered and previous codons are degenerate ones, which encode the same amino acid. This kind of gene mutation is called synonymous mutations, namely the amino acid sequence of the mutant gene encoding is the same to that of the wild type gene.

2) Missense mutation

Missense mutation is a mutation where one or more base pairs are changed to cause substitution of a different amino acid. This in turn can render the resulting protein partially or completely nonfunctional. For example, as the codon for the 6th amino acid (glutamic acid) in the β-chain gene of human hemoglobin is erroneously changed from GAG to GUG, the 6th amino acid on the β-chain polypeptides is incorrectly substituted from glutamic acid to valine to cause the sickle-cell disease (Fig.3.11).

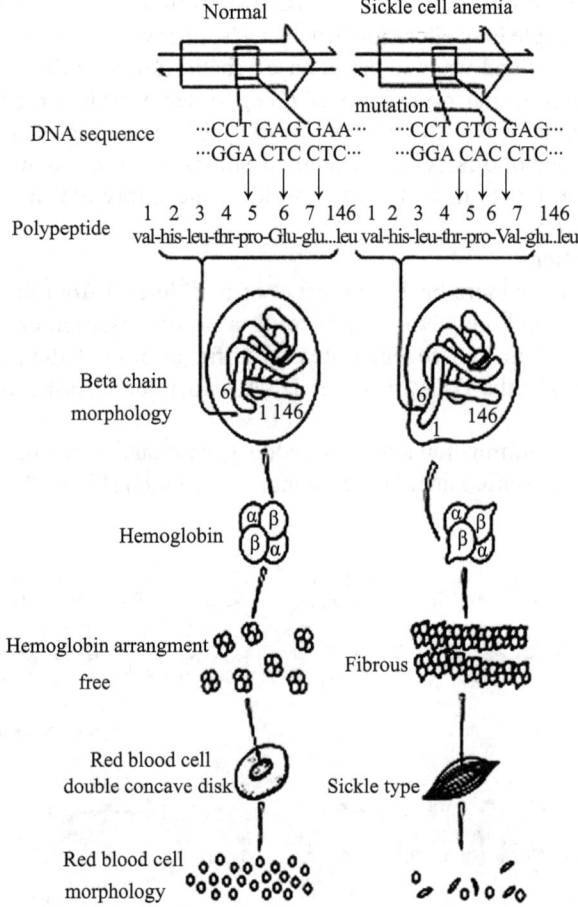

Fig.3.11 Sicklemia caused by a single gene mutation

3) Nonsense mutation

Nonsense mutation is divided into amber mutation, ochre mutation, and opal mutation. Mutations in the structure of genes can be classified as base substitution, frameshift, DNA fragment insertion and deletion. The mutation where one or more base pairs are changed to make the codon for an amino acid become a stop codon is called nonsense mutation. The base is mutated to form a stop codon bring forward terminate protein synthesis. The nonsense mutation where the codon is altered to UAG is called amber mutation, and the nonsense mutation where the codon changes into UAA is also called ochre mutation.

In molecular genetics, auxotrophic mutants refer to the mutant strains or cells which are unable to synthesize some nutritional substances (such as amino acids, biotin) and can grow on the minimal medium (the medium composed of glucose and inorganic salts) only when the corresponding organic substance are added. For example, the wild type strain of *E.coli* can grow on the minimal medium, while histidine-defect *E.coli* (represented as *His⁻*) can only maintain normal growth on the minimal medium by adding an appropriate amount of histidine. The process of recovery of a mutant gene to a wild type one is called reverse mutation. For instance, as a great number of *E.coli* (*His⁻*) cells are inoculated in the minimal medium without histidine, there will be a few cells that can grow. The major reason lies in that the *His* gene in the mutant strain has already been restored to the wild type (recorded as *His*).

As phenotype effect of the mutated gene is restored due to the presence of the second mutated gene, the latter mutated gene is called the suppressor of the former one. The suppressor gene does not change the DNA structure of the mutant gene, but just restores the mutant phenotypes. For example, the codon of tyrosine is UAC, and it is replaced by the nonsense codon UAG to stop translation in substitution mutation. When the gene of tyrosine tRNA has mutation and its anticodon is changed from AUG into AUC, its tRNA can still bind with tyrosine and its anticodon AUC can also match with the nonsense codon UAG. So this mutant tRNA can make tyrosine present at the nonsense mutation codon position to undergo normal translation. Here the mutated gene of tyrosine tRNA is a suppressor gene of the former nonsense mutation.

3.4.3 Rules of Gene Mutation

No matter which kind of mutations, spontaneous mutation, induced mutation, morphological mutation, or biochemical mutation, they all have the following characteristics.

3.4.3.1 Randomness

As a group of organisms is concerned, gene mutation has obvious randomness in time, individual, site and phenotypic changes, etc.

3.4.3.2 Independence

In microbial community, gene mutation occurs individually and one gene mutation and another are irrelevant independent events. For example, the mutation rate of resistance on isoniazide in *Bacillus megatherium* is 5×10^{-5}, that on amino salicylic acid 1×10^{-6}, and that on both of the two substances resistance 8×10^{-10}. The latter mutation rate is approximately equal to the product of the former two kinds. It's demonstrated that occurrence of mutation is random for both cells and genes.

3.4.3.3 Stability

The essence of gene mutation is the result of genetic material alteration. So the mutant gene is the same to the wild type one with relatively stable structure and hereditability. For example, streptomycin resistance of mutant strain is still intact after continuous countless passages in the medium without streptomycin.

3.4.3.4 Reversibility

The wild type gene can become a mutant gene through mutation and *vice verse*. For example, the wild type strain can become the streptomycin-resistant mutant strain through gene mutation and the streptomycin-resistant mutant strains can be restored to the streptomycin-sensitive wild type strain. Generally, the process of mutation of the wild type gene into a mutant gene is called forward mutation, and the reverse mutation from a mutant gene to the wild type gene is called reverse mutation.

The mutant strain obtained through forward mutation is called mutant strain, and the strain obtained by reverse mutation is caller for reverse mutant strain as well. As for phenotype, there is no obvious difference between the reverse mutant strain and the wild type strain. But as for genotype, origins of reverse mutation can be divided into three categories: the first is the real gene reverse mutation, which is reversion of the previous mutation site; the second is the intragenic suppressor mutation caused by mutation at different sites in the single gene; and the third type is the mutation occurring at different sites in different genes which inhibits the original mutant gene expression, and this is also called suppressor mutation.

3.4.3.5 Rarity

Mutation rarity refers to the very low mutation rate under normal circumstances. The mutation rate is defined as the probability of the mutation occurring in a cell in specific environment in a generation or other specified unit time. For individuals, when and where mutation takes place is occasional and random. But for a community, mutation always occurs at a certain frequency in a population and the mutation rate in specific environment is constant. Generally speaking, the spontaneous mutation rate is low from 10^{-10} to 10^{-5}, indicating rarity of gene mutation (Table 3.2). But it can increase through treatment with some physical and chemical factors.

Table 3.2 Spontaneous mutation rate of resistance on different drugs in bacteria or phage

Bacterium	Resistance objects	Mutation rate
Pseudomonas aeruginosa	Streptomycin (1000μg/mL)	4×10^{-10}
E.coli	Streptomycin (1000μg/mL)	1×10^{-10}
Shigella	Streptomycin (1000μg/mL)	3×10^{-10}
Salmonella typhi	Streptomycin (1000μg/mL)	1×10^{-10}
Hemophilus pertusis	Streptomycin (1000μg/mL)	1×10^{-10}
E.coli	Phage T3	1×10^{-10}
E.coli	Phage T1	3×10^{-10}
Staphylococcus aureus	Sulfathiazole	1×10^{-10}
Staphylococcus aureus	Penicillin	1×10^{-10}
Bacillus megatherium	Isoniazid	5×10^{-10}

3.5 Heredity and Human Diseases

In 1900, the average human life span was only 36 years, while in 2000 it has been doubled to 70 years. Although the human life span has been extended, the shadow of diseases accompanies us from birth to death. For a long time, scientists have been committed to solve this difficult problem why the human body suffers from illness. In fact, in addition to trauma and extreme nutrient deficiency, most of human diseases are caused by genetic and environmental factors. Genes are the most unique mark of the individual life, and its abnormal structure and function, is the major origin of diseases, this is called gene mutation. Gene mutation can occur in somatic cells, and also in the reproductive cells. The vast majority of human genetic diseases are caused by gene mutation.

3.5.1 Genetic Diseases and Their Diagnosis and Treatment

3.5.1.1 Genetic diseases

Genetic diseases are illnesses caused by genetic material alteration. At present, it is known that more than 4000 kinds of human diseases are related with genes. Genetic diseases are increasing at the rate of more than 100 kinds each year. With development of medical treatment, the incidence of some infectious and epidemic diseases in the population has gradually dropped, while the relative incidence of genetic diseases is increasing. Human genetic diseases can be classified into three main categories: single genetic diseases, polygenic diseases, and chromosomal diseases. Single genetic disease is the result of alteration of a base pair in a single DNA sequence, including albinism, haemophilia, daltonism, Huntington's chorea, polydactyly, allergic rhinitis, hyperpresbyopia, hypermyopia, and so on. Because the hemophilia gene is passed on with the X chromosome, it can be "lost" during the genetic process. However, compared with the single genetic diseases, polygenic diseases are the most common human genetic diseases in large populations, but its incidence is not as high as but only 1%-10% of that of the single gene diseases (1/4-1/2). These diseases, like congenital heart disease, high blood pressure, diabetes, asthma, schizophrenia, tuberculosis, myasthenia gravis, gout, mild and moderate myopia, psoriasis, rheumatoid arthritis, strabismus, manic depressive psychosis, fall into this class. Polygenic diseases are usually caused by effect of variation of multiple genes in combination with environment. In such diseases, heritability is used to represent effect of genetic factors on morbidity. If heritability is higher than 70%, it indicates that the genetic factors play important roles, like cleft lip, bronchial asthma, nerve schizophrenia, and etc. Heritability of 50%-60% demonstrates that both the genetic and environmental factors are important for these genetic diseases, such as high blood pressure, coronary heart disease and etc. If heritability is lower than 40%, it means the environmental factors are predominant in the diseases, including congenital heart disease, digestive ulcer, maturity-onset diabetes, etc. Chromosomal disease is the result of chromosome aberration (Fig.3.12), including chromosome number and structure change. At present, it's known that there are more than 500 kinds of hereditary diseases, of which 75% are sex chromosome abnormality and 25% are common chromosome abnormality.

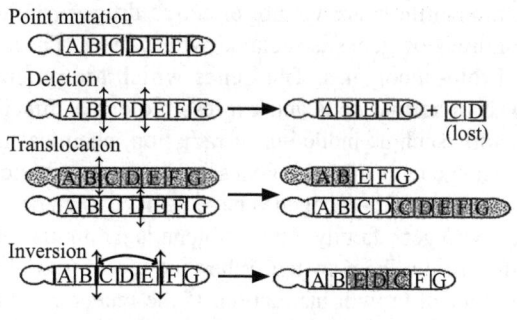

Fig.3.12 Chromosomal mutations

For example, testicular dysgenesis syndrome (47, XXY) and gonadal dysgenesis (45, XO) are the most common genetic diseases of sex chromosome abnormality. The most common genetic disease of chromosome number abnormality is mongolism (trisomy 21 syndrome), and chromosome structure disorder is like cri-du-chat syndrome (see polygenic inheritance). According to statistics, about 10% of people suffer from single genetic diseases and 20% from polygenic diseases. Taking chromosomal disease into account, it's roughly estimated that about 25% of physiological defects, 30% of children deaths, and 60% of adult diseases are caused by genetic diseases and about 1/3 of people are burdened with genetic diseases. In our country, 3% of over 15 million new-born babies every year have birth defect, 80% of which are caused by genetic factors.

Hemophilia occurring in the British royal family and spreading in other European royal families through marriage in the 19th century can be the most famous familial genetic case in history. Hemophilia is originally explained as "hemophilic disease". Due to lack of the blood coagulation factor, the patients bleed easily, must rely on emergency transfusion to maintain life after surgery or trauma, and so have to "take blood as a friend". Hemophilia only occurs in males but is passed to the offspring by females. The British queen Victoria is a haemophilia gene carrier and her third daughter Alice and little daughter Beatrice among her nine children are also carriers. Alice married the grandson of the grand duke, Hessian, in German Empire, and transmitted the hemophilia gene to her daughter Irene and Alexander. Irene married Germany emperor's son and brought the hemophilia gene to the Prussia family. Alexander married the Russian tsar Nicholas II, passed hemophilia to his son, and her four daughters became potential carriers, too. One granddaughter of Beatrice was also a carrier and married Prince Alfonso VIII of Spain to result in hemophilia spreading in the Spain royalty. Disease genes threaten health of many people through heredity, but with further understanding of genes, some genetic diseases can be controlled or avoided through medical means or the weapon of law. Pathologists think that all human diseases have genetic background, but the genetic factors play primary roles only in about 12% of these diseases. In contrast, more human diseases are gradually formed under influence of the acquired environment. Natural living conditions, state of crowd aggregation, dietary components, habits, and so on, can cause single or multiple genes mutation in individuals and bring about disorders of original normal cells in a very strict division procedure so as to result in all kinds of diseases.

Cancer, the most serious threat to human health, is mainly caused by abnormity of multiple genes and chromosomes, and involves in not only cell oncogene activation and suppressor inactivation, but also germinal mutation and somatic mutation. To some extent, cancer is based on gene mutation. Although there are yet not effective drugs or treatment methods for cancer, researchers have found hundreds of genes associated with cancer. These genes are generally divided into two categories. ①Proto-oncogenes. The genes which are potential to induce malignant transformation of cells exist in normal cells and express biological function. However, as the genes are activated by some factors, such as induction of radiation, chemical factors, and biological factors, or translocation and rearrangement of some genes in the chromosome, they may turn into oncogenes. At present, more than sixty proto-oncogenes have been characterized, including *Scr* gene family, *Ras* gene family, *Mgc* and *Myb* gene family. The *Ras* gene is the firstly recognized proto-oncogene and transmits cell growth information like a switch. When the receptor on the surface of cells is stimulated, *Ras* is turned on to send a cell growth instruction. If the receptor is not stimulated, *Ras* will be dormant. About 30% of cancers are caused by *Ras* mutation that the switch protein is permanently opening to instruct reckless growth of cells which may lead to occurrence of tumors. ②Tumor suppressor genes. The human body has some genes just like a brake, they are able to control cell growth and division and so called "tumor suppressor gene". Normally, it inhibits cell proliferation and tumor occurrence and metastasis, and participate in cell adhesion, signal transduction, DNA repair and so on. If these genes have some mutation so as to lose activity and decrease in their functions, cells are propagated repeatedly out of control, finally resulting in tumor occurrence and metastasis. The *Rb* gene, locating in chromosome 13, is the first cancer suppressor gene to be found in a study on children's neuroblastoma and identified. In some common adult tumors, such as bladder cancer, breast cancer and lung cancer, *Rb* gene is also proved to have base deletion or inactivate. The second identified cancer suppressor gene is *P53* which locates in chromosome 17 and its mutation is presented in more than half of human cancers. It is the most common genetic alteration in human malignant tumors and usually looses its function in leukemia, lymphoma, sarcoma, brain cancer, breast cancer, lung cancer, gastrointestinal tract cancer, etc.

In 1994, researchers confirmed that two breast-cancer-causing genes, *BRCA1* located in the 17th chromosome and *BRCA2* in the 13th chromosome respectively, take part in repair of damaged DNA induced with radiation. Normally, the protein products encoded by these genes can help repair DNA replication errors. Once the genes have mutations, they can't work and lose repair function. Consequently, the damaged DNA is continuously replicated to result in breast cancer. In recent years, new cancer suppressor genes have being found constantly, such as renal cell carcinoma-related *VHL* gene, liver cancer-related *M6P/IGF2r* gene, colon cancer-related *DCC* gene, etc. Although until now the determined disease or disease-related genes are still in small quantity, genetic technology as a new method for analyzing specific individual disease has been applied to medical practice. Human beings never stop exploring their own bodies, and such a question will be present with deeper understanding about our own bodies: Can human predict what diseases will take place in them so as to take measures to prevent them from happening? In the early 21st century, we have understood diseases at the molecular level. With further gene researches, scientists have proposed a bold concept - "all diseases are genetic diseases", which was confirmed in the 1970s. Many genetic diseases are direct examples to understand relationship between diseases and genes. But is having a cold or losing bowels by eating food also related with genes? The answer of scientists is yes. When the same flu virus spreads in the same area, why do some people have a cold but others not? Usually we would say different people are different in resistance. In fact, this is the result of effect of your genes. Genes often determines your susceptibility to some kinds of diseases. But remember that it doesn't mean "you must be" but "are easier" to get sick. Like AIDS, there is a gene called *CCR5* presented in the oriental people and it makes them easy to be infected by HIV. The fact is opposite in Europe and the United States, and this is entirely caused by genetic differentiation.

All diseases are related to genes and there are three aspects that the gene controls life activities of organisms, gene structure, gene expression and gene products. Generally speaking, human diseases are not simply dependent on the gene structure, but disorder in gene expression and protein products is often the chief culprit to disease occurrence. And genes with same structure can bring about different consequences under adjustment of external factors. Therefore, from this viewpoint, genes are universal but not everything. There is not a corresponding relationship between genes and human characters, as gene expression is also regulated and controlled by environmental factors.

3.5.1.2 Diagnosis and treatment of genetic diseases

Study of genetic diseases can help us to solve some basic problems. First of all, it can help us decide whether we give birth to a child or not. For many parents who are healthy themselves but pathogenic gene carriers, genetic tests on embryos in pregnant mothers have been able to conduct. Prospective parents can know earlier about fetal conditions so as to make their own decisions. Secondly, gene chip technology lets us master our own genetic information through a slight amount of blood and easily detects what diseases we are susceptible to, and even which kind of medicine you eat will be effective or ineffective. Moreover, through gene drugs to regulate gene expression, congenital immunodeficiency, haemophilia, brain tumor, coronary heart disease and other diseases have achieved a good therapeutic effect. Yet the technology to repair the genes with structure defect is still limited in cultured cells and it's supposed that we can see reports on gene repair in a few years.

1) Gene diagnosis

Gene diagnosis, also called DNA or molecular diagnosis, is a method to determine whether patients have abnormal genes or carry pathogenic microorganism in the samples taken from patients with genetic tests.

At present, the diseases detected with gene diagnosis are basically classified into 3 classes: infectious diseases such as tuberculosis, coxsackie B3 myocarditis, hepatitis c virus (HCV), AIDS, etc.; identification of biological properties of various tumors, such as leukemia; and analysis of abnormal genes for genetic diseases like diabetes. Currently, pathogenic genes can be identified by utilizing gene chips to analyze the human genome. Cancer, diabetes, and many other diseases are caused by genetic defects, medical and biological researchers can identify the mutated genes causing cancer in a few seconds. After 10 years, diagnosis conformability of client's diseases with gene chip analysis will reach above 50%. (See details in Chapter 11)

2) Gene therapy

The basic principle of gene therapy is based on understanding of inheritance mechanism of humans. From the perspective of molecular biology, gene therapy is to use the normal and functional gene to replace or supplement the defect gene. From the point of generalized disease treatment, it can be thought that a new genetic material is transferred into a somatic cell so that the normal gene or the exogenous gene expressing in low quantity which does not exist originally can be normally expressed, so as to endow patients with a new function of resistance on a disease and to achieve the goal of treatment. But it must be pointed out that so-called gene therapy is a procedure for transferring a functional gene into a patient's body to express and make the protein product play a role in disease treatment. This is different from nucleotide drugs application and they can't be confused.

The result of gene therapy is a surgery on a gene, which blocks the source of a disease, and so someone describes it as "molecular surgery". It is a highly integrated, comprehensive and complex biological high technology. It combines the technologies of gene isolation, gene transfer, and efficient expression and regulation of exogenous genes in human body. Because the object of gene therapy is human, it must be effective and safe. So it's more difficult than that of expression *in vivo* of gene engineering technology.

Gene therapy can be divided into sexual cell gene therapy and somatic gene therapy. Because gene therapy of sexual cell involves in the offspring problems and the volunteers willing to receive treatment are very few, it can't enter the clinical trial stage. Somatic gene therapy technology has greater operability, thus it is the mainstream of gene therapy. However, it doesn't change the genetic background of gene defects, which is the drawback of the somatic gene therapy. With regard to the technology, ethics and safety aspects, gene therapy is still faced with many puzzles. Even so, with development for about 10 years, gene therapy has made a good progress, and the general trend is encouraging. It is believed that gene therapy, a new technology, will promote revolutionary medical changes in future.

3.5.2 Prenatal and Postnatal Care

The so-called "eugenics" refers to giving birth to a healthy and intelligent child. In 1883, the English scholars, Galton, studied systematically on anthropology and genetics, and created a word "eugenics". It originally meant that comprehensive researches on the genetic quality of the offspring under the social control were carried out to improve their intelligence and physical strength continuously. In 1960, the American human geneticist, Stem, divided eugenics into preventive eugenics and progressive eugenics. The task of the former is to reduce hereditary diseases in offspring, i.e., to reduce frequency of adverse genes resulting in adverse phenotypes, and so it is passive eugenics. The task of the latter is to make more individuals with excellent mental and physical characteristics present in the offspring, i.e., to increase pathway of favourable gene combination, and so it is active eugenics.

In order to prevent occurrence of genetic diseases, eugenics gives several measures as follows. ①Carrying out premarital examination. ②Avoiding consanguineous marriage because the homozygous probability of harmful genes in the offspring is many times higher than that in random marriage. ③Advocating procreation at child-bearing age. According to statistics, the morbidity of Down's syndrome in the children born by mothers younger than 20 or elder than 40, is ten times higher than those children whose mother's age is between 25-34 years old. ④Carrying out genetic consultation. ⑤Promoting prenatal diagnosis such as amniotic fluid diagnosis to prevent birth defects. ⑥Avoiding contactig teratogens in first trimester of pregnancy. For example, streptomycin can cause fetal acoustic nerve injury, chloramphenicol will result in gray syndrome, and ionizing radiation will bring about slow growth of the fetus.

When looking forward to the future, human beings will realize conscious control of their own evolution for the first time in the natural history with development of life science, which will make our coming generations more beautiful and healthy.

3.6 Genome

3.6.1 Genome

Genome is all of DNA in a cell, including all of genes and interval sequences between different genes. It is well-known that life is determined by genes and each creature has its own genome. The genomes of different kinds of organisms are different. In recent years, scientists have defined genome as a complete set of chromosomes which contains both DNA molecules and genetic information carried on DNA molecules. All the genomes in a life form of cells are composed of DNA, but there are also some organisms such as virus whose genomes are RNA.

The human genome includes nuclear and mitochondrial genome. Nuclear genome has about 3 billion base pairs (bp), and is composed of about 35,000 genes. Mitochondrial genome is actually a circular DNA molecule and more than 16,000 base pairs in length. This genome gets its name as it is located in mitochondrion where the human extranuclear genetic materials are.

3.6.2 Human Genome Project

The HGP was firstly proposed by American scientists in 1985. The aim is to illustrate the sequences of the 3 billion bp of human genome, make clear all human genes, figure out their positions in chromosome, decipher the entire human genetic information, and make the human beings know well of themselves at the molecular level firstly. It's known that human being has 46 chromosomes which contain 3 billion bp encoding about 60,000-80,000 proteins. These coding regions account for only 2% of the genome (the functions of the rest 98% of genome are still to be discovered). In addition, some chromosomes have higher gene density. In the last decade, a lot of efforts have been made to establish a physical map of the human genome in which genes are arranged in the genome according to their sequences by placing road signs. At the same time it also provides an excellent completely-sequenced framework of the human genome. The physical map has helped to determine about 100 pathogenic genes. The greatest challenge is to find genes related with those diseases in a complex genetic model, such as genes involved in diabetes, asthma, cancer, and mental illness. The human chromosome map is shown in Fig.3.13 and the gene map and physical map are shown in Fig.3.14.

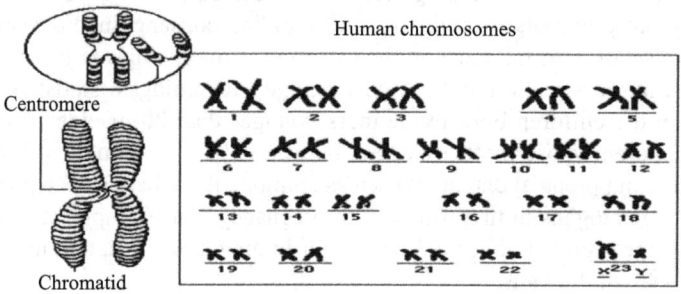

Fig.3.13　Map of human chromosomes

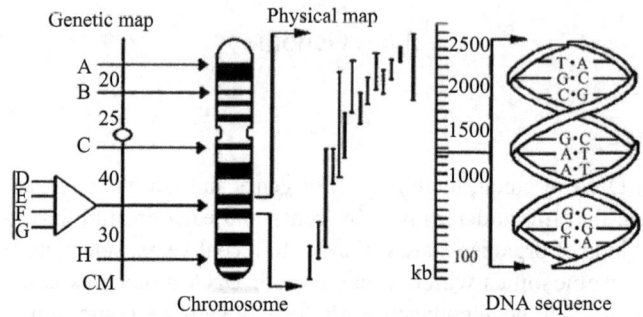

Fig.3.14　Gene map and physical map

　　If analog of the human genome is compared to an encyclopedia with 1 billion words, this book can be divided into 23 chapters and each chapter is just a chromosome. Each chromosome contains thousands of "stories" which are called genes. These "stories" are composed of a series of words with three letters. While each word is four basic chemical "letters", i.e. random permutation and combination of four bases, including A, G, C and T. The human genome project is to "read" the 3 billion chemical "letters", and is so to determine the sequences of the 3 billion base pairs of all the human chromosomes. The human genomes store most of the genetic information related to human birth, aging, sickness, and death. Reading these letters will bring revolution for disease diagnosis and new drug production, so deciphering the human genome is compared to the biological "holy grail".

3.6.2.1　Goal and significance of human genome project

　　The goals of human genome project are as follows: ①constructing the map of human genes in high resolution; ②obtaining physical map of human gene; ③determining complete sequence of the human DNA; ④comparing the sequence information with the selected and representative non-human organism genes; ⑤creating advanced experimental technology to automate the work for acquiring gene map, physical map, and DNA sequencing on a large scale; ⑥developing calculation method and software, establishing the databases that can gather, explain, and send numerous complex gene maps and DNA sequence data from human genome research.

The HGP plays very important roles in life science research and biological industry development, and it speeds up the pace of gene function research. It will make us to have more and more "pathogenic gene samples" so as to let us forecast and treat more diseases. So it has immeasurable impact on human society.

First, obtaining the entire human gene sequence will help humans know about pathogenic mechanisms of many genetic diseases, such as cancer, and provide theoretical basis for molecular diagnosis, gene therapy, and other new methods. In the coming future, the individual's resistance on diseases can be determined according to differences between person's DNA sequences. The 21st century medicine is administrating suitable drug for a specific disease, based on each person's "genetic characteristics", and also named individual medicine. More importantly, it cannot only prevent disease attack in patients in the future, but also keep their offspring away from the same disease through gene therapy.

Second, the human genome program decoding life codes helps people further understand gene expression and regulation. Protein is the real substance taking effect in human body. Human functional genomics actually means learning about expression spectrum that influences development and the whole biological characters of organisms by using genomics knowledge and tools. Some people regard HGP as a table of a life cycle, because it no longer focuses on individual gene but solves the genomic problem at a cellular level and involves in all genes and their expression products so as to establish overall understanding of life phenomenon. At present, researchers have already started comprehensive research on gene expression through DNA chip and other new technologies and on the features and functions of numerous possible proteins or peptides in the human body through analytic means, such as protein chip production, standardized two-dimensional protein gel electrophoresis, chromatography, mass spectrometry, etc. Scientists predict that proteomics research will promote drug development to get a substantial breakthrough so that people can conquer chronic diseases like cancer.

At last, the human genome mapping has great significance to reveal human development and evolution history. Studies on evolution is not based on hypothesis any more. Utilizing comparative genomics to investigate the ancient DNA, it can indicate the mystery of life evolution and the relationship between ancient and contemporary organisms, and help people know better of the human status in nature.

Because gene mapping can be used not only for pathogenic gene research, but also for further research of gene structure, function and expression, the human genome map with high resolution will bring benefits to medical research such as gene diagnosis and therapy, deduction or prediction of infectious potential of genetic disease, diagnostic tests, rational drug design, standardization of gene drug production, etc. Studies on the human genome are not simply purposed to accumulate data, but to discover the inherent law underlying in the numerous data, so as to understand and protect living organisms better.

3.6.2.2 Related maps and test methods of Human Genome Project

Because chromosomes carrying genes cannot be directly used for sequencing, the strategy of HGP is to divide the whole human genome step by step in order from a broad scope to a narrow one, and finally to obtain continuous clonal lines with minimal overlap, used for sequencing. The process that decomposes the genome into small structural regions which are easy to operate is called mapping. According to different signs and ways, the map can be divided into genetic linkage map, physical map and transcription map (also called gene map). The following steps are generally used for sequencing of the decomposed large DNA fragments: ①randomly cutting large DNA fragments into small ones (about 1500bp); ②cloning the small fragments into the sequencing vector; ③sequencing 10-30 subclones for each kilobase DNA at high coverage rate; ④assembling overlapping reading frames into a continuous overlapping line of multiple sequences; ⑤obtaining the final confirmation sequence from the highest quality reading frames.

"Mapping" is determined to be the premise for sequencing in the HGP, and this is purposed to ensure integrity of human genome. However, the sequencing speed of the genomic DNA can be limited by the mapping speed. Therefore, American scientist Dr. Venter put forward an assumption for sequencing the human genome at the same time: the human genome DNA is randomly broken up with mechanical methods and loaded into appropriate vectors, each fragment of 500bp at the end of the inserted DNA fragments is sequences, and then mapping is performed. This method is opposite to the strategy mentioned above. Venter named this strategy as "random whole genome sequencing" or "whole genome shotgun strategy". Success of this method greatly speeds up DNA sequencing and related research. Gel electrophoresis for automatic DNA sequencing instrument has already been unqualified for large-scale DNA sequencing in a short time. Therefore, a capillary gel electrophoresis for sequencing instrument developed by Dr. Venter and the famous PE Company has become the main tool for carrying out the plan, and its efficiency is far higher than the most advanced gel electrophoresis for sequencing instrument at present. Owing to improvement of methods and equipments, the HGP has been completed much earlier than expected.

3.6.2.3 Research progress of Human Genome Project

In 1985, the HGP was put forward by American scientist Dulbecco who was awarded the Nobel Prize. On March 7th, 1986, Dulbecco published an article entitled "A turning point in cancer research: sequencing the human genome" in the journal *Science*. In this article, he pointed out that cancer and other diseases were associated with genes, and described the way and important significance of sequencing the whole human genome. In 1988, Department of Energy of the United States and National Institutes of Health initially headed the HGP in America, which was approved by the congress and sponsored financially by the government. Ever since then, HGP which is an international cooperative organization has been set up. Several nations raised funds and research strength to take an active part in the international research project.

In October 1990, the International HGP was formally launched, and it was expected to take 15 years to sequence 3 billion base pairs, and map and arrange all the genes (estimated 80,000-100,000 genes at that time), by investing 3 billion dollars. Six countries including the United States, the United Kingdom, Japan, France, Germany, and China are responsible for the global HGP, in which the United States undertook 54% of the task, the United Kingdom 33%, Japan 7%, France 2.8%, and Germany 2.2%. In September 1999, China was permitted to participate in the HGP and undertook 1% of the sequencing task.

In 1998, Perkin-Elmer (hereinafter referred to PE) Company which is the largest suppliers of DNA sequencing instruments cooperated with Gene Research Institute led by Venter to establish Celera Genetic Information Company, and then announced that they would complete the human genome sequencing in 3 years by using the latest technology. This made the plan enter a competition between the public institution and the private one, which speeded up the pace of human genome research.

On December 1st, 1999, the International HGP Joint Team announced that it has completely deciphered the genetic codes of the human 22th chromosome. By sequencing 3.35×10^7bp on this chromosome, 679 genes have been found. These genes are related with human congenital heart disease, immune dysfunction, schizophrenia, hypophrenia, and many malignant tumors (such as leukemia, etc.). By means of statistical analysis of the genes on the 22th chromosome, scientists began to suspect the number 100,000 of genes estimated originally.

On May 8th, 2000, Germany and Japanese scientists announced that the gene map of the human smallest chromosome, the 21st chromosome, had been drawn basically. It was found that leukemia, Down's syndrome (congenital dementia), amyotrophic lateral sclerosis, Alzheimer's disease (presenile dementia), manic depression and some cancers were related with the 21st chromosome which contains 225 genes.

On June 26th, 2000, the director of National Human Genome Research Institute, Francis Collins, and the chairman and chief scientist of Celera Company, Craig Venter, announced successful plot of the draft of the human genome. The sequence information in the working draft (namely working frame diagram) is updated every day and released on the internet directly for free reference of the global public in 24 hours. This indicates that the results of human genome research is the common wealth of mankind, and is no longer monopolied by a few people or companies.

Since the human genome draft was announced and the human genome sequencing sketch was published, the research on the human genes had made a series of important findings. ①The total number of human genes is between 30,000 and 35,000, and is less than half of the estimated number. This shows that use of genes in humans is more efficient than that in other species. ②There are "hot spot" regions where the gene density is higher and large "desert" regions without human genes in the genome. The study shows that, the gene density is the highest in the 17th, 19th, and 22th chromosome, and lower in the X, Y, 4th, and 18th chromosome. About more than 1/3 of genomes contain the repeating sequences whose roles need further research. ③Everyone has 99.99% of genes the same with others, and people of different species have higher similarity than those of the same species in genes. Any two different individuals have a different nucleotide in every 1000 nucleotides, and this is called single nucleotide polymorphisms (SNPS). Everyone has a set of his own SNP profile which plays a decisive role in "personality".

3.6.2.4 International institutions of Human Genome Project

Human Gene Organization (HUGO) is an international organization which commits to realize HGP, and is closely related to global human genome sequencing and mapping. HUGO was jointly established by a group of scientists who firstly engaged in human genome research in 1989. Its purpose is to maintain cooperation of research institutions all over the world so as to complete the great project.

HUGO plays a comprehensive coordination role in HGP. HUGO activities include: sorting out data from various research workshop in the organization; constructing the genetic and physical map of the human genome; and taking relevant social ethics, laws, and intellectual rights into account. HUGO is responsible for collecting data and biological materials, encouraging technical communication and sharing, providing information and advice in the HGP, serving as an agency for contact between financial institutions of national governments and genome research institutes. HUGO is the contact interface between HGP and all kinds of related groups or organizations. HUGO currently has more than 100 members on behalf of more than 50 countries and has established three representative offices including HUGO in the United States, HUGO in Europe, and HUGO in Oceania. Their common goal is to complete management responsibilities of the organization. Each member and each country just acquire more information about HUGO activities through the respective representative office.

National Human Genome Research Institute (NHGRI) initially was founded in the United States in 1989 and formerly known as National Human Genome Research Center. Its mission is to lead HGP and National Institute of Health (NIH). NHGRI is one of the 24 associations, centers, or branches of NIH, and is an intermediary organization through which the federal government commits to biological medicine research field. The research system led by NIH owns the largest and most advanced experimental facilities in the world.

3.6.3 Post-Genome Project

After research for more than 10 years, scientists have got an almost complete human genome map, the post-genome era in which the research emphasizes on protein and pharmacogenetics has started and will present a more severe challenge to the scientists.

The human genome map does not tell us "identity" of all genes and their encoded protein. In the human body, proteins play a role as "bricks" in life building, and may contain the "key" that develops disease diagnosis methods and new drugs.

All the proteins encoded and regulated by genes in human cells are called proteome. The chief scientist, Venter, in Celera Company in the United States said at the science annual conference that the human genome, proteome, and drug were the three stages in life science research. Now Celera has entered the second phase, the proteomics research. Celera therefore has bought a large number of protein identification and analysis equipments so as to identify and classify millions of protein fragments every day to draw a proteome map finally. Before Celera turns to proteomics research, a few companies have started. Large-scale biotech companies in the USA now own a database which contains 115,000 kinds of human proteins. Another American company, Cytogen, has drawn the interaction diagram of more than 70 human protein families. In fact, in the last few years, some of the world famous pharmaceutical companies and a large number of biotech companies in small-scale have turned their attention to protein and set off a competition for search for a new protein and determination of their functions. These companies focus on protein isolation and identification, protein interactions, or protein structure analysis.

The latest research results have enhanced scientists' understanding of protein. In the past, scientists thought that a gene was responsible for manufacturing a protein and knowing gene was enough to know protein. But preliminary analysis of the human genome map published recently shows that the human body only contains about 30,000 genes. According to this result, scientists propose that gene may be composed of fragments that can be spliced in different combination ways and one gene can produce various proteins.

The chief scientist of HGP, Collins, said that research on human proteomics would be difficult and now it just took the first step. The reason for this is that DNA and genes are relatively easy to identify and isolate, but for protein, it is time-consuming. Nowadays, the technology for protein determination is far behind the work of deciphering genome, and the best laboratory can only isolate and identify 100 kinds of proteins every day. It is estimated that there may be hundreds of thousand proteins in the human body and it will probably take 10 years to identify them. But with dedication of more and more laboratories to protein isolation and identification, the technologies developed in the next few years it is possible to make ordinary laboratories identify tens of thousands proteins each day.

It can be predicted that proteomics research will make a substantive breakthrough in drug development and make life science research realize its final goal that the drugs for various diseases, including cancer, AIDS and so on, will be developed. For example, scientists have identified the proteins transferring from healthy cells to cancer cells, or the proteins preventing cells from canceration. Based on these findings, the technology for detecting related protein in blood samples can be developed, so as to help diagnose early cancer and eventually develop "switch" drug to control canceration.

3.6.4 Application of Bioinformatics in Human Genome Project

Bioinformatics is a research hotspot in the field of biology, which will be expected to become more and more important and attract more attention in the near future.

3.6.4.1 Recent tasks

As the sequencing data of protein and DNA will exponentially increase in the next few years, bioinformatics will rapidly develop in the following aspects:

1) Information analysis in large-scale genome sequencing

Large-scale sequencing is the basic task of human genome research. Every step is closely related with information analysis. At present, each step from optical density sampling and analysis of sequencer, base readout, carrier identification and removal, splicing and assembly, sequence gap filling, to repeating sequence identification and gene annotation, is closely dependent on genome informatics software and databases. Especially splicing and sequence gap filling even need to combine experiment design with information analysis. The difficulties in splicing and assembly are how to handle repeated sequences. This is especially prominent in the human genome which contains about 30% repeated sequences.

2) Discovery and identification of new genes and new SNPs

The working draft of human genome has been completed, therefore finding new genes becomes an urgent affairs. Genome informatics and the very large-scale computing are important means for discovering new genes, so it is said that most of new genes are predicted via theory. For example, the complete genome of beer yeast (about 13 million bp) contains more than 6000 genes, among which about 60% are acquired through bioinformatic analysis.

After human genes are found, the problems naturally demand to solve as follows: What difference of genes between different races is; and what difference of genes between normal people and patients is, that is usually mentioned SNPs. Constructing SNPs and the related database is the important step for application of genome research results. In 1998, the research has been carried out to find new SNPs which focus on Expressed Sequence Tag (EST) the world. In our country, the Chinese SNPs research is also very important.

3) Comparative studies of the complete genome

Nowadays, bioinformatic experts not only have a lot of sequences and genes, but also have more and more complete genomes. With these data, people can analyze some major biological questions. Such as: Where is origin of life? How does life evolve? How does the genetic code origin? How many genes does the minimum independent living creature need at least? How do these genes make it alive? The size of the mouse genome is similar with that of the human one. They all contain about three billion base pairs, and the number of genes is also similar, but differences of the phenotype between mouse and human are really huge, why is this? Similarly, some scientists estimate that difference of the genomes between races is only 0.1%; difference between human and ape is about 1%. However, their phenotypic difference is very significant. Therefore, phenotypic difference is due to genes and DNA sequences, as well as the whole genome and chromosomal composition. In a word, comparative genomics which results from research of complete genomes will develop a new field.

4) Large-scale analysis of gene expression profile

Along with gradual completion of the human genome sequencing, some scholars have put forward the following questions: Even if we have obtained the complete human genome map, to what extent can we describe people's life activity? Therefore, they put forward a series of questions which cannot be explained from the above data. For example, whether and when do products of gene expression appear? What are concentrations of gene expression products? Does the post-translational modification process exist? If the modification process exists, how does it run? What is effect of gene knock-out or gene over-expression? What is polygenic phenotype? Summarizing these questions, although we know gene and nucleic acid sequences, we don't know how genes play their functions, or how they express in the specific time and space, and what the expression level is? In order to acquire the gene expression profile, new technologies have been developed at the two levels of nucleic acid and protein. DNA chip technology at nucleic acid level, and two-dimensional gel electrophoresis and sequencing mass spectrum technology which are also called proteomic technology at protein level now have been applied in research work.

5) Structural simulation of biomacromolecules and drug design

With implementation of HGP, it is estimated that 35,000 human genes can be located within a few years, and their primary sequences may be determined. However, to understand their functions, one needs to find out the molecular basis of their proteins products. It is necessary to further learn about their three dimensional structure. At the same time, in order to design drugs, one also needs to understand the 3D structure of corresponding protein receptors, which is an urgent task for scientists.

3.6.4.2 Long-term task

Long-term task of bioinformatics is to understand the human genome, find the fundamental law of human genetic language so as to clarify some major natural philosophy questions in biology, such as the origin of life and evolution, etc. The key and core of the research is to understand non-encoding regions.

1) Structural analysis of non-encoding regions

In recent years, studies on the complete genome have indicated that non-encoding regions only account for 10% to 20% of the entire genome sequence of microbes such as bacteria. But non-encoding regions of higher organisms and the human genome account for most of the genome sequence. From the view of biological evolution, with improvement and complication of functions of organisms, non-encoding region sequences significantly increase, and this trend shows that this part of sequences must have important biological functions. It is widely believed that non-encoding regions and genes are relevant with expression regulation in four-dimensional space-time. Therefore, the encoding characteristics of these regions and the rule of gene expression and regulation are hot issues in the future research in quite some time. As the human genome is concerned, people have really mastered the law of encoding regions in DNA (gene) so far. A lot of data suggest that this part of the sequences accounts for only 3%-5% of the genome. That is to say, as much as 95%-97% sequences of the human genome are non-encoding regions. How to further understand functions of the non-coding region sequences is a real challenge which scientists face up to currently.

2) Origin of genetic codes and research of evolution

Since Darwin published *The Origin of Species* in 1859, theory of evolution has been the most important contribution to development of natural sciences and natural philosophy of humans. The core of the theory of evolution is to describe the history of organism evolution (phylogenetic tree) and explore the mechanism of evolution. Since the mid-20th century, along with development of molecular biology, evolution research also deepened to the molecular level. Currently research on molecular evolution is an important means of evolution research, and a set of theory methods depending on nucleic acid and protein sequences have been established. In recent years, with increase of sequence data, the debate on sequence differences and evolutionary relationship have become more and more fierce, because the phylogenetic tree based on a certain molecular sequence can only reflect relationship between such sequence and system development. It does not represent real evolution relationship between species, namely, there might be differences between the genetic tree and the species tree. At the same time, discussion about vertical evolution and horizontal evolution is gradually concerned. The current data give us a more complex and comprehensive evolution mode which enlightens us to understand thoroughly the law of evolution, and it must use the entire genome information and develop accordingly new theory methods.

In a word, the current period is a dynamic new epoch of bioinformatics research. Many scientists also believe that it is the harvest time of the human genome research. It will not only gives people all important achievement of basic research, but also can bring the huge economic benefit and social benefit. In the next few years, DNA sequence data will increase at an unexpected speed. It is a rare opportunity. Our country should utilize these data as soon as possible so as to stand on the frontier of the international scientific area.

Chapter 4

Gene Engineering

Gene engineering is also called genetic engineering. In 1997, the national genetic engineering symposium was held in Beijing, according to the proposal of the symposium, genetic engineering was called gene engineering which referred to DNA recombination technology. Gene engineering is a procedure to modify and recombine genes from different organisms *in vitro* by utilizing DNA recombination technology such as artificial methods for "shearing" and "splicing", then introduce the recombinant DNA into host cells to replicate, transcript, translate and express the exogenous gene, so as to produce the gene product which mankind needs. According to desire of people, with DNA recombinant technology heredity of an organism can be altered to create a new species at the gene level. Furthermore, the technology can provide useful products and services for mankind through an engineering-oriented method.

In 1972, scientists succeeded in recombining the gene fragments of different sources *in vitro*, which are tentative trials about feasibility of gene engineering. In 1973, Herker Boyer (University of California, San Francisco) and Stanley Coher (Stanford University) cut two kinds of plasmids with endonuclease, then connect the two plasmids together with DNA ligase to acquire a hybrid plasmid with two origins of replication, and finally introduced it to *E.coli*, so that a new discipline, gene engineering was founded for oriented modification of organisms. Actually, genetic engineering refers to all of the technologies that it artificially modify hereditary characters of organisms, including gene engineering as well. DNA recombination means that it cleaves and links different genes to construct hybrid DNA *in vitro* with enzymatic method. Molecular cloning mainly is the process that recombinant DNA molecules replicate in the host cell. All the three above have relations but not the same as gene engineering. In 1977, W. Gilbert and F. Sanger established DNA sequencing technology that provided a significant approach for analysis of gene structure, drawing a gene map and gene synthesis. In 1978, the gene of human growth hormone releasing inhibition factor was expressed successfully in *E.coli* via gene recombination. In 1979, gene recombination of human insulin also succeeded and the expressed product was obtained. Hence, gene engineering has progressed into a mature stage.

The birth of gene engineering is based on the facts as follows: different genes have the same material basis; genes can be cleaved; genes can be transferred; peptide and genes have a corresponding relation; genetic codes are universal in the different between organisms; and genes can replicate so as to transmit genetic information to offspring. Gene engineering gradually develops and matures on the basis of discovery of bacterial restriction endonuclease and plasmids, and success in DNA recombination technology and DNA sequence analysis. In 1971, H. O. Smith and D. Nathans discovered the bacterial restriction endonucleases which had extremely strict bases specificity, and could break down phosphodiester bond at the specific site (specific recognition nucleotide sequence) of invading exogenous DNA molecules and to cause "gaps". In further studies, the restriction enzymes, *Eco*K, was discovered, and it could cut DNA specifically at a certain site on the DNA strand, and at the same time methylation modification enzymes were discovered. Thus, these findings provide key tool enzymes for gene recombinant. Now, about 500 kinds of restriction enzymes have been found, and more than 100 kinds of enzymes have been applied in the field of gene engineering.

The vector is used as an important vehicle in gene engineering. An exogenous gene can insert into DNA sequence of a vector, and then replicate in host cells. When vector DNA replicates in host cells, exogenous genes are also amplified and expressed. Currently, the major types of commonly used vectors are plasmid, phage and virus. Plasmids that exist in extrachromosomal state are genetic factors and are capable of automatically replicating in a host cell. Plasmids are generally circular DNA molecules, sometimes linear molecules, and are regarded as an independent replicon. Once they carry new target genes, plasmids themselves may be a stable recombinant DNA. In a receptor cell, it is not necessary that plasmids can conduct self-replication without integration with chromosomes of the host cell and at the same time the heterologous gene replicates as well. Plasmids have obvious phenotypic features (such as drug resistance), and so it is an important vector commonly in gene engineering operation. Bacteriophage is a kind of viruses that infect bacteria, phage λ is often used in gene engineering, and it has a remarkable advantage to carry as long exogenous gene as 23kb. However, plasmid and phage have a disadvantage that they can't be "parastism" in receptor cells of higher organisms, while animal viruses just have such advantage. Therefore, animal viruses usually are used as vectors for gene engineering in higher organisms. For instance, RNA retroviruses can transfect cells from primates at high efficiency, and so are concerned much by researchers in the area of gene therapy.

Gene engineering is the main technology of contemporary bioengineering, and a growing point for contemporary life sciences as well. The main content of gene engineering includes several aspects as the following. ①Isolation, purification of tool enzymes and property research for gene engineering. ②Isolation and purification, and identification and analysis of exogenous DNA and vector DNA, i.e. preparation of DNA fragments containing the target gene (isolation of DNA fragments containing the target gene from genome of complex organisms through a series steps of enzyme digestion, PCR amplification, and etc.), and construction and selection of vectors. ③Cleavage and ligation of exogenous DNA and vector DNA (ligating the DNA fragment with the target gene to a vector with self-replication ability and selected markers, to form a recombinant DNA molecule *in vitro*), and reverse transcription of the exogenous DNA. ④ Analysis of DNA sequence, selection of host cells, and gene transfer, that is, introducing the recombinant DNA molecules to the selected receptor cell, so as to make it replicate and amplify in the cell. ⑤Cloning and screening of engineered cells, and expression and regulation of exogenous gene in the recombinants. The process and its application of gene engineering shows in Fig.4.1.

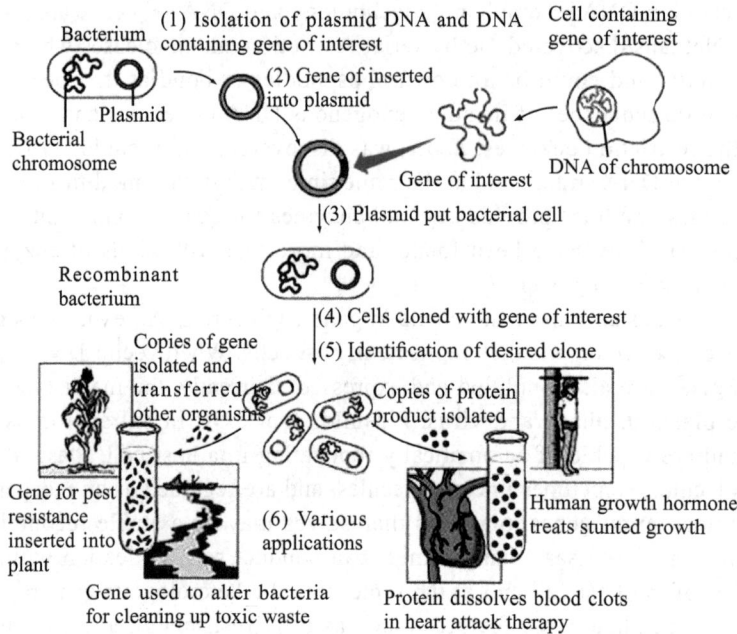

Fig.4.1　Schematic diagram of gene engineering process and its application

Emergence of genetic engineering has important theoretical and practical significance. The most significant feature of genetic engineering is breaking the boundaries between different species, so as to perform gene recombination, exchange and transfer within and beyond species. Hence, gene engineering is the fundamental technology for artificially-oriented modification of heredity of organisms. The purpose of genetic engineering includes several aspects as follows: to utilize DNA recombination technology to create supernatural species; to expand production of useful substances that lack in normal organisms; to develop a new drug; to conduct theoretical research; and to diagnose and treat diseases. Meantime, in biological field, gene engineering provides a new way for theoretical study on differentiation, growth, development, evolution, gene structure and function of one single cell, tissues and organs, and tumor formation mechanism. In medical field, it lays the theoretical and technical foundation for adopting gene therapy to cure hereditary diseases and tumor. In industrial and agricultural production, many precious drugs are obtained through gene engineering, and which cannot be available via traditional technologies. For example, human insulin, human growth hormone, tissue plasminogen activator (tPA), human urokinase, SCP, brain hormone, hepatitis B vaccine (HBV), interferons, and dozens of monoclonal antibodies, and human relaxin, pepstatin, superoxide dismutase (SOD), erythropoietin (EPO), colony stimulating factor, and etc, are all in a clinical trial stage. It is estimated that in the future, gene engineering will play a crucial role in production of enzymes, proteins, amino acids, organic acids, biopesticides, herbicides and antibiotics. Moreover, research on transgenic animals and transgenic plants has made considerable progress and will have a far-reaching impact on development of agriculture and husbandry.

4.1 Tool Enzymes for Gene Engineering

The key technology of gene engineering is DNA recombination technology which means taking out the target gene from chromosomes and then ligating it to vector DNA to form the recombinant DNA molecule, and whose whole process is manipulated by catalysis of enzyme. Gene recombination and isolation is involved in a series of enzyme-catalytic reactions in correlation. There are many important nucleases which have been known, such as endonuclease, exonuclease, and reverse transcriptase which used mRNA as template to synthesize the complementary DNA. These enzymes apply to gene cloning experiments widely (Table 4.1). Especially, discovery and application of restriction endonucleases and DNA ligase make it possible to cut and ligate DNA molecules *in vitro*. There is no doubt that restriction endonucleases and ligase are important enzymology basis for DNA recombinant technology. Therefore, in order to deeply understand the basic principle of gene manipulation, it is obviously essential to discuss common nucleases in the gene recombinant experiment.

Table 4.1 Common nucleases in DNA recombination experiment

Nuclease name	Major Function
Restriction endonuclease II	Cleaving DNA molecule at the specific site of base sequence
DNA ligase	Ligating two DNA molecules or fragments
E.coli DNA polymerase	Filling single-strand gap of double-stranded DNA molecule through adding free nucleotides one-by-one only to the 3'-end
Reverse transcripatase	Using mRNA as template to synthesize complementary DNA
Polynucleotide kinase	Adding a phosphate group only to the 5'-end of polynucleotide chain
Terminal transferase	Adding homopolymer tail to the 3'-end of linear dsDNA or ssDNA molecule
Exonuclease III	Cutting nucleotide residue from the 3'-end of double-stranded DNA
λ exonuclease	Removing single nucleotide from the 5'-end of double-stranded DNA, so as to expose the extended 3'-end of single chain
Alkaline phosphatase	Removing terminal phosphate group from the 5'-end and the 3'-end of DNA molecule
Taq polymerase	Use ssDNA as template to synthesize a new complementary strand in a direction from 5'-end to 3'-end at high temperature (72℃)

All of the enzymes applied in gene engineering are termed generally as tool enzymes. Those cutting phosphodiester bonds between the two adjacent nucleotide residues so as to cause hydrolytic cleavage of polynucleotide chain are called nuclease. Among them, RNase is a type of nuclease that catalyzes hydrolytic cleavage of RNA, but DNase is a kind of enzyme that catalyzes the specific hydrolytic cleavage of DNA molecules. According to different ways of hydrolytic cleavage of nucleic acid, nucleases are usually divided into two types, one is exonuclease that cuts one nucleotides one by one from the end of a polynucleotide chain, the other is endonuclease that cleaves the phosphodiester bond within a polynucleotide chain to form small nucleic acid molecule fragments.

4.1.1 Restriction Endonuclease

Restriction Endonuclease is defined as endonuclease which can identify specific nucleotide sequences of double-stranded DNA molecule and cut DNA at the specific site. Such enzymes are mainly isolated from prokaryotes and purified. For this reason, most of bacteria have some functional obstacles for infection of phages. Up to now, it has not yet been found that one phage can infect two different species of bacteria, even if adsorption and transcription can go smoothly. There is another functional obstacle that is so-called host-controlled restriction and modification controlled. When one type of viruses is isolated from its natural host A, infection rate of the viruses for host B will be only 10^{-4}. Most of them cannot grow, and only one-ten thousandth of viruses can. If the viruses are isolated from host B and then infect host B again, the infection rate will be 100%. But if those viruses are applied to host A, the infection rate will reduce to 10^{-4}, such phenomenon is called host restriction. Research shows that the viral DNA degrades in a new hosts but the host DNA itself is not degraded, this phenomenon is called restriction-modification action, and also refers to the R/M (restriction-modification) system for short. Therefore, restriction phenomenon refers to that bacteria can destroy those foreign DNA molecules when they invade bacteria; while modification phenomenon refers to that when foreign DNA molecules invade bacteria they are modified in current generation and are not destroyed by foreign host bacteria. In addition, restriction enzymes and modification enzymes (also called methylase) have been found in host cells, which identify the same DNA sequence but have opposite role in the same host cells. Methylase have species-specific property and usually modify only host DNA itself, and hence restriction enzyme can't degrade host DNA. However, since methylase can't modify exogenous DNA, but restriction enzyme can do. Therefore, viruses and phages have their own specific host scope. But methylase occasionally makes a mistake to modify exogenous DNA so that exogenous DNA avoids degradation by host restriction enzyme. For this reason, when viruses of host A infect host B, one-ten thousandth of those viruses can inhabit and propagate in a new host. Thus it can be seen clearly that R/M system not only constitutes self-defense system with hereditary stability, but also is useful for species evolution; and it plays the same roles in process of DNA recombination as in microbial cells. To sum up, DNA recombination technology is a simulation of biological function in nature.

4.1.1.1 Nomenclature and types of restriction endonucleases

All of enzymes that can identify and cut specific nucleotide sequence of double helix DNA molecules are generally termed as restriction enzyme. Now more than 500 kinds of restriction enzymes have been discovered from microbes, and so it is necessary to name and classify these enzymes in order to distinguish from one another.

Nomenclature method of restriction enzymes is based on the principle conjunction between genus names and species name that put forward by H.O. Smith and Nathane, that is to say, restriction enzyme is represented by choosing the first letter (capital) of genus name of the bacterium and the first two letters previous varieties (lowercase) to constitute a three-letter symbol. If one strain contains several enzymes, roman numerals will be placed respectively behind the three letters for the difference between different enzymes. For example, *Hae* I, *Hae* II and *Hae* III refer to three different enzymes from *Haemophilus aegytius*. If it has strain or serotype name, the first letters of strain or serotype name will be placed behind the three letters, for example, the three types of enzyme from *Haemophilu sinfluenzae* Rd are respectively Hind I, Hind II and Hind III, wherein d refers to the name the strain. For another example, *Hinf* I is one enzyme from *Haemophilus influenzae* Rf, wherein f means serotype. When the previous two letters of the bacteria names belonging to various species of the genus is completely the same, symbol of one enzyme adopts the first different letter (lowercase) of the species name which replaces the last letter of the three-letter symbol. For example, the two enzymes from *Haemophilus parainfluenzae* are expressed as *Hpa* I and *Hpa* II, and the two enzymes from *Haemophilus parahaemolyticus* of the same genus are respectively described as *Hph* I and *Hph* II.

Now three categories of restriction endonucleases have already been identified, such as Type I, Type II, Type III. These three types of restriction enzymes have different properties (Table 4.2). Since restriction endonucleases activity and methylation enzyme activity of type II are separated, its endonuclease activity has sequence-specific, they have been widely applied in gene cloning.

*Eco*K and *Eco*B, two representatives of Type I, were first identified in two different strains (K and B) of *E.coli* by M. Meselson's and R. Yuan's research group (1968) respectively. The molecular weight of these enzymes is about 300,000 Da. Type I restriction enzymes are single multifunctional enzymes and consist of three different subunits. The cofactors, ATP, Mg^{2+}, and S-Adenosyl methionine, can recognize specific sequence. For instance, recognition sequence of *Eco*B is TGANsTGCT; that of *Eco*K is AACNsGTGC, but the cleavage site locates in the distance 1000bp far from their recognition site, and the methylation site is consistent with the recognition site. Hydroxyl end of DNA fragments, which is formed by removing phosphoryl group from 5'-end of DNA fragments, can't be used as phosphorylation substrates of polynucleotide kinase, thus it is unsuitable as a tool enzyme.

Table 4.2 Types and key properties of restriction endonucleases

Property	Type I	Type II	Type III
Modification and Restriction activity	Single multifunctional	Independent of endonuclease and methylase	Bifunctional enzyme sharing with one subunit
Protein structure of restriction endonuclease	Three different subunits	Single element	Two different subunits
Cofactors	ATP、Mg^{2+}、S-adenosylmethionine	Mg^{2+}	ATP、Mg^{2+}、S-adenosylmethionine
Host-specific site sequence	*Eco*B:TGA(N)$_8$TGCT *Eco*K:AAC(N)$_6$GTGC	Rotational symmetry (except type II s)	*Eco*P1: AGACC *Eco*P15: CAGCAG
Cleavage site	Cleaving randomly at the site far at least 1000bp from host-specific sites	Host-specific sites and nearby	Cleave 24-26bp away from 3'-end of host-specific sites
Enzymatic transition	No	Yes	Yes
DNA translocation	Yes	No	No
Methylation site	Host-specific sites	Host-specific sites	Host-specific sites
Recognition of non-methylation sequence and cleavage	Yes	Yes	Yes
Sequence-specific	No	Yes	Yes
Application in DNA cloning	Useless	Very useful	Rarely useful

Type II restriction enzymes were first isolated from *Hemophilus influenzae* Rd by H. O. Smith, K.W.Wilcox and T. J. Kelly (1970).

Type II enzymes have no abnormal characteristics liking Type I enzymes, and so they are extremely important for DNA manipulation. Type II enzymes have two same subunits and have only restriction function, its molecular weight is 20,000-100,000Da, the cofactor is only Mg^{2+} and the recognition sites are usually rotational symmetry. They cleave DNA at the recognition site with higher specificity; and meanwhile, the methylation sites are consistent with the recognition ones, their substrate is double-helix DNA, they are an ideal tool enzyme for gene engineering. Type II enzymes recognize a special target sequence of double-stranded DNA and break nucleotide chain to form separated DNA fragments with a certain length and sequence. By means of gel electrophoresis of DNA fragments that are produced with restriction endonuclease, activity of restriction endonucleases can be analyzed and studied. Now, a lot of Type II restriction endonucleases have been already isolated from many bacteria. With development of research work, more strains with Type II restriction enzymatic activity are likely to be found.

Type III enzymes contain two type different subunits and have dual function. They require, but not necessariey, ATP, Mg^{2+} and S-AdoMet as cofactors. Type III restriction enzyme have specific recognition sequences. For example, *Eco*P recognizes sequence of AGACC and cuts DNA at the site about 20-30bp far from the recognition site. Its methylation site is consistent with the cleavage site of restriction enzymes, and function of Type III enzymes is yet to be further studied.

4.1.1.2 Nature and Property of Type II Restriction Enzyme

The basic characteristic of Type II restriction enzymes is that recognition sequences are 4-7 nucleotides, recognition sites are cutting ones, and nucleotide sequences of these sites often exhibit dual rotational symmetrical structure. There are three cutting ways as follows: ①cutting phosphodiester bond simultaneously at the symmetry axis of the double chain of recognition sites, in order to form restriction fragments with blunt ends, such as *Hae* III [Fig.4.2 (a)]; ②cutting phosphodiester bond simultaneously from 5'-ends on two sides of symmetry axis of the double strand, in order to form cohesive staggered single-stranded 5'-ends with 2-5 nucleotides, such as *Eco*R I [Fig.4.2 (b)]; ③cutting phosphodiester bond simultaneously from 3'-ends on two sides of symmetry axis of the double chain, in order to form cohesive staggered single-stranded ends of 2-5 nucleotides with 3'-OH, such as *Pst* I [Fig.4.2 (c)].

```
   5'GG ↓ CC3'              5'G ↓ AATTC3'            5'CTGCA ↓ G3'
   3'CC ↑ GG5'              3'CTTAA ↑ G5'            3'G ↑ ACGTC5'
       (a)                       (b)                      (c)
```

Fig.4.2 Main cutting ways of type II restriction enzyme
(a) *Hae* III cleavage site; (b) *Eco*R I cleavage site; (c) *Pst* I cleavage site

Those among Type II restriction enzymes that recognize the same sequence but come from different sources are known as different sources isoenzyme. Those enzymes whose recognition sequence and cutting way is the same are termed as isoschizomers. For example, *Hpa* II and *Msp* I have the same recognition sequence which is CCGG, and the same cutting way, and so both enzymes are isoschizomers. Some enzymes, source and recognition sequence are all different, but restriction fragments have the same sticky end after cutting with these enzymes, and so are called isocaudamer. For example, *Bam* H I, *Bgl* II, *Bcl* I, *Sau*3A I and *Xho* II can cut DNA molecules and then form 4 nucleotide sticky end, GATC, and so they are a group of isocaudarmers.

A restriction fragment is a DNA fragment resulting from cutting a DNA chain with a restriction enzyme, the length of restriction fragments which are formed by digesting DNA with various different restriction enzymes. If x is the A or T occurrence frequency in DNA molecules and y is G or C occurrence frequency as well, cutting frequency (or site frequency) F in DNA molecules of the restriction enzymes is calculated as:

$$F = x^n y^m$$

wherein, n——A-T pairs within recognition sequence of restriction enzymes;
and m——G-C pairs within recognition sequence of restriction enzymes.

If base pairs (B) forming DNA molecule and recognition nucleotide sequence of restriction enzymes are all known, then the number of theoretical cleavage sites (N) for restriction enzymes to cut DNA molecules is as follow:

$$N=BF$$

Assuming that the numbers of four type nucleotide residues in a DNA molecule are equal, then the occurrence frequency of cleavage sites is $(1/4)^4$ when restriction enzymes cut DNA molecules, wherein the recognition sequence is four base pairs. In other words, a cleavage site will emerge once per average 256 base pairs. Similarly, for restriction enzymes recognizing six base pairs, their occurrence frequency of the cleavage site is $(1/4)^6$, or rather one cleavage site occurs every 4096 base pairs. That is, average length of restriction fragments is 256 or 4096 nucleotide pairs, These fragments are all possibly encoding genes, and the latter fragments even may constitute a genome.

It should be noted that all of the known Type II restriction enzymes have specific digestion conditions. Once the conditions change, specificity of enzymes will reduce so as to change recognition sequence. For example, under the conditions of salinity less than 50mmol/L, pH>8.0 and adding glycerol, the recognition sites of *Eco*R I is different from the original site, and this makes *Eco*R I denatured. The number of cleavage sites of the denatured *Eco*R I is 15 times the original of *Eco*R I. *Bam*H I also has a similar denaturation effect. Hence, when all kinds of restriction enzymes perform cutting reaction, it should ensure the process in specific reaction conditions in order to achieve optimal reaction speed.

In all Type II restriction enzymes, now *Eco*R I is widely used in molecular genetics, and comes from the mutant Ry-13 of *E.coli*. Its molecular weight is 58,000Da with two of the same subunits and this enzyme has already been purified 1500 times and achieved gel electrophoresis pure. *Eco*R I can identify six nucleotide base pairs with stricter specificity, and cut natural DNA to form larger fragments with the sticky AATT end. Thus *Eco*R I can be applied in DNA structural analysis and genetic recombination (Fig.4.3).

In addition to DNA recombination, restriction enzymes are also used to construct new carrier, DNA molecule hybridization, DNA sequence analysis, preparation of radioactive DNA probes and DNA base methylation recognition.

There are many factors that influence activity of restriction enzymes. The first is purity of DNA. Generally, the substances that contaminate DNA are proteins, SDS, EDTA, phenol, ethanol, chloroform and so on. Especially, DNase can make DNA degradation, and its influence is most remarkable. And hence DNA purity should be improved as much as possible. In addition, DNA methylation level also affects enzymatic activity. The higher methylation level is, the weaker cutting action is. Meantime, DNA structure also has influence on enzymatic activity. For instance, a cyclic double helix DNA is more stable than a linear DNA. When cutting DNA molecules of the same molecular weight, enzyme amount for cyclic double-helix DNA is 10-20 times than that for linear DNA. Moreover, enzymatic reaction temperature is usually at 37℃, but some enzyme have different, such as *Sam* I at 25℃, *Apa* I at 30℃, *Mae* I at 45℃, *Bcl* I at 60℃, *Taq* I at 65℃ and so on. When reaction temperature is above or below the optimal temperature, enzymatic activity will decrease without exception. Secondly, composition of enzyme reaction buffer also plays an important role. The reaction system usually contains $MgCl_2$, NaCl, KCl, β- mercaptoethanol, dithiothreitol, bovine serum and albumin, the latter three substances can improve stability of enzyme, Mg^{2+} can activate enzymatic activity. Above all, appropriate enzyme reaction conditions should be determined in order to ensure optimal cleavage results.

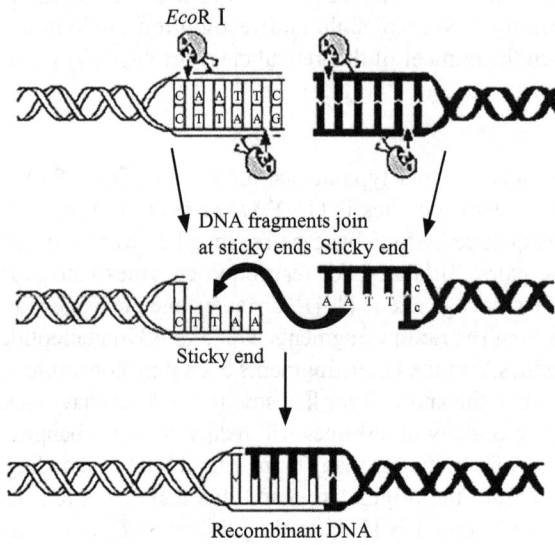

Fig.4.3 Action of restriction enzyme *Eco*R Ⅰ

When digestion reaction comes to an end, we must destroy enzymer activity, such as heating inactivation, usually heat preservation for 5min at 65℃. Some denaturants such as urea, SDS and guanidine, make heat-resistant enzymes inactivate. But before continuing subsequent gene manipulation, it is necessary to purify DNA fragments after enzyme digestion.

4.1.2 DNA Ligases

Just like restriction endonucleases, DNA ligase also has the most important significance for gene engineering. All of them are essential basic tool enzymes for constructing recombinant DNA molecules *in vitro*. Restriction endonuclease can cut DNA molecules into different sizes of fragments, while only DNA ligase is used to link DNA fragments together from different sources and seal nicks in DNA, so as to form new hybrid DNA molecules.

DNA ligase exists extensively in organism cells and at present mostly from *E.coli*, and its molecular weight is 74,000 Da. DNA ligase can catalyze formation of a phosphodiester bond between the 5'-end with a phosphate group on one DNA strand and the free 3'-end with a hydroxyl group on the other DNA chain, so as to ligate the two DNA strand. At the same time, since formation of a phosphodiester bond between a hydroxyl and a phosphate group is an endergonic reaction, energy is required for the ligation reaction. In *E.coli* and other bacteria, NAD^+[(nicotinamide adenine dincleotide) (oxidized form)] serve as energy source for ligation reaction, but ATP (triphosadenine) serve as energy source in animal cells and phages.

DNA ligase is mainly used for normal DNA synthesis and repair of damaged DNA. In gene recombination *in vitro*, DNA ligase is used for DNA ligation, but it can only ligate fragments with sticky ends and cannot ligate those with blunt ends. In addition, DNA ligase can't ligate two single-stranded DNA molecules or cyclic single-stranded DNA molecules, and ligated DNA chains must be part of double-helix DNA. Actually, DNA ligase is used to seal nicks of the double helix skeleton. So called nick is single-strand DNA breakage on some one strand of the double-stranded DNA, which occurs due to loss of a phosphodiester bond between two adjacent nucleotides. But DNA ligase is not used to close gaps, and the so called gaps refer to single-strand DNA breakage on some one strand of the double-stranded DNA, which results from loss of one or several nucleotides.

During the reaction process catalyzed by DNA ligase, at first covalent NAD^+-enzyme complex products, that is to say, a phosphate amide bond between NAD^+ and ε-amino group of lysine residue on DNA ligase forms, through action of the enzyme, NAD^+ will be transferred to DNA nicks to form a pyrophosphate bond with 5'-phosphoryl group, which makes 5'-phosphoryl group activate. With action of the ligase, a phosphodiester bond between activated phosphoryl group and 3'-hydroxyl group forms so as to releases NAD^+, and then the DNA ligation reaction is completed. The optimal temperature of DNA ligation is 37℃, but sticky ends are generally short, and hydrogen bond will be not stable when temperature is too high. And so the optimal reaction temperature often depends on speed of ligation and terminal binding rate. Usually, 4-15℃ is ideal temperature. In the past, the result of ligation was often judged through gel electrophoresis, but now it is considered that the transformation ratio of competent cells is an ideal judgment standard.

T4-DNA ligase is obtained from *E.coli* cells infected with bacteriophage T4, and preparation easy. It is an expression product of gene 30 of bacteriophage T4 DNA. After the bacteriophage infects *E.coli*, lysogenic bacteria of *E.coli* will form, and DNA ligase of high concentration will be obtained by culturing the bacteria at 40℃. In later period, cells produce T4-DNA ligase of high concentration, whose molecular weight is 68, 000Da and cofactor is ATP. The enzyme can ligate both sticky-ends and blunt-ends, but larger dosage of the ligase is required.

4.1.3 DNA Polymerases

Now common DNA polymerases are *E.coli* DNA polymerase, Klenow fragment of *E.coli* DNA polymerase Ⅰ (Klenow enzyme), T4 phage DNA polymerase, T7 DNA polymerase, and thermostable DNA polymerase (e.g. *Taq* DNA polymerase). Different sources DNA polymerases have their respective enzymological characters. Since optimal temperature for *Taq* DNA polymerase is 75-80℃, it is now widely used in PCR and DNA sequencing. All of DNA polymerases catalyze formation of a phosphodiester bond between phosphate group and hydroxyl group on the ends of two DNA fragments. So DNA polymerase can be used for DNA molecule repair and DNA molecules recombination *in vitro*.

DNA polymerases have three different types, that is, DNA polymerase Ⅰ, DNA polymerase Ⅱ, and DNA polymerases Ⅲ, which are also called *Pol* Ⅰ, *Pol* Ⅱ and *Pol* Ⅲ respectively. *Pol* Ⅰ and *Pol* Ⅱ are involved DNA repair, and *Pol* Ⅲ relates to DNA replication. Of the three DNA polymerases, only *Pol* Ⅰ has most closely relation with DNA molecular cloning. Under certain conditions, *Pol* Ⅰ can catalyze hydrolysis reactions of DNA strand. That is to say, *Pol* Ⅰ hydrolyzes DNA chain from 3'-OH end to the 5'-P end to release mononucleotide molecules. Therefore, *Pol* Ⅰ are a kind of exonuclease cutting DNA in the 3'→5' direction, and it's substrate can be double-stranded DNA, as well single-stranded DNA. The main purpose of DNA polymerase Ⅰ in molecular cloning is preparation of DNA probes with radioactive labels for nucleic acid hybridization through DNA nick translation. The characteristics of the polymerases can refer to Table 4.3.

Table 4.3 Properties of DNA Polymerases

DNA polymerases names	Activity of 3'→5' exonuclease	Activity of 5'→3' exonuclease	Polymerization speed	Processivity
E.coli DNA polymerase	Low	Yes	Middle	Low
Klenow enzyme	Low	No	Middle	Low
Reverse transcriptase	No	No	Low	Middle
T4 DNA polymerase	High	No	Middle	Low
T7 DNA polymerase	High	No	High	High
Chemically modified T7 DNA polymerase	Low	No	High	High
Genetically modified T7 DNA polymerase	No	No	High	High
Taq DNA polymerase	No	Yes	High	High

DNA polymerases depending on RNA are also called RNA-guided DNA polymerases or reverse transcriptase which is required for formation of complementary DNA(cDNA) that based on mRNA template, and used for constructing cDNA library so as to isolate genes encoding certain proteins. It is one of the most important tool enzymes in molecular biology. In recent years, reverse transcription polymerase chain reaction (RT-PCR) technology which couples reverse transcription with PCR makes isolation of eukaryotic genes more rapid and effective. Meanwhile, reverse transcriptase can utilize single-stranded DNA or single-stranded RNA as a template to synthesize molecular probes for experiments.

T4 Polynucleotide Kinase can catalyze transfer of a γ-phosphate from ATP to DNA or the 5'-end of RNA, so as to label γ-^{32}P on the 5'-end of nucleic acid molecules, realizing end labeling. Besides, it can also make DNA with loss of the 5'-phosphate end conduct phosphorylation.

4.1.4 Other Enzymes

Other tool enzymes, such as alkaline phosphatase, exonuclease and nuclease, also play a relatively important role in gene engineering. Here we no longer give uncecessary details. Please see other relevant books for further information.

4.2 Gene Engineering Vectors

To fulfill gene recombination really, exogenous genes must be transfered to receptor cells. In this process, exogenous gene must firstly ligate with a certain carrier to form recombinant DNA molecule, and then it can enter the host cell. These carriers which can load exogenous gene fragments and carry them into receptor cells are termed as gene engineering vectors. A useful gene is called a target gene or an insert. A host cell is a receptor cell used to amplify vectors and inserts, and amplification process of inserts is known as molecular cloning. The host cell which carries recombinant DNA molecules is referred to as gene engineering cells.

Gene engineering vectors determine replication, amplification, passage and expression of exogenous genes. There are some essential requirements for gene engineering vectors as follows: ①the vectors should have effective carrying capacity, and can enter host cells; ②the vectors should have a single or a few cleavage sites of several restriction enzymes, that is, the vectors themselves are a replicon, they all can self-replicate before and after carrying exogenous target gene into the host cell or can integrate into chromosome of the host cell; ③in the host cell, they can control exogenous gene expression; ④the vectors should have selective markers in order to identify recombinants conveniently, and loading and unloading procedures should be simple; ⑤the vectors should be safe, reliable, and easily controlled.

At present, such vectors which have already been constructed and applied in gene engineering are plasmid vector, viral vectors, bacteriophage vectors and vectors constructed by combining with each other or other genomic DNA. Now the widely-used vectors are bacterial plasmids. Different vectors have different structures and biological property, and apply to different purposes. The vectors used in DNA recombination include as follows: ①clone vector, it is a carrier for amplifying DNA fragments; ②shuttle vector, it is used for amplification of eukaryotic DNA fragments in prokaryotes, and then the recombinant DNA is introduced into eukaryotic host cells to express; ③expression vector, it is used for gene expression.

With development of molecular biology and DNA recombinant technology, the vectors should not only have the basic properties mentioned above, but also meet special requirements, such as high copy number, strong promoter and stable mRNA, high isolation stability, structural stability, high transformation frequency, wide host scope, high capacity of inserting the exogenous gene, enabling to replicate and transcript wholly, and matching with the host. In addition, the vectors should express target genes at high levels when the host cell does not grow or grows at low rate. But the vectors meeting these requirements completely are few, especially when animal cells are used as the host cell, and now the vectors used for the above purpose are mainly viruses. The target gene that enters the host is generally only one single gene, while it is hard to recombine with one genome or several genes simultaneously, so, it is necessary to do further research and development work for vectors.

4.2.1 Plasmid Cloning Vectors

Those plasmids used for completing gene engineering manipulation are called plasmid cloning vectors. Plasmids are genetic material extrachromosomal and self-replicating in cells. Except yeast killer plasmid which is RNA molecule, the other plasmids are mainly cyclic double-helix small DNA molecules, and their molecular weights have marked difference from less than 1kb to more than 500kb. Every plasmid has a sequence of DNA replication origin, which helps DNA of plasmids replicate in the host cell. Plasmid can replicate along with replication of the host chromosome and are symbiont within the host. Their existence has nothing to do with death of the host. On account of smaller molecular weight, plasmids transfer and migrate easily between different hosts. Whereas, they can encode genes which the host chromosome does not have, express genetic characters uncontrolled by chromosomes, and give host cells extra genetic features, such as antibiotic resistance. In addition they can encode to produce antibiotic enzymes system, glycolytic enzymes system, aromatic compounds degrading enzymes system, enterotoxin and limit-modified enzymes system.

Plasmids have many types. According to host genetic traits, *E.coli* plasmid can be divided into F factor (sexual factor), R plasmid (drug resistance factor) and Col plasmid (colicin producing factor). According to transfer properties, *E.coli* plasmids can also be divided into conjugative plasmid and non-conjugative plasmid. Beside self-replication, the former also has bacteria-matching genes and plasmid conjugation and transfer genes; and the latter is incapable of self-transfer. Those gene that encode barteriocin are called *E.coli* Col plasmid (Fig.4.4); and those genes that encode antibiotic resistance are called *E.coli* R plasmid. Some plasmids that have 10-100 copies in each host cell are termed as high copy plasmid; and others that have only 1-10 copies are called low copy plasmid. The common definition of plasmid copy number refers to average plasmid number each chromosome or each bacterial cell corresponding in the normal growth conditions. When two or more different types of plasmids cannot coexist in the same one host, they belong to a single incompatibility group, but plasmids belonging to different incompatibility groups may coexist in the same cell.

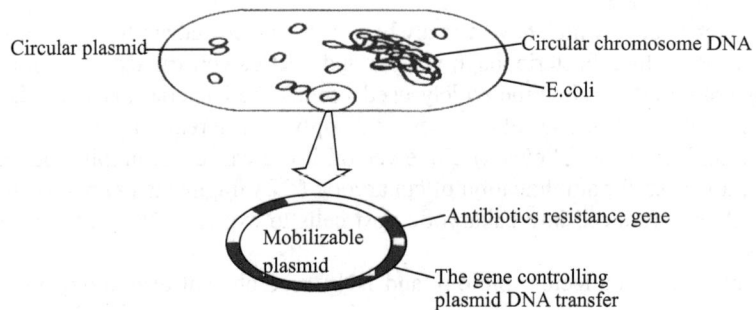

Fig.4.4 Structual map of *E.coli* plasmid molecule

According to copy control types, plasmids can be divided into stringent plasmid and relaxed plasmid. Replication of the former synchronizes with the host chromosome, and relates to protein synthesis of the host. But stringent plasmid is irrelevant to activity of DNA polymerase I, and hence when protein synthesis terminates, plasmid and host chromosome replication will also stop. As a result, there are only one or several copies of such plasmids in each host. On the contrary, replication of the latter is not synchronous with the host chromosome replication, but relates to activity of DNA polymerase I Additionally, relaxed plasmid is irrelevant to protein synthesis, and hence when protein synthesis terminates, plasmid also can replicate. And so, there are 10-20 copies of such plasmids in each host. If needed to extract plasmid, chloramphenicol is usually adopted to treat the host cell so as to stop the protein synthesis and chromosome replication, which makes plasmid copy number in the host cell increase to thousands of copies. In order to improve expression efficiency of engineered bacteria in gene engineering, relaxed plasmid is generally adopted as a vector.

Essential conditions for plasmid cloning vectors include: ①they must have many unique restriction sites in the molecular structure, and had better locate in selected marker; ②they must have transformation function after constructing recombinant plasmid; ③they should have two or more strong selected markers; ④their sizes are relatively smaller, and they should be relaxed plasmid in order to control and manipulate easily; ⑤they should have a narrow host scope, not be infectious, not be mobilized by other plasmids.

According to the principles above and the specific conditions, a suitable plasmid vector would be selected in gene engineering. At present, many artificial plasmids, such as pBR322, pBH10, pBH20, pTR262, pAT153, pX~3, pMK16, pKC7 and so on, have already been constructed through DNA recombinant technology. Because of limitation on contents in this book, we introduce only pBR322, pUC 19 and Ti plasmid in the following text.

4.2.1.1 Plasmid pBR322

The plasmid pBR322, so-called universal plasmid, is an important artificial plasmid with drug-resistant site and is one of the widely-used vectors currently (Fig.4.5). Usually, the lowercase p represents plasmid, and abbreviation initials or numbers are used to describe the plasmid. As for example pBR322, the BR stands for Bolivar and Rogigerus who constructed this plasmid, and the 322 for digital numbers relates to two scientists. Plasmid pBR322 which belongs to relaxed has 4363 base pairs in length and contains two antibiotics genes (Amp^r gene and Tet^r gene).The plasmid has unique *Bam*H I , *Hind* III and *Sal* I restriction sites inside the Tet^r gene, and has a unique *Pst* I restriction site inside the Amp^r gene. Plasmid pBR322 which belongs to relaxed contains a replication origin site in order to ensure the plasmid's performing replication function and its high copy numbers in *E.coli*. Plasmid pBR322 not only cannot transfer between host cells in the nature, but also won't cause transmission of antibiotic resistance gene. With double antibiotic resistance genes, pBR322 is very useful in manipulation of gene engineering.

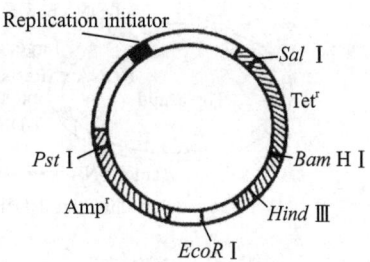

Fig.4.5 The structure of plasmid pBR 322

4.2.1.2 Plasmid pUC19

Although pBR 322 is a widely-used plasmid vector, it contains lesser unique clone sites and its screening procedure is time-consuming. Therefore, people modified pBR322 by using multiple cloning sites (MCS) technology, and common pUC series of plasmid vector are a typical example. Based on the pBR322 plasmid vector, researchers integrate *lacZ'* gene with multiple clone site at the 5'-end, and then develop a series of new plasmid vectors with a double-functional assay character.

The size of plasmid pUC19 is 2686bp. The plasmid carries four elements that one *ori* site of pBR322, one Amp^r gene, a regulation fragment of β-galactosidase gene (*lacZ'*) of lac operon in *E.coli*, and one repressor gene *lacI* which regulates expression of *lacZ* gene. Meanwhile, it also has one multiple cloning site, one Amp^r gene which allow to perform double screening of recombinants through a colour reaction and ampicillin resistance.

The screening process of cells containing plasmid pUC19 is relatively simple. If the cell contains plasmid pUC19 without inserted target DNA, blue colonies will form when culturing *E.coli* cells on agar media containing both IPTG (an inducer of lac operon) and X-gal. If the cells contain plasmid pUC19 with inserted target DNA, white colonies will form when culturing *E.coli* cells on the same media. Hence, recombinants may be selected conveniently according to colour of bacterial colonies.

4.2.1.3 Ti plasmid

In the nature, when soil agrobacteria invade into plants via a vulnus, Ti plasmid in soil agrobacteria integrates into plant chromosomes. Genes of Ti zone have two functions that decide formation of crown gall tumor in plants, and control synthesis of opines. Thus, Ti plasmid can induce formation of plant tumor. T-DNA sequence of Ti plasmid can promote Ti plasmid transfer to Ch-DNA of plant cells genome.

Based on the function that Ti plasmid can enter plant cells and integrate into DNA molecules of plant chromosome, scientists load exogenous genes onto Ti plasmid to form heterozygous Ti plasmid. When such plasmid is introduced to *Agrobacterium tumefaciens*, and then this *Agrobacterium tumefaciens* infects plant cell so as to form the transgenic plants. Discovery of Ti plasmid is a milestone which opens a new way for plant gene engineering. This procedure is shown in Fig.4.6 below.

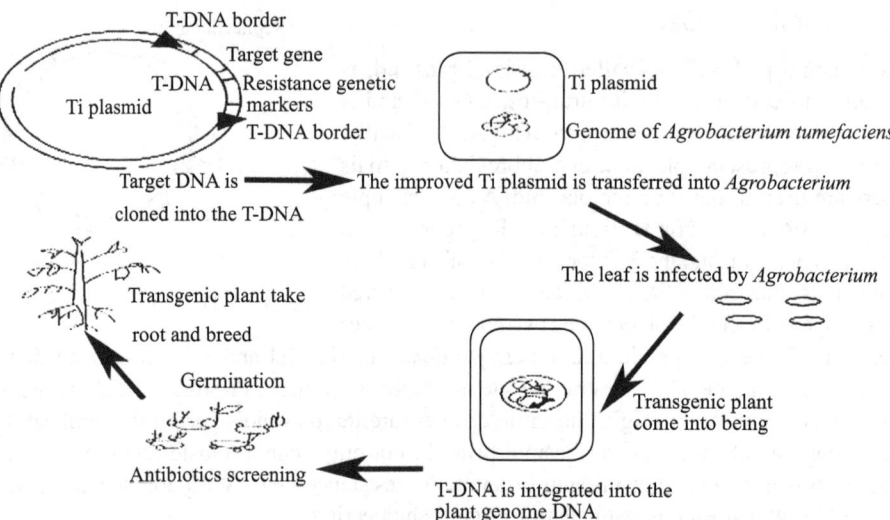

Fig.4.6 Introduction of exogenous gene into plants with Ti plasmid

In addition, a shuttle plasmid vector is constructed. It is an artificial vector with two different replication origins and selected markers, so that it can survive and propagate in two different hosts. Since this kind of plasmid vectors can carry exogenous DNA sequence to shuttle between different species cells (especially between eukaryotes and prokaryotes), they are very useful in research work for gene engineering.

4.2.2 Phage Vectors

Plasmid vectors can carry the largest DNA fragment less than 10kb, but genomic library construction often needs to clone much bigger DNA fragments in order to reduce the amount of clones in the library. So phage vectors are applied in gene engineering as a cloning vector.

4.2.2.1 Phage structure and nucleic acid types

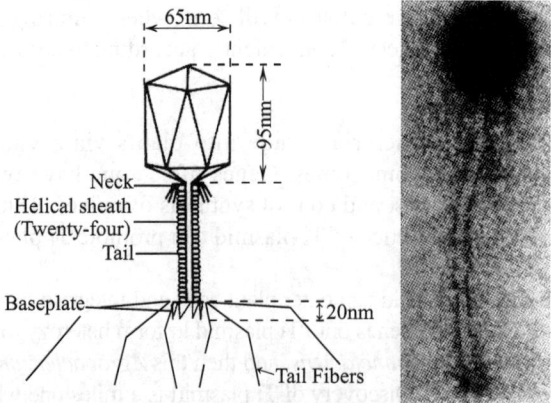

Fig.4.7 Structure model of T4 phage and electron microscopy photos

Different kinds of phage particles have great difference in structure. Structures of phages are generally icosahedron without tail, icosahedron with tail and linear shape. The typical structure is shown in Fig.4.7.

As for nucleic acids of phages, the most common is linear double-stranded DNA. Besides, several other forms, such as circular double-stranded DNA, linear single-strand DNA and single-strand RNA, etc., have been found. The molecular weight difference among various kinds of phages can reach one hundred times, and DNA bases of some phages are not made up of the standard four bases A, T, G and C. For example, the DNA bases of phage T4 do not contain base C, and 5-hydroxyl methyl cytosine (HMC) substitutes for base C.

4.2.2.2 Phage λ vector

Phage λ is a double-stranded DNA virus, and its host is *E.coli*. This virus exists in two ways of bacteriolysis and lysogeny in the host. In the lysogenic pathway, after infection, phage λ DNA can integrate into chromosome DNA of the host, the prophage replicate along with chromosomal replication of the host. In the bacteriolytic pathway, phage λ controls the biosynthesis in the host cell after infection, terminates information expression controlled by the host DNA and degrades the host DNA, and synthesizes a large number of phage DNA and the coat proteins to envelope to form the whole phage particles. Finally, the host dissolves and many viable whole phage particles release.

Phage λ DNA is a linear double-helix molecule consisting of 48,502bp, and its molecular weight is 3.1×10^7 Da. It has so far been found that the DNA encodes 61 genes. Half of these genes control the life activity cycle, the rest can be replaced by exogenous target gene with no effect on phage life. In the meantime, it can control exogenous gene expression. The DNA of wild type phage λ has a 12-base sticky end at both ends, and these two ends are completely complementary and are easy to form circular structure in the host.

4.2.2.3 Cloning vectors based on phage λ DNA

The phage λ DNA of wild type has excess restriction sites for most of restriction endonucleases which are often used in the gene cloning at present, such as five *Eco*R I restriction sites and seven *Hin*d III restriction sites. It is obviously not suitable to be used as a vector for gene cloning, so those genes must be modified via genetics methods in order to eliminate some extra restriction sites and remove non-essential region. So far two types of λ DNA vectors have been constructed. One is insertion vector, and the other is substitution vector. When the exogenous DNA is cloned to λ insertion vector, it will make λ phage lose certain biological function, which is so-called insertional inactivation effect. According to specificity of insertional inactivation effect, λ phage insertion vector can be further divided into two subtypes, inactivation of immunity function and inactivation of *E.coli* β-galactosidase. The vector can be inserted by an exogenous target gene at a single restriction enzyme site. Substitution vector is a cloning vector which is constructed on the basis of phage λ DNA. For this vector, DNA fragments can be replaced with exogenous inserted DNA in its central part. The size of exogenous target DNA which the vector can accommodate is generally about 15kb.

4.2.2.4 *In vitro* packaging of λ recombinant molecules

The recombinant DNA molecules constructed with phage λ DNA as a vector must be packaged into complete virus particles. Only in this way they just have capacity to infect the host. The process is mixing phage λ head proteins, tail proteins and recombinant DNA molecules under appropriate conditions, and then automatically packaging into complete phage particles. To provide phage packaging protein, head part, and tail proteins, two mutants of the phage with amber mutation should be prepared. One is Dam λ phage, which infects the host and then produces a large number of phage head proteins in the lysate. The other is Earn mutant of λ phage which infects the host and then produces phage tail proteins at high concentration in the lysate. When these lysates above-mentioned mix with recombinant DNA, the product of gene E (mainly coat protein) and that of gene D together package recombinant DNA into the phage head under the action of the product of gene A. And then, under the action of the products of gene W and gene F, the head and the tail will be connected into the complete phage particles. After packaging into the complete phage particles, infection efficiency of every microgram of recombinant DNA for host is plaque of 10^6, which is 100-10,000 times higher than that of unpackaged recombinant DNA.

4.2.3 Cosmid vector

Cosmid vector is a special type of artificial plasmid vector that contains *cos* sequence of λ DNA and plasmid replicon. The word "cosmid" derives from abbreviation of *cos* site-carrying plasmid, and its original meaning is plasmid carrying the sticky end *cos* site. Since cosmid includes *cos* sites, origin of plasmid replication, an antibiotic selection marker and one, or more restriction sites, cosmid therefore is a special vector with duplex properties of plasmid and *cos* phage DNA vector. The size of DNA fragments cosmid clones is 40kb or so, they can duplex and conserve in *E.coli* cell as well. Because of having both advantages of plasmid vector and phage vector, cosmid is often used to construct genomic libraries.

4.2.4 YAC Vector

YAC is an abbreviation of yeast artificial chromosome, and it is a cloning system used to clone large exogenous DNA fragments in yeast cells. It is isolated from yeast chromosome and is a linear cloning vector with capacity of self-replication. It consists of origin of replication (*ori*), centromere element (CEN), telomere (TEL) and yeast selection marker. In brief, YAC has two arms, at the end of each arm there is one telomere, and all of the elements essential for artificial chromosome exist on the arm. Besides, there are also selectable markers that provide for screening of recombinant artificial chromosome from yeast cells. In fact, YAC vector emerges in the form of plasmid, and such plasmid size is 11.4 kb. In process of cloning, YAC vector firstly is digested with appropriate enzyme and the two arms are recovered, then the two arms are ligated with large exogenous DNA fragments. Finally, a genuine artificial chromosome is formed. The experimental results show that each YAC vector can load DNA fragments of over one million bases, and its carrying capacity is several times higher than that of cosmid. YAC cannot only ensure integrity of the gene structure but also decrease dramatically the clone number for a genomic library, thus it reduces difficulty in construction of a genomic library. Such plasmid which can assemble large DNA fragments has a vital significance in current development of the human genome project. YAC vector is shown in Fig.4.8.

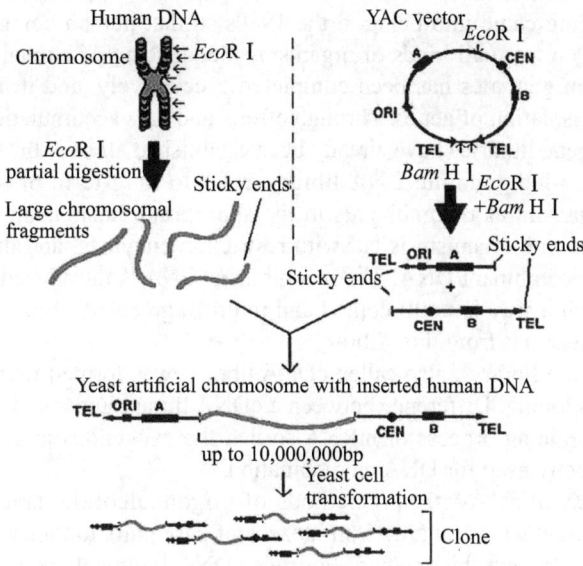

Fig.4.8 Sketch diagram of YAC vector cloning

4.3 Target Genes

The constructive process of gene engineering cell must have corresponding target genes, and such genes are generally structural genes. The purpose of gene engineering is to acquire new species with high application values through recombination of genes related with fine traits. Accordingly, specific target genes should be isolated from current organism populations. Gene is a fragment of DNA with heredity, and its average size is about 1000bp. As early as in 1946, Beadle and Tatum put forward a theory of "one gene, one enzyme", that is, one gene will be expressed to form one protein molecule via transcription and translation. A complete gene must include structural gene, regulator gene, operator gene and promotor gene, among which structural gene contains all information of protein.

The required exogenous genetic information is transmitted into host cells through the target gene, and it makes the host to show the required traits. Therefore, the desired target gene should possess such features: free of excess interference composition, high purity and appropriate size for recombinant manipulation. The methods for obtaining exogenous DNA mainly rely on continuous improvements of DNA sequencing, gene discovery and artificial gene synthesis, also benefit from understanding and utilizing rules of gene expression in organisms.

Since the 1960s, scientists have carried out research work about determination of nucleotide sequences of DNA, but unfortunately, they made little progress. In 1975, the plus-minus method was invented by Sanger and other scientists, and with this method DNA fragments of 100-500 nucleotides could be directly analyzed, which was a significant breakthrough. After the chemical degradation method was introduced by Gilbert Maxam and others in 1977, DNA sequence could be analyzed more quickly. In the same year, Sanger and coworkers put forward the double deoxidizing termination method in which DNA sequence can be determined rapidly, accurately, and reliably, and it is one of the important methods for DNA sequencing by now. With rapid development of computer technology, automatic DNA sequencing came into being in the 1980s so that human beings could acquire large-scale sequence information from all kinds of organisms. Sequencing of human genome, rice genome and many microorganism genomes has been completed successively, and it laid the foundation for discovery, synthesis and isolation of genes. Through efforts and data accumulation for dozens of years, several gene banks and gene libraries have already been established all over the world.

Gene banks, also called genomic DNA library, refer to long-term preservation of an entire genome of a certain organism as recombinants in the appropriate hosts using cloning method. Cell genome DNA from some one organism is cut with restriction enzymes, and then recombines with a suitable vector to form recombinant DNA. The recombinant DNA is introduced to a host cell, and the genome preserved in such a way is multi-copied and multi-fragmented. When we need some certain DNA fragment, we can search it from this "library".

Genomic library (gene library), also called cDNA library, was formed from reverse-transcription mRNA of organisms by cloning. Difference between a cDNA library and a genomic library lies in that introns are removed in splicing process of mRNA so that the exogenous target DNA fragments in a cDNA library can be directly used for DNA recombination.

From the 1970s, chemical synthesis methods of oligonucleotide have also been improved gradually. The sizes of target genes ranging from dozens of base pairs to thousands of base pairs have already been synthesized through this method. For those DNA fragments more than 200bp in length, section synthesis is needed, and then various sections of fragments are ligated into a complete gene with DNA ligase. Therefore, if the sequence of a target gene can be found from the current cDNA library, it can also be synthesized using chemical methods for gene recombination.

4.3.1 Target Genes Obtained via Physical Method

Since different genes have certain difference in structure and physical properties, the biomacromolecule, chromosomal DNA, are cut with restriction enzymes or broken mechanically to produce DNA fragments encoding different genes. If the content of G-C base pairs and that of A-T base pairs in different genes differ, those genes with high content of G-C have higher density than the total DNA or other DNA fragments, and vice versa. Consequently, the corresponding band is taken out to identify the gene by utilizing extremely precise density centrifugation or gel electrophoresis and EB staining, and observing position of DNA bands with a fluorescence microscope. Nucleic acid hybridization technology and Southern blotting technology can be also applied in isolation and identification of genes. By far, many genes, such as arbasin gene, crystal protein gene from *bacillus thuringiensis*, polygonal protein gene from polygonal virus and β-galactosidase gene, have been successfully acquired with physical isolation technique.

4.3.2 Target Genes Obtained via Chemical Synthesis

Chemical synthesis of DNA molecular plays a vital role in development of molecular cloning and DNA identification method. Synthetic DNA fragments can be ligated to form a long integral gene, serve as target gene for PCR amplification, introduction of mutation and even nucleic acid hybridization, and also can serve as sequencing primers. Synthesis of short single-stranded DNA (ssDNA) fragments has become a conventional technology for molecular biology and in biotechnology labs. Now, DNA synthesizer can automatically synthesize DNA fragments. As cells from various organism species have their own preferential codons which can be re-designed in chemical synthesis of DNA fragments to fit a special host cell. The procedure for polymerizing 5'-or 3'-deoxyribonucleotide or 5'-phosphoryl oligonucleotide fragments as raw materials condensation of genes one by one with chemical method is named as chemical synthesis method. At present chemical synthesis has phosphodiester method, phosphotriester method and solid phase phosphite triester method. Nevertheless, pure chemical synthesis reaction has low specificity and more side reactions. The longer the synthetic fragments are, the more difficult isolation and purification of the fragments are and the lower the yield is. Since discovery of DNA ligase, genes have been synthesized usually by adopting chemical synthesis combined with enzyme method. At first, using chemical synthesis, 10-15 nucleotide fragments in a certain base pair sequence are synthesized, and their 5'-ends are phosphorylated with polynucleotide kinase. Base sequences of these fragments should belong to the two strands of DNA respectively to make the fragments arrange alternately, i.e., when the corresponding parts of two nucleotide fragments match, both of the sides have sticky ends. And then, the third fragment arranges at an appropriate site by matching with sticky ends. Finally, the two fragments in the same strand are ligated with DNA ligase, and a complete gene in full-length is formed. So far, genes of nearly one hundred kinds of products, such as bacterial tyrosine tRNA, human growth hormone, interferon, angiotensin, human insulin, human growth hormone releasing factor, lysozyme, tPA and so on, have been synthesized with the method above. As for advantages, this method possesses feature of randomness, and we can fulfill artificial design, synthesis and assemble of non-natural genes, so as to provide a powerful approach for protein engineering.

4.3.3 Obtaining Target Genes from Gene Library

Construction of a genomic library is one of the effective methods for obtaining target genes encoding a protein from the macromolecular DNA. In prokaryotic cells, structural genes usually produce a continuous encoding region in genome DNA. But in eukaryotic cells, exons tend to be separated by introns. So, with regard to isolation of prokaryotic genes and eukaryotic genes, different cloning steps should be adopted.

4.3.3.1 Isolation of prokaryotic target genes

In prokaryotic cells, the target DNA usually accounts for 0.02% of the total chromosomal DNA. To clone procaryotic genes, firstly, enzymolysis of the total DNA is performed with restriction nuclease, and then these different sizes of DNA fragments are ligated with vectors to form recombinant DNA molecules. Finally, the recombinant clones with exogenous DNA are identified, isolated, cultured and further evaluated. The whole process is called genomic library construction. According to definition, a complete genomic library should include all the genomic DNA of target organisms.

Constructing a genomic library adopts mostly the restriction endonucleases (such as *Sau*3A1) that recognize sequence of four bases and decompose genomic DNA. By adjusting the condition of enzymolysis reaction, DNA can be digested partly to produce fragments in all the possible sizes. Since the restriction endonuclease sites are not arranged randomly in the genomic DNA, some enzymolytic fragments may be too big to clone, and then the genomic library is not complete. Therefore, it is difficult to find a specific target fragment.

After constructing a gene library, the clones with the target DNA sequence need to be identified. There are three general identification methods that include DNA hybridization with a labelled DNA probe, immune hybridization between the antibody and the gene product, and identification of the protein activity.

Success or failure in DNA hybridization depends on base pairing stability between the probe and the target DNA sequence. Double-stranded DNA molecules can produce single-stranded DNA molecules through heat treatment or alkali denaturing.

In DNA hybridization experiment, the target DNA firstly denatures, and then the single-stranded target DNA binds with cellulose nitrate membrane or nylon membrane at high temperature. The single-stranded DNA probe is labelled with radioactive isotope or other substances, and then incubated with the membrane binding with the target DNA. If the nucleotide sequence of samples complements with the probe, hybrid molecules can be formed through the base pairing effect (Fig.4.9). Finally these hybridized molecules are detected via autoradiography or other ways. The conditions of hybridization reaction are very important, and stable binding always requires at least 80% among 50 bases of the fragments which completely matching pair.

People do not possibly get the complete sequence of some genes in a genomic library. It is necessary to use another restriction endonuclease to construct another library, and then use the original probe to screen. We can also construct a genomic library in which length of insert fragments is larger than the average length of the original karyogenes so as to increase possibility to obtain the complete target gene.

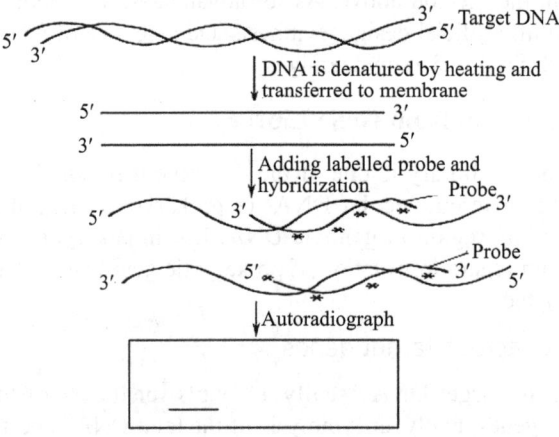

Fig.4.9 Schematic diagram of DNA hybridization

In the event without DNA probe, other methods for screening in a library may be chosen. For example, if one target DNA can be transcripted and translated, once this kind of protein appears, even a small amount of the protein, immunological method can be used for the target clone detection. This procedure has much in common with DNA hybridization technically.

All of clones in a gene library are cultured in the suitable medium and then transferred to a membrane which is treated to lyse the bacteria and simultaneously released proteins are attached to the membrane. Then antibody corresponding a protein encoded by a certain target gene is added (called the primary antibody), and redundant impurities are eluted to remove after reaction. The second antibody labelled with a special enzyme, such as alkaline phosphatase, corresponding the primary antibody, is added, after elution again, a colourless substrate of such enzyme is added. If the secondary antibody binds with the primary antibody, colourless substrate will be hydrolysed with the enzyme binding on the secondary antibody, and so a coloured product will be produced (Fig.4.10).

According to location in which clones manifest colour change, the corresponding clones in the original culturing plate can be found out. These clones may contain a complete gene, also only a part of the gene. In spite of a part of the gene, it also can produce a protein domain enough to be identified by the primary antibody. In addition, integrity of the gene in positive clones still needs further identification.

Clones in gene library
↓
Culture in medium
↓Transmembrane
Colony transfer onto membrane
↓Lysis
Naked protein releasing from colony
↓Add primary antibody
The primary antibody binds with the naked target protein
↓Elute free primary antibody, add the secondary antibody
The secondary antibody binds with the primary antibody
↓Elute free secondary antibody, add coloring reagent
Visualization
↓
Locate positive clones

Fig.4.10 Screen of positive clones via immunoblotting

If the target gene encodes one enzyme which can't be produced by the host cell, then the target gene can be screened through existence of such enzyme activity. For example, those genes which encode α-amylase, endoglucanase and β-glucosidase in various organisms are isolated via this method. i.e., the nuclear genomic library firstly is transferred into a selective bacterial strain which is then cultured in medium with a particular substrate, and those clones that can utilize such substrate are screened. If protein encoded by the searching gene which plays a very vital role in growth of the host cell, after transferring the genomic library into these mutant cells, those who can grow in the minimal medium without required substrate must have a functional target gene. Many genes with important function including those genes associated with antibiotic biosynthesis and nitrogen fixation of plant rhizobium have been successfully obtained by using genetic complementary method.

4.3.3.2 Isolation of eukaryotes target genes

Target genes of eukaryotes can also be obtained from a genomic library. But genomes of eukaryotes are bigger than prokaryotes, and so the genomic library is more complicated than the cDNA library. An established genomic library also contains non-transcription and non-translation sequences. Eukaryotes genes contain intron sequences which can transcribe but cannot translate. RNA of transcription needs transfer from nucleus to cytoplasm where RNA must remove introns in order to synthesize mRNA. And a clone containing a target gene in a genomic library may contain any inserted sequence or non-transcription sequence near the gene. Therefore, the recombinant genes isolated from a genomic library have many problems with expression in the hosts. Whereas, a cDNA library contains only transcription and translation sequence, because a cDNA library is a gene population synthesized with reverse transcriptase by adopting mRNA in the cytoplasm as a template. Therefore, as for eukaryotes, firstly a cDNA library is often constructed and then a target gene is obtained from it. In establishing a cDNA library, if the selected cell or tissue type is suitable, it will be easy to screen the needed gene from a cDNA library.

Construction of a cDNA library includes the following steps: ①obtaining the cell or tissue that contain the target gene; ②extracting the total RNA and mRNA from the tissue or cell; ③synthesizing the first cDNA strand using predesigned the primers, reverse transcriptase, four kinds of dNTP and the corresponding buffer (Mg^{2+}), with mRNA as a template; ④synthesizing the second cDNA strand; ⑤methylating the cDNA and ligating an adaptor; ⑥linking the double-stranded cDNA with a suitable vector.

In organism cells, specific protein and its mRNA template can form a complex in the cytoplasm. If mixing the corresponding antibody against the protein with cytoplasm synthetizing this protein, precipitation that contains the template mRNA of this protein will form. After purification of the mRNA, the corresponding gene of this protein can be synthetized by means of the principle of construction of a cDNA library. Furthermore, in a cDNA library, if a target gene expresses, its corresponding protein product can be identified using immunological method, or detected through activity of the expressed protein. So the plaques on the culturing plate of a cDNA library can be transfered to a duplicating plate, in which screening of the target plaque may be performed, then on the culturing plate the colony corresponding to the target plaque is found so as to amplify it to isolate the corresponding target gene. So far, the genes of human growth hormone, interferon, urokinase and etc have been obtained by using cDNA library technology.

4.3.4 PCR Amplification of Target Genes

In 1985, Mullis and others, from Amercian Cetus Corporation, established a system to quickly amplify a specific DNA fragment, i.e. technique of polymerase chain reaction (PCR) (Fig.4.11). Such practical invention brought a dramatic revolution in the field of molecular biology. In 1993, Mullis was awarded the Nobel Prize, and meanwhile PCR has already been a fundamental technique to quickly amplify a specific DNA fragment via enzymatic reaction *in vitro*.

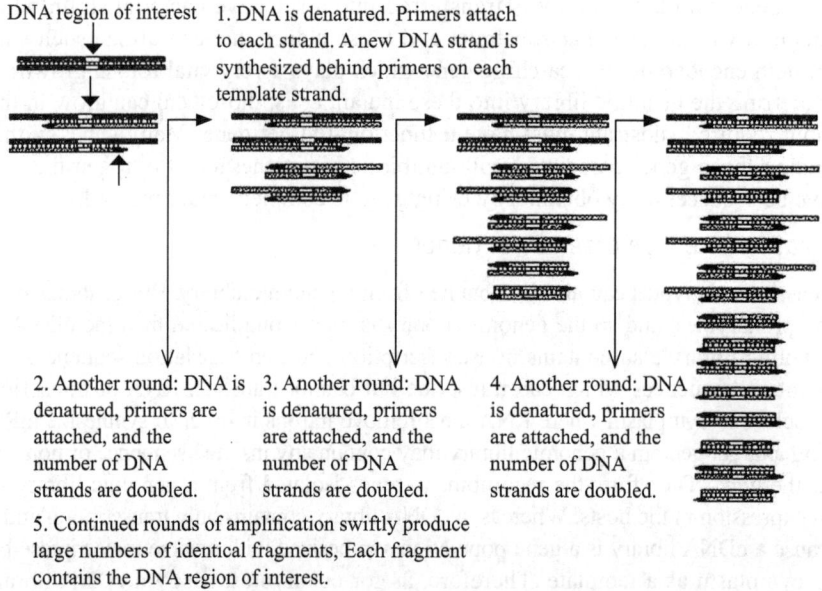

Fig.4.11 Diagram of principle of PCR chain reaction technology

PCR technique requires a reaction system with several components as below: ①the primers (about 20bp) which are complementary with either end of both strands of the target gene; ②thermostable enzyme, such as *Taq* polymerase; ③dNTP; ④DNA template containing the target DNA sequence. Generally, the target gene of 100-500bp can be emplified by PCR technology. The procedure of PCR includes three steps as below. ①DNA denaturation. It is the process in which double stranded DNA unwinds to change into single-stranded DNA by breaking hydrogen bonds between the two complementary bases at 95°C. ②Annealing. Temperature of the reaction system is lowered to 55°C so as to make a pair of primers match with the complementary sequences of the single-stranded DNA. ③Elongation. Temperature of the reaction system is adjusted to the optimal temperature of *Taq* polymerase, 72°C, a new DNA strand can be synthesized with the target single-stranded DNA as a template using DNA polymerase.

PCR technique has two features. Firstly, it can guide synthesis of a specific DNA sequence, because the starting point of a new DNA strand is determined by the annealing sites of the primers at either end of the DNA template. Secondly, the specific region of DNA can be amplified quickly and massively. Because the primers are designed according to the base pairs complementary with both ends of the amplification region, every new synthetic DNA strand has a new primer binding site and it involves in the next cycle of PCR. After twice cycles, the number of double-stranded DNA molecules contained in the reaction mixture, i.e. the copy number of the DNA region that locates between the two primers binding site, can reach 2^2 theoretically. After n cycles, the DNA copy number can reach 2^n. PCR technique can amplify a target gene on a large-scale in a short time, which makes PCR technique have extensive application in the fields of biology, medical science, anthropology, forensic medicine and so on. In other word, PCR technique brings a profound change to molecular biology.

Beside general PCR, there are some improved PCR methods for obtaining target genes, such as retrotranscription PCR(RT PCR), anchor PCR, reverse PCR, and etc.

4.4 Ligation of a Target Gene and Vector DNA

DNA fragments containing target genes must ligate with other appropriate DNA molecules which can conduct self-replication, such as plasmid, virus and etc., so as to form recombinant DNA which can be introduced into host cells via transformation or other ways and can proliferation normally, so that target genes can be expressed.

In vitro DNA recombinant technology mainly depends on restriction endonucleases and DNA ligases. DNA molecules are cleaved with restriction endonuclease to form cohesive ends with 1-4 single-stranded nucleotide. When the same enzyme is used to cleave vector DNA and exogenous DNA molecules, or those enzymes which can produce the same sticky end are used, the target gene with complementary cohesive ends and vector DNA form covalent bond under action of DNA ligase, and then recombinant DNA is formed.

4.4.1 Insertion Inactivation Method

An exogenous gene can be inserted into a selective marker of a vector to inactivate the marker gene, and this method is known as insertion inactivation. For example, when *Bam*H I cut a circular plasmid and an exogenous gene both with only one cleavage site locating in a marker gene, it will lose expression ability after the target gene inserts into its sequence. With catalysis of T4 ligase, the complementary ends of the plasmid and the target gene can form covalent bond, and a recombinant plasmid is constructed to reform a circular plasmid. For example, pBR322 is a plasmid containing single cleavage sites of *Hin*d Ⅲ、*Bam*H Ⅰ and *Eco*R Ⅰ If pBR322 and exogenous DNA are cut simultaneously by *Hin*d Ⅲ and *Bam*H Ⅰ, the vector and exogenous DNA have one *Bam*H Ⅰ end and one *Hin*d Ⅲ end, respectively. After annealing they can be ligated in a certain direction. But the obtained exogenous DNA fragments with this method may have two opposite insertion directions, which is inconvenient for gene cloning. Besides, vector DNA and the target gene are prone to conduct self-cyclization and self linear polymerization between vector molecules and between target gene molecules, so as to exert influence on heterozygous recombination rate.

4.4.2 Directional Cloning

Directional cloning refers to that a target gene inserts into a vector in the correct designed direction. According to properties of restriction endonuclease, when a specific DNA molecule is digested with two different restriction enzymes simultaneously, DNA fragments with two different cohesive ends will form. Thus, according to the unique orientation, the vector and the exogenous DNA fragment will form recombinant DNA molecule through annealing. Adopting directional cloning technique, the exogenous DNA fragments can insert into the vector molecules in a specific direction(Fig.4.12).

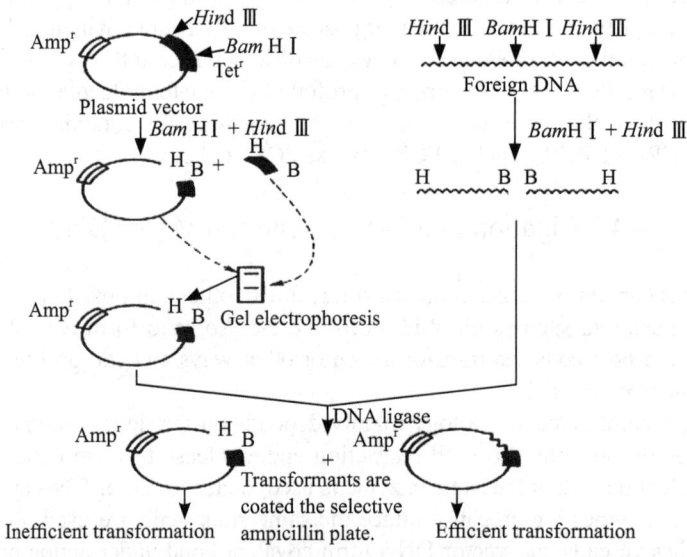

Fig.4.12 Directional cloning of an exogenous DNA fragment

When the vector molecules and target genes are digested with different restriction enzymes, they can form complementary sticky ends, but sometimes non-complementary sticky ends or blunt ends. In the given conditions, blunt ends of DNA fragments can be ligated with T4 DNA ligase. Additionally, non-complementary sticky ends of DNA fragments need to be treated by single-stranded DNA specific S1 nuclease to form blunt ends, and then are ligated effectively by T4 DNA ligase. Efficiency of ligation between blunt ends of DNA fragments generally is significantly lower than between sticky ends, and moreover the recombinant DNA formed with this method can't be cut at the original cleavage site.

Common ligation methods for blunt ends of DNA fragment mainly include homopolymeric tailing, linker ligation and adaptor ligation.

In order to make exogenous DNA fragments as much as possible insert into vector DNA so as to form recombinant DNA, efficiency of ligation reaction must be increased. For this purpose, several aspects generally are taken into account as below. ①Alkaline phosphatase treatment, homopolymeric tailing or cos plasmid may be employed to prevent non-recombinants from self-cyclization, and reduce emergence of non-recombinant "cloning". ②The total DNA concentration and the proportion of the between the vector and the exogenous DNA can be adjusted reasonably, and so the efficiency of ligation reaction can be improved. ③According to different reaction types, reaction temperature and time may be controlled reasonably, so as to increase the efficiency greatly.

4.5 Introduction of Recombinant DNA into Receptor Cells

After recombinants with exogenous DNA fragments are constructed *in vitro*, they must be introduced into appropriate host cells for reproduction and massive and consistent recombinant DNA molecules can be obtained. This process is called gene amplification. Therefore, the selected host cell must have replication capacity of exogenous DNA, and can also express phenotypic features of recombinant molecules, so as to select and identify transformants.

If recombinant DNA molecules *in vitro* aren't introduced into an appropriate host cell, they cannot display its life vitality, and will degrade gradually with time. Accordingly, appropriate technology should be adopted to introduce recombinants DNA into host cells, so as to make the target gene to amplify and express massively. With development of gene engineering, from lower prokaryotic cells to simple eukaryotic cells, further to higher plants and animals with complex structure, they all can serve as receptor cells for gene engineering. But whether exogenous recombinant DNA molecules can effectively be introduced into receptor cells and expressed at high efficiency or not, is related with various factors, such as receptor cells, cloning vectors, gene transfer methods and other factors. According to gene transfer principle and method of genetic material transferring among organism cells in the nature, several gene transfer techniques, such as the transformation, transduction, transfection, hybridization, cell fusion and liposome mediated transfection have been developed.

4.5.1 Receptor Cells

Receptor cells applied in DNA recombination is also called host cells or gene expression system, which provides gene replication, transcription, translation, post processing and secretion with essential conditions. Selection of receptor cells depends on the used vector system and various genotypes of receptor cells. The requirements for receptor cells include high transform or transfection efficiency of recombinant DNA, genetic stability, match between receptor genotypes of receptor cells and selective markers of vectors, easy selection of recombinants, and expression of exogenous genes at high efficiency and stable accumulation.

Receptor cells can absorb recombinant DNA molecules (genes) and keep them stable. Prokaryotes are an excellent type of receptor cells, these cells have many advantages, such as superior absorption to exogenous DNA molecules, fast-breeding, simple genome, easy to cultivate and genetic manipulation. Therefore, prokaryotes were often used to construct engineered bacteria for expression of target genes, or used for construction of a cDNA library and a genomic library as receptor bacteria. Currently prokaryotes used as receptor cells for gene cloning are mainly *E.coli* and *B.subtilis*.

In recent years, eukaryotic cells, such as yeast and some plant and animal cells, used as gene cloning receptor cells are concerned more and more. Some properties of yeast are similar to prokaryotes, so yeast has already been used as receptor cells for gene cloning for early. Since the passage number of somatic cells of animals is limited, generally germ cells, spermatovum cells, embryonic cells or hybridoma are used as receptor cells for gene transfer.

4.5.1.1 Microbes

E.coli, *Bacillus subtilis*, yeast and mould have been widely used as host cells in gene engineering. But expression products of *E.coli* often form insoluble inclusion body in which protein folds in an abnormal form. Hence, inclusion body will need to dissolve, in order to acquire bioactive target proteins after renaturation. Purification and isolation of expressed products still have many problems with long procedure, complex technique and low yield of bioactive protein. Host cells, genetically modify *E.coli*, can enhance their capacity to secret and express the target product.

Bacillus subtilis is mainly used for secretory expression. Its defect is that expressed products are easy to be hydrolyzed by protease secreted by itself, and stability of recombinant plasmids is poor. *Streptomycete* has strong secretion capacity of expressed products, and it is cultured conveniently and is often used in expression of antibiotic resistance genes and biosynthetic genes. With lactose metabolism genes, citric acid absorption gene or protease genes on the plasmids, *lactobacillus* can be used for food gene engineering. *Pseudomonas* can be used to construct engineered bacteria which have a variety of degradation ability required for environmental protection. *Corynebacterium* is mainly used for amino acid gene engineering. *Cerevisiae fermentum* is nonpathogenic eukaryotic cell produce no endotoxin, has been already widely used in eukaryotic gene expression after its peptide chain glycosylation system is genetical modified.

4.5.1.2 Animal and Plant Cells

The vector used in plant cells is very limited. These vectors are generally used in transgenic plants, but rarely in plant cell culture engineering. At present target genes are transferred into plants mainly with agrobacterium-mediated method, and so dicotyledonous plants are more useful in expression system.

Insect cells can express both prokaryotic genes and mammalian genes. And these cells have a stronger secretion and modification ability, but the glycosylated oligosaccharide chain is very different from that of human glycoprotein. Currently they are used to produce antibodies. Mammalian cells have very stronger modification ability after protein synthesis and can secrete expressed products to outside of cells. These cells can be used to express various human glycoproteins. Whereas, their culture condition is very strict, culture cost is higher and they are easy to be polluted. Now, common animal receptor cells are L cell, HeLa cell, vero cell and CHO cell.

4.5.2 Recombinant DNA Transfers into Receptor Cells

Among many methods for introducing recombinant DNA into receptor cells, transform (transfection) and transduction are mainly applicable for prokaryotic cells and lower eukaryotic cells (yeast); while micro-injection and electroporation are mainly used in higher plants and animals.

4.5.2.1 Transformation

As to prokaryotic cells, transformation is often used to introduce target genes to receptor cells.

Transformation of prokaryotic cells is a procedure that one recombinant DNA molecule with a target gene enters into a cell by binding with host cell, and then replicates and so the target gene expresses in the host cell. This procedure includes preparation of competent cells and transformation. Competent cells refer to a physiological status of receptor cells that can absorb exogenous DNA. An exogenous DNA molecule with certain genetic information is introduced into a host cell, and this cell acquires a new genetic trait through homologous recombination between different DNA molecules by transformation. The procedure for absorbing directly recombinant DNA molecules formed with lysogenic bacteriophages or viruses DNA as vectors is termed as transfection. Therefore, transfection is a special form of transformation.

In 1928, Griffith used *Diplococcus pneumoniae* infected mice to lead death in mice, and the used *Diplococcus pneumoniae* was a pathogenic bacterium whose external cell wall was coated with polysacchride, had smooth surface of the cell wall and so was called S type. Later he isolated a mutant strain with rough surface of the cell wall which was so called R type whose external cell wall was not coated with polysacchride. This R type strain would not cause death when infected mice. When the pathogenic S type strain was heated and killed, and then was mixed with the R type strain, injection of mice with the mixture caused death of mice and the S type strain was isolated from the heart blood of the killed mice. Hence, it was considered as the killed S type strain led to transformation of the R type strain. Subsequently, it was found that adding extract liquor of the S type to the culture of the R type strain also caused transformation from R type to S type. In 1942, Avery extracted and isolated DNA from S type strain, and adding this DNA to the cell culture of R type strain at a dose of $1/6 \times 10^{-8}$ still resulted in transformation effect. So DNA is thought to be the transforming factor of S type strains. Thus, it can be seen that transformation is a natural phenomena in objective existence in the biosphere. The transformation experiment of *Diplococcus pneumoniae* is shown in Fig.4.13.

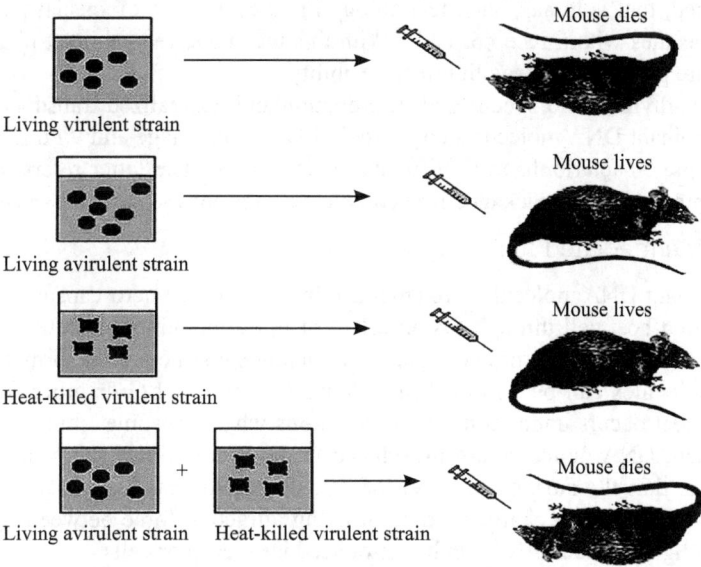

Fig.4.13 Schematic diagram of transformation of *Diplococcus pneumoniae*

Transformation can make exogenous DNA enter host cells, but it is necessary to change host cells into competent cells. Some scientists think that competent cells lose part of cell wall or are damaged, exogenous DNA can be absorbed through the cell wall. Others also think that a certain enzyme, i.e. a competence factor, may exist on cell surface, and it can bind with DNA and promote cells to absorb DNA. The competence factor has already been extracted from competent cells of *Diplococcus pneumoniae*, *Bacillus subtilis* and *Streptococcus* spp. If incompetent cells get this factor, they will become competent. About 90% of *Diplococcus pneumoniae* can be induced by the competent factor. Such factor is a protein whose molecular weight is 500-10,000Da, and has species specificity. Once the donor cells DNA is absorbed by competent cells, the DNA will exchange with homologous DNA of a receptor cell, which is called homologous recombination. The result is that the donor DNA integrates into the receptor genetic material, so as to replicate and express along with the receptor DNA, and so is transferred to all the progeny cells.

For preparation of competent cells the following aspects should be noted: ①under optimal conditions, receptor cells should grow to the exponential phase in culture medium so as to make the receptor cells density (OD_{600}) reach 0.4 or so; ②calcium choloride ($CaCl_2$) solution should be selected, and temperature should be controlled at 0–4℃ in the whole process of preparation.

E.coli is the most widely used receptor cell for gene cloning, and can change into competent cell via induction. But some cells can change into competent cells by changing culture conditions and media.

4.5.2.2 Transduction

With the aid of lysogenic phage or virus, exogenous DNA is transferred to host cells in gene engineering, and this process is called transduction. In order to make phage particles infecting receptor cells, recombinant phage DNA must be packaged firstly *in vitro*. Beside phage DNA molecules, phage particles with infections ability also include coat protein. Accordingly, packaging technique *in vitro* should be established, that is to say, such technology imitates a series of special packaging process of phage DNA molecules within receptor cells. With this technique recombinant phage DNA can be packaged into mature phage particles with infection ability.

Transduction is divided into specialized transduction and generalized transduction. The former refers to that recombinant DNA molecules composed of lysogenic phage and viral DNA as vector are packaged into complete bacteriophage or virus particles *in vitro*. The latter refers to that any DNA fragment or recombinant DNA is packaged into complete bacteriophage particles *in vitro*.

4.5.2.3 Liposome mediated fusion effect

When recombinant DNA molecules are embed in a liposome micro-capsule, exogenous target gene is transferred to a host cell through fusion effect of liposome with host cell. A liposome which composed of a lipid bilayer is an artificially-prepared membrane structure. In formation of liposome, the target DNA molecules can be wrapped in it. With the aid of PEG or other fusion promoting materials, fusion effect occurs under controlled conditions when liposomes mix with receptor cells, and then recombinant DNA molecule are introduced into receptor cells. The principle of liposome mediated method is that the surface of receptor cell membrane carries negative charges, while liposome particles have positive charge, and so utilizing attractive force between opposite charges, DNA, mRNA and single-stranded RNA can be introduced into receptor cells.

4.5.2.4 High-voltage electroporation

Exogenous DNA molecules can also be transferred into receptor cells via electroporation. So-called electroporation refers to a procedure that host cells are put in external electric field, perforation occurs on the cell wall through the electric pulse, and as a result DNA molecules can enter host cells through the pores. By adjusting the electric field strength, the electrical impulse frequency and DNA concentration, exogenous DNA can be introduced into eukaryotic cells. The basic principle of electroporation is that under an external electric field, due to big electrical potential difference, cell membrane (compose of phospholipids) appears unstable so as to produce pores which allow macromolecular material (such as DNA fragments) and micromolecular material to enter cytoplasm, but not causes lethal cellular damage. If the external electric field switches off, the formed pores will recover themselves. When the voltage is too low, DNA cannot enter cell membrane; and when voltage is too high, cells will be damaged irreversible. So it is recommended that the voltage should be controlled in the range of 300-600V and temperature at 0℃. Lower temperature makes the pores recover slowly, so as to increase opportunity of DNA entering cells. Electroporation used for gene transfer is more convenient and has higher transformation rate than calcium chloride method, but this method needs special electroporation apparatus.

4.5.2.5 Other gene transfer techniques

When metal microparticles reach a certain speed under external force, they can enter a plant cell, whereas does not cause lethal cellular damage and cells can still maintain normal life activities. Using this characteristic, exogenous DNA containing target genes firstly is mixed with metal microparticles such as tungsten, gold and etc., to make DNA absorb on their surface, and then a gene gun is used to bombard the particles to make DNA along with the metal microparticles enter plant cells through the helium shockwave. Particle bombardment method is generally applied to transgenic plants, either plant organs or tissues.

By utilizing micromanipulation system and micro-injection technique, exogenous genes will be injected directly into fertilized eggs of experimental animals, to make the exogenous gene integrate into animal genome. And then, the fertilized ovum integrated with the exogenous gene is transplanted into the uterus of a receptor for further development, so as to obtain transgenic animals. This procedure is shown in Fig.4.14.

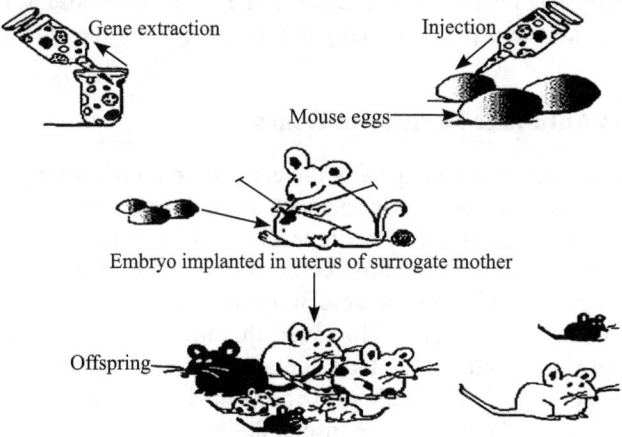

Fig.4.14 Construction of transgenic mice

The transfection method mediated by calcium phosphate or DEAE-glucan also can be used in transfection of animal cells. This is a conventional method for transferring exogenous genes to mammalian cells to conduct instantaneous expression. Mammalian cells can capture DNA-calcium phosphate deposits attaching to the cell surface, so as to fulfill the introduction of exogenous gene.

DEAE (diethylaminoethyl glucan) is a polymeric cation material of macromolecule, and can promote mammalian cells to capture exogenous DNA molecules. Its mechanism may be that DEAE can inhibit nuclease activity or promote endocytosis when binding with DNA.

4.6 Screening of Recombinants

When recombinant DNA formed by ligating a target gene with vector DNA is introduced into host cell, because of interference of unpredictable factors and operating errors, recombination and expression of the target gene will not be done in all the ways designed in advance. The clones that acquire target gene and express effectively only account for a small part of all the clones, and most of them are still original receptor cells which do not contain the target gene. Therefore, in order to isolate the genuine positive clone containing the target gene from many transformants, recombinants must be screened through various approaches. At present, a series of elaborate and reliable detection methods have been established, including nucleic acid hybridization method based on specific probes, Southwestern blotting, immunochemical method, genetic detection method and physical detection method.

To sum up, recombinant screening methods can be divided into two types, screening at the level of nucleic acid and screening at the protein level. The former refers to nucleic acid hybridization method which is based on base pairing principle of DNA-DNA and DNA-RNA, using probe technique as the core. Now *in situ* hybridization, Southern hybridization, Northern hybridization and etc have been developed. The latter mainly include antibiotic resistance and auxotrophy detection, plaque formation observation, target enzymatic activity, and immunological characteristics and bioactivity of target proteins.

No matter what screening method will be adopted, the ultimate purpose is to verify whether gene exists in host cells and is able to express normally in accordance with the order and the way designed by people.

4.6.1 Screening via Antibiotic Resistance Genes

Screening via antibiotic resistance gene is the earliest and most widely-used method in gene engineering. Antibiotic resistance gene markers such as Tet^r, Amp^r and Kan^r have been assembled in a vector when DNA recombinant was designed(Fig.4.15). After the plasmids encoding the resistance gene carry target genes into host cells, the cells obtained are corresponding drug resistant. When such drug is added in the screening medium, only the clones that contain the plasmid with resistance markers survive. However, this method can only prove that cells have plasmid, but can't determine whether the clone contains the target gene. In order to prevent false results, people further develop a new method of insertional inactivation. One plasmid often has two kinds of resistance genes, the target gene inserts deliberately into a resistance gene *in vitro* so as to make inactivation of the gene. So host cells obtained by this method can only survive in the medium with another antibiotic, but can't grow in the medium with the two antibiotics. The strain screen out in this way is guranteed to contain the target gene. However, since screening should be done twice, operation is more troublesome.

For example, plasmid, pBR 322 contains two antibiotic resistance genes. Amp^r gene has unique restriction sites Pst I, and Tet^r gene has two restriction sites Sal I and BamH I. When exogenous DNA inserts into Sal I /BamH I sites, Tet^r gene inactivates. And then the strain with recombinant DNA changes from $Amp^r Tet^r$ into $Amp^r Tet^s$. Therefore, the bacterial colony that can grow in Amp^r medium but can't in $Amp^r Tet^r$ medium may be the required recombinant.

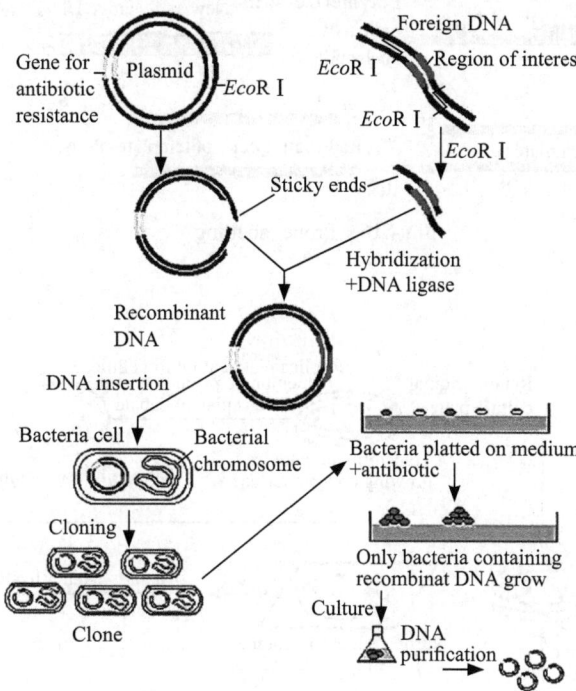

Fig.4.15 Scrnning of recombinants via antibiotic resistance markers

4.6.2 Screening via Nucleic Acid Hybridization

Molecular hybridization utilizing the base pairing principle is an important approach for nucleic acid analysis, and also is used for identification of gene recombinants. The key of nucleic acid hybridization method is to acquire probes with radioactivity or nonradioactive but with other similar radioactivity. Importantly, DNA or RNA sequences of these probes are known. According to experimental design, the probe containing the target DNA fragment is prepared firstly, and then the target gene can be identified by using hybridization method.

The approaches of nucleic acid hybridization include *in situ* hybridization, Southern blotting and spot hybridization. *In situ* hybridization is a type of hybridization that recombinant colonies or plaques are transfered onto filter membrane and DNA releases, and then hybridization is performed with DNA probe. Probe Labelling is shown in Fig.4.16, the procedure of *in situ* hybridization is shown and Fig.4.17.

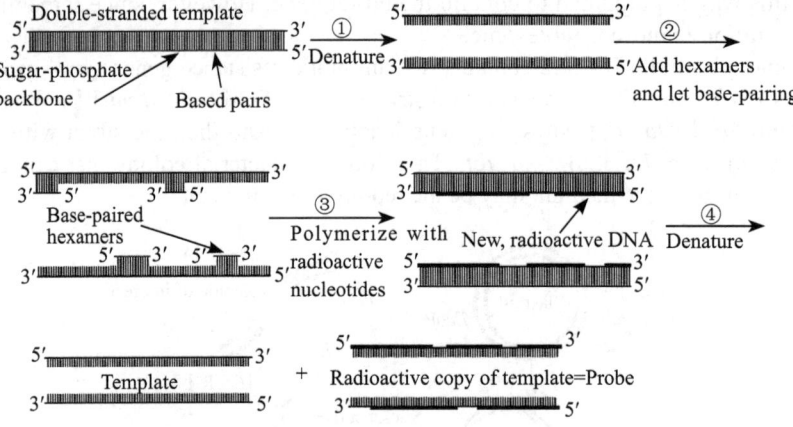

Fig.4.16 Probe labelling

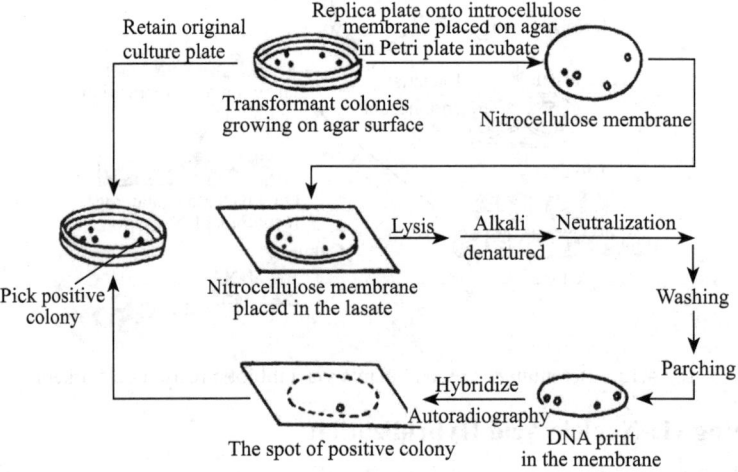

Fig.4.17 *In situ* hybridization

Southern blotting is a typical ectopic hybridization, and it was developed by Southern in 1975 and named after his name. The procedure of this method is that recombinant DNA is cut by restriction enzymes, the target gene is isolated, after electrophoresis is done and then transferred to a film, and finally after fixation hybridization is done with the probe.

4.6.3 Screening via β-Galactosidase Chromogenic Reaction

β-galactosidase chromogenic reaction is an assay method that utilizes β-galactosidase activity both in the recombinant DNA and in the host cell, which shows auxotrophic complementation, i.e.

alpha complementation, and thus recombinants can be screened through the intuitive chromogenic reaction.

Auxotrophy refers to those microorganisms that lose the ability to synthesize one or some growth factors. If one host cell belongs to an auxotrophy type, the cell can grow only if culture medium contains the nutrients. Beside a target gene, the recombinants containing such exogenous DNA which express the nutrients will show auxotrophic complementation, so that the recombinant cells possess complete metabolism ability and can grow even if the medium does not contain the nutrient. For example, some host cells lack leucine synthase gene and some lack the gene of tryptophan synthase, then recombinants will be screened out from host cells through the selective medium. This screening method is also known as auxotrophic complementation.

For example, pUC plasmid vector contains the regulation fragment of β-galactosidase gene(*lacZ'*). The strain that carry the complete *lac* operon can produce β-galactosidase. If such cells contain plasmid pUCl9 without a target DNA, the product of repressor coding genes(*lacI*) can't bind with the *lacZ'* promoter in the culture medium which contains IPTG(Isopropy lthiogalactoside). Therefore, when *lacZ'* can transcript and translate, the produced *lacZ'* protein binds with one protein encoded by chromosomal DNA of the host cell to form heterozygous β-galactosidase with activity. If the chromogenic substrate, X-gal(5-bromide-4-chloro-3- indole- β-D-galactose glycosides) exists, it will be hydrolyzed by heterozygous β-galactosidase to form the blue product, i.e., those colonies containing plasmid pUC19 without exogenous DNA fragment show blue. If an exogenous target gene inserts pUC19, the *lacZ'* gene structure will be destroyed, and so the recombinant cell cannot produce functional LacZ' protein. Thus recombinants can't form heterozygous β-galactosidase, either, and the recombinant colonies show white. According to the colour of the colony, so the recombinant containing the target gene is screened. This method greatly simplified the process to screen recombinants among the clones containing these plasmids.

4.6.4 Screening via Immunoassay

Immunoassay is a detection method with high specificity and sensitivity (Fig.4.18). The basic principle of this method is that expressed products (protein or peptide) of target genes in host cells serves as antigen, immunized serum of expressed products serves as antibody, and hence expressed proteins are detected through antigen-antibody reaction and existence of target genes may be further judged. If the target genes in recombinants can be transcripted and translated, according to location of the clones in which immunoreaction and colour change, the corresponding clones on the original culturing plate will be determined, and therefore recombinants can be screened.

Immunoassay method can be divided into radioactive antibody assay (RIA), immuno- precipitation test and enzyme linked immunosorbent assay (ELISA). The most outstanding advantage of these methods is that one can detect with them those cloned genes which do not provide any selectable phenotypic trait for their hosts. However, these methods need to use specific antibodies.

Recombinants also can be screened through enzyme activity and protein gel electrophoresis. If the target gene encodes one enzyme that it cannot be encoded by the host cell, then the recombinant can be screened according to existence of such enzyme activity. If the target gene product can be detected, the results show that the transformant is a recombinant with the target gene. In addition, if the objective enzyme plays a very important role in cell growth, by designing a suitable selective medium, transformants can also be identified via growth of clones in such medium.

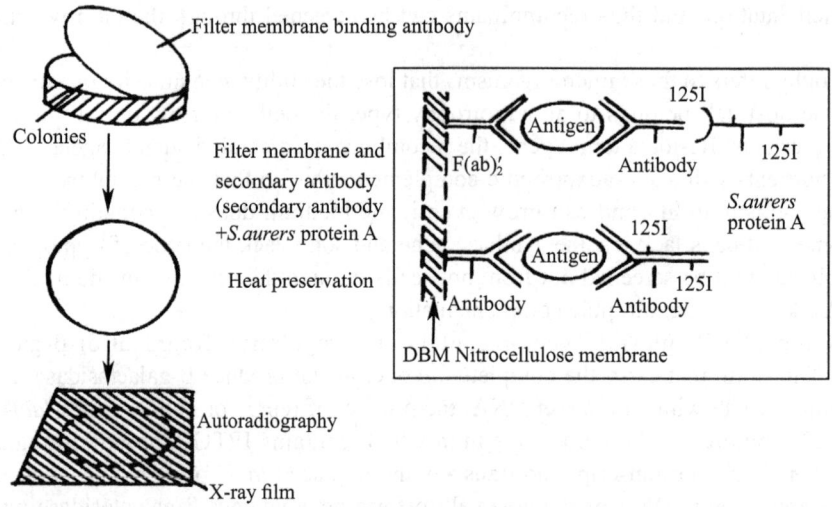

Fig.4.18 Immunoassay method

If recombinants containing an exogenous gene can express the target gene correctly, the peptide (protein) expressed by the exogenous gene will increase in total proteins. Recombinants can also screen through protein gel electrophoresis. When total protein extracted from a recombinant conducts gel electrophoresis, a new protein band will appear on the electrophoresis map. According to this phenomenon, recombinants can be preliminarily identified.

There is also Western blotting method. It is a detection method which combines protein electrophoresis with immunoblotting. Total protein is extracted from the transformed colonies; with SDS-polyacrylamide gel electrophoresis various proteins are separated in different bands; the separated bands are blotted or transferred to a solid membrane; then the antibody against the target protein (primary antibody) and the secondary antibody with a specific label are added onto the membrane successively to conduct antigen-antibody reaction and colour change detection; and the positive reaction result represents that the detected transformant is a recombinant.

4.7 Achievements and Application of Contemporary Gene Engineering

4.7.1 Achievements of Contemporary Gene Engineering

In the early 1950s, Watson and Crick put forward DNA double helix model and the central dogma of genetic information transmission, which not only confirmed that the genetic information material is DNA, but also illustrated relationship between biomacromolecules such as DNA, RNA and protein, and their biological functions. Until the early 1970s, a series of tool enzymes for genetic engineering were found, and cleavage, ligation, transformation and expression of DNA were fulfilled to achieve the purpose to modify organism heredity artificially in an oriented way. As a result, the gene engineering was born. It gave life science new vitality, and became the backbone technology of contemporary bioengineering. It provides a powerful new approach for illustrating life mysteries and life activity laws, and opens a new window to explore major problems, such as differentiation, growth, development and evolution mechanism of cells, tissues or organs. It also propels immeasurable development of life science. In addition, it is a very useful technique for elucidation of pathogenic mechanisms of cancer, AIDS and hereditary diseases. Now cause of tumors has made breakthrough and the oncogene has been found. AIDS is a fatal disease caused by human immunodeficiency virus type I (HIV-1). Up to now, anti-HIV vaccine is still in shortage, and an effective drug also is not designed. But the HIV life cycle and its pathogenic mechanism have been discovered through gene engineering technology, which lays the theoretical basis for design and manufacturing of therapeutic drug. In clinical diagnosis, gene engineering technology also plays an important role in epidemic, cancer, genetic diseases and some parasitic diseases. Radioactive isotope probes have been used in detection of more than 60 kinds of genetic diseases and more than 40 oncogenes, and this provides a favourable approach for accurate treatment of these diseases. Nowadays scientists at home and abroad have been carrying out efforts that introduce normal genes into patients in order to replace those pathogenic genes or supplement those deficient genes. These studies aim at difficult diseases including cancer and genetic diseases. Besides deep studies in labs, some clinical trials have been performed. Undoubtedly, with further research on gene therapy, a series of important gene drugs will come into being, and the third revolution in the medical field will be coming. So gene prevention and gene therapy for cancer and hereditary diseases, are also an important development trend for gene engineering.

4.7.2 Application of Contemporary Gene Engineering

In practical application, gene engineering has invaluable promise and huge potential in pharmaceutical, food and chemical industry, environmental protection, mining, energy, husbandry and other aspects. Especially in the pharmaceutical industry, it has not only changed the traditional technology, but also has upgraded the traditional technology. More importantly, via gene engineering people can obtain many new drugs which cannot be obtained with the traditional technology. For instance, insulin, growth hormone, brain hormone, α-interferon, urokinase, hepatitis vaccine, domestic animal vaccine, cattle chymosin, protein C, htPA, and dozens of McAb have all been commercialized. Besides, there are a lot of precious drugs, such as EPO, β-interferon, λ-interferon, IL-2, SOD, colony stimulating factor, human proinsulin, prourokinase, rennin, ANF, β-endorphins etc., are on trial or clinical trial. In a word. from the beginning of gene engineering, the first beneficiary industry is the pharmaceutical industry, and it has great potential in this industry. So gene engineering must have great influence on development of the pharmaceutical industry in the future. Moreover, in medical area, gene therapy technology in gene engineering will play an important role. Gene therapy refers to recombining human normal DNA with lysogenic virus DNA to form recombinant DNA hybrid, then introducing the recombinant DNA into the human body via viral infection to integrate with human chromosomes. It can be used to replace a mutated gene, supplement missing genes or close abnormal genes. Thus, it can radically treat congenital genetic diseases, malignant tumors, AIDS, cardiovascular diseases and diabetes.

Beside production of therapeutic drugs and diagnosis reagent, gene engineering also can be applied directly to diagnosis and treatment of diseases. Gene probes have the advantages of high accuracy and sensitivity in diagnosis and investigation of many diseases, such as infectious diseases, epidemic disease, tumors and human hereditary diseases, so it is an advanced technology with promising application.

In the chemical and food industry gene engineering also shows its huge advantages. It not only changes the traditional technology, but also provides many new chemicals and food grade enzymes for human. At present, industrial products, such as chymosin, SCP, α-amylase, glucanase, L-tryptophan, L-threonine, L-proline, L-homoserine and so on, have been commercialized. Other more than one hundred genes, such as vitamin C, riboflavin and cellulase etc. have been cloned successfully, and are in research and development. Breakthrough progresses have been made in husbandry industry, transgenic animal and plants have been obtained. These achievements have tremendous promotion for husbandry industry in the future. Because DNA recombinant technology breaks through incompatibility of distant hybridization between various species, thus its application is very wide. In agriculture, beside direct application to modification of plant genes to acquire stable high-yield crops, it can be used to breed the crops with resistance against pathogenic microorganism and pests so as to conduct biological control. For example, *Bacillus thuringiensis* insecticide can kill more than 150 kinds of lepidoptera caterpillar, and is widely applied to control pest of cereal, fruit trees, vegetables, corn, tobacco and various forestry plants. Its effective component is δ-endotoxins (a kind of protein). Hence, *Bacillus thuringiensis* δ-endotoxins gene has recently recombined with suitable plasmid expression vectors to form recombinant DNA, and the recombinant DNA has been transferred into *pseudomonas* cells to express the gene effectively. Such genetic engineering bacteria are cultured, and then are inactivated. The latter has been approved as insecticide by Enviromental Protection Agency (EPA) of the USA. There are also many different uses in livestock husbandry. A specific gene is recombined with lysogenic viral DNA to form recombinant DNA. Through infection or microinjection technique, it is transferred into fertilized eggs of animal, after culture and development, beside animals with high quality, animals with high farrowing rate, dairy cows, horses with high milk yields, and animal with high quality furry can be obtained. Moreover, non-natural animals that do not exist in natural world can also be obtained. Giant goat, giant rat and giant fish are typical examples. Now the giant pig has been bred out. In addition, if introducing the recombinant DNA into cells of adult animal such as dairy cattles or milk goats, animals with special expression performance will be obtained, and may be used as a bioreactor to express particular exogenous gene. For example, mammary glandular cells can be secreted corresponding products, such as hormones, antibodies and enzymes. From these achievements it can be well proved that gene engineering has wide application prospect in animal husbandry.

Moreover, gene engineering has also made remarkable achievements in energy source, mining and environmental protection. Nowadays, due to population growth, industrial and agricultural production expands rapidly, and so industrial wastes increase greatly. Pesticide abuse, and oil exploiting, transportation and processing cause severe contamination of air, soil and water such as rivers, lake and so on, in which their harm level has already reached the marginal value of human tolerance and it's time to harness these problems. In the past, physical, chemical or traditional biological engineering technologies were adopted to remove pollutants. Now, gene engineering technology has become an effective means to control environment pollution. Chakrabarty AM transferred four plasmids with capacity of decomposing different hydrocarbons to the same *pseudomonas* cell, and obtained a super engineered strain. This strain is immobilized in straw, and the oil ingredients absorbed in the straw can be degraded by the engineered bacterium. In the same conditions, the several-hour effect of an engineered strain is equal to one-year effect of a natural strain. A patent of this technology has already been approved, and it creates a precedent that gene engineering controls environmental pollution. In addition, YOL plasmid from *Pseudomonas taetrolens* with ability to decompose methyl aromatic compounds is transferred to *pseudomonas* cells with capacity to decompose halogenated catecholamine to obtain an engineered bacterium. The engineered bacterium can decompose all kinds of chlorinated benzene acid ester compounds which are environmental pollutants, so as to purify environment. There are also engineered bacteria to eliminate mercury pollution. To sum up, gene engineering also has a great potential in environmental protection.

With development for over past 40 years, gene recombinant technology has already become an important technology for gene manipulation and has played a key role in life science. Meanwhile, since the day of its birth, gene engineering sets application as the research objective, and focuses on research and development of new protein drugs. The 21st century is a century of life science and biotechnology. Bioengineering industry will become a pillar industry in the new century. With final completion of human genome project and development of genomics, proteomics, bioinformatics and gene therapy, life science and biotechnology will leap up to a new development platform. It can be believed that gene engineering will soon make a breakthrough in new gene screening, new drug (vaccine) development, design new drug and gene therapy. So, it brings hope to conquer AIDS, cancer and other refractory diseases. The biopharmaceutical industry will be a sunrise industry with rapid development in the post-genome era.

Chapter 5

Cells and Cell Engineering

Cell engineering refers to a technology of contemporary bioengineering, which applies theories and methods of contemporary cell biology, developmental biology, genetics and molecular biology. It perform genetic operation and recombine structures and inclusions of cells at cellular level according to demand and design of people, in order to change structures and functions of cells, i.e., by means of cell fusion, nucleoplasm transplantation, chromosome or gene transfer and tissue and cell culture to propagate and culture new species rapidly that people need. The most striking advantage of cell engineering is that the genetic material of a cell is transferred to the receptor cell to form the hybrid cell, so as to avoid many gene manipulations including gene isolation, purification, cleaving and splicing, and to increase gene transfer efficiency. But generally speaking, cell engineering is an operation at cellular level, and is also called cell manipulation technology. It includes cell fusion technique, organelle transplantation, chromosome engineering and tissue culture technique. New species can be bred through the cell fusion technique, which break the barriers of traditional hybridization within the same species and realize interspecies hybridization. This technology can make plant cells or animal cells of different species or sources fuse together. This has unprecedented significance for creation of new varieties of plants, animals and microorganism.

5.1 Basic Concept and Technology of Cell

The living things have diversity on the earth, while various cells have one thing in common, i.e. all the living things are made up of cells.

5.1.1 Basic Concept of Cell

In 1855, Virchow proposed that new cells were formed in the existing cell by means of cell division, that is to say, cell can't be produced spontaneously by inanimate matters. Cell theory can be summarized in the two following points: ①all the organisms are made of cells and cell products; ②the new cells must be obtained through division of existing cells.

5.1.1.1 Cells are basic unit of life activities

Cells are basic unit of life activities in all living things, except virus. All living things are made up of cells. The cell is an independent, orderly, self-controlled system of structure and function. Every cell is a complete device in order to meet the needs of their own metabolism. Even multicellular organisms, various tissues perform a specific function with cells as the basic unit. On the other hand, between different tissue cells have extensive relations and signal communication, which show the relationship of division of labour cooperation. All kinds of fine division of labour and clever cooperation between cells ensure performance of various metabolic activities of complex multicellular organisms in an orderly manner.

In addition, cells will respond promptly to changes of environment. Long-term response and adaption of cells to environment may be modified gradually, which ultimately embody evolution of an entire organism.

Cells are the basis for growth and development of organisms. Growth and development of organisms can be achieved partially through volume increase of cells, but it is not without limit. Multicellular organisms are can grow through cell division and cell differentiation to increase the number of cells. Although morphology and function of various cells in multicellular organisms are different, they all are formed by division and differentiation of a zygote. Therefore, cells are the basic unit of heredity, all kinds of cells contain a full set of genetic information. Plant cells have totipotency, and the Dolly's birth proves that animal cells also have genetic totipotency.

The shape of cells varies and the size is also different. Their shape and size have close relation with their functions. In order to carry out normal growth, cells must obtain a variety of substances from environment so as to meet their own needs. At the same time, metabolites of cells must be discharged outside through the cell membrane. In multicellular organisms, some metabolites of cells are also transferred to other cells.

Unicellular organisms, such as bacteria, chlamydomonas, grass bug, etc., have both trophic function and reproductive function. As for multicellular organisms, different cells or cell groups often perform different functions. Some specially perform trophic functions, and some specially perform reproductive function. The more complex biological structure is, the more accurate cell division is.

It will not be a complete life without cells. If the structural integrity of the cell is damaged, it is impossible to realize complete life activities of the cell. The cell nucleus and the mitochondria and the chloroplast isolated from cells all contain genetic information, but they cannot be cultured *in vitro*.

5.1.1.2 Elemental composition of cells

Cells have extremely complex chemical composition, which constitutes a sophisticated cell structure system at all levels. Thus the basic unit of the life activities, the cell, is formed.

The basic elements constituting cells are carbon, hydrogen, oxygen, nitrogen, phosphorus, sulfur, calcium, potassium, iron, sodium, chlorine and magnesium, and these elements constitute many inorganic compounds and organic compounds, which are needed for cell structure and function. The most basic small biological molecules are nucleotide, amino acid, fatty acid and monosaccharide, which constitute nucleic acid, protein, lipid and polysaccharide and other important biological molecules as well. Those biological macromolecules, such as lipoprotein, glycoprotein etc, generally exist in a compound form. There are two basic structural systems needing for any type of cell. One is the biological membrane system that is composed of phospholipid bilayer with mosaic lipoprotein, and the other is the genetic information replication and expression system which is made up of nucleic acid and protein.

Cells constituting various living organisms have wide varieties, and their morphological structures and functions are different. Their diversity can't be calculated, but they have common basic points as the basic unit of life activities.

(1) There is biological membrane, the cell membrane, on the surface of all cells, which is made of phospholipid bilayer and mosaic protein. It makes cells maintain relative independence from

surrounding environment so as to form the relatively stable internal environment. Cells perform substance exchange and signal transduction with surrounding environment through the cell membrane.

(2) All cells have two kinds of nucleic acid, namely DNA and RNA. But virus, organisms of acellular morphology, has only one nucleic acid, namely RNA or DNA. Now, RNA plays a leading role in the origin process of lives. Its has been proved so by increasing evidence. Double-stranded DNA may have higher stability for permanent storage and precise replication of genetic information, and enhances their repair capability.

(3) The ribosome exists in all cells, and is an indispensable basic structure for any cell (except very specialized individual cells). When translating polypeptide chains, the ribosome will form polyribosome with mRNA.

(4) Proliferation of all cells is performed in a mode of division from one cell to two ones. Genetic materials replicate double before cell division, and assign equally to the two daughter cells. This is foundation and guarantee for life reproduction.

5.1.1.3 Cell categories

Cells can be divided into two categories in biosphere, prokaryotic cells and eukaryotic cells. Bacteria, actinomyces, etc. belong to the prokaryotic cells and they lack true nucleus. Their size usually is smaller than that of eukaryotic cells, and their DNA is circular and cannot bind to proteins and exposes in the cytoplasm. There are no intracellular organelles with membrane system structure. Their cell wall is composed of peptidoglycan, and it is the main obstacle of cell fusion. Nevertheless, prokaryotic cells grow rapidly, DNA without binding to proteins is easily manipulated by people in genetic fields, and so they are also good material for cell transformation.

Yeast, animal cells and plant cells belong to eukaryotic cells. They have larger size, the nucleus and the organelles made of many membrane systems. Plant cells have the cell wall made of several layers of cellulose and hemicellulose. Fungal cell wall is a reticular fiber made primarily of glucan, and additionally mannan, protein and lipid fill in it. Eukaryotic cells usually have obvious cell cycle. In the mitosis period, chromosomes appear highly spiral tightening state, which is not conducive to isolation of genes and brings about difficulty in insertion of exogenous genes. So it is very important for successful cell fusion and cell metabolites production to take some measures to induce synchronous growth of eukaryotic cells.

Eukaryotic cells have many varieties, plant cells, animal cells and special fungi are all eucaryon.

5.1.2 Regulation of Intracellular Enzymes Activity

Metabolism is the most important traits of the life activities, and all life activities are maintained via normal metabolism. Various chemical reactions of metabolism in organisms are performed under action of enzymes. Without enzyme, metabolism will stop, and life will stop as well. If enzyme is lost, the whole biosphere would disappear. Contemporary life science has developed to the molecular level, so as to explain essences and laws of life phenomena from relationship between structures and functions of biological macromolecules. It has great scientific significance for exploration the relationship of enzymes with life activities, metabolic regulation, diseases, growth and development, etc. at the molecular level.

Most metabolic reactions in cells are catalyzed by enzymes. Adsorption of extracellular substances and intracellular metabolic reactions both need enzymes catalysis.

Regulation of enzyme activity is such a process that a certain number of enzymes regulate the speed of their catalyzed reactions through alteration of their molecular conformation or molecular structure. Regulation of enzyme activity is more timely, rapid and effective than regulation of enzyme synthesis, and it is an economical regulation mode for organisms under the circumstance lack of nutrients. By means of change of activity of one or more key enzymes, it can influence the flow of intermediate products in metabolism pathways. Such activity regulation usually is driven by reversible bind of enzymes to specific metabolites of small molecules (allosteric effectors such as end-product, etc.).

5.1.2.1 Allosteric regulation

Recent studies suggest that regulatory enzymes regulated via feedback inhibition are generally allosteric enzyme. The essence of enzyme activity regulation is allosteric regulation of allosteric enzymes. Besides the active centre binding to zymolyte (also named the catalytic site or the active site), allosteric enzymes still have a site which can bind to end product and is called as regulatory center (or allosteric site). When it binds to the final product, the conformation of the enzyme molecule will change so as to influence binding of the substrate to the active center. The binding of the regulatory center to the final product is reversible. When concentration of the final product reduces, the final product is dissociated from enzyme molecule, and the enzyme restores original conformation and then binds to the substrate, and catalysis occurs. In a metabolic pathway of many steps of reactions, allosteric enzyme is the one that can alter the conformation under influence of the end-product. In the process of allosteric regulation, enzyme molecules only happens conformation changes. All proteins that have two or more binding sites may change their conformation when one site binds to its effector, those proteins are called allosteric proteins and have quaternary structure of multiple subunits.

Allosteric enzyme is allosteric protein, and usually the first enzyme, or the one which catalyzes the key reaction in some metabolic pathway. For example, threonine deaminase is the first enzyme in isoleucine synthesis, the enzyme can be inhibited by the end product, isoleucine, in a feedback way, so as to reduce the intermediate metabolites α-ketone butyric acid, and avoid excessive accumulation of the end product.

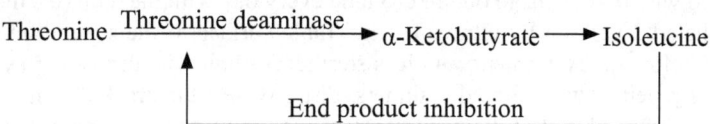

Regulation of the glycolytic pathway and the tricarboxylic acid cycle in cells is done by means of feedback inhibition.

5.1.2.2 Modified regulation

Modified regulation is accomplished through covalent regulatory enzymes which carry out reversible covalent modification by catalysing some groups of the polypeptide chain (Covalent bonds of enzyme molecules will change, namely enzymatic primary structure changes.). This makes regulatory enzymes lie in a state of interconversion between the active enzyme and the inactive one, resulting in activation or inhibition of regulatory enzymes so as to control speed and direction of metabolism. At present there are many reversible covalent regulatory proteins, which usually have the following interconversion modes:phosphorylation/dephosphorylation, acetylation/deacetylation, adenylylation/deadenylylation, methylation/demethylation, etc. For example:

In prokaryotic cells:

Low activity state←adenosine acylase←glutamine synthetase→deadenylylation enzyme→ highly active state

In eukaryotic cells:

Low activity state←dephosphorylase ←pyruvate dehydrogenase→phosphorylase→highly active state

Modified regulation is an important way for enzyme regulation in organism body, because there are many covalent regulatory enzymes in organism metabolic branch way, which are the key enzyme that regulates the metabolic flow.

5.1.3 Cell Culture Techniques

Since it serves as the basic structural and functional unit of organism, cell must be studied inevitably in the research on mechanism and control of life activities, Therefore, it is necessary to seek for approaches which can maintain growth, and proliferation, as well as keep their structure, differentiation and cell functions *in vitro*. This results in birth of cell culture techniques. Whereas cells cultured *in vitro* lack the ability of anti-infection, so prevention against contamination is the primary condition which determines success or failure of cell culture. Even if the labouratory with perfect equipments is used, improper operation of experimental technicians may also lead to contamination. Hence, experimental technicians must establish the sense of strict aseptic operation, and aseptic operation *should* be followed through the whole process of cell culture.

5.1.3.1 Preparation before culturing

Good experimental plan and operating procedures should be made at the beginning of experiments. According to the requirements of experiments, the needed equipment and materials are prepared, checked to ensure no error and placed at operation places (e.g., culture room and superclean bench), and then disinfection starts. This process can avoid increase in contaminated probability when the materials go there and back in experiments.

5.1.3.2 Disinfection before operation

There is a buffer room outside the aseptic room. The ground of sterilizing room and buffer room needs to be cleaned with 0.2% benzalkonium bromide every day with the mop (the mop is of exclusive use), and be irradiated 30-50min by ultraviolet ray. Table surface of the superclean bench is cleaned with 75% alcohol before an experiment, and it is sterilized 30min via ultraviolet ray. Some operation appliances, such as pipettes, waste liquid cylinders, dirt box, test tube rack etc., need to cleaned with 75% alcohol and then are placed at the bench to sterilize via ultraviolet radiation. Don't expose cell and culture solution to ultraviolet rays.

5.1.3.3 Washing and dressing

Just like a surgery in principle, experimen tal technicians should change shoes, clothes, and wear a hat in buffer room, and then enter the sterile room. It is necessary to wash their hands thoroughly when they enter the sterile room, and disinfect with 5% alcohol before operation. If touches the polluted items or goes in and out the culture room during the experiment, one must wash his/her hands with disinfectant. One must wear long-sleeved clean work clothes because the forearms must stretch into the cabinet of the bench when he/she works in the superclean bench.

5.1.3.4 Flame sterilization

When culturing tissues or cells in a sterile environment or doing other aseptic operations, we should ignite an alcohol lamp or gas lamp at first. All the subsequent operations, such as opening or closing a bottle, should be done near flame or by roasting with flame. But after sucking the culture solution the pipette should not be roasted with flame, because residual culture liquor on the pipette tip may be charred to form a carbon film, and the harmful substances will be brought into the culture medium when reusing the pipette. When opening or closing the bottles containing cell culture, the flame sterilization time should be short so as to prevent from burning cells due to high temperature. In addition, sterilization of rubber plugs through flame cannot last for a long time, in order to avoid burning them to produce toxic gases endangering cultured cells.

5.1.3.5 Operation technique

When doing some culturing operations, technicians' movements require accuracy and agility, but do not need to be too fast in order to prevent from causing air flow which may increase the chance of contamination. We should not use hands to touch the disinfected vessels. If so, they need to be disinfected by roasting with flame or substitute alternative items for them. Cells or tissues should not get exposure in the air untimely before treatment. The bottles containing unused nutrient solution should not be opened too early, either. If the used culture media are not reused, the bottles containing them should be immediately closed to reduce contamination risk. When sucking the nutrient solution, PBS, cell suspension and other solutions, the special pipette for each solution should be used separately and should not be mixed with others to avoid expanding contamination or leading to cross contamination. During operation technicians should not face the operation area to speak or cough in order to prevent bacteria or mycoplasma from spitting onto the working table. In the process of operation, hands or relatively dirty items should not pass over the mouths of the opened bottles. Because the bottle mouth is easy to contaminate. If the pipette tip contacts the bottle mouth, the pipette should be discarded and replaced with a new one.

5.1.4 Cell Fusion Technique

Cell fusion is also called cell hybridization, i.e., it is a process where two or more cells are merged to form one cell in the natural conditions or with artificial methods (biological, physical or chemical method). Cell fusion induced artificially was developed as an emerging technique in the 1960s. It can accomplish fusion between cells from the same species, as well as fusion between cells from different species. During cell fusion, molecular rearrangement and merging occur to the cell membrane, and chromosomes and genetic materials recombine with each other to form new living cell. Somatic cell hybridization is a method for producing hybrid cells only by somatic cell fusion instead of sexual process. Cell fusion technique has been widely used in cell biology and medical research fields. Cell fusion is an important technique of cell engineering, and its main process includes the following steps.

5.1.4.1 Preparation of protoplasts

Due to the tough wall of plant and microorganism cells, enzymes are usually required to degrade the cell wall. Due to no cell wall, animal cells have no need of this step.

5.1.4.2 Induction of cell fusion

The process of plant cell fusion is as follows: The suspensions of two parental cells (protoplasts) are adjusted to a certain cell density, mixed according to the proportion of 1:1 or the other specific proportion, and high concentration of polyethylene glycol (PEG) or other inducers are gradually dripped into or electric shock is used to promote cell fusion (Fig.5.1).

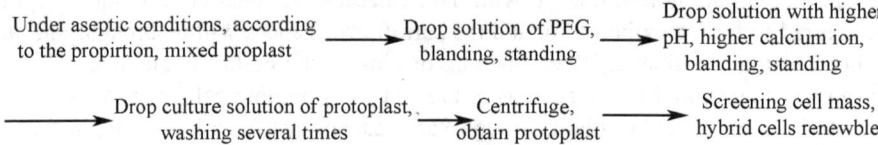

Fig.5.1 Fusion process for plant cells

The inactivated Sendai virus (HVJ, a paramyxovirus) can induce fusion of *Ehrlich* ascites tumor cells to form a coenocytic cell. In addition, paramyxoviruses, smallpox virus and herpes virus can also induce cell fusion. The process of animal cell fusion induced by Sendai virus is showed in Fig.5.2.

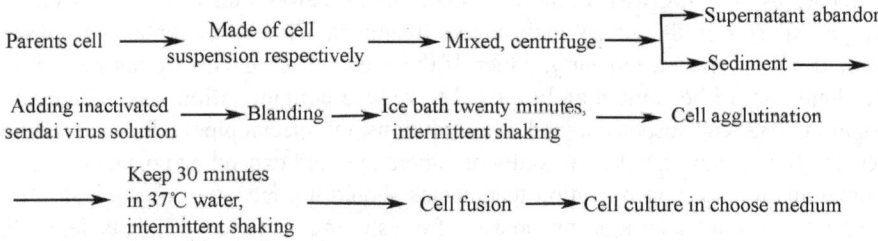

Fig.5.2 Fusion process for animal cells

5.1.4.3 Screening hybrid cells

The mixtures mentioned above are transferred to a specific screening medium so as to let the hybrid cells selectively grow, and those which are not fused cannot grow. So the hybrid cells with the parental genetic characters are obtained.

5.2 Culture Characteristics and Nutritional Requirements of Animal and Plant Cells

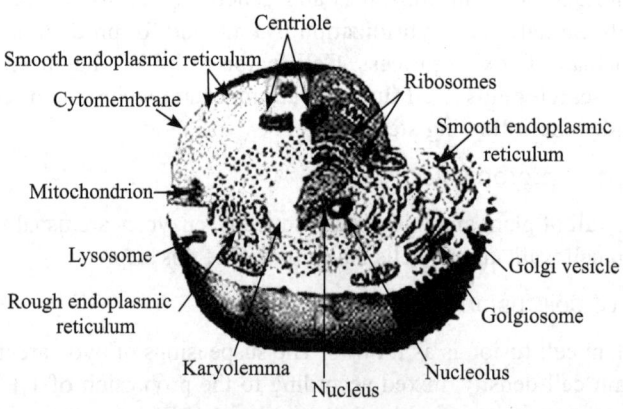

Fig.5.3 Structure model of animal cell

The structural system of cells constituting animals and plants is almost identical to their functional system. Many important organelles and cell structures, such as cell membrane and nuclear membrane, chromatin, nucleolus, mitochondria, the Golgi apparatus, endoplasmic reticulum and ribosome, microtubules and microfilaments, etc., have the same morphological structure and composition as well as the same functions in different cells. Plant cells have some special cell structures and organelles, which animal cells do not have, such as cell wall, vacuole and chloroplast and other plastids. Structures of animal and plant cells are shown in Fig.5.3 and Fig.5.4.

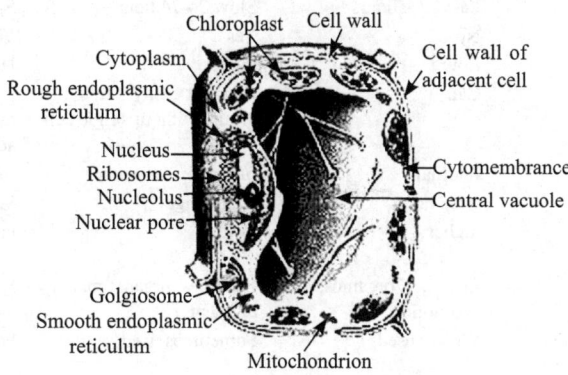

Fig.5.4   Structure model of plant cell

5.2.1   Culture Characteristics of Animal and Plant Cells

5.2.1.1   Culture characteristics of plant cells

The size of plant cells is far bigger than that of microbial cells, and they have cell wall mainly composed of cellulose so that the cells can endure pulling force rather than twisting and shearing force. In the culture process plant cells grow slowly, culture of plant cells is contaminated to microbe and so needs to use antibiotics in the media. Oxygen should be supplied during culture, but due to high viscosity the culture media cannot withstand ventilation stirring with high intensity. Culture of plant cells has the population effect and the contact inhibition phenomenon. Cell products retain in cells, and the yield is low. During culture plant cells exhibit totipotency in structures and functions.

5.2.1.2   Culture characteristics of animal cells

The difference between animal cells and plant cells is cell wall. Without cell wall, animal cells have low mechanical strength and poor ability to adapt to environment. Culture of animal cells needs less oxygen. The products of culture distribute in extracellular environment, and the cost of reaction process is higher, but the price of products is also very high. Mass culture of animal cells should not make use of experience from microbial reaction. Culture characteristics of animal cells and plant cells and microbial cells are shown in Table 5.1.

Table 5.1 Structures and culture characteristics of animal cells, plant cells and microbial cells

Categories / Comparative items	Microbial cells	Plant cells	Mammalian cells
Size	1-10μm	10-100μm	10-100μm
Cell wall	Yes	Yes	No
totipotency	Yes	Yes	Yes
Cell differentiation	No	limited	Yes
Nutritional requirements	Simple	More complex	Very complex
Growth rate doubling time	Fast 0.5-5hours	Slow 24-74 hours	Slow 15-100 hours
Group effect	No	Yes	Yes
Shear stress sensitivity	Low	higher	Highest
Suspended growth	Can	Can, but easily into the group, without single cell	Some cells can be, but the majority of cells required adhesion surface growth
Metabolic regulation	Internal	Internally, hormones	Internally, hormones
Environmentally sensitive	wide range	wide range	Very sensitive, narrow range
Cells or product concentration	higher	low	low
Product type	Enzymes, organics, antibiotics	Enzymes, organic matters, pigment	Hormones、enzymes、MAb
Traditional variation and screening techniques	Widely used	Sometimes used	Few used

5.2.2 Nutritional Requirements of Animal Cells and Plant Cells

5.2.2.1 Nutritional requirements of plant cells

Compared with animal cells, the most striking advantage of plant cell culture is that plant cells can grow in simple synthetic media. The ingredients of media are composed of inorganic salts, carbon source, vitamins, plant growth hormones, organic nitrogen source, organic acids and some composite materials.

1) Inorganic salts

The optimum concentration of inorganic salt is not the same for different culture forms. Usually, concentration of inorganic salt is about 25mmol in the medium. Nitrate concentration generally adopts 25-40mmol/L. Although nitrate can be used for inorganic nitrogen source alone, it is helpful for cell growth to add a small amount of ammonium salt. If adding some amber acid or other tricarboxylic acids, ammonium salt can also be nitrogen source alone. Potassium must be added in the medium, and its concentration should be 20mmol/L or higher. The concentration of phosphate, magnesium, calcium and sulfur is 1-3mmol/L. Sodium element such as NaCl or KCl, need not to be added in the medium, but they have non-toxic effect on cells even though their concentration reaches as high as 60mmol/L.

2) Carbon source

Sucrose or glucose is the regular carbon source. Fructose is worse than those two just mentioned. Other carbohydrates or organic carbides are not suitable as a single carbon source. Usually, increase in sucrose content in the medium can raise yield of secondary metabolites of cultured cells.

3) Plant growth hormones

Most plant cell culture media contain natural and synthetic plant growth hormones. The hormones can be divided into two categories: auxin and cytokinins. Auxin can promote cell division, and the most effective and the most common auxin is indole acetic acid (IAA), 2, 4-dioxygen phenoxyacetic acid (2, 4-d) and naphthyl acetic acid (NAA). Auxin can make plant cells and tissues in culture to form the roots, and most effective and most common ones are indolebutyric acid (IBA), IAA and NAA. Cytokinins usually is derivatives of adenine. Most common mitogen is 6-amino purine (BA), 6-chaff amino purine (K), zeatin (Z). Cytokinins and auxin usually are used together to promote both cell division and growth. The amount of plant hormones is 0.1-10mg/L, and varies according to different cells.

4) Organic nitrogen source

Organic nitrogen source adopted usually includes protein hydrolysate (including casein protein hydrolysate), glutamine or amino acid mixture. Organic nitrogen is good for early growth of plant cells in the primary culture. L-glutamine can replace or supplement some protein hydrolysate.

5) Organic acids

Adding pyruvic acid or intermediate products of the tricarboxylic acid cycle, such as citric acid, succinic acid and malic acid, can guarantee plant cells grow on the medium with ammonium salt as a single nitrogen source, and increase resistance to potassium salt at least up to 10mmol. The intermediate products of the tricarboxylic acid cycle can also increase the growth rate of the protoplasm and the cells inoculated at low density.

6) Composite materials

Composite materials, whose components are not determined, such as yeast extraction liquid, malt extraction fluid, coconut milk and fruit juice, usually serve as cell growth regulators. Now these biological preparations usually have been replaced by others nutrients whose components have already been known. It is also shown in many cases that some of the extraction liquids are harmful to cell growth. At present, the widely used composite material is coconut milk, and its concentration in the medium is 1-15mmol/L.

5.2.2.2 Nutritional requirements of animal cells

The nutritional requirements for animal cell culture are higher than other cells. The carbon source is mostly glucose, and the nitrogen source is various amino acids. Animal cells can passage and propagate *in vivo*. The discovery spurs people to find the medium of defined chemical composition that can maintain continuous growth of animal cells and replace "natural" media, such as embryonic extraction, protein hydrolyzate, lymph fluid, etc. Eagle's basic medium, more complex DMEM medium, 199 medium, RPMI 1640 medium and CMRL 1066 medium are commonly used. Although they are chemosynthesis media, 5%-20% serum is added in order to replace uncertain components. There are also developed complex medium, such as $NCTC_{109}$, $NCTC_{135}$, F_{10} and F_{12}, etc.

Components of the culture medium for animal cell are as follows:

1) Amino acids

Essential amino acids are a kind of amino acids which cannot be synthesized by animals themselves but cells require. Furthermore, cysteine and tyrosine are also needed. Because the cell lines are different, the desired amino acids also vary. Sometimes non-essential amino acids are also added, and amino acid concentration often limits available maximum cell density. The balance between them affects the growth rate of survival cells. In cell culture, most cells need glutamine as energy and carbon source.

2) Vitamins

Eagle's minimum medium contains only vitamin B, and others are obtained from serum. When serum concentration decreases, requirements for other vitamins increase sharply. Vitamin restriction can be seen from the cell survival rate and the growth rate rather than the maximum cell density as an index.

3) Salt

It mainly refers to metal ions and acid radical ions, such as Na^+, K^+, Mg^{2+}, Ca^{2-}, Cl^-, SO_4^{2-}, PO_4^{3-} and HCO_3^-. They are main ingredients which determine osmotic pressure of a medium. As to suspension culture, reducing calcium content can make the cell aggregation and adherence minimize. Sodium carbonate concentration relates with concentration of gas phase CO_2.

4) Glucose

Most media contain glucose used as energy. It can form pyruvic acid through the metabolic pathway of glycolysis, which can be converted to lactic acid or acetoacetic acid into the Krebs cycle to form CO_2. As to embryonic cells and transformed cells, lactic acid accumulates obviously in the medium, which indicates that the Krebs cycle does not play the same role *in vitro* as *in vivo*.

5) Organic additives

Complex media contain nucleotide, intermediates of the citric acid cycle, pyruvic acid, lipid and other organic compounds. Likewise, when serum concentration decreases, these compounds must be added because they are beneficial to cloning and maintenance of these special cells.

6) Serum

In tissue culture, the most common natural medium is serum. The reason is that serum contains a lot of proteins, hormones and other nutrients, which promote cell growth and propagation, and cell adhesion, and neutralize toxicity of some toxic substances. The most common serum is calf serum and calf fetus serum. Human serum is used for culture of human cells.

In most of medium for culture of animal cells serum must be added, but in many cases, cells can maintain life activities and proliferate in the medium without serum.

5.3 Plant Cell Engineering

Plant cell engineering is an experimental subject which is based on operation of plant tissues and cells *in vitro*. It is a procedure where plant tissues and cells as a basic unit are cultured and propagated, or they undergo factitious refined manipulations under the *in vitro* conditions, to make some biological characteristics of cells change according to people's intention, so as to modify or create new varieties, accelerate reproduction of individual plants, or obtain useful substances.

The theoretical basis of cell engineering is cellular theory and cellular totipotency theory. In 1902, Haberlandt, a German scholar, published the famous paper *"Experiment of plant cell culture in vitro"*. The article pointed out that cells as a basic unit in higher plant organs and tissues may conduct division and differentiation *in vitro*, even to form embryo and plants. This opinion was later called cellular totipotency theory. Due to the limitations of the technology at that time, the experiment failed, but it opened a new area of botany, plant tissue and cell culture. In the middle of the 20th century, with plant cell tissue culture combined with cellular genetic manipulation, plant cell engineering developed. It not only promoted the development of plant science, but also had important influence on agriculture and forestry, even development of the pharmaceutical industry.

Soon after, Hannig (1904) cultured successfully the embryo of radish and scurvy grass with normal developmental ability in his experiments, and became the founder of plant tissue culture. In 1933, the founders of plant physiology in China, Li Jidong, Luo Zongluo and Luo Shiwei successively found that extracted solution of ginkgo endosperm and young mulberry leaves could promote growth of ginkgo embryo and maize root *in vitro*. Other scholars during same period also established separately the continuous cultures of plant tissues, and successfully realized continuous growth of excised plant tissues in artificial media. These research fruits laid the foundation for modern tissue culture. In 1957, Skoog and Miller discovered kinetin, and pointed out that kinetin can strongly induce differentiation of callus tissues to grow out the bud in tissue culture, which was important progress in plant tissue culture. From then on, exploration to obtain a complete plant with the tissue culture methods was carried out vigorously worldwide. Now more than 600 kinds of plants can be rapidly propagated via tissue culture method, some cereal crops, vegetables, flowers, fruit trees and medicinal plants, etc., with important economic value, have realized large-scale industrialized and commercialized production.

Although in China the research work on plant tissue culture started relatively late, with Chinese people's diligence and wisdom, in just for more than 20 years, we have made great progress in many aspects. Chinese scholars have developed new products and new strains through unicellular culture of anther and pollen, such as tobacco, rice, wheat, barley, rape and sugar cane, etc., whose total cropping area is more than 1 million hectares. The main crops obtained through detoxication and rapid propagation include banana, potato, sugar cane, cassava, vanilla, strawberry, orange, apple, grape, flowers and ornamental plants. Cells of plants such as lithospermum, pseudo-ginseng and so on have already been cultured on a large scale in a fermentation tank. China's traditional medicine involves about 5000 species of plants, and cell culture is one of the most important aspects in development of traditional Chinese medicine resources.

5.3.1 Significance of Plant Cell Culture

Plants are important resource for human survival. They provide food, medicine, spices, pigments, and so on for human beings. In terms of drug, plants contain considerable secondary metabolites which can be used as drug for us (including our ancestors). 75% of the population on the earth takes drugs made from plants to treat and prevent diseases. Unfortunately, excessive growth of the population and rapid increase in demand for plant medicines cause predatory development of the natural resources of plant medicines. Many natural resources of plant medicines have been exhausted, and it is not feasible to satisfy the market demands through artificial plantation in a large area. In the late 1980s, plant cell culture technology alleviated the contradiction and provided the opportunity and the method for producing more useful substances. Plant cell culture technology has opened up a new approach for rational utilization of plant resources, and has become a main research content of plant cell engineering, as well as is one of the most active research fields of contemporary biotechnologies.

Plant cell culture is a technology where cells and cell metabolites are obtained under sterile condition and artificial control via culturing plant cells at high density. Secondary metabolites refer to the products which are produced in metabolic process but are not essential for general life activities, such as pigments, toxins, antibiotics, auxin, alkaloids, etc.

Now some developed countries deploy a large number of human and financial resources on exploiting the field with huge economic potential. Currently the world's biggest batch of cells (tobacco cell) obtained through industrialized culture has amounted to 20,000L. In "Eighth Five-Year Plan" and "Ninth Five-Year Plan" and "863 Project" our country allocated funds continuously funding to sponsor the research on production of the antitumor drugs taxinol through industrialization culture of taxus cell. At present the yield of taxinol has reached the advanced world level of 60mg/L.

Industrialized plant cell culture mainly adopts suspension culture system and immobilized cell culture system. The former is suitable for rapid mass proliferation of cells, but is not beneficial to secondary metabolite accumulation frequently; on the contrary, cells grow slowly and content of secondary metabolites is relatively higher in the latter system.

5.3.2 Methods for Plant Cell Culture

The mode of large-scale culture of plant cells is mainly suspension culture method and immobilized culture method. Suspension culture is divided into batch culture method, semi-continuous culture method and continuous culture method.

5.3.2.1 Batch culture method

After all the medium is added into a reactor at one time, plant cells are inoculated and cultured under the appropriate conditions for a certain time to complete all of the fermentation process. This manipulation process is called batch culture of plant cells. For this culture method, plant cell growth rule is similar to that of microbes. Nutrients in the culture solution decline with cell growth. Cell growth process is also divided into lag phase, logarithmic growth phase, conversion period, stationary phase, and attenuation stage.

Studies show that the maximum accumulation stage of secondary metabolites is the later period of cell growth, and so is called the production period. Two-stage culture method is established in production, and it includes two culture stages. At the first stage, the medium is suitable for cell growth and so is used to reproduce cells. At the second period, the medium is suitable for synthesis of secondary metabolites and so is used to produce the target product.

5.3.2.2 Semicontinuous culture method

Semicontinuous culture method refers to the culture mode based on batch culture in which some part of the culture liquid for batch culture is taken out and the same amount of the fresh medium is supplemented, so as to maintain the constant volume of the culture liquid. With this method nutrients in the medium can be continually supplemented, the number of times of inoculation reduces, and the surroundings of cell culture change with time, just like batch culture method.

5.3.2.3 Continuous culture method

At some time after feeding and inoculation, cells and the culture liquid are collected continuously at a specific speed, and the fresh medium is supplemented at the same speed, which is called continuous culture. This culture mode can make the environment of cell growth maintain constant for a long time.

5.3.2.4 Immobilized culture method

By adopting immobilized reactors, cells are immobilized on the nylon mesh and then loaded in a packed bed, or fixed on the surface of the hollow fiber reactor membrane or fixed on the porous mesh plate, so as to make cells locate in a reactor with both the gradient distribution and multiple growing points. The culture medium is added to culture the inoculated cells, or the fresh medium is supplemented continuously to realize continuous culture. If necessary the sterile air should blow into the reactor. The advantages of immobilized culture lie in fixed position of the cultured cells, which makes it easy to obtain high-density cells and populations and establish the physical and chemical links between cells, and is beneficial to differentiate into tissues from cells, control the culture conditions and obtain secondary products. For example, cayenne pepper cells are immobilized on the polyurethane acetate foam, and the yield of capsaicin is 1000 times higher than that in suspension culture. If the precursors, such as phenylalanine and capsaicin isomer are added, the yield of capsaicin will be much higher.

5.3.3 Artificial seeds

Plant cell engineering covers many areas, but this book can't introduce more details. Hence, artificial seeds are taken for example to illustrate the status of plant cell engineering in contemporary agriculture and its production process.

Artificial seeds are also called synthetic seeds or somatic seeds. Plant propagation needs to take up a great amount of cultivated land, and propagated seeds also could spread diseases and insect pests. And there are a lot of difficulties in transplantation and long distance transportation of the tube seedlings. In the 1970s, Muralshige put forward that a very few explants could be cultured synchronously to form many embryoids which are embedded in a capsule to make them have functions of the seed, and can be applied directly for sowing in the field. Compared with the tube seedlings technology and natural sexual reproduction of seeds, the advantages of artificial seeds are as follows. ①A lot of excellent hybrid seeds can quickly be obtained without sexual breeding, and can keep original characters. Especially, for those plants which are hard to get seeds, artificial seeds have more important practical significance. ②For those special plants which can't normally produce seeds, such as triploid, aneuploid, gene engineering plants, it is possible to reproduce increasingly in a short period through application of artificial seeds. ③Compared with production of seeds in fields, the process for making artificial seeds can be controlled artificially, and cannot be affected by seasonal limits, avoiding adverse factors of natural disastrous climate and the risk of pathogenic bacteria. In addition, it saves both land and labour, and can realize the industrialization and intensive production. ④In production of artificial seeds, nutrients, plant growth regulators, azotobacter, pesticides can be added, which is different with microbe culture. ⑤Compared with the tube seedlings, artificial seeds can avoid transplanting, are easy store and transport. Hence, artificial seeds are concerned more in the biological area, and many meaningful research fruits have been made in the field. So far, preparation of artificial seeds of somatic embryos of many plants, such as carrot, celery, American ginseng and alfalfa and so on has been studied.

5.3.3.1 The composition of artificial seeds

Artificial seeds consist of three parts, embryoid, artificial endosperm and artificial testa. Embryoid is a structure similar to natural seeds with germ and radicle, embryonal structure which is similar natural seeds, has ability to grow up into a plant. Artificial endosperm is a solution of nutrients prepared artificially to meet the requirements for growth and development of embryoids. Generally, with the culture medium for formation of embryoids as the main component, a certain quantity of plant hormones, antibiotics, pesticides and herbicides are added according to people's need so as to provide the conditions required for normal germination of embryoids. Artificial testa is a colloid compound film that wraps on the outer layer of artificial seeds. The film can allow smooth exchange between internal and external gas, as well as can prevent water in artificial endosperm and all kinds of nutrients from leakage. In addition, it still should have certain mechanical pressure resistance.

5.3.3.2 Preparation of artificial seeds

1) Preparation and synchronous growth of embryoids

Non-synchronism of embryoid growth is one common phenomenon in somatic embryogenesis.

The embryoids in different developmental periods in different sizes are often found in the same explant. Because of asynchronous, lots of mature embryos cannot be obtained to make artificial seeds at a time. Therefore, induction of synchronous growth of embryoid has become the core issue of preparation of artificial seeds. The following measures can be taken to promote synchronous growth of embryoids. ①At the early stage of cell culture, selective inhibitors of DNA synthesis such as 5-amino uracil are added, which can make cells stop DNA synthesis temporarily. When inhibitors are eliminated, cells will carry out synchronous division. ②Low temperature treatment can inhibit cell division, tubulin synthesis and spindle formation. In due time, the temperature increases to normal culture temperature and partial synchronization growth can be accomplished. ③In an appropriate period of cell suspension culture, the suspending culture is filtered with nylon or wire mesh with a certain aperture, or undergoes density gradient centrifugation. And then embryonic cell mass at a certain stage of embryonic development are collected to obtain embryonic cell mass in the uniform size. The cell mass is transferred to the medium without auxin, and most of the embryoids develop synchronously. ④In some plant cell culture, there is ethylene synthesis peak before cell growth reaches the peak period. So ethylene synthesis has close relationship with cell growth. When ventilating intermittently nitrogen or ethylene gas to the suspension culture, the mitotic synchronous rate increases significantly. ⑤Embryoids at different development stages have different osmotic pressure, in order to meet requirements, the media with certain osmotic pressure can be prepared to control development of embryos, so as to make the embryonic development stay at the specified stage (e.g., the osmotic pressure of q globular embryo of sunflower is about 17.5%), thus the embryoid can start the stage of synchronous development.

There are many factors that control synchronous growth of cells and embryoids. In addition to appropriate regulation through the physical and chemical factors mentioned above, genetic factors including cells sensitivity of test material and embryogenesis potential, have great influence on synchronous growth of cells and embryoids. For studies on embryogenesis and synchronous control, many aspects such as material selection, culture procedure and embryogenesis rules should be considered comprehensively. For example, embryoids just harvested contain high moisture, are not enough mature, and also are hard to store. Generally, embryoids should be subject to natural drying for 4-7 days, till they turn opaque.

2) Preparation of artificial endosperm

Some nutrients, plant hormones and antimicrobial substances are added in the embedding colloid of artificial seeds to constitute artificial endosperm which ensures supply of the nutrients embryoid growth needs. Generally, with the culture medium for formation of embryoids as the main component, a certain quantity of plant hormones, antibiotics, pesticides and herbicides are added according to people's need so as to provide the conditions required for normal germination of embryoids as far as possible. It should be noted that artificial endosperm composition of different plants should be different, so it needs test to determine.

3) Artificial testa materials

After getting a large number of mature embryoids in synchronous development, artificial testa should be used to wrap up embryoids so as to prepare artificial seeds. The most ideal artificial testa is sodium alginate which has advantages of non-toxicity, high coherency, water-holding capacity and ventilation, easy use and low cost. Other materials, such as agar, gelatin, carrageenan and locust bean gum, etc., can also be chosen.

4) Preparation process of artificial seeds

①Artificial endosperm solution is mixed with 2% sodium alginate, and then embryoids in a certain proportion are added to mix well. ②Using a dropper with a certain aperture, the mixture is dropped into 2.0%-2.5% water solution of calcium chloride. Due to exchange of calcium ion with sodium ion, capsule of sodium alginate may be produced outside of the embryoid to form artificial testa. With ion exchange complexation for 10-15min, round, rigid artificial seeds are formed. The size of the capsule can be controlled via change of the dropper diameter, and the embryoid numbers in every seed depends mainly on the density of embryoid in the embedding agent. Every seed had better contains only a reproductive body, therefore, according to the size of reproductive body, the appropriate dropper diameter can be chosen. ③When artificial testa has appropriate hardness, artificial seeds must be immediately put in the sterile water to rinse for 20 minutes in order to terminate the ion exchange. Thickness of artificial seeds can be controlled according to duration of ion exchange complexation. If the testa is too thick, artificial seeds are hard to germinate, on the contrary. Testa may influence on storage, transportation and sowing of artificial seeds. ④After termination of reaction, artificial seeds are picked up to dry up naturally. Adhesion between the capsules occurs easily when they contact each other, and meanwhile it is easy to dehydrate quickly which causes capsule cracking. In order to overcome the above disadvantages, Dupont Corporation in the United States adopts a painting called Elvax4260 to treat surface of artificial seeds, and this has good effect. In addition, 5% $CaCO_3$ or talcum powder also has certain adhesion resistance.

5.3.3.3 Storage and germination of artificial seeds

Because of seasonal reason, storage of artificial seeds has always been a difficult problem to solve. But due to high content of water, artificial seeds processed with sodium alginate are easy to germinate under normal temperature and shrink resulting from dehydration, and so they have big difficulty in storage. Liquid paraffin method, low temperature method, drying method, plastic and aluminum foil packing method are all adopted, but none of them are ideal. The reason is that selected somatic embryos are just in the vigorous growth periods. Generally speaking, mature somatic embryos are cultured for 5-6 days and then grow up into a little plant with root and bud. The above used storage methods are hard to inhibit germination of the embryo embedded in the colloid ball.

According to physiological and biochemical features of natural seeds during seed formation, scientific researchers have obtained the embryo in static state with high activity in suspension culture of carrot cells, by adopting regulation culture method. The so-called static state are the state that growth of the radicle of somatic embryos is controlled. These embryos are stored without special conditions to inhibit their germination. Embryos in the static state are shorter and thicker than normal ones, and they have low content of water, and accumulate more dry substances and active substances. Therefore, they have some characteristics of dehydration resistance of and high stress, so they can be stored for a long time. If this method can be used to somatic embryo culture of other plants, maybe it will open a new approach for storage of artificial seeds.

At present, scientific researchers have made model artificial seeds of more than 10 species of plants, but actual application has many problems. Three main problems should be solved: ①high quality and quantity somatic embryos cannot be formed by culturing the explants from many important plants; ②present artificial endosperms and testa are not good enough, and they cannot effectively prevent against microbial contaminate; ③artificial seed storage and germination have not yet got a breakthrough for research.

5.4 Animal Cell Engineering

Animal cell engineering is an advanced technology that combines life science with engineering. It is widely used in the basic disciplinary area of life science so as to provide effective approaches and methods for exploring genetic control of lives at the molecular level, the cellular level to the overall level for research on life science. Meanwhile, utilizing engineering technology, mass of animal cells was cultured to obtain its metabolites, even animals themselves (cloning technology). Hence, it shows a wide application prospects for human beings to modify nature, create excellent species, produce biopharmaceuticals and guarantee human health.

In 1885, Roux from Germany put the medullary plate tissue of chick embryo in physiological saline at constant temperature and made the tissue survive for several days, and firstly put forward the concept of tissue culture. This is the first record of organ survival *in vitro*. In 1907, neural cells from frog embryo cultured by Hanrison, not only survived for weeks, but also developed the axon, his work was accepted as the start of the animal culture technology *in vitro*. The 1950s was a rapid development period of *in vitro* culture technology; at the same time, a variety of culture application technologies came into being by combining culture technology *in vitro* with other technologies of life science. So far, there have been over 5000 cell lines (strains) cultured successfully in the world. In 1958, Japanese scholar Okada found that inactivated Sendai virus could induce cell fusion of Ehrlich ascites tumor, and then created a new field of animal cell fusion. In 1975, Kohler and Milstein established lymphocyte hybridoma technique, obtained the precious monoclonal antibody and achieved great breakthrough in the field of immunology. In 1997, British scholar Wilmut and others obtained the cloned animal, Dolly, by transferring the nucleus of goat embryo to unfertilized oocytes whose nucleus was removed. This is a great scientific breakthrough, and this result has been repeated in several animals.

Chinese scholars also made important contributions to animal cell engineering. For example, they carried out studies on nucleus-cytoplasm recombination between fishes with different genetic relationships, accomplished obtained nucleus-cytoplasm recombination between varieties, genuses and families, and obtained recombination fishes with special characters. The goats were obtained successfully by means of somatic cell cloning in Shanghai and Yangling of China.

5.4.1 Significance of Animal Cell Culture

Success of cell culture *in vitro* created the conditions for studies on cell structures and functions. Various cell lines and strains from lower animals, higher animals, the highest mammals and even human beings have been established, and include normal cell strain, viruses or other factor-transformed cell strain, gene mutations cell strain, hybridoma cell strain, and so on. These cultured cells have become the important study model in cellular biology, and developmental biology, genetics, immunology, tumor biology, neurobiology and other disciplines, play a very important role in science research and development, and also become the most important foundation for rapid development and application of cell engineering (Fig.5.5).

With further development of genetic engineering technology and cell fusion technology, many foreign protein genes can be transferred into animal cells and be amplified, so as to make animal cells express valuable protein with high quality. At the same time, hybridoma technology makes hybridoma cells secrete all kinds of monoclonal antibodies. Therefore, production of various biological products with in large-scale animal and plant cell culture technology has made great progress.

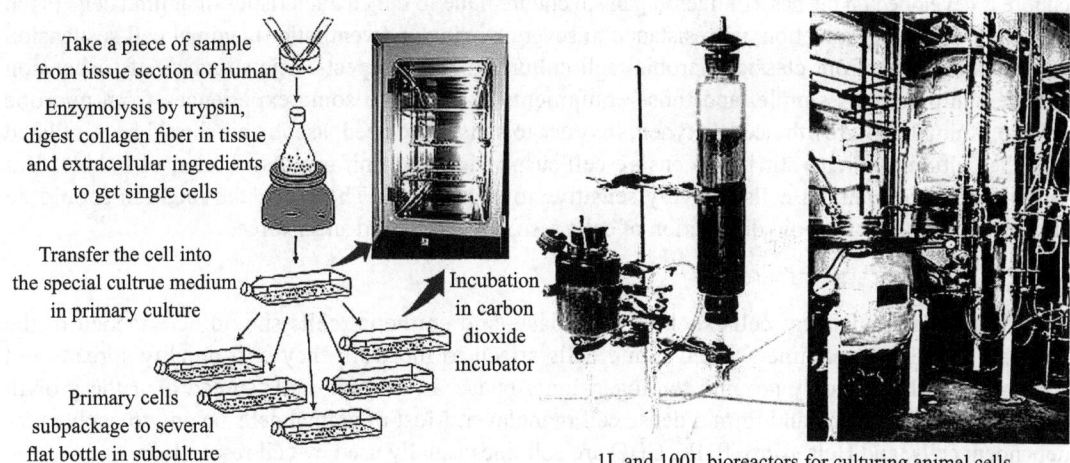

Fig.5.5 Animal cell culture

In recent years, many countries have already invested great manpower and financial resources in development of large-scale animal cell culture technology, so as to further utilize genetically modified animal cells to produce monoclonal antibodies and other fine glycoprotein. Getting drug proteins (such as virus antigens, mammal antibodies and enzymes, etc.) is a very complex process. It requires that molecules should have accurate folding and glycosylation. These requirements in bacteria and yeasts are difficult to satisfied, whereas hybridoma and recombinant DNA technology can often make animal cells produce and secrete a certain amount of useful proteins. Just for this reason, large-scale animal cell culture is of great value in next 10 years in drug production.

5.4.2 Culture Methods of Animal Cells

Beside components of basal medium, animal cell culture medium *in vitro* still needs serum, and the most common serum is calf serum. Serum can provide many hormones, trace elements, minerals and lipid for cell growth. It is an indispensable material for cell mitosis. But serum is not the ideal medium ingredient due to its flexibility and indefiniteness, which bring difficulty with experimental work in precise quantitative analysis and repeatability. So research and further development of alternatives of serum media have important significance.

Animal cell culture *in vitro* has two types. One is anchorage-independent cells which derive from blood, lymphatic tissue cells, many tumor cells (including hybridoma cells) and some transformed cells. These cells can be cultured through suspension culture which is similar to microorganism culture methods. The other is anchorage-dependent cells, and most of animal cells, including non-lymphoid tissue cells and many heteroploid somatic cells belong to this type. They must stick to the solid or semisolid surface with positive charges to grow.

5.4.2.1 Suspension culture

So-called suspension culture refers to the process that cells freely suspend to grow in liquid media. It is mainly used for culture of anchorage-independent cells such as hybridoma cells, etc. Animal cell suspension culture is developed on the basis of microorganism culture. Due to the characteristics of animal cells (such as lack of cell wall protection, no resistance to severe stirring and ventilation), animal cell suspension culture is different from classic microbial cell culture in many aspects. The structure of suspension culture equipments is simple, and those equipments can refer to some experiences from microbe culture. Animal cell growth needs oxygen, in order to satisfy the need, aseptic air should be ventilated into the culture medium, stirring to ensure cell suspension and uniform distribution of bubble. But without cell wall, animal cells are very sensitive to shear force. Therefore, the medium should be stirred appropriately to avoid destruction of cells resulting from fluid shear force.

5.4.2.2 Anchorage culture

Anchorage-dependent cells such as fibroblasts and epithelial cells should be attached to the wall of the culture container. When round cells attach to the wall, they will quickly spread, and start mitosis, then quickly get into the logarithmic phase. In general, cells spread over the growth surface after a few days, and form a dense cell monolayer. Most of animal cells belong to anchorage-dependent cells, and Hela, Vero, BHK, CHO are cell lines usually used by cell researchers.

Anchorage-dependent cells were cultured by roll bottle system in the earlier period, and this system has simple structure, low investment, mature technology and good reproducibility. But when using the roll bottle system, labour intensity is high, and the cell yield expressed as the volume is low. In order to overcome these disadvantages, Van Wezel developed micro-carrier system to culture anchorage-dependent cells in 1967. Micro-carriers refers to a kind of beads whose diameter range is 60-250μm. It is made from natural glucan or various kinds of synthetic polymer composition. For use of the micro-carrier system, animal cells attach to the micro-carrier suspending in the medium, and gradually grow into a monolayer. This model combines monolayer culture with suspension culture, and possesses advantages of the two culture method. ①The superficial area/volume ratio is greater. If 1mg Cytodexl is added into 1mL medium, superficial area will be up to $5cm^2$ which is great enough for growth of $(75-100) \times 10^7$ cells. Due to the greater specific surface area of the micro-carrier, cell productivity of unit volume medium is high. ②The micro-carrier suspends in the medium, and cell growth environment is uniform, which simplify monitoring and control of the environmental factors. Simple microscope can be used to observe cell growth. ③The medium has high availability. ④The labour intensity is small, and the occupied space is small. Since the micro-carrier culture system has many advantages, it has been widely used in mass culture of animal cells to produce all kinds of biological products.

5.4.2.3 Immobilized culture

Suspension culture is designed for anchorage-independent cells, but anchorage culture is applicable for anchorage-dependent cells. In addition, another embedding culture method is applicable for both types of cells. For this method, the density of cells is high, and cells' ability of resistance to shear force and contamination are high as well. Calcium alginate embedding method is often used for culture of anchorage-independent cells, and collagen embedding method for culture of anchorage-independent cells.

5.4.3 Application of Animal Cell Mass Culture Technology

With the development of hybridoma technology and DNA recombination technology, various bioreactors used animal cell have been designed and developed, large-scale cell culture technology has been widely used in research and production of contemporary biopharmaceuticals. Large-scale cell culture technology refers to that animal and plant cell were cultivated to produce a mass of bioproducts at high density under the artificial conditions. Its application greatly reduces the quantity of experimental animals used for research on disease prevention, treatment and diagnosis. So it provides a powerful tool for production of bioproducts such as vaccine, cytokines and artificial tissues. Many animal cell large-scale culture systems have been established, and are used to produce bioactive substances. The representative systems include submerged air lift culture system, micro-carrier culture system, micro-capsule culture system, macro-carrier culture system and hollow fiber carrier culture system. These five culture systems mentioned above can all produce different bioproducts, and have different technical features. According to incomplete statistics, there are more than ten thousand laboratories and more than 400 factories in the world, which are engaged in research and development of biotechnology products. They have made great breakthroughs, some examples are as follows.

5.4.3.1 Monoclonal antibody

Monoclonal antibody is applied widely to diagnosis *in vitro*, body imaging, disease treatment and many industries (such as immune purification). The traditional method (culture method of ascites tumor of mice or rats) for producing monoclonal antibody has not already met the actual needs. It is an economical and reliable approach to produce all sorts of different monoclonal antibodies with an animal cell mass culture system. Hybridoma cells belong to semi-suspension culture cells, suspension culture method can be used for large-scale culture of hybridoma cells, and micro-capsule immobilized culture method is also applicable for them. Now in production of monoclonal antibodies *in vitro* with bioreactors, two basic strategies are generally adopted. ①Large volume and high-density suspension culture, with air lift or agitating bioreactors, in a way of batch culture. The main features of this strategy that have relatively simple engineering structure of bioreactors, proportional amplification of stirring power and ventilation quantity in the aspect of dynamics can be amplified proportionally, low shear force and high oxygen transmission rate. ②Relatively small volume and high density culture systems can be used, such as hollow fiber perfusion culture, fixed bed and macro-carrier perfusion culture and micro-capsule culture.

Similarity of these systems is that the cell density is high up to 5×10^7-5×10^8/mL, which means that hybridoma cell concentration per volume has been close to that the cell concentration in the mice ascites. Adopting the continuous perfusion culture system, cells will be in a more stable environment. Fresh culture solution is perfused continuously, and products are recovered at the same time. And hence, little harmful metabolic waste is accumulated. Usually cells can be cultured continuously for several months. Moreover, monoclonal antibody 0.1-0.5g/L cultured supernatant can be harvested. The main advantages include four aspects as follows. Firstly, unit operation method may be adopted to culture and amplify step by step, and the cost can decline with expansion of culture scale. Secondly, the pollutants of rat transmitted diseases and the non-mouse source antibodies reduce so as to control the quality of monoclonal antibodies easily. Thirdly, a variety of bioreactor and culture systems can be chosen to produce all kinds of human source antibodies expressed by various host cells. The last but not least, the whole operation process can be engineered and controlled automatically, so as to increase reproducibility of amplification process.

Adopting the serum-free culture medium, monoclonal antibodies of top quality in a 1000L culture system has been succeeded in British company, Celltech. In 1988, in China Zhu Dehou and others cooperated with Shanghai Blood Center to produce anti-blood type A and anti-blood type B monoclone antibody as a reagent for determining blood types successfully by applying a cell mass culture system (hollow fiber culture system and macro-carrier culture system). The products of these two monoclonal antibodies with the titer unit higher than the required standard have the yield of 50L. Among 110,000 clinical cases of identification of blood types, none of them has false results, and this attains the same level as that in the 80s of the 20th century and can save 100L of human blood.

5.4.3.2 Vaccine

Vaccine is a biological product made from pathogenic microorganisms, such as bacteria and virus. Due to the effect to raise immune level of animals and enhance the body's resistance to distress, vaccine is widely used for prevention and control of infectious diseases in humans, poultry and livestock. Traditionally, cell culture is always used to produce virus vaccine for human and animals, and some vaccines have been produced on a large scale. The vaccine for foot-and-mouth disease is one of the main products manufactured with cell large-scale culture method. In 1983, British Wellcome Corporation produced the vaccine for foot-and-mouth disease by using the cell culture solution whose volume was high up to over $2100m^3$. Other products, such as vaccine for rabies, polio and cattle leukemia, have been explored to develop with cell mass culture technology.

5.4.3.3 Interferon

Human body can produce a kind of material when infected by virus, and the substances interfering with (holding back) the second infection of the same virus, this substances are called interferon (IFN). Therefore, IFN is a natural antiviral protein. It cannot only resist virus directly, but also enhance the body's immune function so as to promote the body to eliminate virus ultimately and conquer disease. After the war between humans and "SARS" virus in 2003, IFN aroused people's great interests again. Since the 1970s, many companies and research institutes have used cell mass culture technology to produce IFN via continuous culture of cell strains. For example, British Wellcome Corporation cultured Namalwa cell to produce α-IFN with a 8000L bioreactor, in order to meet the needs for clinical practice. And another British company, Celltech, produces α, β and γ-IFN used automatic air lift culture system, and the products have been sold all over the world.

5.5 Chromosome Engineering

Chromosome engineering refers to a procedure in which the aim to directionally change hereditary characters of organisms or breed out new varieties is achieved, by adding, deleting or displacing all the chromosomes or part of them from the same or different species, in accordance with the people's preliminary design. It is the cell engineering technology for changing genetic composition of cells at the chromosomal level. Chromosome engineering is different from transgenic engineering. The former focuses on integration of completely artificial chromosomes into the host's genome, but in the later a single gene fragment is implanted into the host's genome. Although the purpose of transgenic engineering is clear and the background is simple, it is not easy to isolate target genes. But, a single gene, a single or several chromosomes or fragments of a chromosome are relatively easy to transfer to receptor cells. But biological effects of the two have fundamental difference. Transgenic engineering can only change some one link or metabolite in physiological activities of host organisms, while chromosome engineering can modify overall physiological functions of hosts, which is the wonder of chromosome engineering.

Although the term "chromosome engineering" was proposed in the 1970s, as early as in the 1930s, American E.R.Sears and his students began the related research. It is not only concerned in modification of genetic basis of plants and breeding new varieties, but also an effective approach for basic research such as gene positioning and chromosome transfer. Chromosome engineering can be divided into animal chromosome engineering and plant chromosome engineering.

5.5.1 Plant Chromosome Engineering

The main purpose of plant chromosome engineering is to add, remove or replace chromosomes with traditional hybridization and backcross method. Its basic procedure includes artificial hybridization, cytological identification and screening the needed materials among hybrids or hybrid progenies. Taking common wheat as an example, the used materials are as follows: monosome and nullisome system, teisome systems, alien addition lines, alien substitution lines, translocation lines.

Higher plant cells are generally diploid ($2n$), i.e. cells contain two sets of chromosomes from parents. And gametophytes of higher plants, such as pollen and embryo sac, contain only one set of chromosomes, and are called haploid (n). Hence, the chromosome number in haploid is half of that in diploid. For example, the diploid rice has 24 chromosomes, and its haploid only 12 chromosomes. With regard to their origin, some species are formed through hybridization between over two species, and their cells contain more than two genomes, and are so called multiple diploid. Their haploid is multiple haploid. For example, common wheat cells contain six sets of chromosomes ($6\times$), and it is hexaploid (three times of the diploid). Its chromosome composition is $2n = 6\times = 42$. From the viewpoint of evolution, common wheat is formed through twice hybridization between three diploid species ($2n$) and evolution. Its haploid belong to multiple haploid, and the chromosome number n is $21(3\times)$. But people still call it haploid conventionally.

After colchicine was used for biological research in 1937, research work of polyploid has made rapid progress. The chromosomes engineering-based breeding scheme put forward by Chase in 1963, is also called ploidy manipulation breeding, i. e., tetraploid changes into diploid through ploidy manipulation, firstly selection, hybridization and screening are performed at the level of diploid, and then chromosomes are doubled to make the hybrid restore tetraploid level. It shows good application prospect for utilization of wild species. Now in China scientists have succeeded in induction of double haploid and single haploid, chromosome doubling and $2n$ gamete utilization, and got some "double haploid, wild species", tetraploid hybrids which have been wildly used in breeding, such as tetraploid wheat, octaploid triticale, etc. Triploid watermelon is seedless, and so has no progenies. The alien addition line of Agropyron intermedium-added obtained by scientific researchers through chromosome engineering can resist to stem rust and leaf rust. The alien substitution line of Agropyron intermedium-substituted wheat chromosome 3D has resistance to 15 physiological strains of stem rust. The alien substitution line of triticale 6 R has resistance to powdery mildew. Xiaoyan No. 6 is one translocation wheat line possessing chromosomes from two oat grass species, and can resist to various rust diseases and dry and hot wind, has high yield. It has been applied widely in agricultural production in a large area. These results show that chromosome engineering has important significance in breeding new disease-resistant varieties.

5.5.2 Animal Chromosome Engineering

Animal chromosome engineering mainly aims at creating a new species with all the genetic traits human needing by conducting animal chromosome transfer with cellular micro-manipulation methods (such as micro-cell transfer method, etc.). Because polyploid or the individual with severe chromosome deletion and repeated chromosomes is not easy to survive in the higher animals, chromosome engineering in higher mammalians have adopted cell fusion technique at cellular level.

In 1965, Harris and Watkins first prepared the interspecific heterokaryon containing mice and human cells. They succeeded in performing cell fusion between Human HeLa cell cultured from cervical cancer tissue and mouse cells (Ehrlich mice tumor cells grown in the peritoneal cavity) with the aid of Sendai virus. The fusion cells had two nuclei from different species firstly, and were called heterokaryon. After fusion between two nuclei of a heterokaryon, the heterokaryon conducted cell division to form a mononuclear cell line. The hybrid cells were bigger than normal cells and identified easily. In subsequent cell division, both of the two sets of chromosomes from human and mice were not replicated with cell division. In normal conditions, the human chromosomes were lost randomly with cells division. So the final hybrid cells often had a complete set of mouse chromosomes and different quantity of human chromosomes with different serial number (or with silk sticky fragments). Development of chromosome banding technique makes people distinguish the human chromosomes from the mouse chromosomes easily according to the chromosome of zone characteristics, and determine which one or several chromosomes are left in the hybrid cells.

People have 30,000-40,000 genes which control all life activities of people. Some parts of specific life activities are controlled by specific genes. If the human chromosomes or fragments of one chromosome are transferred to other organism cells to form "humanized organisms", they will express some physiological activities and the phenotype of human. Scientists conceive that if a group of related human genes are transferred to another organism, then the organism should be able to produce biological molecules, cells, tissues and organs with human characteristics. From this humanized organisms, we can obtain blood cells, muscle cells, nerve cells, even the heart, the liver, etc, which can be accepted by the human body when transplanting them. Of course, we can also get human antibodies, serum albumin and insulin used as drug. Thus, human beings can establish a "production factory" for cells, tissues and organs through chromosome engineering.

At present, artificial chromosome and chromosome engineering have been researched in Shanghai, China, and the research mainly aims at producing human antibodies with mice. Researchers removed the gene cluster for producing antibodies in mouse cells, and then transferred the human chromosome fragments including the complete antibody gene cluster to mouse cell. And the cultured humanized mouse can produce a complete set of the human antibodies.

Once chromosome engineering achieves success, human can create new organisms, which can change garbage into hydrogen, degrade oil pollutants on the sea surface, and convert cellulose into glucose and alcohol, in order to realize human's dream. This is the charm of "chromosome project" and the main trend of development of international biological technologies.

5.6 Research of Stem Cells

Stem cells, i.e. undifferentiated cells, are a kind of cells with self-renewal and differentiation potential. They include embryonic stem cells (ES) and somatic tissue-derived stem cells (STDSC). All kinds of cells, tissues and organs in the body, and even a whole organism are all developed from stem cells.

5.6.1 Stem Cell Technology

Research of stem cells generally includes the following three stages. First step is obtaining stem cell lines, which is the most important. They can be isolated from early human embryo or organs and tissues and identified and can keep characteristics of stem cells *in vitro* for a long time (general should undergo over 25 passages). The second step is to induce stem cells differentiation *in vitro*. Specific inducers are used to make the induced cells differentiate into the expected cell types, so as to form particular tissues and organs. This technology is known as directional differentiation. With genetic engineering approach, exogenous genes may be introduced into stem cells to induce directional differentiation, so as to displace original pathogenic genes. The third is induce differentiated cells immortalization. The final undifferentiated cells from inducted differentiation ES cells or their progenitor cells only conduct primary culture or short-term culture, cannot proliferate *in vitro* or conduct passage culture, and so cannot become immortalized cell lines. Therefore, the differentiated cells of ES cells induced must be immortalized, and should be implanted relevant animal or human organs or tissues to proliferate. The above research of stem cells requires both trivial operation and high experimental skill. In China, Professor Xu Rongxiang has made a major originative breakthrough in skin stem cell in situ regeneration. After properly treating the burned skin, they successfully induced stem cells in the base layer of epithelial tissue differentiation to form skin cells, and the injured skin recovered rapidly. This technology shows that our research work on stem cell takes the lead in situ stem cell repair and regeneration of tissues and organs.

Huge potential application value of stem cells arouses much concern in many countries and institutions. They invested heavily in research of stem cells and have made a series of important discoveries. In December, 1999, research of human stem cells was evaluated as the first of the ten great scientific accomplishments by American journal *Science*.

In fact, so far, people have not yet known about many problems about stem cells. In early 2000, researchers in the United States founded stem cells in the pancreas accidentally. Canada researchers founded "dormant state of stem cells" in the retina of human, rat and cattle. Some scientists demonstrated that stem cells from the bone marrow could develop into liver cells and stem cells could develop into blood.

With expansion of the research field of stem cells in depth and breadth, people will have more comprehensive understanding about stem cells. The 21st century is the era of life science, and also the era for creating world wonders for human health and longevity. Hence, application of stem cells will have broad prospects.

5.6.2 Embryonic Stem Cells and Tissue Stem Cells

Embryonic stem cells originate from stem cells of early embryo, primitive germ cells or teratoma tissue. The cells in the inner cell mass in the blastula developed from the zygote during embryonic development are known as embryonic stem cells (Fig.5.6). Embryonic stem cells have "totipotency", i. e. the ability of forming all tissues and organs; therefore it is called the "omnipotent" cells. With progression of directional differentiation, embryonic stem cells loose "totipotency" gradually to become more specific and develop toward specific tissues and organs such as heart, lung, skin, and bone marrow, blood vessels, skeletal muscle and liver. They can only develop and differentiate into several kinds of cells in a system, but can't form cells in other systems, so these stem cells are called pluripotent stem cells, also known as tissue stem cells. Pluripotent stem cells go on with differentiation and development to form more specialized cells, namely committed progenitor cell, which can only proliferate to differentiate into only on type of cells, such as red blood cells, muscle cells, nerve cells, etc. Terminal cells loose division ability, and only can differentiate into cells or tissues with specific functions according to a certain procedure, completing the special physiological function, such as oxygen transportation, muscle contraction, signal transmission, etc. When terminal cells age and die gradually, pluripotent cells can produce new terminal cells in order to supplement and substitute for died terminal cells, so as to maintain dynamic balance between growth and decline of tissues and organs. After birth, the body still retains a small amount of pluripotent stem cells beside committed progenitor cells, and the pluripotent stem cells can continue proliferation and differentiation in the growth process.

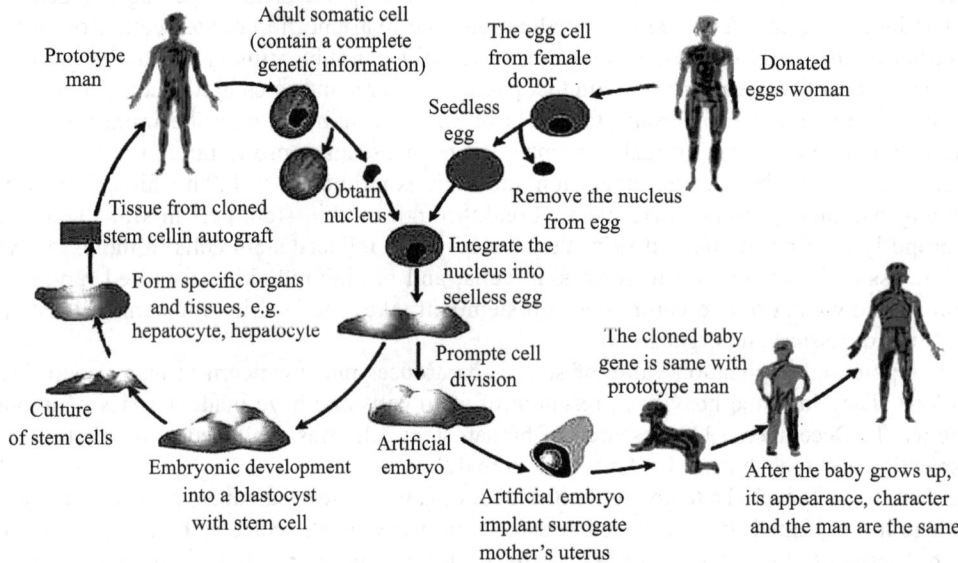

Fig.5.6 The preparation and application of embryonic stem cells

Embryonic stem cells are shown in Fig.5.7. Pluripotent cells and committed progenitor cells are called tissue stem cells or adult stem cells, such as hematopoietic stem cells (HSC) from bone marrow (Fig.5.8), mesenchymal stem cells (MSC) and neural stem cells (NSC) in the nervous system (Fig.5.9), etc.

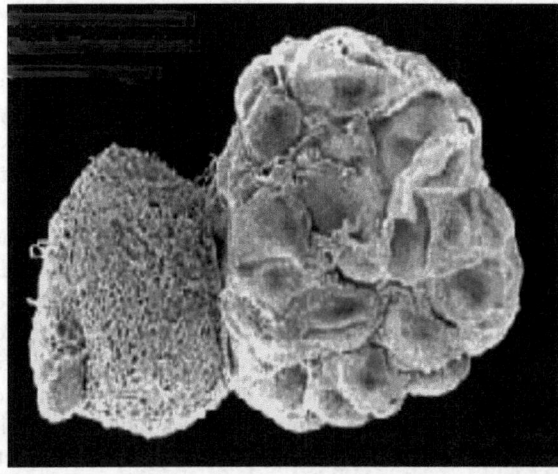

Fig.5.7 Micrograph of embryonic stem cells

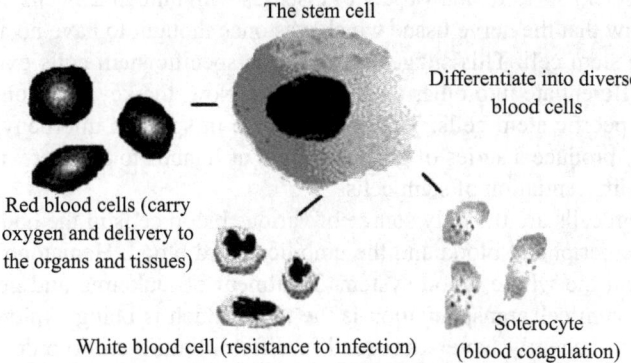

Fig.5.8 Hematopoietic stem cells

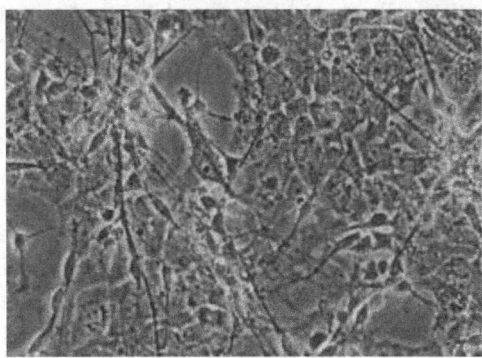

Fig.5.9 Micrograph of neural stem cells

5.6.3 Research Progress of Stem Cell

Stem cell research started with embryonic stem cells. In the 1950s, scientists first found embryonic stem cells in teratoma, from then on developed the research course of stem cell biology. In the 1980s, British Evens and Kaufman and American Martin individually isolated the embryonic stem cells from the early embryo of mice and cultured *in vitro* them independently. They established embryonic stem cell line with development totipotency, and promoted booming development of ES cell biology research. Meanwhile, this also solved model problem in the research on development and genetics and cell differentiation in mammals. Since then, the ES cell lines from some animals such as hamster, rat, rabbit, cattle, sheep, pigs and primates have been established. For example, German and American medicine research teams transplanted the neurogliocyte differentiated from ES cells by culturing into the test mice successfully, and researchers in Missouri made paralyzed cats restore the ability of some body movements through the mouse embryo cell transplantation technology. In 1998, Thomson and others from Wisconsin University in the United States cultivated human ES cells successfully and made them keep totipotency of ES cells to differentiate into all kinds of somatic cells.

Thus, it is possible for scientists to treat a variety of diseases with human ES cells.

Recent studies show that the nerve tissue which was once thought to have no ability to regenerate still contains the nerve stem cell. This suggests that tissue-specific stem cells exist widely and also have the potential to differentiate into other cells or tissues. Now, the key question is how to find and isolate various tissue-specific stem cells. They often locate in specific microenvironment in which mesenchymal cells may produce a series of growth factors or ligands to interact with stem cells, so as to control renewal and differentiation of stem cells.

Hematopoietic stem cells are the only source of various blood cells in the body, and mainly exist in the bone marrow, the peripheral blood and the umbilical cord blood. Hematopoietic stem cells can be differentiated to form the whole blood system. Treatment of leukemia and genetic hematonosis through hematopoietic stem cell transplantation is the topic which is being explored in the medicine area. American scientists recently have succeeded in differentiating embryonic stem cells into the hematopoietic precursor cells in the human bone marrow, and further cultured them to form red blood cells, white blood cells and platelets. These results suggest that humans may obtain the inexhaustible blood source.

The research of neural stem cells started later than hematopoietic stem cells. Because the fetal brain tissue which isolation of neural stem cells needs is more difficult to obtain, and the dispute on research of embryonic stem cells has not stopped, the research work is still at the starting stage. Recently, Polish scientists successfully have obtained the brain cells by culturing stem cells in the umbilical cord blood and inducing them to differentiate, and they may be used for treatment of Parkinson's disease, cerebral concussion, etc., and repair of brain damage.

For the first time, an Israel scientist cultivated embryonic stem cells to obtain the human heart tissue, which had electrical and mechanical properties of the new cardiac tissue and could beat normally. Chinese scientists have successfully cultured stem cells in vitro to form the mucosal tissues of the stomach and the intestine. This is another major research fruit in the area of regeneration *in vitro* of human organs, following the all-around repair of the skin tissue with the stem cell in situ culture technique. At the beginning of 2005, Pang Wenxin working with Institute of Hematology of Union Medical University in Peking found the stem cells in the muscle tissue which had hematopoietic potential.

However, research work of human ES cells aroused a big dispute all over the world. For the reasons of social ethic consideration, some countries issued some laws to prohibit research of human ES cells. From the view of either basic research or clinical application, research of human ES cells bring much more benefits than negative influence in the ethical aspect, so many people are calling for research of human ES cells passionately.

5.7 Embryo Engineering and Animal Cloning Technology

Embryo engineering is a comprehensive breeding technology, and is complementary to stem cell bioengineering. It is mainly engaged in embryo transfer and splitting, ovocyte maturation *in vitro* and transplantation, embryonic mosaic, embryo freezing preservation and clinical embryo transfer for treatment of diseases. Adopting contemporary embryo bioengineering technology is to an effective way of development of animal husbandry toward high-yield, high-efficiency, high-quality and intensification. The main technical procedure of embryo engineering includes embryo *in vivo*, embryo *in vitro* and cloned embryos.

Since the 1970s, application of biotechnology breeding system with embryo engineering as the core has replaced the traditional breeding method with selection and hybridization as the main approaches, so as to result in a profound revolution of animal breeding technology. As a series of experimental techniques, embryo engineering provides technical guarantee for further research of basic theories in life science.

In China embryo engineering technology research started in the mid 1970s, and was gradually developed on the basis of tracking, digesting and absorbing international advanced technology. After decades of efforts, Chinese scientists have made remarkable achievements in the high-level theoretical research of embryo engineering technology, and their research results have been listed in the world's advanced rank. In application practice, cattle and sheep embryo transfer technology and freezing technology have made success and gradually become the routine technology in animal husbandry industry. "Test tube animals" obtained through *in vitro* fertilization of isolated embryos and "cloned animals" obtained through nuclear transplantation give birth successively; and research of human embryonic stem cells also has made great progress.

5.7.1 Embryo Engineering

5.7.1.1 Embryo transfer technique

Embryo transfer (ET) refers to a procedure where the early embryo in the female livestock which has been mated is isolated and transferred into the female individual with the same physiological state, in order to make it continue to develop into a new individual. Animal embryo transfer-related technical system is mainly used for rapid propagation, introduction and germplasm resource conservation of good varieties or individual animals. Main technical links of embryo transfer include that synchronization of estrus of a donor and receptor, superovulation of the donor, donor's estrus identification and hybridization, embryo collection, embryo test and identification and embryo transfer.

People have studied embryo transfer for more than 100 years. The first success was rabbit embryo transfer in 1890 performed by Walter Heap from University of Cambridge, UK. In 1930s, embryo transfer gradually attracted attention from animal husbandry and veterinary professionals, and embryo transfer in many animals, such as sheep (in 1934) and goats (in the late 1950s) and so on, succeed in succession. In the 1970s, frozen embryo transfer succeeded. Later, people realized that superovulation and embryo transfer technique had huge potential in improving reproductive capability of excellent female livestock and speeding up modification of varieties, embryo transfer began to be applied in practical production. In 1971, the first commercial embryo transfer company, Alberta company, was founded, and later International Embryo Transfer Society (IETS) was founded in 1974. Now, embryo transfer technique has become the basic technique and approach in the area of embryo engineering technology, and superovulation and embryo transfer technique has become the main approach for rapid propagation of male livestock of excellent varieties and excellent females of core livestock groups. Application research of embryo transfer are developed rapidly in some developed countries such as the United States, Canada, France, Japan, Britain, Netherlands, Australia and Germany. They have established complete cow embryo transfer business companies to sell embryos to other countries. In these countries, breeding plans are fully implemented, and progress in genetic improvement is 10% greater than expected.

In China, livestock embryo transfer technology started later than in other countries. Embryo transfer in sheep and nanny goats succeeded respectively in 1974 and in 1980; since then its application research developed more quickly in cattle and sheep production. The development history of embryo transfer in cattle and sheep in our country is divided into three stages, experimental research stage (from the early 1970s to the mid 1980s), development acceleration stage (from the mid 1980s to the mid 1990s) and application stage (from the mid 1990s to now). Now embryo transfer has started full industrialization stage. After the 1990s, embryo transfer played an important role in rapid breeding of good varieties of livestock, and embryo transfer in cow, beef cattle, Angora goat, Boer goat and sheep has already realized industrialized application in our country.

The significance of superovulation and embryo transfer technology in production of animal husbandry includes the following points. ①Reproductive potential of females of excellent varieties may be fully played to make their progenies increase rapidly. Through superovulation and embryo transfer technique, the normal ovulation number and the embryo number increase so as to obtain a large number of embryos from outstanding females. The embryos are transferred to surrogate receptors, relieving excellent females from the task of embryonic breeding so as to save the long gestation period for excellent females. Thus, the number of excellent females' progenies increases rapidly, and their pregnancy task is removed to shorten the reproduction interval, so as to raise the reproductive ability of excellent females, The role of excellent females is fully played in breeding, so as to accelerate the breeding process. ②In the breeding, this technique can shorten the generation interval and increase selection intensity. With this technique the number of progenies meeting the requirement for identification can reach in a shorter period, finishing the progeny identification work in advance so as to speed up genetic modification and shorten the breeding process. ③Embryo transfer is also the foundation for other embryonic biological technologies (such as embryo splitting, embryo mosaic, gender identification and nuclear transplantation, etc.), and it is an important approach for life science research.

5.7.1.2 Embryo cryopreservation technique

The cryopreservation of embryo and oocyte refers to one long-term embryo and oocytes preservation technique, adopting special protective measures and freezing process to make embryos or oocytes stop metabolism at-196℃. After a period of heating, the frozen embryos and oocytes may still restore metabolic capability. The technique has laid a solid foundation for mammal breeding, reproduction and factory production, and has great theoretical and practical significance. Embryo cryopreservation research began in the 1950s, and Smith used glycerol as an antifreezing agent to store rabbit embryo successfully for the first time. In 1971, mouse embryos were cryopreserved with slow freezing method successfully by Whittingham. In 1978, Willadsen established for the first time the "conventional freezing method" in the freezing research of cow embryos, which could realize rapid freezing and rapid thawing, and so now has been widely used in freezing process of commercial embryos. In 1984, Takeda and others adopted one step freezing method for the first time to freeze mouse embryos successfully. In 1985, Fahy Rall applied vitrification freezing method to perform one step freezing experiment on mouse embryos in the eight-cell phase without the aid of any freezing instrument, getting the post-freezing development rate of 87.5%. Since then, vitrification freezing method has used successfully in preservation of embryos of cattle, sheep, goat, rabbit and rat. At present, the embryo freezing technique has been widely used in animal husbandry production, and common protective agents are glycol, sucrose, serum PBS, etc.

Commercial application value of embryo cryopreservation mainly includes the following points: simple and cheap long-distance transportation of embryos so as to replace transportation of live animals and save transportation and quarantine expenses; reducing the risk of disease transmission resulting from livestock transportation in order to promote international exchange of excellent varieties; and not requiring for feeding donors and receptors in the same place, synchronization of estrus, recovery at the same date and embryo transfer as required in fresh embryo transfer. In the frozen embryo transfer, what is needed is only to observe the estrus date of receptors. When receptors are in the corresponding embryonic developmental period, the frozen embryo may be thawed at any time to perform embryo transfer. Embryo cryopreservation also has the significance of protection of excellent breeds in scientific research. Establishment of excellent animal variety embryo library or gene pool can preserve the varieties, the lines and the rare mutants which may be lost as a result of diseases, natural disaster and war, and can control genetic stability within a variety and maintain genetic consistence among animals of some inbred lines, so as to save a lot of time and fund.

5.7.1.3 Embryonic splitting technique

Embryo splitting refers to a biotechnology using micromanipulation system to split mammal embryos into several parts with developmental potential. Embryonic splitting technique is a method of embryonic cloning for obtaining progenies of monozygotic twins or multiple births. Most of early mammal embryos belong to adjustment development type. After removing half of an early embryo, the rest can still develop into a complete embryo. For early embryos at the cleavage stage, each blastomere has the same developmental ability before at least 8-cell stage. At the stage of morula, adjustment development ability of one single blastomere is weakened, and the success rate of embryo splitting greatly reduces. With development of embryo splitting technique, its manipulation procedure gradually is simplified. Embryo splitting methods include microscopic manipulation instrument splitting and unarmed splitting, and the former include micro-needle splitting and microscopic knife splitting.

In 1904, 2-cell embryo split test in frogs was first done by Spemann, and progenies of monozygotic twins were obtained. In 1930, Pincus proved firstly that single blastomere of 2-cell rabbit embryo in pseudopregnancy body can develop into smaller blastocysts. But until 1970, Mullen for the first time obtained progenies of monozygotic twins of mice by splitting 2-cell mouse embryo blastomere and then culturing *in vitro* and transferring the split blastomere. In 1974, Trounson successfully obtained sheep of monozygotic twins from the split embryo. After the 1980s, mammal embryo splitting technique has developed rapidly. Based on summarization of the experiences of predecessors' researches, Willadsen established the systematic embryo splitting method, and obtained progenies of sheep quarter and eighth embryo, and those of cow quarter embryo with this method. At present, the animals whose progenies of half-embryos have been obtained are mouse, rabbit, sheep, goats, cows, horses and human; those whose progenies of quarter embryos have been obtained are rabbit, sheep, pigs, cattle and horses; and progenies of eight embryos also exist.

In 1986, Chinese scientist Zhang yong got mice of monozygotic twins using embryonic splitting method for the first time, and in 1987 obtained goats of monozygotic twins, In the same year, Tan Liling got calf of monozygotic twins. In 1986, Dou Zhongying obtained calf from split embryos and in 1990 obtained progenies of cow quarter-embryos.

In animal husbandry production, embryonic splitting can be used to increase the number of animals with excellent genetic characters, and raise efficiency and benefit of embryo transfer technique. In scientific research, individual animals of monozygotic twins or multiple births can provide valuable experimental material for animal genetics and developmental biology, animal breeding science, which can eliminate genetic difference and increase accuracy of experimental results. Embryonic splitting is the basis for bioengineering technologies such as embryonic mosaic and nuclear transplantation. After gender identification, according to intention, determine whether the other part should be transferred, so as to control the animal gender.

5.7.1.4 Embryo *in vitro* production technique

Embryo *in vitro* production technique system is based on research results of *in vitro* fertilization (IVF) technique. IVF, also called test tube embryo technology, refers to that the fertilization process of sperms and eggs *in vitro* were completed through artificial manipulation, the fertilized eggs are *in vitro* cultured and transplanted to obtain various animals. The basic principle of IVF is that primary oocytes in cumulus oocyte complex (COC) become mature in the artificially simulated environment (nutrition, temperature, humidity, gas, osmotic pressure, pH value, etc.) and at the same time the sperms obtain energy to complete fertilization.

Researches of mammal IVF has a history of 100 years. In 1878, German scientist Sehenk first did experiment of IVF in rabbits and guinea pigs, but until 1951, Zhang Minjue and Austin found that sperm have capacitation phenomenon in mammals simultaneously in rabbits and rats, and it should be said that this is the milestone for research of mammal IVF. In 1954, Zhang Minjue obtained the first IVF mammal in the world, "test tube rabbit". Rapid development of IVF since the 1960s, especially many important modifications of in sperm capacitation technique and cell culture methods have made IVF technique step into the livestock research field. In 1978, the world's first test-tube baby also was declared to give birth, and later a large number of test-tube babies obtained with microscopic fertilization of sperms came into being successively in many countries. This technique has become an ideal approach for treatment of male infertility. At present, test-tube animals have been obtained by means of ovarian oocyte *in vitro* maturation, frozen sperm *in vitro* capacitation, *in vitro* fertilized egg, and embryo cryopreservation and transfer. According to special needs of all kinds of embryo related techniques, oocytes, fertilized eggs and pre-transfer embryos in different cell division phases can be provided accurately.

In 1990, Chinese scientist Qian Jufen and others obtained the first test tube goat. In 1994, Liu Ling obtained the first batch of goats with oocyte follicle *in vitro* maturation technique, i.e., IVF tube goats. Xu Rigan (1989), Fan Biqin (1989), Zhu Yuding (1989), Lu Kehuan (1990), Qin Pengchun (1990), and others successively obtained test tube calves, which made cow IVF technique in China reach the advanced world level.

The main technical links of *in vitro* embryo production include a series of steps, including oocyte acquisition and *in vitro* maturation, sperm capacitation, IVF of mature oocytes and capacitated sperm, fertilized egg *in vitro* culture and transfer.

Significance of embryo *in vitro* production include the following: producing a large number of high-quality and cheap embryos within a short time, so as to accelerate the breeding process of livestock of excellent varieties; providing appropriate experimental materials for the rest research of embryo engineering such as embryonic gender identification and embryonic mosaic discovering fertilization mechanism and nature of early embryo development with the aid of IVF technology, so as to provide important theoretical basis for animal embryology and developmental biology. In addition, the technique also has important practical value in saving endangered animals, and development and utilization of rare animals.

5.7.2 Animal Cloning Technology

5.7.2.1 Introduction to animal cloning technology

Cloning refers to a procedure for obtaining a cell group or an individual animal population with the same genetic background through asexual reproduction. In the broad sense, animal cloning refers to animal asexual reproduction. Many animals with the same morphology and functions and genotype are produced from a single individual by means of asexual reproduction method. The procedure mentioned above that after performing embryo splitting the split embryo parts are transferred into receptors respectively to conduct gestation for production many animals with the same genetic characters, i.e. embryonic cloned animals, belongs to the simplest artificial animal cloning way.

Animal cloning technology mainly refers to nuclear transplantation, including embryonic splitting technique, and is now the current hot spot in research of embryo engineering and life science all over the world. Nuclear transplantation refers to that the nucleus of a mammal cell (nuclear donor) is transplanted to another mature oocyte (nuclear receptor) with the nucleus removed through special external manipulation (microinjection, electric fusion, *in vitro* culture, etc.), so as to propagate mammal embryos and their species with the same genotype without sexual reproduction process. The animal individual developed with this technique is called nuclear transplantation cloned animal or karyoplasm hybridization animal.

With development of research of limitations of embryonic splitting and cloning technology, people started to research nuclear transplantation technique and made success. Nuclear transplantation technique greatly increases the production efficiency of cloned animals higher than embryonic splitting technique, and so basically substitutes for embryonic splitting method for research and development of production of cloned animals. Establishment of the nuclear transplantation technique system is an important milestone in the development history of reproductive biology. Now, the animal obtained by means of nuclear transplantation is called clone animal, and this is really clone animal. This technique is the main method for animal cloning currently.

According to different between nuclear donor cells in nuclear transplantation, nuclear transplantation can be divided into embryonic cell nucleus transplantation, embryonic stem (ES) cell nucleus transplantation, fetal fibroblasts nucleus transplantation and adult somatic cell nucleus transplantation.

In 1938, German scientist Spemann first put forward nuclear transplantation test in amphibians, and had a try. Briggs and King injected germ cell nucleus to oocytes without nucleus, and then got normal tadpole, which shocked the whole biological area. In China Professor Tong Dizhou also carried out basic theory research and achieved good results in fish nucleus transplantation. In 1981, Illmensee and Hopper for the first time reported that ICM cell nucleus was transplanted to zygotes without nucleus, and by culturing *in vitro*, 34% of the recombinant embryos developed into morulae. These embryos were transferred to receptors, and 19% of them development and gave birth. In the development history of nuclear transplantation cloning technology, there are two landmark events. One is that in 1986 Willadsen reported that the first mammal animal developed from the recombinant oocyte without nucleus was obtained by transferring one single 8-cell blastomere into the mature sheep oocyte without nucleus to produce the recombinant embryo. The method used in that study basically constituted the procedure of mammal nuclear transplantation, and subsequently a variety of nuclear transplantation cloning technology animals such as cattles, pigs, rabbits, monkeys, goats and so on were obtained. The other event is that British scientist Wilmut and others successfully obtained the cloned sheep "Dolly" by using somatic nuclear transplantation technique on February 17th, 1997; and it was the cloned progeny which human beings first used somatic cells from an adult animal, breast epithelial cells, and once aroused much attention all over the world. It had important theoretical significance that it indicated that after treatment differentiated cells could restore totipotency in the environment of cytoplasm of mature oocytes with nucleus removed. In 1998, the United States produced "cloned-from-clone" mice (i.e. Honolulu technique). Japanese Kato, Chinese American scientist Yang Xiangzhong and others reported the birth of somatic cloned cattle subsequently. In 2000, five cloned pigs were produced for human organ transplantation in the United States. In November of 1999, Genetics and Developmental Biology Research Institute of Chinese Academy of Sciences and Yangzhou University first reported that somatic cloned goats (two) were born. Research Institute of Zoology of Chinese Academy of Sciences reported that development of the heterogeneous recombinant cloned embryo obtained by means of recombination between somatic cells from a panda and oocytes from a rabbit had progressed to the blastocyst stage, in order to save the rare animal resource in the world, which gained international attention and support. At present, animal cloning technology has begun to enter the production application stage.

In accordance with purposes of animal cloning, it can be divided into reproductive cloning and therapeutic cloning. The former aims at rapid propagation or "replication" of excellent animal individuals; and the latter recombines with embryonic stem cell technology to solve the problems in autologous transplantation of tissues and organs in the human medicine (See the section "embryo stem cells" in this chapter). In February 2004, South Korea and the United States scientists successfully cloned the early human embryo and extracted embryonic stem cells, and first proved the feasibility of the human therapeutic cloning. This was a significant step for application of cloning technology, and once again aroused much discussion on cloning.

5.7.2.2 Significance of Animal Cloning Technology

Reproductive cloning (hereinafter referred to as the animal cloning) has extensive practical value in livestock biotechnology breeding, rapid breeding of good varieties, embryonic industrialized production, rare animal protection, etc. It is the new approach for interdisciplinary life science research, and has important theoretical significance in the following aspects. ①Clone animals have high economic value. Cloning production of progenies of excellent animal individuals can greatly improve survival performance of livestock population, avoid separation and recombination of excellent genes in sexual reproduction, rapidly increase the frequency of excellent genes and their combination in the population, and as far as possible increase genetic contribution of excellent livestock breeds, so as to greatly speed up the breeding process. ②Animal cloning is a powerful measure for rescue and protection of rare animals. Scientists will propel rapidly present homologous cloning into heterogeneous cloning, i.e. "borrowing pregnant belly", and this undoubtedly will greatly promote the work for protection of endangered mammals. ③Transgenic clone mammals will provide cheap drugs, health care products for human body-tolerable transplant organs. ④The animal cloning technology is an important research method in embryology, developmental biology and cell biology, genetics, molecular biology. It can combine with embryonic stem cells, molecular biology technology to discover various genes and their activities in animal development, so as to provide a new idea to solve so many complicated problems in biology.

5.7.2.3 Existing problems

At present there are mainly four problems in animal cloning technology. ①Progenies of somatic cell cloning animals likely appear aging phenomenon. For example, the cloning sheep "dolly" underwent "euthanasia" because of progressive pneumonia. Its lived for less than seven years (sheep can usually live for 11-12 years). Its death has led to scientists further debate about premature senility of cloned mammals, and the aging cause of cloned animals is to be further studied. ②The overall success rate of cloned animals is 1%-5% and at a lower level. ③The study of theories of cloned animals can't keep up with the pace of technological development, and researchers engaged in theories of other emerging disciplines such as cell biology and molecular genetics need to do some deeper work. ④Cloning technology itself needs more extensive permeation and integration with other disciplines and technologies.

From the current development situation, these difficulties and problems can be overcome. The biggest problem in animal cloning technology is ethics problem.

In November 1997, a document titled "Declaration of World Human Genome and Human Rights" was adopted at United Nations Educational, Scientific and Cultural Organization Conference 29th in Paris. This document explicitly pointed out that reproduction of humans with cloning technology should be banned because it violated the dignity of human beings. In January 1998, 19 European countries such as France, Denmark, Sweden, Italy, Norway, Portugal, Romania, Spain and etc signed a agreement to strictly prohibit human cloning, European Protocol on Banning Human Cloning. This is the first international legal document on banning human cloning in the world. The agreement prescribes that any research institutions or individuals in each contracting country it be prohibited to use any technique to create one person with genes similar to a living or defunct, otherwise they be punished severely. In September 2003, international conference "Opposition to Human Cloning" held in New York, more than 16,000 scientists as representatives from 63 international famous science organizations signed a statement to urge the United Nations to ban human cloning worldwide. China and the governments of the United States, Britain, Germany, Japan and other countries have also clearly opposed to human cloning. Chinese Ministry of Health announced that China would not approve, participate in, sponsor or accept research of human cloning from Chinese and foreign scientists. So far, only 30 countries in the world officially have banned human cloning, but in many countries a lot of scientists are studying on human cloning secretly, and it is claimed that their cloned human has been born.

Various countries have not yet come to an agreement with the key problem whether research of embryonic stem cells on earth conform to with medical ethics or will destroy human lives. Many scientists point out that human cloning is a great threat to human dignity, and also may cause serious social, ethical, moral, religious and legal problems. At the same time scientists generally believe that we should distinguish reproductive cloning, i.e. breeding cloned human, from therapeutic human cloning, i.e. embryonic stem cell cloning, and avoid confusion. Therapeutic cloning is a technology from which a lot of people can benefit, and embryonic stem cell research brings hope for curing many diseases. About 100 million patients with alzheimer's disease, cancer, diabetes and spinal cord disease will not be able to be healed on the globe if therapeutic cloning is banned. Although countries all over the world have different history, culture, religion, ethical tradition, and there are different views on the problem of the origin of life, but scientific research that can serve humans should be protected. However, most of the debates have focused on issues related with human life reproduction. Some scientists and ethicist think that human embryos used for cloning are a form of life, although they do not necessarily represent a natural person individual, but extraction of stem cells from embryos need completely kill the embryonic tissue, and this is equivalent to carrying off a life in fact.

On March 8th, 2005, at the United Nations General Assembly Session 59, The United Nations Declaration about Human Cloning, adopted by Law Committee of UN General Assembly, was approved by voting by ballot. The Declaration called for member nations to ban all the forms of human cloning, including therapeutic cloning for embryonic stem cell research. The nations agreeing on therapeutic cloning, such as China, the United Kingdom, Belgium, France, India, Japan, Singapore and others, vetoed. The Chinese delegation said, the Chinese government would continue to stand the ground against reproductive cloning, but the reason for China's vetoing was the Declaration did not reflect different opinions on the issue from different nations. Since a declaration is not an international legally binding convention, vetoed nations stated that they would not restrain their doctors to do embryonic stem cell research and therapeutic cloning. The ethical problems will be detailed in Chapter 13 "On Human Cloning".

5.8 Research Progress and Future Prospects of Cell Engineering

Cell engineering has permeated into many fields of human life, and achieved many developmental research results, among which some results have been popularized in production and gained obvious economic and social benefits. With further research of cell engineering technology, its prospects and the impact will emerge increasingly.

5.8.1 Biopharmaceutical Engineering

In the past 20 years, animal and plant cell engineering technology combined with the latest theories and technologies of cell biology, biochemistry, molecular biology and engineering to develop into an advanced comprehensive high technology. Biopharmaceutics based on the animal and plant cell engineering and other bioengineering technologies has created great economic benefits and social effect.

After the 1990s, the global sales amount of biopharmaceuticals grows at an annual rate of 30%, and this speed is much higher than the average annual growth speed of less than 10% in the whole pharmaceutical industry. Biotech drugs account for 9% currently in the global pharmaceutical market, and have become the key of development of novel medicine industry. At present, there are about 1400 biopharmaceutical companies in the United States, among which more than 20 companies have formed the scale production and their total capital is more than 40 billion dollars. Until April 2003, more than 370 varieties of biopharmaceutical products have been approved to use for clinical diagnosis and treatment of 200 kinds of diseases by the United States FDA. Japan is second to the United States with respect to development of biological technologies. There are now about 600 biopharmaceutical companies, and they are constantly strengthening development of the world market and have launched into the European and Asian markets. Europe lags behind Japan in development of biotechnologies. In Europe there are about 300 biopharmaceutical companies. The scale of the biotechnology-derived market relating to production of pigments, toxins, antibiotics, interferon, human insulin and human growth hormones has increased from 2.34 billion dollars in 1995 to 4.15 billion dollars in 2002.

At present, high-tech bioproducts are used in the medical area develop rapidly in the world, and the number of approved new bioengineering pharmaceuticals and vaccine products increases quickly and from annual 5-7 products in the early 1990s to annual 32 ones in 2000. Recently five kinds of important monoclonal antibody products have appeared in the United States market, and the market demand is 100kg/a for every kind of them. The demand for other several monoclonal antibodies and recombinant protein products also reaches 1-20kg/a. In the coming five years, the United States FDA may be expected to approve 5-10 kinds of new antibody products annually. Among these bioengineering products, half of them are produced through animal cell culture technology. Based on the hot spot of future biopharmaceutical research, vaccine, monoclonal antibody, recombinant human protein, gene therapy, cellular therapy and IFN are the most popular field of research and development. The statistical data from international authority institutions show that the antibody market for global treatment and diagnosis is up to 7.3 billion dollars in 2002, is expected to grow at the annual rate of

16.6% and will be up to 15.7 billion dollars till 2007.

In China biotechnology research started in the mid of 1970s. After 20 years of efforts, it has initially formed a complete basic framework including research, development, production system, and has got a series of biotechnology achievements and products with high level and huge market potential. In 1986, the sale amount of contemporary biotechnology products was only 200 million yuan, and till 1996 it has increased to 11.4 billion yuan. Production of natural products, such as panaxoside, vinblastine, paclitaxel and so on with cell engineering made great progress. It cannot only avoid industrial pollution, but also reduce resource destruction resulting from extraction of natural products from natural resources. Hence, it is helpful for protection of natural resources.

Now, scientists at home and abroad are studying vaccines for "SARS" and "avian flu", and the competition for the leading position of the technology has intensified increasingly.

Large-scale culture technology of animals and plant cell is an important approach for accomplishment of industrialization of cell engineering products. By studying the environmental conditions for target products, a series of downstream process and manufacturing technologies with high expression, low consumption and stable uniformity have been established, so as to complete the whole production process from the lab to scaled production, realizing standardization and industrialization of target products in order to meet the big needs for market application. Secondary metabolites of plant cells (such as pigment, toxin, antibiotics, auxin, alkaloids, etc.) can be obtained with plant cell mass culture, opening up a new reasonable way for plant resources. Production of enzymes, growth factors, vaccines and monoclonal antibodies with important medical value through animal cell large-scale culture technology has become an important trend in the pharmaceutical high biotechnology industry.

Large-scale culture of animal and plant cell can be performed in a special bioreactor, since the 1970s bioreactor research has got great development. According to the characteristics of different cells and designed functions of different bioreactors, the bioreactors for animal and plant cell suspension culture include air lift bioreactor and agitating bioreactor; the ones for animal cell suspension culture and anchorage-dependent culture, hollow fiber bioreactor and rectangular channel cellular ceramic bioreactor; the ones for animal cell anchorage-dependent culture, micro-carrier agitating bioreactor; the ones for cell embedding culture, fluidized bed bioreactor and fixed bed bioreactor. Research and development large-volume bioreactors have become the goal for bioengineering companies in countries such as America and Britain. Wellcome Corporation has established a 8000L agitating bioreactor to produce vaccines and other products; Sumitomo Corporation has also established a 8000L bioreactor to produce tPA; and Celltech Corporation has developed and established a 10,000 L air lift bioreactor to produce MAb. In addition, most of companies are now designing, developing and testing larger-volume bioreactors. According to the related statistical data, at present mammalian cell culture ability of bioengineering pharmaceutical companies in the world is 400,000 L, and the scale can meet the annual productivity of about 2500kg MAb. The demand for clinical diagnosis and treatment with antibodies and vaccines is promoting development of animal cell mass culture technology.

5.8.2 Stem Cell Research, Reproductive Cloning and Therapeutic Cloning

The differentiation potential of embryonic stem cells and tissue stem cells has provided great hope for many patients that need organs for transplant therapy and cell substitution therapy. In stem cell therapy application, hematopoietic stem cells and skin stem cells have been used earlier, and other stem cells, including embryo midbrain cells for Parkinson's disease, pancreatic duct cells for diabetes, etc. are also in clinical research. Scientists have been able to use donor cells at different development stages to perform somatic cell cloning successfully in many animals, which lays the foundation for further development of reproductive cloning and therapeutic cloning.

With development of stem cells and therapeutic cloning, regenerative medical engineering which has the purpose and the task to repair and replace morphology and functions of human tissues and organs has developed rapidly. Haseltine, the chairman and chief executive officer of American Human Genome Science Corporation thought, regenerative medicine includes four stages: simulating the effect of growth factors to stimulate self-repair function of the organism; culturing tissues or organs *in vitro* for transplantation after identifying the essential growth factors; making old tissue regenerate by reconstructing the biological clock of cells; and exploring new development of nanotechnology and material science. These developments of new technologies will make humans construct new components of cells, organs and tissues to obtain a new combination integrating the human body and natural composition. It shows good prospects for human to conquer diseases and maintain youth.

But to make it become a reality, a lot of basic applied research work has to be done, and only if these problems are completely solved, stem cells may be used in clinical practice. Because embryonic stem cells can only be taken from embryos, we must take into account ethical problems that may occur when we use human embryonic stem cells.

Chapter 6

Enzymes and Enzyme Engineering

6.1 Overview of Enzymes and Enzyme Engineering

6.1.1 Concept of Enzyme and Enzyme Engineering

Enzyme is a kind of indispensable biocatalysts for self-replication and metabolism of substances in organisms. Enzyme can specifically and effectively catalyze substrate reaction in the mild conditions such as room temperature, normal pressure, neutral pH, and so on. So development and application of enzymes are very important for the contemporary new technology revolution. Metabolism is the most important feature of life activities, all life activities are carried out through normal run of metabolism, and various chemical reactions related with organism metabolism are all controlled by enzymes. No enzyme, no metabolism, no life, either. If there was no enzyme, the whole biosphere would not exist. Most of enzymes are located in cells, and some others are secreted outside cells. Contemporary life science has been developed into the molecular level, so that we can utilize relationship between structures and functions of biological macromolecules to explain the nature and laws of lives phenomena, and explore relationship between enzymes and life activities, metabolic regulation, disease, growth and development, this undoubtedly has great scientific significance. The research of physicochemical properties and action mechanism of enzymes has very important significance for clarifying the nature of life phenomena.

Enzyme is also an important tool in the molecular biology field. The presence of some specific tool enzymes makes that determination of the primary structure of nucleic acid have some important breakthrough. Discovery of restriction enzymes facilitated the birth of DNA recombinant technology, and promoted development of the genetic engineering. Enzymes reflect obviously the wonderful functions of recognition, catalysis and regulation in organisms system.

Enzyme engineering is one of the main contents of contemporary bioengineering. With rapid development of enzymology, especially application of enzymes, enzymology combine with engineering interinfiltration and development into a new technology science, i.e. enzyme engineering, which is an edge discipline and technology formed by combining basic principles of enzymology and microbiology with chemical engineering. Usually, the technical procedure for enzyme production and application is called enzyme engineering. It studies enzymes for the purpose of application. In a certain biological reaction device, enzyme's catalytic properties are utilized to convert corresponding materials to useful products. So enzyme engineering is an important part of bioengineering.

6.1.2 Research Content of Enzyme Engineering

Enzyme engineering consists of four parts: ①production, isolation and purification of enzymes; ②modification of enzyme molecules; ③enzyme and cell immobilization and enzyme application; ④enzyme reaction kinetics and bioreactors. Generally, enzyme engineering is thought to date from the World War II. Since the 1950s, enzymes were isolated from microbial fermented broth to make enzyme preparations. After the 1960s, presence of enzyme and cell immobilization technology made application technology of enzymes to take on an entirely new look. Since the late 1970s, development of microbiology, genetic engineering and cell engineering brought vitality for further development of enzyme engineering, and had great effect from enzyme preparation methods and application scope to post-treatment technology. Although about 8000 kinds of enzymes have so far been found and identified, commodity enzymes preparation in large scale production and application have only dozens of kinds.

Therefore, according to different means for researching and solving the above problems, enzyme engineering is divided into chemical enzyme engineering and biological enzyme engineering. The former refers to research and application of natural enzymes, chemically modified enzymes, immobilized enzymes and chemical artificial enzymes. As a product of combination with contemporary molecular biology, the latter focuses on enzymology and gene recombinant technology and mainly includes three aspects: ①large-scale production of enzymes with genetic engineering technology (cloning enzyme); ②modification of enzyme genes for production of genetically modified enzymes (mutation enzyme); ③design of new enzyme genes for the purpose of production of synthesis of new enzymes which have never existed in the nature.

In 1971, the first International Enzyme Engineering Conference was held in Hennileer, and at that time enzyme preparations had been widely used in industrial and clinical practice. Chihata Ichiro and others used the amino acid resolution technique of immobilized amino acylase for industrialized production of L-amino acids, and hence created new field of application immobilized enzymes. At International Enzyme Engineering Conference held in 1983, Chihata Ichiro won the prize. Later, production of L-aspartic acid with immobilized aspartase and production of high fructose syrup with immobilized glucose isomerase both succeeded. So immobilized enzyme research is still the research center of enzyme engineering, and it is applied more and more widely. In addition to application in traditional food industry (like lactose decomposition, cheese manufacturing, milk disinfection, alcohol production, etc.), immobilized enzymes are also widely applied in other areas, such as organic synthesis reactions, analytical chemistry, medical treatment, waste liquid treatment, affinity chromatography and so on. Contemporary biotechnologies have been widely sued in industry, agriculture, pharmaceuticals and food, and play a very important role in contemporary resource, energy, environmental protection and so on. Several new biotechnology industries have become the high-tech fields for prior development. As an important part of bioengineering, enzyme and enzyme engineering not only are stressed attended by both biochemical workers and those in industry, agriculture and medicine health care areas.

Cell immobilization (immobilized cell) technique is developed gradually based on immobilized enzymes. In recent years, cell immobilization technique has developed more rapidly and has been applied more widely than enzyme immobilization technique. In industrial application, studies on production of alcohol and beer through fermentation of immobilized yeast cells are more remarkable. A Japanese scientist named Toshio Onaka obtained high quality beer within one day with yeast cells embedded with calcium alginate gel. In France, Corriell and others immobilized yeast cells on the carrier of PVC pieces and porous brick carrier to carry out batch test of beer fermentation, and the production process could run continuously for 8 months. Previously most of companies focused much more on immobilized bacteria and yeasts, however, many metabolites with industrial value (such as enzymes, antibiotics, organic acids and steroid compounds) are made by filamentous fungi. At present, the method for immobilizing filamentous fungi mainly includes adsorption method and embedding method. But the embedding method limits sufficient supply of oxygen for cells, so as to reduce the production efficiency of metabolites products with immobilized filamentous fungi. Swedish Mosbach, et al put forward a general immobilization method for producing metabolites, which utilizes polymers to embed all kinds of cells, such as bacteria, yeasts, plant and animal cells and artificially constructed cells. In 1980, Wagner and others reported, the gene of penicillin acylase in *E.coli* ACTT11105 cloned to a plasmid, to get the hybrid strain *E.coli* 5 K (PHMl2) with higher enzyme-yield activity which was immobilized and used for production of penicillin acylase, this is the first case of combination of gene engineering with enzyme engineering. In 1980, Lim and Sun reported, embedded islet cells with calcium alginate could be used in the treatment of rat diabetes.

In research of enzyme engineering, bioreactors of enzyme and enzyme inhibitors are two important parts. Enzyme bioreactors often can increase catalytic efficiency, simplify process and increase economic benefits. By combining with the immobilization technique, we have developed many new techniques such as enzyme electrode, enzymatic membrane reactors, immune sensor and multiple enzyme reactors which are very valuable in chemical analysis, clinical diagnosis and monitoring of industrial production processes. Enzyme inhibitors play a special role in metabolic control, biopesticides, bioherbicides, and their low toxicity is a welcomed advantage. Developments of the enzyme inhibitors have been stressed increasingly by international industry institutions.

From progress and situation of enzyme engineering, it can be expected, in the future, a number of genetic engineering enzyme preparations will emerge, and an application upsurge of molecular modification enzyme preparations will come. Enzyme inhibitors and activators will still attract great attention from human, and play an important role in clinical, industrial and agricultural production, so as to make greater breakthrough in control of enzyme activity. Enzymatic method production has major contributions to the chemical synthesis industry. Mimic enzyme, artificial designed synthesis of enzymes, antibody enzyme, and hybrid enzyme will become active research fields. The non-aqueous system enzyme reaction technique (of enzymatic reaction in reverse micelles, enzyme reaction in organic solvents) will be one of the hot spots.

6.1.3 Classification, Naming and Structural Characteristics of Enzymes

6.1.3.1 Classification of enzymes

In 1961, based on reaction types, the Commission on Enzymes of International Union of Biochemistry classified enzymes into six classes, each of which was further classified into sub-classes, and assigned each enzyme a systematic serial number. The six classes of enzymes are as follows. ①Oxidoreductase. This class of enzymes involve in energy production, detoxication and synthesis of some physiological active substances in organisms. They catalyze hydrogen atom transfer, electron transmission, oxygen atom adding or introduction of hydroxyl group of substrates. This class of enzymes includes oxidase, dehydrogenase, reductase, peroxidase, oxygenize and cytochrome oxidase. ②Transferase. This class of enzymes can transfer some atom groups from one substrate to another one, and the transferred groups include amino, carboxyl, methyl, acyl and phosphate group. They involve in metabolism and synthesis of nucleic acid, protein, sugar and fat. Important transferases are acyltransferase, glycosyl transferase and ketoaldehyde transferase. ③Hydrolase. This class of enzymes catalyzes hydrolysis of substrates, and the hydrolyzed chemical bonds include ester bond, glycoside bond, ether bond and peptide bond. They play hydrolysis action *in vivo* and *in vitro*, and are also the enzymes applied the most widely in humans. Important hydrolases include lipase, glycosidase and peptidase. ④Lyase. This class of enzymes catalyzes removing or adding reactions of chemical groups on substrate molecules, including formation of double bonds and addition reaction. ⑤Isomerases. This class of enzymes catalyzes spatial isomerization reactions of substrate molecules, including racemization, epimerization, cis-trans isomerization, intramolecular transfer, etc. ⑥Ligase. This class of enzymes catalyzes cleavage of ATP and other high-energy phosphate bonds, and at the same time makes condensation reaction between other two molecules, so they are also called synthetases. This class of enzymes has relation with synthesis of many life substances. Their characteristic is that high-energy phosphate ester such as ATP is needed to act as energy and some enzymes need metal ions to act as cofactors.

6.1.3.2 Nomenclature of enzymes

1) Common nomenclature

According to the rule of common nomenclature, many enzymes are named according to their substrates, and their names are the substrate name with addition of the suffix "ase". So urease is the enzyme that catalyzes hydrolysis of urea. Some are named according to properties of catalyzed reactions. For example, oxidase and transaminase belong to this. Additionally, some are named by combining the two above methods, and cholesterol oxidase and alcohol dehydrogenase belong to this type. On this basis, some are named by adding their source or other characteristics, and for example, diaphorase and iron enzyme belong to such type.

Common nomenclature is simple and has been used for a long time. However, it is not systematic, and some confusion phenomena such as one name referring several enzymes or many names referring to one enzyme often appear. So in 1961, the Commission on Enzymes of the International Union of Biochemistry (I.U.B) formulated systematic nomenclature of enzymes and the classification rules, at the same time the enzymes accepted were tabulated, and recommended biochemical workers in various countries to name and classify enzymes in accordance with these rules.

2) Systematic nomenclature

In accordance with this method, the name of each enzyme consists of the substrate name and the reaction type name. For example, alcohol dehydrogenase catalyzes the following reaction:

$$CH_3CHO + NAD^+ + H^- \rightleftharpoons CH_3CH_2OH + NAD^+$$

In here, the substrate is ethanol and NAD^+, and the reaction type is oxidoreduction reaction, and so this enzyme is named as alcohol: NAD^+ oxidoreductase. If one of substrates is water, the word of water will be omitted, for example, acetyl CoA hydrolase is such case. But for exceptions, some enzyme names have been widely adopted, accepted, will not lead to confusion, chaos, and so can remain in vogue. For example, peptide-peptide hydrolase is one such enzyme.

Systematic nomenclature is very clear. From the name, both the substrates and the reaction type can be known, but the name is complex and not convenient to use. So the Commission on Enzyme recommended a common name for each enzyme, which was placed in a parentheses bracket [ ], such as alcohol: NAD^+ oxidoreductase [alcohol dehydrogenase].

In the systematic nomenclature, the classification number for each enzyme consists of four numerals, and in front of them is the two capital letters "EC" (Enzyme Commission). The first numeral in the number represents the class of the enzyme and the second numeral the major group of the class, for example, for oxidoreductase, the numeral represents the type of the electron donor group; for transferase, the property of the transferred group; for hydrolase, type of the hydrolyzed chemical bond; for lyase, the type of the cleaved bond; for isomerase, the type of isomerization; and for ligase, the type of the formed bond. The third numeral represents the number for each subgroup in each group, and every numeral has different meaning in different groups of different classes. The fourth numeral represents the serial number for each enzyme in each subgroup. For example, the numbers EC1.1.1.1 for one enzyme means that the enzyme is oxidoreductase, the electron donor CH-OH, the receptor NAD^+, and the serial number 1, namely ethanol dehydrogenase. For another example, in the enzyme number EC3.4.4.4 (trypsin), "3" represents hydrolase; the second numeral "4" the enzyme's action on peptide bond; the third "4", the enzyme's action on peptide-peptide bond but not on the peptide bond at both ends of a peptide chain. So the Commission on Enzyme prescribes, in those papers mainly involved in enzymes, the number, the systematic name and the source of the used enzyme should be given clearly in the first narration, and thereafter, according to personal habit, the common name or the systematic one may be used.

Whether an enzyme catalyzes a positive reaction or a reverse reaction, the same reaction type should be adopted in its name. When the reaction in only one direction can be proved or the reaction in only one direction has significance, naturally the reaction type should be adopted to name the enzyme. Sometimes nomenclature of enzymes is habitual. For instance, in all the reactions including mutual conversion between NAD^+ and NADH, ($DH_2 + NAD^+ \rightleftharpoons D^+ + NADH$), the enzymes catalyzing the reactions are traditionally all named as DH_2: NAD^+ oxidoreductase. In addition, for various classes of enzymes, there are some special nomenclature rules. For example oxidoreductase can often be named as donor: receptor oxidoreductase, and transferase as donor: receptor transferase.

It is worth noting that although those enzymes with the same catalysis function from different species, or different tissues or different organelles of the same species catalyze the same biochemical reactions, their primary structures are not exactly the same and sometimes the reaction mechanism may be have difference. For example, according to different metal ions contained in enzymes, superoxide dismutase (SOD) can be divided into three categories: CuZn-SOD, Mn-SOD and Fe-SOD. They have not only different primary structures, but also different physical and chemical properties. Even if belonging to the same category of CuZn-SOD, the primary structures of enzymes from red blood cells of a cow and from those of a pig are different. However, neither common nomenclature nor systematic nomenclature distinguishes these features, and these enzymes are named as the same name and so are called isoenzyme. Therefore, when an enzyme is discussed, usually both its source and name should be described.

6.1.3.3 Composition and structural characteristics of enzymes

Although a few RNA molecules have been demonstrated to have catalytic activity, most of enzymes are protein. Thus, enzymes have the structural characteristics of proteins inevitably, and have spatial structure forms at four levels. The primary structure refers to a covalent skeleton of polypeptide chain with the certain amino acid sequence. The secondary structure contains a variety of fine structure forms, helix, sheets, turns and coils formed similar amino acid residues in the primary structure under the action of hydrogen bonds. The tertiary structure is formed by further coiling molecules on the basis of the secondary structure, and includes including specific three-dimensional arrangement. The quaternary structure refers to specific three-dimensional arrangement of the folded polypeptide chain. It should be indicated that the disulfide bond formed through dehydogen of sulfydryl groups from two cysteine residues has important effect on structure of enzyme protein. For example, human epiderm growth factor with six cysteine residues formed three disulfide bonds. Insulin has one disulfide bond in the chainA, and two disulfide bonds are formed between the chainA and the chainB. Those enzymes with activity are all globulin, i.e. the folded to form polypeptide chain with tight structure, whose hydrophilic groups of amino acids locate on the surface and hydrophobic groups in the center. Enzymes protein have three composition forms. ①Monomeric enzyme, it is composed of the polypeptide chain with only one active site, its molecular weight is 13-35kDa, and the majority is hydrolase. ②Oligomeric enzyme, it is composed of several same or different subunits which are bound together, the single subunit has no activity, and only if binding each other the subunit has activity. its molecular weight is over 35kDa. ③Multienzyme complex, it refers to the system in which various enzymes catalyze consecutive reactions. The product of a previous reaction is the substrate of a next reaction. Only the minority of such enzymes is composed of a single protein, and most of the enzymes are complex protein or called holoenzyme which is composed of the protein part (enzyme protein) and the non-protein part. Such enzyme protein itself has no activity, and can activate in the presence of cofactors. Cofactors may be inorganic ions and also organic compounds, and all belong to small molecules. Some enzymes only need one of them, and some need both. About 25% of enzymes contains tightly-bound metal ion or catalytic process need metal ions, including iron, copper, zinc and magnesium, calcium, potassium, sodium and so on, and they play a part in maintaining the activity of enzyme and finish catalysis process of the enzyme. Organic cofactors are divided into coenzyme and prothetic group, according to extents of their binding to enzyme protein. Binding of the former to enzyme protein is loose, and that of the latter is tight. But sometimes both of them are referred to as coenzyme. Most coenzymes are nucleotides and vitamins or their derivatives (Table 6.1). They are often essential components of food for organisms and their supply shortage can cause some diseases for human body. Among the above six classes of enzymes, except hydrolase and ligase, other enzymes all need specific coenzymes in biological reactions.

Table 6.1 Some common coenzymes and their precursors and deficiency diseases

Coenzymes	Precursors	Deficiency diseases
Coenzyme A	Pantothenic acid (VB_3)	Dermatitis
FAD, FMN	Riboflavin (VB_2)	Growth retardation
NAD^+, $DADP^+$	Nicotinic	Pellagra
Thiamin pyrophosphate	Thiamine	Beriberi
Four hydrogen folic acid	Folic acid	Anemia
Deoxyadenosine	Cobalamin (VB_{12})	Pernicious anaemia
Pyridoxal phosphate	Pyridoxine (VB_6)	Dermatitis
Proline hydroxyl substrate	Ascorbic acid (VC)	Scurvy

Enzyme catalysis reaction has the following features. ①Enzyme catalysis reactions have strong specificity. Chemical catalysts are poorly specific to reactants, and have no strict requirements for substrates, however, biocatalysts, enzymes, have strict requirements for substrates. One kind of enzymes usually catalyzes one kind of substances or one substance to produce chemical reactions or result in particular chemical bond change to produce special products. ②Enzyme catalysis reactions have high catalytic efficiency. Activation energy of enzymatic reactions is very low, and the catalytic efficiency usually is 10^6-10^{13} times that of chemical catalyst. ③Enzyme catalysis reaction needs mild reaction conditions. The reaction conditions of chemical catalysts are usually high temperature, high pressure, strong acid or alkali, but those of enzymatic reactions are usual normal temperature, atmospheric pressure and neutral pH values. ④Enzyme activity regulatory mechanism is complex. Various regulation forms of enzyme activity exist in organisms, at first, enzyme quantity and degradation process are regulated and controlled at different levels. The regulation forms include proenzyme activation, hormone action, and covalent modification or isomerization. In a multienzyme reaction system the regulation mechanism is more complicated. Hence, the industrialization process should be controlled according to detailed reaction situation so as to gain the optimum conversion efficiency.

6.1.4 Enzyme Activity and Activity Units

The determination of enzyme activity is an essential job in enzymology research, enzyme preparation production and application. In enzyme preparation production, fermentation process control, evaluation of extraction and purification methods, and storage and application of enzymes are based on determination of enzyme activity. In production of alcohol and distilled spirit, quality and amount of koji are determined through determination of enzyme activity in koji. In production of beer, quality of malt is judged by determination of enzyme activity of malt. In other product fermentation processes, determination of enzyme activity is all involved. This indicates that determination of enzyme activity has great significance for instruction of production practice.

Because enzymes are difficult to purify and very unstable, quality or volume cannot be used to represent when it is necessary to quantitative description of biocatalysts quantity, usually, enzyme activity is adopted to represent the existing amount, based on the catalytic specificity of enzymes. The so-called enzyme activity refers to the ability for one enzyme to catalyze a chemical reaction. The strength of enzyme activity is expressed as the unit number of enzymatic activity per unit enzyme preparation. For liquid enzyme preparations, enzyme activity is expressed as the activity unit number per milliliter enzyme liquid (U/mL); and for solid enzyme preparations, as the activity unit number per gram enzyme (U/g). Under certain conditions, enzyme activity expresses as the reaction speed. The higher the speed of enzymatic reaction is, the higher enzyme activity is. So, through measurement of the enzymatic reaction speed, we can know about the strength of enzyme activity.

Enzyme activity unit (U) is one basic measurement unit formulated artificially to describe enzymes quantitatively. It means the enzyme amount which is needed to complete the designated reaction amount (the reaction amount may be expressed as the decreasing amount of one substrate or the increasing amount of one product) in the certain reaction conditions (the optimal enzymatic reaction conditions), within the unit time (1min or 1h). Under the prescribed conditions, the prescribed reaction amount completed within the unit time represents that the actual enzyme amount of the enzyme preparation participating in the reaction is one unit; and thus, if 10 such prescribed reaction amounts are completed, there are 10 units of enzyme amount in the enzyme preparation.

In order to eliminate the confusion phenomenon of enzyme activity unit, in 1961, the Commission on Enzymes of International Union of Biochemistry formulated the consistent enzyme unit, in the optimal reaction conditions (the optimal substrate, the optimal pH, the ion strength of the optimal buffer and 25℃). The enzyme amount needed that catalyze 1.0mmol substrate to convert to products is one international enzyme activity unit (IU). Although the international unit can be used as a unified standard for comparison of enzyme activity, this unit often appears too complicated in practical application. So, in general, we adopt the respective prescribed units for different enzymes. For example, in Chinese standard QB546-80 the activity unit of a-amylase is provided as the enzyme amount needed to decompose 1g soluble starch per hour, and there is another provision that the enzyme amount needed to decompose 1mL 2% soluble starch solution into colourless dextrin per hour that should be one enzyme unit. The latter obviously is small than the former. Saccharifying enzyme activity unit refers to is one enzyme unit of the enzyme amount needed which converts soluble starch to 1mg reducing sugar per hour under the designated conditions. For protein enzymes, it is provided, under prescribed conditions, the enzyme amount needed to decompose the substrate casein to produce 1mg tyrosine is one enzyme unit. Because one kind of enzyme often has various determination methods and different enzyme units may be adopted in practice, when applying any enzyme preparation the unit number cannot be considered only. It should be also noted how the adopted enzyme unit is defined and in what conditions the reaction is performed and what method is used to determine the enzyme activity.

In addition, there is a concept of specific enzyme activity that refers to the enzyme activity unit number per unit enzyme preparation. Here "enzyme preparation" can be understood as all kinds of enzyme products in a broad sense such as animal and plant tissue homogenate as enzyme source, microbial materials, enzyme extraction liquor or preparations made through purification.

In isolation and purification of enzymes, we need to do follow-up determination of specific activity so as to evaluate purification methods in each step. With purification treatment, impurity are removed, specific activity of the enzyme will increase gradually. The specific activity which does not increase any more when continuous purification is done is called constant specific activity. It indicates that enzyme preparation is most pure that. The specific activity at that time is thought to be the activity unit number per mg enzyme protein.

6.1.5 Source of Enzymes and Production Methods

Enzyme production refers to the procedure for obtaining the needed enzyme through preliminary design and artificial manipulation and control. The production methods of enzymes can be divided into extraction method, fermentation and chemical synthesis. Among them, the extraction method is used earlier and is being used even today. Fermentation has been the main method for enzyme production since the 1950s. And the chemical synthesis process is still at the laboratory stage.

Enzymes as biocatalysts are widespread in animals, plants and microbe cells. In earlier times, production of enzymes was accomplished directly through extraction and purification of tissues of organisms such as animals and plants as the main source. Extraction method is simple and convenient, but tissues or cells containing enzymes must be obtained. Thus, this method is affected by climate and geographical environment, or extraction from cells may be performed after culturing microbial cells. But this makes the process become miscellaneous, and product contains more impurities and purification will become more difficult. But animal and plant materials has long period, limited source, influenced by geography, climate and seasons, and at the same time they are limited by technology, economy and ethics and so on. So, with increasing expansion of the application scope of enzyme preparation, those enzymes relying solely on plant and animal sources can't meet the demand, and so many traditional enzyme sources cannot adapt to the demand for enzyme application in today's world. Therefore, after the 50s of the 20th century, fermentation method for enzyme preparation was adopted. However, extraction method of enzyme from plant and animal tissue cells still has the applicable value in the area with rich animal and plant resources. For example, trypsin, pancreatic amylase and pancreatic lipase or the mixture of these enzymes, pancreatin, is extracted from the pancreas of animals; alkaline phosphatase from the animal small intestine; papain from papaya; bromelin from pineapple bark; pectinase from the waste *Aspergillus* cells in citric acid fermented broth; and kallikrein from the submandibular gland in animal.

In theory, like other proteins, enzymes can be produced through chemical synthesis. In the middle

Chapter 6 Enzymes and Enzyme Engineering

60s of the 20th century, the new technique of chemical synthesis appeared. In 1964, Chinese scientists synthesized complete molecule of cattle insulin with bioactivity with chemical synthesis, based on the amino acids sequence, this opened up a new era of chemical synthesis of proteins. In 1969, the United States scientist Gutte and Merrifield also obtained RNA enzyme (ribonuclease) containing 124 amino acids for the first time with the chemical methods, and developed a set of automation techniques for solid phase synthesis of polypeptide, accelerating the synthesis speed. Now peptide synthesizer can be used to accomplish chemical synthesis of enzymes. However, chemical synthesis of enzymes requires many kinds of amino acids with high purity as a single substrate, the synthesis cost is very high, and the enzymes having clear chemical structure can be synthesized with this method. Because of more steps, the chemical synthesis reaction is only applicable to production of short peptides. Due to limitation of reagents, equipments and cost, artificial synthesis of macromolecular enzyme proteins containing more amino acid residues with chemical synthesis method is still very difficult, from the economic and technological viewpoint.

All kinds of microbes are widespread existence in the nature, and these microorganisms possess the secretion ability of some enzymes. At present, enzymes in industry are produced generally through liquid submerged fermentation or solid fermentation with microbes. So, fermentation method has become the main method of enzyme production since the 1950s. It Utilizes life activities of microbial cells, required enzymes can be obtained in the fermentation. According to the different culture methods of microorganisms, fermentation can be divided into solid fermentation, liquid submerged fermentation, immobilized cell fermentation and immobilized protoplast fermentation, etc. Since the 1980s, in addition to the microorganism fermentation method, enzymes have been produced through animal and plant cell fermentation. Currently, most of commodity enzyme preparations are produced with microorganism fermentation method, and most of enzyme-producing microorganisms are obtained through isolation and screening from the nature. Soil, surface water, deep sea, hot springs, volcanoes, forest in the nature are the main sources of enzyme-producing microorganisms. The procedure for screening enzyme-producing microorganisms includes sample collection, separation and initial strain screening, determination of enzyme-producing performance and further screening. Enzyme production with microorganism has prominent advantages as follows. ①There are many different kinds of microorganisms, almost all of enzymes can get from microorganisms fermentation. ②Due to fast microbial propagation, the enzyme production cycle is short, microorganism culture is simple, and enzyme yield can increase by controlling culture conditions. ③Microbes have strong adaptability and strain capacity, and new high-yield strains may be bred out through adaptation, induction, mutation and DNA recombinant technology. Enzymes produced by microbial cells can be divided into two types that include structural enzyme and inducible enzyme. The former will express under the circumstance of need in the cell growth process, and the latter will express only after adding corresponding inducers to fermentation media. The inducers are generally substrates or products of enzyme-catalytic reaction. Generally, the amount of enzyme which the cells express can be controlled by various factors, synthesized enzyme amount is limited. According to the principle of economics of cells, synthetic enzyme quantity meets mainly the needs for cell growth and metabolism, when enzymes as the final target product, wild type microbes cannot meet the needs for a large scale enzyme production. Therefore, in industrial production of enzyme preparations, all microorganisms are high-yield enzyme-producing strains genetically modified. Routine physical or chemical mutagenesis breeding methods can be used for breeding of high-yield strains and make important contributions to establishment and development of the enzyme preparation industry.

In recent years, with development of gene recombinant technology and research progress of microbial genomics, genetic engineering method has been adopted more and more to construct high-yield enzyme-producing strains in the areas of academy and industry, and this technique has been used in large scale industrial production. Some more efficient new methods such as DNA shuffling and genome shuffling also have started to be used for breeding of high-yield strains.

6.2 Fermentation Production of Enzymes

All organisms can produce a certain amount of enzymes in certain conditions. The process that enzymes are produced in the body is called biosynthesis of enzymes. The fermentation production of enzymes is the procedure for producing the needed enzymes by utilizing life activities of cells (microbial cells, animal cells and plant cells) through artificial manipulation and control, based on preliminary design.

6.2.1 Basic Theory of Enzyme Biosynthesis

Enzymes have catalytic activity, but it has been also already found that in addition to "classic enzyme", some other biological molecules also have catalytic activity. For instance, riboenzyme, that is to say, RNA itself is a biocatalyst. So, the synthesis of enzymes mainly means biosynthesis of RNA and protein.

The genetic information carrier in some organism cells, DNA molecules contains the gene corresponding to certain enzyme, this gene will be expressed by kind of cells to synthesize the enzyme protein. DNA can transcript into the corresponding RNA and then peptide chains is produced through translation to form enzyme molecules with complete spatial structure through further process. Under the action of RNA polymerase (transcriptase), RNA can be produced with DNA as templates and nucleotide triphosphate as the substrate.

Translation is the process that the base sequence of RNA molecule is converted to the amino acid sequence of peptide chains, this process is different in different organisms. Peptide chains are synthesized in the ribosome through action of various tRNAs, enzymes and accessory factors with mRNA as the template, amino acid as the substrate.

Enzyme biosynthesis is influenced by many factors and is also regulated in various forms, just like as protein synthesis. Regulation at the transcriptional level is crucial for enzyme biosynthesis. According to the theory of gene regulation, four kinds of genes in the DNA molecules are related with enzymatic synthesis, among which structural genes and enzymes have their own corresponding relationship. The genetic information of structural genes can transcribe into the genetic code on mRNA which translate into peptide chains of enzyme protein. Therefore, synthesis of enzymes is regulated by genes, and there are three kinds of regulation modes. ①Catabolite repression. It refers to the phenomenon that rapid utilizing carbon sources and nitrogen source which repress biosynthesis of some enzymes (mainly inducible enzyme) in fermentation. For example, glucose represses biosynthesis of β-galactosidase; and fructose, biosynthesis of α-amylase. ②Induction of inducers. This is a procedure that biosynthesis of enzymes starts or accelerates by adding some one substance, and it is also called induction of enzyme biosynthesis. The substance playing the action of induction is called inducer. For example, lactose induces biosynthesis of β-galactosidase, and starch induces biosynthesis of α-amylase. ③Feedback repression. This is also called product feedback repression. It refers to the phenomenon that enzyme biosynthesis can be repression by the product of enzyme catalysis or the end product of one metabolic pathway. The substance which causes feedback repression is called corepressor. For instance, as the end product of histidine biosynthetic pathway, excessive accumulation of histidine in turn gives feedback repression of 10 kinds of enzymes in its synthetic pathway.

Enzyme biosynthesis undergoes double regulation of genes and metabolism. Microbe enzyme biosynthesis and activity regulation mechanism is shown in Fig.6.1.

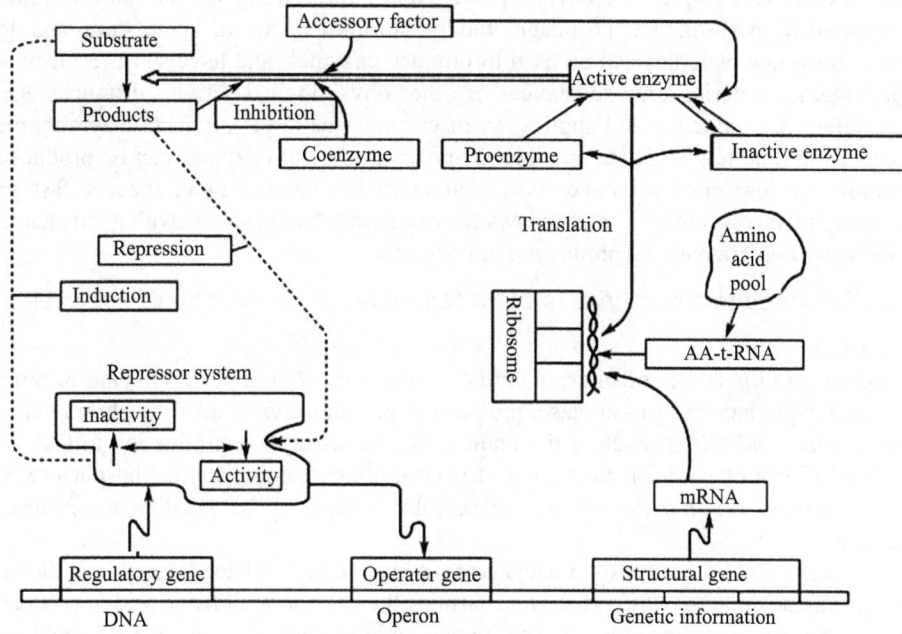

Fig 6.1　Regulation of microbe enzyme biosynthesis

Just like protein synthesis, synthesis of enzymes is controlled by genes, which determine chemical structure of enzyme molecules. But from the view of enzymes, only some genes can't guarantee production of a large number of enzymes. Synthesis of enzymes also is regulated by metabolites (substrate and product or analogue of enzyme reaction). When inducers exist, the enzyme yield can increase by several times and even hundreds of times. Instead, the product of some enzymatic reactions, especially the end product can play repression effect to decrease the number of synthesized enzyme. According to the operon theory, the operon in gene is made up of the operator gene and several adjacent structural genes. Structural genes can transcribe genetic information to synthesize corresponding mRNA, which then can translate into specific enzymes. The operator gene can control activity of structural genes. There is a kind of regulatory genes in cells which can produce repression protein which can bind to repressor to result in isomerization effect and increase in its affinity with the operator gene, so that RNA polymerase can't get to the position of the structural gene. So, DNA can't be transcribed, mRNA can't be synthesized. Therefore, synthesis of enzymes is repressed. Inducer can also bind to the repressor protein to change its structure and reduce the affinity with the operator genes, making the operator gene become free again. Then the structural genes start to transcribe into mRNA, and then translate into specific enzymes.

6.2.2 Enzyme Fermentation Production

One of the premises for enzyme fermentation is breeding the enzyme-producing cells with excellent characters. A good enzyme-producing strain should meet the following requirements: ①fast propagation, high enzyme yield, properties of enzymes conforming to the usage requirements, and preferably exoenzymes which are easy to purify; ②no variation or degradation of strains, stable enzyme-producing performance, no phage contamination or invasion; ③the strain should be easy to culture, cheap raw materials can be used to produce enzymes, and have short fermentation cycle; ④nonpathogenic bacteria, toxic substances or other physiological active substances cannot form to ensure safety of production and application of enzyme; the target product, enzyme protein, other byproducts should be few as far as possible. At present, most of enzymes can be produced through fermentation, because microbes have many characteristics, such as more species, fast grow, easy culture, strong metabolic ability and so on. Many enzyme-producing strains with good characters have been widely used in fermentation production of enzymes.

6.2.2.1 Common microorganisms for in fermentation production of enzymes

1) Bacteria

Bacillius subtilis is one of microorganisms most widely applied to enzyme production, and can be used for production of α-amylase, protease, β-glucan enzyme, alkaline phosphatase, etc. For example, *Bacillus subtilis* BF7658 is the main strain for α-amylase production in China. *Bacillus subtilis* As 1.398 can be used for production of neutral protease and alkaline phosphatase. α-amylase and protease produced by *Bacillus subtilis* are exocellular enzymes, but alkaline phosphatase exists in the cytoplasm.

E.coli can produce a variety of endoenzymes, which can be isolated through cell disruption. For example, glutamate decarboxylase is used to determine the content of glutamic acid or produce γ-amino butyric acid. Aspartase catalyze fumaric acid to add amino group to form L-aspartic acid. Ampicillin acylase is used to produce new semi-synthesized penicillin or cephalosporin. β-galactosidase is used to decompose lactose. Restriction endonuclease, DNA polymerases, DNA ligase, exonuclease etc play an important role in genetic engineering.

2) Yeast

Yeast is widely used to produce invertase, pyruvate decarboxylase, alcohol dehydrogenase, etc. *Candida* yeast is used to produce lipase, uricase, invertase, alcohol dehydrogenase. In addition, it has the enzyme system for decomposing alkane to ferment petroleum, and can be used for oil fermentation; and 17α hydroxylase produced by it with strong activity can be used for conversion of steroid to produce testosterone.

3) Mold

Aspergillus niger is a mould and belong to the group of *Aspergillus niger* of *Aspergillus*. It can be used to produce various enzymes, which may be extracellular enzymes and also intracellular enzymes, for example, saccharifying enzyme, α-amylase, acid protease, pectinase, glucose oxidase, catalase, RNA enzyme, lipase, cellulose enzyme, hesperidinase, naringinase, etc. *Aspergillus oryzae* can be used for production of saccharifying enzyme and protease, which is widely applied in traditional production distiller's Qu and soy sauce Qu in China. In addition, *Aspergillus oryzae* is also used to produce amino acylase, phosphodiesterase, pectinase, etc.

Penicillium chrysogenum among the *Penicillium* is used for production of glucose oxidase, phenoxymethyl penicillin acylase (mainly acting on penicillin Ⅴ), pectinase, cellulase Cx, etc.

Penicillium citrinum is used for production of 5′-phosphodiesterase, lipase, glucose oxidase, chymotrypsin, nuclease S_1, nuclease P_1, etc.

Trichoderma can produce the cellulose containing C_1 enzyme, C_x enzyme and cellobiase, etc. In addition, *Trichoderma* contains 17α hydroxylase with strong activity, which is often used for steroid conversion.

Rhizopus is used for production of saccharifying enzyme, α-amylase, invertase, acidic protease, lipase, pectinase, cellulase, hemicellulase, etc. *Rhizopus* can produce 11α hydroxylase with strong activity, and so is the important strain for steroidal conversion.

Mucor is used for production of protease, saccharifying enzyme, α-amylase, lipase, pectinase, chimosin, etc.

4) *Actinomyces*

Streptomyces is one of the most important strains among *Actinomyces*. It is often used in the production of glucose isomerase, penicillin acylase, alkali protease, neutral protease, chitinase, etc.

In addition, *Streptomyces* used in production 16α hydroxylase which can be used for steroid conversion.

6.2.2.2 Process of Fermentation Production of Enzymes

When the excellent enzyme-producing strains are obtained, it will be the key of production how to realize microorganism mass culture and enzyme production through fermentation? Production of enzyme with fermentation method is a very complicated process. Because detailed production strains and target enzymes vary, seed preparation and fermentation methods and conditions are different. The main factors that affect enzyme production include culture medium design, selection of fermentation ways and fermentation condition control. Since protein synthesis needs to consume a lot of ATP, aerobic microbes are adopted and fermentation is done in a ventilation agitation tank.. In addition to nutrition conditions, environment conditions such as dissolved oxygen concentration, temperature, pH values and so on also have important influence on microbial growth and enzyme production, and all these need to adjust and control. In addition, protein is very easy to inactivate in the conditions of high shear force, therefore shear force of the fermentation system should be controlled properly. Protein is a kind of natural surfactant, and under the bubbling conditions a large number of proteins accumulated in the fermentation are easy to form foam, which affects the normal operation of fermentation tanks. So defoaming device should be designed in design of fermentation tank, and defoaming agents should be added in time in the process of fermentation. In operation of the fermentation, flow adding supplementary method is often adopted to increase the enzyme yield.

Because enzymes are special proteins, synthesis of a large number of proteins needs rich nutrients and energy. Like culture media of other fermentation products, the media for production of microbial enzymes contain carbon source, nitrogen source, inorganic salt and growth factors. In the production process of enzyme preparations, enzyme-producing promoters often are added in culture media, i.e., enzyme yield can be increased by adding some substance, such as tween-80, phytin, cleaning agent LS, polyvinyl alcohol, ethylenediamine tetraacetate (EDTA), etc., but its mechanism has not been clarified. For inducible enzymes, enzyme yield can be increased remarkably by adding proper inducer in the medium, general substrates of enzymes or substrate analog. For example, resveratrol or benzyl alcohol as the inducer must be added in the medium in production of lignin peroxidase with white-rot fungus. At the same time, many enzymes can be used as industrial catalyst, and the selling price is not high, thus, those cheap, abundant agricultural side products which meet the requirements for cell growth and enzyme synthesis should be chosen as raw fermentation materials, common carbon source includes starch, dextrin, molasse, sucrose, glucose, etc.; and nitrogen source fish powder, bean cake powder, peanut meal and urea. Inorganic ions such as calcium and a small amount of growth factors such as vitamins, amino acids, purine base, pyrimidine base, etc. are used.

Production of enzymes with microbial fermentation has two kinds of main ways, solid fermentation and liquid submerged fermentation. Solid fermentation is also called surface culture or bran koji culture, which utilizes wheat bran, rice bran, etc., as basic raw material, to perform microorganism culture after adding suitable amount of inorganic salt and water to prepare the medium. Common culturing modes of solid fermentation include shallow dish culture, drum culture and multiple ventilation thick-layer culture. The characteristics of solid fermentation are simple equipment, easy popularization, and favour for mold culture and enzyme production. Its disadvantages are difficulty in control over fermentation conditions, incomplete material availability, big labour intensity, easy bacteria pollution, etc. This method is not suitable for production of intracellular enzymes. Liquid submerged fermentation technique is also called liquid submerged culture, and is a kind of stirring ventilation culture mode that makes use of liquid media to conduct agitating ventilation culture in a fermentation tank. The fermentation process needs certain equipment and technical conditions, and power consumption is also higher than liquid fermentation, but availability of both the raw materials and the enzyme yield are higher, and culture conditions are easy to control. At present, enzymes are produced mainly with the liquid submerged fermentation technique in industry. But distiller's yeast (containing a great amount of amylase and saccharifying enzyme) culture, production of food grade enzyme and feed additives still adopt solid fermentation technique.

6.2.3 Enzyme Biosynthesis Patterns

In certain culture conditions, the growth process of enzyme-producing cells in batch culture also includes lag phase, exponential phase, stationary phase and death phase. Through analysis the relationship between enzyme production and cell growth, enzyme biosynthesis patterns can be divided into the following three types.

6.2.3.1 Synchronous synthesis pattern

Synchronous synthesis is also called growth coupling pattern, i.e. synthesis of enzyme is synchronous with cell growth. A great amount of enzyme is produced when cells enter into the exponential phase, and synthesis of enzyme will stop when cells grow into the stationary phase. For this type of enzymes, their biosynthesis can be induced, but is not repressed by catabolism metabolites and reaction products. And after removing the inducer, or when cells enter into the stationary phase, synthesis of enzyme will stop immediately, which shows that the mRNA corresponding with this type of enzymes is not stable. For example, tannase which is induced by tannin in *Aspergillus oryzae* belongs to synchronous synthesis type.

Some enzymes start to synthesize after cells grow for a period of time, and enzyme synthesis stops after cells enter into the stationary phase. This is a special type of growth coupling model, which is also called mid-term synthesis type. This type of enzymes which its synthesis has phenomenon of feedback repression, and their corresponding mRNA is not stable. For example, biosynthesis of alkaline phosphase in *Bacillus subtilis* is repressed, and phosphorus is necessary for cell growth, and hence, phosphorus must exist in media. When cells grow for a certain time, inorganic phosphorus in the medium is almost exhausted (less than 0. 01mol/mL), repression is removed and enzymes are synthesized in large quantity. Because the corresponding mRNA for alkaline phosphatase is unstable, its life span is only 30min or so. So, when cell growth enters into the degenerating phase, synthesis of enzymes will be also stopped.

6.2.3.2 Continuous synthesis pattern

Synthesis of enzymes begins with cells growth, but when cells enter into the stationary phase, enzyme synthesis can continue for a longer time. Cell growth has partial correlation with synthesis of enzymes. This type of enzymes can be induced, but not repressed catabolites and products. The corresponding mRNAs of these enzymes are relatively stable, and enzymes can be synthesized for quite a long time after the stationary phase. The enzyme synthesis pattern also can be changed based on different inducers. As for production of β-galactosidase with *Aspergillus*, β-galacturonic acid or pure pectin should be used as the inducer, and synthesis of this enzyme belongs to continuous synthesis pattern. If crude pectin (containing glucose) is used as the inducer, synthesis of the enzyme will delay. If glucose content is high, enzyme synthesis will begin after cells enter into the stationary phase when glucose is exhausted in medium. Under this condition, the synthesis of enzyme pattern will become delayed synthesis type.

6.2.3.3 Delayed synthesis pattern

After cells enter into the stationary phase, enzymes begin to synthesize and accumulate. This type of enzymes is not synthesized in the exponential phase, this may be due to catabolite repression effect, when repression is relieved, enzymes begin to synthesize. Besides, their corresponding mRNA molecules have high stability, so the cells can continue to utilize accumulated mRNA to translate to enzymes after they stop growth. Many hydrolases belong to this pattern in which synthesis of enzymes is not related with cell growth. For example, acidic protease is produced by *Aspergillus Niger*, after the cells enter into the stationary phase, the enzyme will begin to synthesize and accumulate more.

From the patterns of enzyme synthesis (Fig.6.2), we can know that the main factors of enzyme biosynthesis are stability of mRNA and repressor in the medium which influence patterns of biosynthesis. If mRNA has high stability, enzyme can continue to synthesize after cells stop growth; and if mRNA has poor stability, enzyme synthesis will terminate with cells' stopping growth. When some substances in the medium cannot influence enzyme synthesis, it will begin with cell growth. On the contrary, enzyme synthesis will begin after the repressor is removed, when cells grow for a period of time or cell growth enters into the stationary phase. Although microbial growth has certain relation with enzyme production, strain variation or medium change may alter the period of enzyme synthesis. *Bacillus* spp. have stronger ability to form extracellular enzymes than other non-spore-formation microorganisms, production of extracellular protease has close relationship with formation of spores. Generally, the mutants that can't form spores cannot synthesize a great amount of alkaline protease, and those which loose the ability to synthesize protease can't form spores. Amylase production has no direct relationship with formation of spores, and in some strains amylase activity is the highest when their growth mass reaches the maximum. Some strains (such as *Bacillus subtilis* and *Bacillus stearothermophilus*), have the highest amylase activity in the exponential phase. The strains which are most sensitive to repression or cannot produce a great amount amylase when sugar is not exhausted and before cells enter into the stationary phase. Amylase is produced more in the stationary phase when using crude raw materials in industry, and the enzyme activity increases with bacterial autolysis. *Bacillus subtilis* BF-7658 has the highest amylase activity in the degenerating phase.

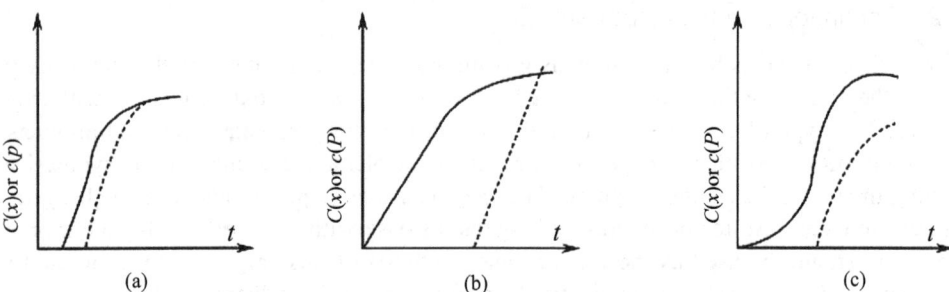

Fig.6.2 Some kinds of patterns of enzyme biosynthesis
(a) The formation of enzymes are related to cells growth; (b) The formation of enzymes are partly related to cells growth;
(c) The formation of enzymes are unrelated to cells growth

6.2.4 Growth Dynamics of Enzyme-Producing Cells

In a certain medium and under certain culture conditions, the growth rate of enzyme-producing cells is influenced by all kinds of intracellular and extracellular factors, and its change is very complex. However, cell growth has its own law, we can control the production conditions, and maintain the cell growth rate within a certain range according to the need, in order to achieve the ideal effect after mastering the law.

Cell growth dynamics is mainly engaged in the law of cell growth rate and effect of environmental conditions. For decades, many scholars have studied these aspects. Among these studies, in 1950, French Monod first proposed the dynamics equation to express microbial growth.

In the process of cell culture, the cell growth rate is proportional to cell concentration:

$$r_x = \frac{dX}{dt} \mu X \tag{6-1}$$

In the above equation (6-1), r_x is cell growth rate, X is cell concentration, and μ is specific growth rate.

When the medium only has a kind of restrictive matrix, μ is the function of concentration of the restrictive substrate, and this is the Monod growth dynamic model:

$$\mu = \frac{dX}{dt} \cdot \frac{1}{X} = \frac{\mu_m \cdot S}{K_s + S} \tag{6-2}$$

In this equation (6-2), S is concentration of restrictive substrate, μ_m is the maximum specific growth rate, and refers to the specific growth rate when concentration of restrictive substrate is excessive, that is, when $S \gg K_s$, $\mu_m = \mu$, K_s is the Monod constant, and refers to the concentration of restrictive substrate when the specific growth rate reaches half of the maximum specific growth rate, i.e. when, $\mu = \frac{1}{2}\mu_m$, $S = K_s$.

The Monod equation is the basic equation of cell growth dynamics, and has important application value in respect of optimization of fermentation process and process control. In addition, many scholars put forward many different dynamics models, and modified the Monod model in different situations or with different methods, which will not be described in details here.

For the fermentation process in a continuous stirred tank reactor, while the culture solution is added constantly, the same volume of fermentation solution is discharged. In the steady-state, growth dynamics of continuous fermentation of free cells can be expressed as the following equation:

$$\frac{dX}{dt} = \frac{\mu_m \cdot S \cdot X}{K_s + s} - DX = (\mu - D) \cdot X \qquad (6\text{-}3)$$

In this equation (6-3), D is dilution ratio and refers to the ratio of the volume of added culture liquid to that of the fermented liquid per unit time, and its unit is generally h^{-1}. When the specific growth rate (μ) is greater than the dilute rate (D), $\frac{dE}{dt}$ is positive, which indicates that the cell concentration increases. When μ is equal to D, cell concentration will be constant, when μ is smaller than D, $\frac{dE}{dt}$ is negative, which indicates that cell concentration decreases, S rises correspondingly again, so as to build a new steady-state by making μ equal to D. But when D is equal to μ_m, X will tend to zero. So, in the continuous fermentation process of free cells, to make the cell concentration keep constant, the corresponding dilution rate must be controlled well so as to make it equal to the cell growth rate.

The Monod equation is similar to the Michaelis-Menten equation of enzyme reaction kinetics. The maximum specific growth rate μ_m and the Monod constant K_s are determined with double reciprocal mapping method.

6.2.5 Dynamics of Enzyme Biosynthesis

Dynamics of enzyme biosynthesis is mainly engaged in the synthesis rate of enzymes in cells and effect of various factors on it. Research on dynamics of enzyme biosynthesis may be based on the whole fermentation system to study on the synthesis rate of enzymes in cell population, which is called macroscopic enzyme synthesis dynamics. While in the research the synthesis rate of enzymes in cells based on cell individuals is called microscopic enzyme synthesis dynamics or structural dynamics. Macroscopic enzyme synthesis dynamics is associated with cell growth rate, cell concentration and enzyme synthesis patterns.

Generally, the dynamics equation of enzyme synthesis can be expressed as follow:

$$\frac{dE}{dt} = (\alpha\mu + \beta) \cdot X \qquad (6\text{-}4)$$

In this equation: X—Cell concentration (g cell / L)
  μ—Specific cell growth rate (h^{-1})
  α—specific growth-coupled enzyme production coefficient (IU/g cell)
  β—specific non-growth-coupled enzyme production rate ($h^{-1} \frac{IU}{g \cdot cell}$)
  E—enzyme concentration (U / L)
  t—time (h).

Because of different enzyme synthesis patterns in cells, the relation between enzyme synthesis rate and cell growth dynamics also varies.

For synchronous enzyme synthesis pattern, enzyme synthesis rate is coupled with cell growth rate. For non-growth-couple enzyme synthesis, special enzyme production rate b is equal to zero, and the dynamics equation of enzyme synthesis is:

$$\frac{dE}{dt} = \alpha\mu X \qquad (6\text{-}5)$$

Those enzymes of middle-phase synthesis pattern are special growth-coupled type, and their dynamics equation of enzyme synthesis is the same as synchronous enzyme synthesis pattern. Nevertheless, when some repressor exists in the medium, α is equal to zero, which indicates no enzyme is produced, and when the repressor is removed, enzyme synthesis will start.

Delayed enzyme synthesis pattern belongs to the non-growth-couple model. The special enzyme production rate α with growth-coupled model is equal to zero, and the dynamics equation of enzyme synthesis is as follows:

$$\frac{dE}{dt} = \beta X \tag{6-6}$$

For delayed enzyme synthesis pattern, enzymes can be produced in the stationary phase of cell growth, and they are partial growth-coupled model. The dynamic equation of enzyme production is as follows:

$$\frac{dE}{dt} = \alpha \mu X + \beta X \tag{6-7}$$

The related model parameters are μ_m, K_s, α and β, after a model is validated and confirmed, the relevant parameters are usually calculated with linearization treatment and trial and error method.

6.2.6 Measures for Increase of Enzyme Yield

Enzyme fermentation production mainly aims at high enzyme yield. In addition to breeding excellent enzyme-producing stains and ensuring the appropriate fermentation conditions, many measures can be taken, adding inducers, controlling concentrations of repressors, and adding surfactants or other enzyme-producing promoters, all can promote production of enzymes in cells so as to obtain the maximum enzyme yield.

6.2.6.1 Adding inducers

Adding appropriate enzyme-producing inducers to the medium, enzyme yield of inducible enzymes can significantly improve. However, different enzymes have different inducers. For example, lactose can induce synthesis of β-galactosidase, cellobiose can induce synthesis of cellulase, sucrose glycerin monopalmitate can induce synthesis of invertase, etc. But sometimes a kind of inducers can induce production of many enzymes of the same enzyme system. For instance, β-galactosidase, permease, β-galactose acetylase can be induced by β-galactose at the same time. One enzyme often has different inducers. In practical application, according to characteristics of enzyme, inducing effect and inducer sources, an appropriate inducer is selected.

Generally inducers can be the substrate, the substrate analogue or the product in enzyme reactions. Among these three kinds of inducers, enzyme substrate analogues are the most effective inducer, also known as gratuitous inducer which can induce cells to synthesize a particular enzyme, but not really substrate without combining with the enzymes, therefore, gratuitous inducers have no metabolic change, its induction effect is very obvious. For example, the inducing effect of isopropyl-β-D-thiogalactoside (IPTG) for β-galactosidase is higher hundreds of times than that of lactose, and the inducing effect of sucrose glycerin monopalmitate for invertase is higher a few times than sucrose.

6.2.6.2 Control of Repressor Concentration

Biosynthesis of some enzymes can be repressed by repressors, in order to increase enzyme yield, we must try to eliminate the repression. The repression can be divided into product repression and catabolite repression. The repressors may be products of enzyme-catalyzed reaction, end products in one metabolic pathway and glucose catabolites. Control of repressor concentration is one effective measure to eliminate repression and increase enzyme yield.

For example, glucose can repress synthesis of β-galactosidase, when glucose exists in the medium, even though the inducer also exists, β-galactosidase cannot be synthesized more. Only in the medium without glucose, or when glucose is all used up by cells, the inducer can induce production of a large quantity of the enzyme. The similar cases may occur in production of many enzymes. In order to reduce or eliminate glucose catabolites repression, the concentration of glucose and other substrate, which are rapidly-utilized carbon source, should be controlled in the medium. The other carbon source hard to utilize (such as starch, etc.), feeding method or batch carbon source adding method may also be adopted to raise enzyme yield. In addition, when the catabolite exists in the medium, adding a certain amount of cyclic adenosine monophosphate (cAMP) can remove the repression effect, if the fermentation medium contains some inducer at the same time, enzyme can be synthesized rapidly.

If some enzyme is repressed by the end product of one metabolic pathway, the end product concentration may be controlled to remove the repression, and the end product analogue may also be added to eliminate the feedback repression.

6.2.6.3 Adding Surfactants and Promoters

Non-ionic surfactants, such as Tween, Triton and so on, can increase yield of enzyme, because of these surfactants can deposit in the cell membrane to increase cell permeability, and promote the secretion of enzyme. For example, enzyme yield can increase 1-20 times by adding 1% Tween in the medium for mold cellulase fermentation broth. We should pay attention to amount of surfactant adding, in addition, adding surfactant is beneficial for increase of stability and catalytic ability of some enzymes.

Adding enzyme-producing promoters has significant effect for increasing enzyme yield, and this mechanism has not yet been clarified. For example, phytate can increase the yield of Mould protease and phosphodiesterase in *Penicillium citrinum* by 20 times; and polyvinyl alcohol and sodium acetate have effect to raise the yield of cellulose. Enzyme-producing promoters have different effects on different enzymes in different cells. We must determine the optimal concentration of a promoter through experiments to ensure the optimal effect.

6.3 Production of Enzymes with Immobilized Cell Fermentation

6.3.1 Preparation and Characteristics of Immobilized Cells

6.3.1.1 Preparation of immobilized cells

Immobilized cells are developed on the basis of immobilized enzyme technique in which enzymes are enclosed in a certain space to catalyze bioreactions, the immobilized enzyme can be recovered and reused after reaction. Compared with free enzymes immobilized enzymes have the following advantages: ①most easy separation from substrates and products, no enzyme residue in the product solution, and simplified extraction process; ②longer enzyme reaction duration (batch fermentation and continuous fermentation), and easy accomplishment of continuous and automatic fermentation; ③higher stability of the enzymes in most cases; ④strict control of the enzymatic reaction process; ⑤increase of the utilization rate of enzymes is high, production cost can be reduced; ⑥increase of the yield and the quality of the products.

Immobilized cells refer to the living cells that are fixed in a certain space to conduct life activities and can be used repeatedly. In 1978, Japanese scientists produced a-amylase with immobilized cells successfully. Because of barrier effect of cell wall, some substances can't be secreted to outside of cells, scientists imagine, if the cell wall is removed, it is possible to make more intracellular substances secreted into the environments extracellular, this is the purpose of research on immobilized protoplast conducted by scientists. In 1986, alkaline phosphatase and glucose oxidase was produced with immobilized protoplast fermentation by Chinese scientists and made success, which opens a new way for further development of enzyme engineering.

The preparation methods of immobilized cells are the same as those of immobilized enzymes generally, which include embedding method, adsorption method and selective thermal denaturing method. The preparation methods of immobilized enzymes are shown in Fig.6.3. The principle of embedding method is entrapment of microbial cells or enzyme in a network space of water insoluble porous gel polymer, through aggregation function or ion network formation, or through precipitation or change of solvents, temperature and pH values. Gel polymer network can prevent leakage of cells, and can let matrix infiltrate and make the product spread out. Embedding method can be divided into grid type and microcapsule type. The former refers to that entramption of cells or enzymes in a subtle grid of polymer gel, and the latter refers to a semipermeable polymer membrane. Adsorption method is also called carrier binding method, with which microbial cells or enzymes can be immobilized on the water-insoluble carrier through physical adsorption, chemical or ions binding. Its operation method is simple, and it has little effect on activity of microorganisms. But the quantity of bound microorganisms is limited, and it has poor reactivity and repeated usability. Adsorption method is divided into physical adsorption, covalent binding, ion binding and biological specific adsorption. Selective thermal denaturing method is dedicated to cell immobilization, and it is a procedure that cells are treated at appropriate temperature to make the cell membrane protein denature but not to make cells inactivate due to protein denaturing.

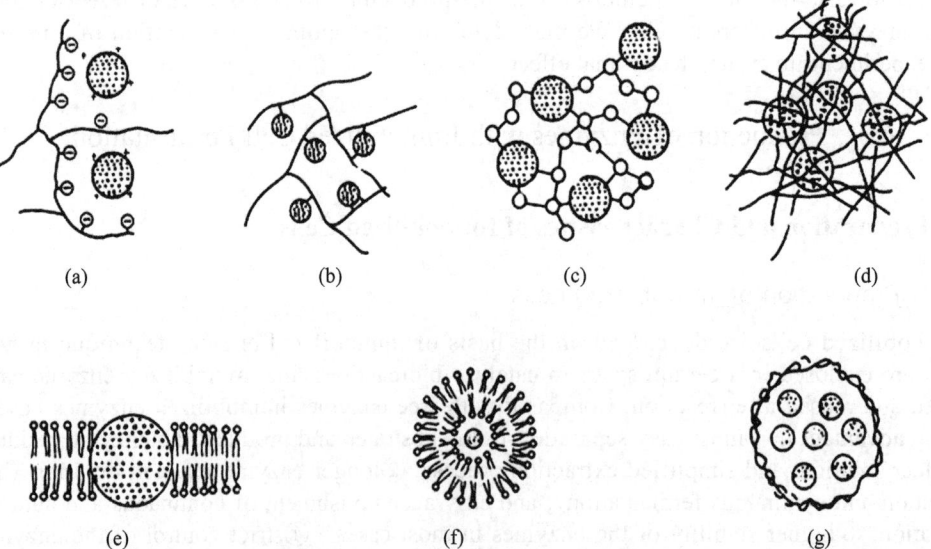

Fig.6.3 Methods for enzyme and cell immobilization
(a) Ion absorption method; (b) Covalent coupling method; (c) Cross-linking method; (d) Polymer-embedded method;
(e) Hydrophobic interaction method; (f) Liposome-embedded method; (g) Microcapsules-embedded method

Cross-linking method refers to immobilization method of intermolecular cross-linking in which double-function reagents combine with enzyme molecules. Because of the functional group of enzymes participate in the reaction, the structure of the activity center structure of enzymes may be influenced, so as to make the enzyme inactivate significantly. In addition, cross-linking agents, such as glutaraldehyde are expensive, which limits its application.

6.3.1.2 Characteristics of Immobilized Cells

Combination of gene engineering with enzyme engineering will possibly solve the problem of stability of recombinant microorganism in the large scale culture. Recently, some reports on application of immobilized gene engineering strain to industrial production have shown, compared with the traditional culturing methods, this technique exhibits certain advantages. We discuss characteristics of immobilized gene engineering strain.

1) Increases the yield of target products

Immobilized *E.coli* BZ18 (pTG201) produces the target product 20 times higher than free cells without selective pressure. In the immobilized system, cells grow fast until reaching the stationary phase, and compared with the relatively free system, the number of living cells with immobilized system can reach more than ten times than with it.

2) Improves expression level of cloned genes

Immobilization methods have significant effect on the cells cultured for several passages in the expression level of cloned genes. In a Hollow fiber membrane bioreactor running continuously, high yield of β-lactamase can be obtained and maintain for more than 3 weeks. Compared with a suspension system, high yield of β-lactamase can be obtained selectively in the immobilized system, and when the immobilized reactor runs till the third and one hundredth day, the yield figures of this enzyme are 100 times and 1000 times those with the suspension system, respectively. In addition, human interferon can be produced steadily for a month by immobilizing China hamster cells on the microcarrier.

3) Increases genetic stability of plasmids

Genetic stability of plasmids is the most important factors in engineered cell fermentation, because plasmids are the carrier of target genes. In comparison of genetic stability of plasmids between in free cells and in immobilized cells in the basic or LB medium, it is found that genetic stability of plasmids in the two media is higher in immobilized cells.

6.3.2 Characteristics of Enzyme Production with Immobilized Cell

Based on a lot of experiment results, immobilized cell fermentation production of exoenzymes has the following characteristics:

6.3.2.1 Increase of the enzyme production rate

After immobilization, cells grow and propagate within certain space and cell density increases. Hence, biochemical reactions can be accelerated, so as to increase the enzyme production rate. For example, immobilized *Bacillus subtilis* can produce α-amylase through batch fermentation, and the volume enzyme production rate (also called enzyme production strength) reaches 122% of that of free cells. For continuous fermentation, the enzyme production rate gets higher.

6.3.2.2 Continuous fermentation at high dilution rate

Immobilized cells in a carrier are not easy to fall off and lose, and can be used repeatedly several times. They can be used in both fed-batch fermentation and continuous fermentation at high dilution rate. For example, immobilized cells for anaerobic fermentation of alcohol and lactic acid can be used continuously for half a year or longer. Immobilized cell for fermentation production of enzymes can be used continuously and stably for more than 30 days.

6.3.2.3 Fermentation stability is high

Due to protection of the carrier, immobilized cells have a wider range of adaption to pH values and temperature, and can steadily produce enzymes through fermentation. This characteristic is beneficial to operation control of fermentation, and helpful for automation production.

6.3.2.4 Fermentation period shortens and equipment utilization rate increases

Immobilized cells may maintain the good enzyme-producing performance for a long time growing to a certain degree, to shorten the fermentation cycle. For example, immobilized *Aspergillus Niger* in semi-continuous fermentation produce glucoamylase, the cycle of the first batch fermentation is 120h, which is the same as free cell fermentation. But after the second batch fermentation, the fermentation cycle reduces to about 60h. If continuous fermentation is adopted, immobilized *Aspergillus Niger* can produce the enzyme steadily at high dilution rate. This can raise the utilization rate of fermentation equipment much more.

6.3.2.5 Products are easy to isolate and purify

Immobilized cells are easy to isolate from the fermented broth, because if there are less free cells, it is helpful for isolation and purification of products from the fermented broth, so as to improve the quality of products.

6.3.3 Technical Conditions for Enzyme Production with Immobilized Cell

The basic technical conditions for enzyme production with immobilized cells are the same as those with free cells. Here only several problems about control over process conditions which should be stressed during enzyme production through immobilized cell fermentation are discussed

6.3.3.1 Pre-culture of immobilized cells

After the cells were immobilized on carrier, in order to make cells grow and propagate in the carrier, pre-culture is an indispensable process. During pre-culturing the growing medium special for cell growth and the process conditions should be adopted, because if cells are growing well, they can be used for fermentation. The fermentation medium for enzyme production and the optimal fermentation process conditions should be used to replace the original medium and conditions. Sometimes, the same culture medium and technical conditions can be also used in pre-culture and fermentation.

6.3.3.2 Control of composition of culture media

Compared with culture media for free cell fermentation, the culture media for immobilized cell fermentation does not have much difference generally. But the structure of some immobilization carriers may be affected by ingredients of media, so we pay attention to this problem in preparation of culture media. For example, prepared immobilized cells with calcium alginate, excessive phosphate may damage their structure, so phosphate concentration must be limited in the medium, and at the same time, certain concentration of calcium ions must be added to the culture medium so as to keep stability of the immobilized cells. In addition, in order to promote transmission and dissolution of oxygen, culture medium concentration should not be too high, especially medium viscosity should be low as far as possible.

6.3.3.3 Supply of dissolved oxygen

During culture and fermentation immobilized cells may be influenced by carriers. So transfer and dissolution of oxygen will be hindered to some extent, especially for the embedding immobilized cells. Oxygen needs to diffuse into inside of gel particles through the gel layer so as to supply for the cells, and so oxygen supply becomes the main restrictive factor. Therefore, it is necessary to increase the amount of dissolved oxygen so that the need for cell growth and enzyme production can be met. Since the reactor for immobilized cells cannot adopt operation of fierce agitation so to avoid damage of cells, increase of ventilation will be the main method for increasing dissolved oxygen. Adding some material in gel is beneficial to oxygen transmission. Co-immobilization of hydrogen peroxide enzyme and cells together is adopted, and then appropriate amount of hydrogen peroxide is added to the medium to produce oxygen through action of the enzyme for cells. The concentration of culture media may be reduced, so as to reduce the viscosity of culture media as far as possible. Supply of dissolved oxygen is a key restrictive factor in aerobic fermentation of immobilized cells.

6.3.3.4 Control of temperature

Immobilized cells adapt to the temperature in a wider range, so the temperature is not hard to control in batch fermentation and fed-batch fermentation. But due to higher dilution rate in the continuous fermentation process, the temperature within the reactor changes greatly. If temperature is adjusted only in the reactor, when the temperature has great difference between the added culturing liquid, it is difficult to meet the requirements for enzyme production. Therefore, generally the culturing liquid must be adjusted to the appropriate temperature before entering the reactor.

In addition, in order to meet the requirements for enzyme production with immobilized cell fermentation technique, we must further study bioreactors for immobilized cells, immobilization carriers and immobilization technique.

6.4 Application and Molecular Modification of Enzymes

6.4.1 Application of Enzymes

Thousands of enzymes in the nature have been found, but only dozens of then have been used in industry wildly, and 80% of the industrial enzymes are hydrolase, which are mainly used for degradation of polymers in the nature, such as starch, protein, fat and other substances. Thus, protease, amylase and lipase are the present three main enzyme preparations in industrial application. Protease can be used in detergent, dairy products industry, leather industry, etc., amylase is used in baking, brewing, starch saccharification and textile industry and, lipase is used in detergents, food and fine industry. In recent 30 years, application of new enzymes has led to emerging of completely new type products, even some unexpected products to replace chemical products of petroleum. Enzymes have been widely used in all walks of life, such as enzymatic diagnosis, enzymatic analysis, enzyme catalysis, enzyme detection, etc.

β-mannanase can be widely used in food, pharmaceutical, paper making, feed, oil exploitation, and fine chemical industry, furthermore, this enzyme has great application potential in conversion of biomass enriching in the nature to oligomannose with important commercial value, as a novel enzyme preparation, development of β-mannanase is an important topic in the field of biotechnology, it has great significance for high added value development, modification of traditional chemical industry through development of biotechnology, and meeting demands in the fields of resources, environment, energy and pharmaceuticals. Now, with new methods, the basophilic strain producing alkaline β-mannanase has been screened and isolated, scientists have completed the systematic project involving in interdiscipline of strain identification and systematic development analysis, determination of enzyme-producing conditions, enzyme purification and property analysis, cloning and expression, sequence analysis of the enzyme gene, small-scale trial and medium-scale trial of enzyme production, component, structure and physiological function analysis of oligomannose, and industrial production tests of the enzyme and oligomannose. In conjunction with characteristics of extreme bacteria and extreme enzymes, with some new methods scientists have solved the difficult problems of alkaline, salt and thermal stability of the enzyme, enzyme activity, oligosaccharide production rate, etc. in this field, and have realized production and application of β-mannanase and oligomannose for the first time.In our country researchers have used alkaline β-mannanase to convert konjac powder, sesbania gum, etc., and take the lead in realizing industrial production of oligomannose in the international area, which opens up a new field of industrial application based on plant hemicellulose resources. Instead of chemicals, alkaline β-mannanase is used for gel in the fracturing process in an oil field. This new process has low cost and good effect, and has exhibited an important role in oil recovery and protection of earth layer environment.

Application of enzymes may refer to the sections of application of contemporary bioengineering technologies in light chemical industry, environment, food, material, and pharmaceutical industry.

6.4.2 Modification of Enzyme Molecules

In application, as macromolecular active substance, enzymes often exhibit unstable, especially under extreme condition of high temperature, strong acid, strong alkaline and high permeability, which thus limits industrial application of enzymes. Modification of enzymes with chemical method can expand the application scope of enzymes.

Enzymes are polymer compounds with complete chemical structure and spatial structure which determines their properties and functions. As long as the structure of enzymes has some fine changes, i.e. enzyme molecules are modified some of their properties and functions will be changed correspondingly. The procedure for changing the structure of enzyme molecules with various methods in order to change some of their properties and functions is called modification of enzyme molecules.

Trough modifying, application scope and application value of enzyme can significantly expand. For example, molecular modification can enhance enzyme activity, increase stability of enzymes, eliminate or reduce enzymes antigenicity. Therefore, molecular modification of enzymes becomes an important field in enzyme engineering. Especially in recent 20 years, with rise and development of protein engineering, modification of enzyme molecules has combined with genetic engineering technology. Through site-directed mutant technology we can store the information of modified enzyme molecules in DNA, by means of gene cloning and expression, the enzymes with new features and functions can be obtained continuously through biosynthesis, so as to make modification of enzyme molecules show broader prospects.

Modification of enzyme molecules includes amino acid substitution modification, zymoprotein side chain group modification, and limited hydrolysis of peptide chain modification, macromolecular binding modification and metal ion modification.

Enzyme protein is formed by combining various amino acids through the peptide bonds. Each enzyme has its spatial structure corresponding with activity and function. Different sequences of various amino acids are the basis for chemical structure and spatial structure of enzymes. If an amino acid is replaced with another amino acid in the peptide chain, some spatial conformation of the enzyme protein may be changed, which leads to changes in some properties and functions of the enzyme, and such modification is called amino acid substitution method which can also be used to modify other functional proteins and peptides. In recent years, rapid development of protein engineering has provided effective modification methods for enzyme.

Side chain groups of enzyme protein can be modified with all sorts of small molecules substance, which is called zymoprotein side chain group modification. The main functional side chain groups of zymoprotein include amino, carboxyl, sulfhydryl, imidazolyl, indole, phenolic hydroxyl, hydroxyl, guanidyl, methyl sulfenyl, etc. These groups play an important role in formation of the stable spatial structure of enzymes, and constitute various auxiliary bonds in which changes can result in changes in the spatial structure of enzyme molecules, finally causing changes of the properties and functions of enzymes. Various modifying agents may modify amino acids far away from the active site, may also modify amino acids located in the active site to lead to covalent change, causing change of enzyme activity due to change of protein conformation that disturbed the exquisite structure of the active site. According to different properties of reactions between chemical modifying agents and enzyme molecules, the modified reactions are mainly divided into acylation, alkylation, oxidation and reduction, and aromatic ring substitution.

Macromolecule binding modification refers to a method that change properties and functions of enzymes utilizing combine of all water-soluble macromolecule substances with enzyme molecules to cause subtle change of the spatial structure of enzymes. Usually, water-soluble macromolecular modifying agents may be dextran, PEG, heparin, sucrose polymers, etc. When using these molecules, they need to activate at first, and then bind to enzyme molecules through covalent bonds, in order to modify enzyme molecules. Macromolecule binding modification is the most widely used method for modification of enzyme molecules. The enzymes which are modified with this method can significantly improve the enzyme activity, increase stability or reduce the antigenicity of the enzymes.

For some enzymes containing metal ions, in order to change properties and functions of enzyme, we should change the metal ion type and quantity by metal ion replacement modification method which is only applicable for those enzymes containing metal ions in the original structure. In these enzyme molecules, metal ions are often part of the active center of enzymes. If metal ions are removed in enzyme, it will often become inactive, and if the original metal ions are back again to the enzyme molecules, the enzyme activity can recover. Adding different metal ions, the enzymes take on different properties. In the process of metal ion replacement modification, a certain amount of EDTA should be added at first in the enzyme solution to make metal ions in enzyme molecules and EDTA form chelates, and at this time the enzyme becomes inactive, then, EDTA-metal chelates are removed from the enzyme solution with dialysis or ultrafiltration, molecular sieve chromatography method. Finally, different metal ions are added to the enzyme solution to make the metal ions bind to the enzyme protein. Because of different kinds of ions by ion replacement, modified enzyme will appear to have different characteristics. As long as the suitable metal ions are selected to replace the original ions, it is possible to improve the enzyme activity, and to increase stability of the enzyme. For example, if Zn^{2+} is removed from zinc type protease, and Ca^{2+} is used to replace for Zn^{2+} to form calcium type protease, the enzyme activity will increase by 20-30 times. Most of α-amylase molecules contain Ca^{2+}, and others contain Mg^{2+}, Zn^{2+} or other ions, so the common α-amylase is mixed ionic type. If other ions in α-amylase molecules are replaced by calcium ion, activity and stability of the α- amylase can be improved. So adding a certain amount of calcium ion is beneficial to improvement of stability and activity of α-amylase in storage and application process of α-amylase.

6.5 Research Progress in Enzyme Engineering

With development of contemporary biotechnology, people have more and more understanding about enzymes, research content of enzyme engineering has been expanding constantly and new types enzyme emerge constantly. At the same time, enzyme application has been expanded. With application of a large number of enzymes, all kinds of enzyme reactors arise at the historic moment, which further promotes development of enzyme engineering.

One of the development trends of contemporary enzyme engineering is to look for extreme enzymes, such as high temperature-resistant enzyme, acid and alkaline-resistant enzyme, salt- resistant enzyme, and so on. These enzymes may exist in bacteria in extreme conditions of survival. In recent years, with research progress of these bacteria, new enzymes have emerged continuously in the enzyme industry.

6.5.1 Artificially Synthesized Enzyme and Simulated Enzyme

Besides enzyme production, in recent years, a new hot topic has arisen in enzyme engineering, it is artificially synthesized new enzyme and simulated enzyme, because people have found that production of enzymes only with microbial fermentation method cannot yet meet the growing demand, and it is necessary to develop new ways and methods for new type of enzymes.

Artificially synthesized enzymes are catalytic substance with functions similar to natural enzymes obtained through chemical synthesis. It may be protein, and also may be quite simple macromolecular substances. Artificially synthesized enzyme requires that people make clear some problems, such as, how such enzymes play a catalytic role, where the key part is, and what the characteristics of the key part are. In addition, synthesis method of artificial enzyme should be simple and economical. In 1977, Dhar reported that the artificially synthesized sequence, polypeptide Glu-Phe-Ala-Glu-Glu-Ala-Ser-Phe had the activity of lysozyme, and its activity was 50% of the natural enzymes. In 1990, Stewart constructed a polypeptide, which consisted of 73 amino acid residues, and whose active site is composed of histidine, aspartate and serine, using chymotrypsin substrate tyrosine ethyl ester as a template and using computer to simulate of the active site chymotrypsin. This peptide had the activity for alkyl ester substrate which was 1% of the natural chymotrypsin, and exhibited the substrate specificity and chymotrypsin inhibitor sensitivity. Another macromolecule which consisted of 34 amino acid residues was synthesized, and it had the same catalysis as RNA enzymes. However, people thought that was too complicated, and continued to look for more simple and more stable, smaller, more economical artificial enzymes. Although benefits of artificial enzymes are not obvious, the research team engaged in the artificial enzyme is growing. Perhaps, in the near future, the artificial enzyme will get a place officially and a rising status in the production field of enzyme engineering, even overwhelming natural enzymes.

The so-called simulated enzyme refers to simple non-protein molecules synthesized with organic chemical method. It can simulate the processes of complexation and catalysis of enzymes on substrates, in order to achieve high efficiency of enzyme catalysis and to overcome the problem of instability of enzymes. Enzyme simulation work can be divided into three levels: ①synthesizing simple complex with similar enzyme activity; ②simulating the active center of enzymes;

③simulating the whole enzyme, i.e. chemical simulation of the whole active site of enzymes including micro-environment. At present, the simulation enzyme work mainly concentrates on the second level. For example, some active groups are introduced into some natural or synthetic compounds, so as to make them have the enzymatic behavior. Currently this kind of enzyme model molecules used to build simulated enzyme includes cyclodextrin, crown ether, cryptand, cage ether, porphyrin, large cyclophanes, etc. Cyclodextrin has been used to simulate successfully chymotrypsin, ribonuclease, aminotransferase, carbonic anhydrase, etc. In 1985, Ben-Der and others made a simulated enzyme of chymotrypsin, utilizing the cavity of β-cyclodextrin as the binding site of the substrate, using carboxyl and imidazolyl linked on the side chain and one hydroxyl of cyclodextrin to constitute the catalytic center.

Artificial enzyme or simulated enzyme generally has some features, high catalytic efficiency, high adaptability, high selectivity and high stability. It is simpler than natural enzyme in structure, and usually has two special loci, one is the substrate binding locus, the other the catalytic locus. In contrast, constructing the substrate binding site is relatively easy, and constructing the catalytic site is more difficult. In practice, the two sites usually stand alone. At the same time, studies show that if artificially synthesized enzyme has a binding site of reaction transition state, the site has double functions of the catalytic site and the binding site. Therefore, in constructing simulated enzyme, macromolecular polymer or macromolecular polymer complexing metal can be used as matrix generally, and the corresponding hydrophobic group is added at the appropriate site, to form a cavity, which can accommodate the substrate and is suitable for binding to the substrate, and of which at the suitable site the catalytic group can be introduced. Since the simulated enzymes do not contain amino acids, their thermal stability and pH stability are better than natural enzymes.

The most simply simulated enzyme is obtained undoubtedly by utilizing the existing enzyme or protein as matrix, and introducing the corresponding catalytic group on the matrix. However, in some sense, this kind of simulated enzyme can be regarded as enzyme modification. According to the active structure of enzymes, some simple small peptides can be synthesized to be mimic enzyme. However, more mimic enzymes utilize synthetic polymer as matrix, and at present the existing artificial enzymes are mainly prepared with molecular blotting method and cyclodextrin.

6.5.2 Ribozyme

In the early 1980s, Cech and Altman independently found that RNA had biocatalytic function, which altered people's traditional concept that enzyme was special protein with catalytic function. Because of discovery of nucleic acid enzyme (ribozyme) with catalytic function, Cech and Altman won the 1989 Nobel Prize for Chemistry. Ribozyme is a kind of multifunctional biocatalyst, and it can act on RNA and DNA, as well as polysaccharide, amino acid ester and other substrates, the reactions catalyzed by ribozyme also has strong substrate specificity and the reaction rate complying with the Michaelis-Menten equation. Ribozyme exists in many organisms, such as protozoa, fungus cell mitochondria, algae chloroplast and phage.

Ribozyme has two models, namely "hammerhead model" and "hairpin model", and needs specific secondary structure to shows its catalytic activity. Compared with the common enzyme, ribozyme has disadvantages of difficulty in introduction to the human body and low stability, due to low cleavage efficiency. Therefore, although it is necessary to conduct further research, ribozyme still has promising potential.

6.5.3 Abzymes

Abzyme, as the result of crafty combination of high selectivity of antibody with efficient catalytic ability of enzyme, appeared in the late 1980s. In essence, abzyme is a class of immunoglobulin with both catalytic activity and enzyme properties in the variable area, and so also known as catalytic antibody. It may be obtained according to the common preparation procedure of monoclonal antibody by using the antigen (hapten) designed in advance. The preparation methods of abzyme include induction method, introduction method, duplication method, etc. So far, in addition to ester hydrolysis, carboxylic acid hydrolysis and amide bond hydrolysis, abzyme catalyzes more than 10 kinds of reactions such as amide formation, light-induced cleavage and polymerization, ester exchange, etc. The specificity of these abzyme-catalyzed reactions is equal to, even higher than that of enzyme-catalyzed reactions, and the catalytic velocity sometimes can reach the level of enzyme catalysis.

As well as known, antibody and enzyme are both protein molecules, and the binding both between antibody and antigen, and between enzyme and substrate has high specificity. However, both of them have a fundamental difference. Enzyme binds to the high-energy transition state molecules, while antibody to antigen (the ground state molecules) combination. Antibody and enzyme are high polymer material, and play their individual different missions in the long evolution process. Although their structure is different, but they have two major points in common. Both are proteins, and both bind to target molecules highly selectively. Antibody binds to antigen specifically to help macrophage ingest and destroy antigen, and enzyme binds specifically to the substance with specific structure selectively in chemical reactions, so as to greatly reduce the activation energy of the reactions, and catalyzes the reactions selectively and efficiently to make them complete under mild conditions.

Antibodies are a kind of protein that animals resist on invasion of foreign matters. Abzyme refers to the antibodies which are prepared with a series of chemical and biological methods with catalytic activity. Enlightened by Pauling's transition state theory and prediction, in 1969, Jencks put forward an opinion that if antibody could bind to the transition state product of chemical reactions it had catalytic performance inevitably. This means that once antibody binds to the transition state product it has the property to catalyze chemical reaction specifically and selectively of enzyme, such as high efficiency and specificity under mild conditions. By analogy, if antibody can bind to the analogues of the transition state product, it will also bind to the transition state product in chemical reaction process, and such antibody has catalytic performance of enzyme. Undoubtedly, endowing antibody with enzyme activity has great significance, and this work has big difficulty. At the first glance, with Pauling's theory and Jenck's theory, antibody with catalytic properties can be obtained easily, but whether the theoretical prediction comes true depends on technological breakthrough. Success of monoclonal antibody technique promotes the birth of catalytic antibody. In 1986, Lerner and Schultz successfully got the catalytic antibody.

Gene engineering and protein engineering can also be used in preparation of catalytic antibodies. First of all, the genes expressing catalytic antibody can be synthesized with artificial method; secondly, the encoding genes are transferred to the bacteria or yeast expression system; finally, the expressed products are screened and purified to prepare catalytic antibodies. Of course, based on chemical modification, enzyme catalytic group can also be implanted to antibody, transformed the antibody into catalytic antibodies. So far, common antibody may also be modified into catalytic antibody by introducing the catalytic group to one common antibody with chemical modification method. So far, hundreds of catalytic antibodies have come into being. Obtained abzymes have successfully catalyzed the reactions which all the six classes of enzymes can catalyze which are studied frequently ester hydrolysis, amide hydrolysis, cyclization, amide formation, decarboxylation, triphenyl hydrolysis, peroxidation, olefins isomerization, oxidation-reduction reaction, etc. Since the birthday in 1986, this new substance with properties of both enzyme and antibody has always been concerned in the scientific world. The research of catalytic antibody is one of interdisciplinary intersections at the scientific frontier, and attracts synthetic chemists, biologists and immunologist. In only over ten years, the research scope has been expanded, the research results have been emerging one by one, and the technology adopted in the research has become advanced increasingly. Abzyme integrates biology, immunology and chemistry, and breaks through the frame of traditional research on simulated enzymes of macromolecules and complex compounds by adopting monoclonal and polyclonal technique, gene engineering, protein engineering and other high technologies. It has opened a new field for biological catalyst research and preparation to make the research level of simulated enzyme have a qualitative leap, and indicate application prospect in the catalytic chemistry, reaction kinetics, medicine, pharmaceutics and many other fields.

6.5.4 Extreme Enzymes

As macromolecular bioactive substance, enzymes often appear unstable in the application process, especially more easily inactivate under the extreme conditions of high temperature, strong acid, strong alkaline and high permeability, so application of enzyme preparation is restricted in the industry and other fields to some extent. But in the long-term production practices, people gradually realize that some microbes can survive in the extreme ecological environment, and are known as extremophiles. According to different resistant environmental conditions, extreme microorganisms can be divided into thermophilic, psychrophilic, halophilic, acidophilic, basophilic, barophilic microorganisms. Many research results show that it is because extremophilic microorganisms contain a large number of enzymes adapting to extreme conditions that makes them survive in the unusual ecological environment condition. Usually, these enzymes which can play biocatalysis under extreme conditions are called the extreme enzyme. Corresponding with classification of extremophiles, extreme enzymes can be divided into thermophilic enzyme, psychrophilic enzyme, halophilic enzyme, acidophilic enzyme, basophilic enzyme, barophilic enzyme and so on. Since the first extreme enzyme, thermophilic *Taq*DNA polymerase was successfully used in genetic engineering and subsequent in the PCR technology, people have begun to constantly explore structure, properties and application prospect of various extreme enzymes.

Extreme enzymes coming from extremophiles can play function in extremely strict conditions, which will greatly expand application scope of such enzymes, and lay a new basis for biotechnology process with high efficiency and low cost. Application of extreme enzymes can change the whole situation of biocatalysts. With diverse regional environments and organism resource advantage in China, systematic investigation, protection, development and sustainable utilization of extreme microorganisms, from resource exploration, mechanism research to application development, will be one precious opportunity for realization of leap development of biotechnologies in China.

6.6 Protein Engineering

Protein engineering is also called the second generation gene engineering. This concept first was put forward by Ulrner the United States Gene Corporation in 1981. The purpose of protein engineering is to produce proteins with higher bioactivity or unique properties, by means of gene engineering on the basis of rational design of structure of protein molecules. With rapid development of molecular biology, crystallology and computer technology, protein engineering has got considerable progress in recent ten years, becomes the important approach for research on protein structure and function, and has been widely used in pharmaceutical production. Protein engineering mainly includes two aspects, analysis of protein structure and prediction of the proteins structure. Analysis of protein molecule structure includes two methods. One is crystal structure analysis, i.e. three-dimensional structure of protein is determined with monocrystal diffraction method in order to study relation between its structure and function; and the other is research of protein conformation, i.e. the spatial structure of protein or peptide molecules is studies with nuclear magnetic resonance (NMR). Therefore, the research target is to obtain structure information of protein molecules as much as possible to find correlation of structural characteristics of protein molecules with their particular functions, which is most important for design and prediction of structure of the protein molecules, and subsequent construction and expression of new protein molecules by means of genetic engineering. Structural prediction of protein refers to prediction of senior structure of protein molecules according to the sequence of amino acid. It aims at predicting the spatial structure model by utilizing known primary structure, i.e. amino acid sequence, so as to research on structure and function of protein molecules and accomplish design of protein molecules.

Gene engineering is a key technology to achieving the aim of protein engineering. This step includes all aspects of gene engineering, from gene separation, cloning and expression, mutation, to analysis of properties of engineered protein. Therefore, protein engineering is often known as the second generation gene engineering.

Purification, and functional and structural analysis of protein are assessment of quality of protein engineering implementation, also are to provide more direct background materials for further structural prediction and molecular design. On the basis of purification of engineered protein, its properties can be analyzed through various physical, chemical and biological approaches. These techniques include protein structure analysis (such as crystal structure, solution structure) and protein function analysis (such as enzymology method, immunology method, protein-protein interaction method, protein-nucleic acid interaction methods). Analysis of structure and function of engineered protein can provide some information for reference for design of protein molecules.

With further understanding about relationship between protein structure and function, protein engineering enters into realm of freedom of human knowledge inevitably.

6.6.1 Design and Modification of Protein Molecules

Molecular design of protein is to provide the design solution for protein engineering modification. Molecular design of protein has the main basis from three aspects.①Evolutional relationship among protein molecules. From microscopic differences between homologous protein sequences, we can find effect of the spatial structure on biological function of protein, in addition, research of evolution of protein also provides information for the structure law of protein. ②Relationship between the primary structure and the spatial structure of protein. From this relationship, senior structure of protein can be predicted. ③Relationship between protein structure and function. From this relationship we can find influence of structural change on function. At present, protein engineering mainly focuses on the field of modification of the existing protein.

Protein molecules can undergo "intensive modification". It refers to intentional modification of several amino acid residuals in protein molecules in order to research and improve properties and functions. Site-directed mutation is the most common method in protein engineering. Generally there are two approaches for protein modification, introducing disulfide bond into protein molecules and changing amino acid sequence of protein molecules. After such modification, we can attain the goal to expand the thermostability range of protein, increase specificity, oxidation resistance and bioactivity, and alter the pH value of protein. For enzyme protein, protein modification can change the K_m and V_{max} of enzymatic reaction, increase the catalytic efficiency of reactions and affinity of enzyme on substrate, and enhance specificity of enzyme.

Protein molecules can conduct "moderate modification". It refers to substitution for a peptide fragment or a certain structural domain of protein molecules. The three-dimensional structure of protein may be considered to be assembled with structural components, which can be exchanged between different protein molecules so as to transfer the corresponding function. In this aspect deletion mutation technique and structural domain exchange technique are often used in protein engineering.

Protein molecules can conduct "extensive modification", which is just the so-called *de novo* design of protein molecules. Based on the sequence of amino acids, the new protein that does not exist in nature is designed and manufactured, to make it have particular spatial structure and expected function. It must be pointed out that the example of *de novo* design of protein is not yet much, and it only relates to a few small peptides, whose structure consists of α-helix or β-folding. Only when people master the law that the primary structure decides the senior structure, and understand relationship between the senior structure and biological function, *de novo* design of protein molecules can really be realized. But the embodiment that restriction endonuclease is modified through protein engineering indicates that enzyme that doesn't exist in the nature can be created by people through protein engineering.

A very successful example of protein engineering is artificially synthesized insulin. In 1953, the whole structure of bovine insulin was decoded by British biochemist Singer, which consisted of the two peptide chains composed of 17 and 51 amino acids respectively. This was the first time that human beings made clear the detail structure of an important protein. In 1958, Biochemical Research Institute, Chinese Academy of Sciences, Shanghai Research Institute of Organic Chemistry and Biology Department of Peking University jointed together to establish a cooperation research group and started to explore synthesis of insulin with chemical method. On September 7th, 1965, the cooperation research group completed complete synthesis of crystal bovine insulin, which became a stunning world event. After strict identification, the synthetic crystallized bovine insulin is the same as the natural bovine insulin in structure, bioactivity, physicochemical properties and crystal shape. Artificially synthesized crystallized bovine insulin is the first artificial synthetic protein in the world, and makes human beings progress much more on the road to discover the mystery of life. So far, insulin is one of a few proteins whose primary, secondary and tertiary structure have been clarified most. This can lay a good basis for modification of insulin through protein engineering. In normal human body, insulin is secreted from the pancreas continuously at a low level so that the blood glucose level in normal persons can be regulated strictly. But for various reasons, insulin cannot be supplied adequately, or insulin cannot play normal physiological function in a target cell, which leads to disorder of metabolism of sugar, protein and fat in the body, so as to cause diabetes. With development of diabetes, metabolic disorder in the body cannot be regulated well, which may lead to chronic complications of the eyes, the kidney and the nerve, the cardiovascular tissues and organs and eventually, to blindness, lower limb necrosis, uremia, cerebral apoplexy or myocardial infarction, or even there is threat to lives. Diabetes is a common disease. With improvement of living standards, the incidence of diabetes is increasing year by year, and prevalence of diabetes in developed countries has already been as high as 5%-10%. Generally speaking, after injection into the human body, insulin may accumulate in the subcutaneous tissue, and can reach the highest concentrations in the blood after 30min or even a longer time. Insulin in the blood may be broken up constantly, and can't maintain the relatively stable concentration. At the same time, diabetes patients need injections of insulin 2-3 times every day, which brings a great deal of pain and discomfort to patients. Therefore, obtaining insulin with high stability and long-term efficacy, even oral insulin through protein engineering, is urgent demand for diabetes patients.

6.6.2 Research Content of Proteomics

The concept, proteome, put forward in 1994 by Australia scholars, and was published in the journal Electrophoresis for first time in July, 1995. It referred to all the proteins expressed by a genome. Corresponding with genome, proteome is also a holistic concept. The function of a gene is realized mainly through its encoded protein, so protein is the real executor of life activities. Proteomics is a new discipline with proteome as the research object and is the new technology system for analysis of the whole set of proteins expressed by the genome in a cell and determination of their functions. It is deemed to be the main part of research of post-genome. Compared with genome, proteome has more complex composition, more active function, and is closer to the essence of life activities. More and more studies have proved that a gene will produce several proteins. The more complex cellular structure is, the more obvious this quantity difference is. In *E.coli*, a gene can encode average 1.3 proteins, but a human gene can encode average 10 proteins. Besides, between proteins there exists active and extensive interaction, and to a great extent we can say that "no protein plays its biological function alone". The research content of proteome mainly includes: ①protein identification; ②post-translation modification; ③the determination of protein function; ④looking for the target protein. If scientists find some one gene, it is not enough to link amino acids one by one according to the gene-determined sequence. Appropriate modification of lipid and sugar must also be ensured. At the same time, if the behaviour of some one protein is to determined, it must also be considered if dissolving in water or embedding in the cell membrane filled with lipid some proteins can exhibit normal function. Therefore, we can be closer to the peak of mastering the essence of life phenomena, so as to find the essence and the law of life activities by researching the whole of proteins, namely proteome.

6.6.3 Application of Proteomic Research in Medicine

In respect of medicine, proteomics research focuses on pathogenesis, early diagnosis and treatment of human diseases, pathogenic mechanism of pathogens, drug resistance and discovery of new antibiotics.

Mutation of multiple genes leads to inactivation of tumor suppressor genes which causes colon cancer. Researchers have found that some proteins appear in the tissue of colon cancer, which shows that appearance of these proteins is a strong indicator for detection of early colon cancer. In addition, some proteins have a wide application in early diagnosis of liver cancer, bladder cancer, dilated cardiomyopathy, and so on. In January 2002, two groups of scientists announced that the discovered mutual relationship between all the proteins in yeast and drew the relevant map. Later, another group of scientists announced that they designed a set of accurate methods for detection of early ovarian cancer with proteomic techniques.

In recent years, WHO has paid more and more attention to the effects of infectious diseases on human health. Besides *Tubercle Bacillus* and multiple drug resistance *coccus* infection, some new infection factors such as HIV virus, Ebola virus and so on have appeared, and analysis of proteomes of these pathogenic microorganisms are very important for understanding their virulent factors, antigenicity and preparation of vaccine. In addition, diagnosis, treatment and prevention of diseases are also important. Now, we have obtained all of the whole genome sequences of nearly 20 kinds of microbes and more than 60 kinds of gene sequences are under investigation. Disease fingerprinting with proteome as the main research content has become a hot spot in the field of disease diagnosis.

6.6.4 Research Progress in Protein Engineering

Emergence of protein engineering has marked that human beings' conquering the nature enters a new stage of development. Protein engineering helps us make full use of genes and proteins in nature. Through protein engineering, artificial design and modification of genes and proteins at the molecular level may be realized, so as to create the genes and proteins which do not exist in nature, in a short time and thus completing the process which needs to take hundreds of years to complete. Protein engineering not only improves the traditional one, but also opens a new research area.

Protein research mainly includes the following several aspects:①analysis of protein structure, discovery of the structural law of protein molecules, and prediction of the spatial structure based on the primary structure; ②establishment of relationship between protein molecules to find out the amino acid residuals which are involved in the interaction; ③explanation of relationship between protein structure and function in order to understand effect of the active site and the primary structure of protein on protein function.

In application of protein engineering, all of proteins and polypeptides with development prospects relate to trials of modification, in which different effects are obtained. Protein engineering of biotechnology drugs and industrial enzymes have been the most significant achievements. Protein and polypeptide drugs include hormones, cytokines, enzymes, enzyme activators or inhibitors, receptors and ligands, cell toxins, antibiotic peptides and antibodies. As a drug, they are expected to increase activity, specificity and stability, control molecule aggregation, reduce immunogenicity and side effect, prolong the half-life in the body, and enhance orientation of the target site through modification.

For example, blood coagulation process is a complex and highly coordinated process. Many diseases result from formation of embolism to clog the blood vessels, namely thrombus. If blood thrombi are formed in the coronary artery, this will cause myocardial infarction; if in the passage that takes blood to the heart, it will cause myocardial infarction; if blood thrombi are formed in the brain or in the pulmonary artery, it will lead to apnea and pulmonary embolism. Usually thrombus is decomposed by plasmin, which exists in an inactive form, i.e. plasminogen form existence in the normal case. Plasminogen is activated by serine protease which is called tPA. When blood fibrin binds to plasminogen to form the complex, its affinity with tPA is enhanced, therefore tPA can only activate plasminogen in the form of complex. Although blood coagulation is important in the early period of wound healing, clinically in most cases the blood coagulation extent should be reduced, especially for those patients who have just recovered from thrombosis and had some microscopic surgery.

Blood thrombi can be dissolved by tPA clinically and treat myocardial infarction and pulmonary embolism; but when patients take tPA, more than 50% of the drug will be cleared away by the body after five minutes. In order to solve this problem, tPA must be delivered through long venous transfusion, tPA is a glycosylated protein, some oligosaccharide bonds of many glycoproteins in the plasma can be recognized specifically by the liver receptors, in the glycosylated tPA molecules, aspartic amide at the 120th site is a recognition sites. If glutamine (Gln) is used to replace for asparagine (Asn), half-life of tPA in the blood circulation is greatly prolonged.

Leech element secreted by the leech salivary gland is a protein which is composed of 65 amino acids, and is a very good thrombin inhibitor. Scientists want to modify leech element, make it become an efficient anticoagulant. So, Asn at the 47th site is changed into lysine (Lys) or or arginine (Arg), its anticoagulation efficiency in the test tube increases by four times. In validation of its effect of thrombus formation resistance in animal models, it is found that the efficiency increases by 20 times, and is 5 times higher than that of heparin.

Growth hormones must bind to specific receptors to get into cells for playing action, and play an important role in human growth and development. Studies show that growth hormones can bind to their receptors, as well as to the receptor of prolactin in many different cells. If you want to avoid side effects during treatment, growth hormones must bind only to its receptor, but try to minimize its binding to other receptors from other cells. Through structure analysis it is found that the region of binding between growth hormones and their receptors overlaps partially with that between growth hormones and the receptor of prolactin, but the two are not identical. Therefore, the activity of binding between human growth hormones and the receptor of prolactin may be reduced selectively, but it has no influence on that between growth hormones and their receptors. This can be accomplished through protein site-directed mutagenesis. Since high affinity binding between human growth hormones and the receptor of prolactin needs zinc ion, while the binding between growth hormones and their own receptors does not need. Hence, scientists want to modify these ligands so as to make their side chain without ability to bind zinc ion, finally making human growth hormones only bind to their own receptors, but not the receptor of prolactin. This work has important meanings and makes great success, and preliminary results fully comply with the design requirements; but much more detection work should be done so as to make the modified human growth hormones change into commercialized drug.

Protein engineering has a broad application prospect, is an irreplaceable approach research and discovery of the law on relationship between protein structure and function in the research of basic theories. It has application in various fields such as agriculture, medicine and so on. For example, protein drugs with high bioactivity, high stability and low toxicity, novel antibiotics and directed immunotoxins are produced through protein engineering. In bioengineering, the unique catalysis and molecular recognition characteristics of engineered proteins are utilized to construct biosensors. By changing structure of proteins, industrial enzymes to catalyze reactions in the organic media are produced. The genes of engineered proteins are introduced to plants to change or improve quality of crops and design new biopesticides. In short, protein engineering has stepped into the mature stage, and its development in the future is promising.

Chapter 7

Microorganism and Fermentation Engineering

7.1 Introduction to Fermentation and Fermentation Engineering

Fermentation originally derived from the Latin word "fervere", and refers to the phenomenon that yeast utilized carbohydrate substances or sprouted grains to produce CO_2. Louis Pasteur thought that fermentation was anaerobic respiration of yeast from ethanol and CO_2 which give energy for organism growth. Generalized fermentation concept refers to that aerobic or facultative anaerobic, anaerobic microorganisms convert organic substances into useful metabolites through metabolism in certain conditions, thereby obtaining fermentation products and industrial raw materials. Now, fermentation refers to culture of microbial or other organism cells (animal and plant cell) culture, or accumulation of metabolic products through bioreactors. The whole fermentation process is also called as bioreaction process.

Common microorganisms used in fermentation production are bacteria, actinomycetes, yeast, mold, basidiomycetes, single cell algae, etc. People began to use metabolic products as food and medications thousands years ago. In the forties of the 20th century, industrial production of penicillin was realized due to application of liquid submerged fermentation technology, which was a new milepost for fermentation industry. As a result of application of excellent strains or gene engineering bacteria, advanced technology and equipment utilization, the traditional fermentation technology(natural fermentation) developed into the contemporary one, fermentation engineering. Fermentation engineering (also known as microbial engineering) refers to the process for production of commercial products that produced various physiological active substances for humans needing in metabolism and growth of microorganisms. In fermentation production, many unit operations are applied including isolation of microorganism, purification of fermentation products, heat, mass and momentum transmission and so on. Therefore, fermentation engineering has become an interdisciplinary science closely related with microbiology, biochemistry and chemical engineering, and it is also an important part of biotechnology.

7.1.1 Development History of Fermentation Engineering

Thousands of years ago, humans used metabolites of microorganisms as food and medicines. Brewing is one of the earliest biotechnologies obtained through practice by humans. In about 6000 BC, western Sumer and Babylon began to use malt to brew beer. In about 2000 BC, ancient Greeks and Romans used grapes to produce wine. According to archaeological results, drinking vessels appeared in our country in the Longshan culture period 4200-4000 years ago. In 221 BC (the later Zhou Dynasty), people knew how to make sauce, vinegar and bean curd in our country. In the tenth century, the live vaccine was prepared to prevent smallpox in our country. Soy sauce, pickles, cheese, milk, wine, fermented dough, manure and retting wheat straw all belong to products of the traditional fermentation technology. However, at that time people did not know the relationship between microbes and fermentation, and so it was difficult to control the fermentation process. People only could rely on experience of production and oral teaching that inspired true understanding. Therefore, that period was called as the natural fermentation period.

In 1680, Holland's Anthony Leeuwenhoek (1632-1723) made the microscope, first observed microorganisms that the naked eye couldn't see. In 1857, the famous French biologist Louis Pasteur (1822-1895) proved that the alcohol fermentation was caused by yeast with Pasteur flask experiments. In 1897, the German Eduard Buchner (1860-1917) found that the crushed yeast cells could still ferment sugar solution to produce alcohol, and called the substance with fermentative capacity as zymase. By then, people had really understood the fermentation phenomenon. In 1905, the German Robert Koch (1843-1910) won the Nobel Prize for his outstanding work on tuberculosis of lungs. He invented the solid medium, got the pure cultures of bacteria, and established a microbial pure culture technique. Thereafter fermentation could be controlled in a man-made way. At the end of the nineteenth Century to the 20th- 30th of the twentieth Century, there were many fermentation products, such as baker's yeast, ethanol, acetone, butyl alcohol, organic acids (lactic acid, citric acid), enzymes (amylase, protease), etc. They mainly were primary metabolites through some anaerobic fermentation and surface solid fermentation.

The British bacteriologist Fleming found that *Penicillium notatum* could restrain the growth of *Staphylococcus aureus* in 1928 (Fig.7.1). Its products were known as penicillin, but Fleming's research results did not arouse attention of people at that time. In the early 1940s, because of the great demand for antibacterial drugs during the Second World War, people began to study on penicillin. Initially using wheat bran as fermentation medium, with surface culture method, the fermentation titer units was about 40U/mL, the purity 20%, and the yield 30%. In 1943, American and British scientists developed a $5m^3$ fermentation tank with mechanical ventilation. The fermentation titer units increased to 200U/mL, the purity 60%, and the yield 75% through submerged ventilation fermentation in this fermentation tank. Therefore the microbial submerged culture technology was established. After more than half a century, the penicillin fermentation level was greatly improved, so that the fermentation titer units was up to 90,000U/mL, the yield 90% and the purity was 99.9%. Due to adoption of the submerged culture technology, the antibiotic industry and even the whole fermentation industry made great progress. Soon other antibiotics, such as streptomycin, novobiocin, chloramphenicol, aureomycin, oxytetracycline etc came into being in succession. The development of antibiotic industry promoted production of other fermentation products. the most prominent progress emerging was of amino acid fermentation industry in the 1950s and enzyme preparation industry and organic acid industry in the 1960s. In this period there were more types of products, such as antibiotics, amino acids (glutamate, lysine), nucleotide, enzyme preparations, organic acids (citric acid), polysaccharide, SCP, vitamins, the products obtained through bioconversion and enzyme reaction, which were mainly primary metabolites and secondary metabolites with aerobic fermentation. It is reasonable to say, this is a heyday of contemporary fermentation industry. New products, new technologies, new processes, new equipments are emerging constantly, the production scale is constantly expanding, and the scope of application is also gradually broadened.

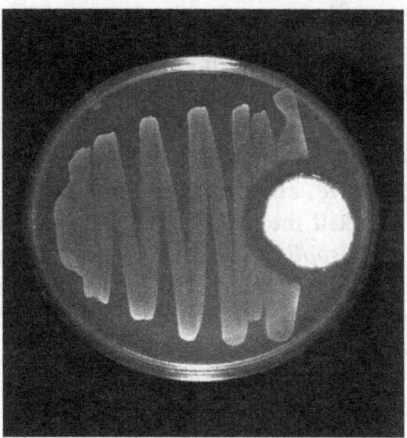

Fig.7.1 Bacteriostasis of *Penicillium notatum*

In 1953, Watson and Crick from the United State discovered the DNA double helix structure model. In 1973, Stanley Cohen group from Stanford University in the United States of America found that two plasmids were cut by *Eco*R I, and then were connected by DNA ligase, and the hybrid plasmid had two replication origin sites and was used to transform *E.coli* successfully. Although their experiments did not involve any target gene, they laid the foundation for establishment of the genetic engineering theory and application of the technology. Since the discovery of DNA double helix structure model and the realization of gene transfer in laboratory, it was possible to follow the human intention to design a new organism. Genetic engineering is recombination of exogenous (target) genes (specific DNA fragments) *in vitro* with vector DNA (plasmids, bacteriophages and so on), and then transferring recombinant DNA into host cells, so as to form the clones (clone, an asexual propagation line or recombinant) that can replicate and express of exogenous genes. We cannot only construct high yield gene engineering bacteria, but also obtain the target products, exogenous proteins, which microorganisms cannot produce through culture of these recombinants. These products include various proteins with physiological activity from plants, animals and human beings, such as insulin, growth hormone, cytokines and monoclonal antibody. These are products manufactured by using the contemporary biotechnology, DNA recombinant technology and protoplast fusion technology. In addition, mass culture of plant cells had a longer history than that of animal cells; some precious secondary plant metabolites can be produced through culture of plant cells, such as alkaloids and steroids, which belong to contemporary biotechnology products. Although contemporary biotechnology products are not of many kinds, but they have high big value and great social benefit, so contemporary biotechnology industry is the ascendant high-tech industry.

7.1.2 Types of Fermentation Products

The major types of fermentation products are microbes, metabolites, enzymes and microbial conversion products. These products are widely used in food, pharmaceutical, agriculture and animal husbandry, light industry, chemical industry, energy, environmental protection and other fields.

7.1.2.1 Microbial cells

The traditional microbial body products include bread, edible and feeding microbial cell proteins (such as *Candida utilis, Candida tropicalis, Saccharomyces cerevisiae, Chlorella* sp., *Arthrospira* sp., *Endomycopsis* and etc.). Contemporary microbial body, include medicinal macrofungi, such as mushrooms, *Cordyceps sinensis, Armillaria, Poria cocos, Ganodorma lucidum, Tremella fucitormis, Agaric*; and biological pesticides, such as parasporal crystal of *Bacillus thuringiensis* which can kill the pests of *Lepidoptera* and *Diptera*. The filamentous fungus *Beauveria, Metarhizium anisopliae* can prevent and treat pine moth, and therefore they can be made into a novel microbial insecticides. Active lactic acid bacteria (LAB) as probiotics preparation, can improve the microecological environment of the human intestinal, and enhance immunity and other physiological functions.

7.1.2.2 Microbial enzyme

As microorganisms have characteristics such as many species, enzyme production variety, convenient production and low cost, with enzyme method, production process can be simplified; equipment investment, the energy and raw material consumption can be reduced; product quality and labor condition can be improved; and wastes can be reduced. Production of microbial enzymes with fermentation method, and isolation and purification of microbial enzymes with contemporary biotechnologies manufacture enzymes preparations are the important parts of the fermentation industry.

7.1.2.3 Microbial metabolites

This type of products can be divided into two categories: ①primary metabolites, such as amino acids (e.g. glutamate, lysine), organic acids (e.g. citric acid, lactic acid), organic solvent (e.g. ethanol, glycerol), nucleotides, proteins, nucleic acids, vitamins, etc.;②secondary metabolites, such as antibiotics, alkaloid, bacteriocin, plant hormones, pigments, microbial polysaccharide, etc.

7.1.2.4 Microbial conversion products

The final products of bioconversion is produced by enzymes or enzyme systems from microbial cells through chemical reactions at specific sites, and they are not microbial cells metabolite utilizing nutrients, i.e. they are produced by biocatalysts' acting on one site of the substrates. The most striking characteristic of bioconversion is strong specificity include reaction specificity, structure and site specificity and stereo specificity. With bioconversion method, glycerol can be converted into dihydroxy acetone, glucose into glucose acid, sorbitol into L-sorbose, and so on. The most important bioconversion in the fermentation industry is conversion of the steroid hormones (such as hydrocortisone, dexamethasone, fluocinonide, etc.). It has so many characteristics, compared with the chemical synthesis method, such as convenient operation process, high yield and low cost. In addition, the conversion efficiency can be improved and the cost can be reduced with bioconversion of immobilized cells or immobilized enzymes.

7.1.3 General Fermentation Process and Its Characteristics

7.1.3.1 General fermentation process

The essence of biological reaction process is production of biological products with biocatalysts, i.e. the fermentation process, as shown in Fig.7.2.

Chapter 7 Microorganism and Fermentation Engineering

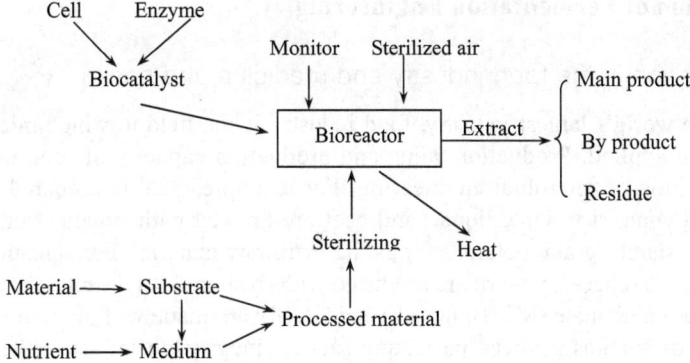

Fig.7.2 Schematic diagram of general fermentation process

As can be seen from Fig.7.2, the general biological reaction process consists of the following four parts: ①pretreatment of raw materials and preparation of fermentation medium; ②preparation of biocatalysts; ③selection of biological reactors and reaction conditions; ④isolation and purification of products.

7.1.3.2 Characteristics of fermentation

Whether culture of microbe cell, or culture of animal and plant cells, treatment of wastewater and production of biopesticides with microorganisms, all of these are called bioreaction processes. If the cells are to be used as the biocatalyst, the process is called as fermentation. If the enzyme is to be used as the biocatalyst, the process is known as enzyme reaction process. The characteristics of biological reaction process are summarized as follows. ①The raw material used in fermentation is mainly carbon source, adding a little organic and inorganic nitrogen source, without toxic substances. ②The bioreaction process is usually carried out at normal temperature and pressure, and in a way of self-regulation of organisms. Various reactions are like a reaction in single equipment, and so the equipment has a variety of uses. ③Complex polymeric compounds, such as enzymes and optical active substances, are easy to be produced. ④The reactions such as oxidation and reduction can be performed in a highly selective way at specific site (group) of complex compounds. ⑤Organisms themselves can also be used as fermentation products, such as SCP with rich in protein, enzymes and vitamins. Except special circumstances, general culture fluid does not cause harm to human and animals. ⑥During the fermentation process contamination of other microorganisms especially phage invasion, should be noted to be prevented strictly, in order to avoid greater harm. ⑦Under the condition that equipment investment does not increase, production capacity can be raised through improvement of performance of the producing organisms. In addition, in actual fermentation, production efficiency can be enhanced by modifying the process technology and equipments to improve product quality and to increase production benefit.

7.1.4 Application of Fermentation Engineering

7.1.4.1 Application in the food industry and medicine and health

As one of the world's largest industry, food industry is the field in which microbial technology was developed and applied. Production value and production capacity of food industry still ranks the first in application of microbial engineering. For example, SCP is prepared with a variety of raw materials; and wine, rice wine, liquor, and beer are brewed with sugars (fruit, tree sap, honey, molasses, etc.) and starch (grain, potato, etc.) as the main raw material. Fermentation dairy products, such as fermented milk, cheese, yogurt, are produced with fresh milk as the raw materials. Condiments such as monosodium glutamate (MSG), inosinic acid, GMP are made with the starch as raw material; and fermented foods such as soybean paste, soy sauce, vinegar, lobster sauce, bean curd, maltose, pickles, etc, with beans and grains as raw material. In addition, there are glucose, maltose, fructose syrup, sweet peptides, sweet protein peptides and other sweeteners, and lysine(food fortifier), citric acid (sour agent), pigments (colouring agent), dextran and Xanthan Gum (thickening agent), glucose oxidase and vitamin C (fresh keeping agent), Nisin peptide and pimaricin (antiseptic), and other food additives, are all microbial fermentation products.

Fermentation engineering is the most widely used in medicine and health which is also the field with the most rapid development and the greatest potential. Fermentation engineering may be utilized to improve pharmaceutical production from various aspects, and to develop new drugs, so as to enhance the human health level. Now there are more than 6000 kinds of antibiotics, most of which are produced by microorganisms. Medical products and veterinary drugs (including semi synthetic antibiotics) have more than 100 kinds, and they mainly include antibiotics resistant on bacteria, such as bacitracin, cephalosporin, aureomycin, chloramphenicol, streptomycin, kanamycin, spiramycin etc.; antifungal antibiotics (amphotericin B, griseofulvin, nystatin); and antitumor antibiotics (actinomycin, mitomycin, intraepithelial DX, mithramycin, bleomycin etc.). In addition, amino acids, vitamins, human insulin, hepatitis B vaccine, interferon, steroid hormones etc. also are microbial fermentation products.

7.1.4.2 Application in chemical and energy products

Chemical raw materials (alcohol, glycerin, isopropanol, acetone, ethanol, etc.), some surfactants, organic acids (acetic acid, propionic acid, lactic acid, butyric acid, succinic acid, fumaric acid, malic acid, tartaric acid, salicylic acid etc.) and polysaccharide (dextran, xanthan gum, pullulan, alginic acid etc) can be produced by microorganisms. Biogas (methane) and clean energy (hydrogen, microbial fuel cell) also can be produced by microbes. In addition, oil can be made with algae.

7.1.4.3 Application in agriculture

To a great extent, developed agricultural economy relies on progress of science and technology. Fermentation engineering technology and its products can provide powerful support for the agricultural development. Biological pesticides including viral insecticides (nuclear polyhedrosis virus, cytoplasmic polyhedrosis virus, granulosis virus, etc.), bacterial insecticides (*Bacillus thuringiensis*, *Bacillus popillia*, etc.), fungi insecticide (*beauvericin* insecticide, insect fungus insecticide, etc.), and protozoa pesticide (microsporidia insecticide, new nematode insecticide, etc.), can all be produced with microorganisms. Some antibiotics used in agriculture (such as blasticidin, blasticidin S, kasugamycin, qingfengmycin, etc.), biological herbicide (weed pathogens include rust, fungi, *Fusarium*, anthrax and nematodes and viruses), and biological yield increasing agent (such as *Rhizobium* and *Freund*

Actinomyces symbiotic nitrogen-fixing bacteria, *Azotobacter chroococcum Beijerinck* and gum yield of *Azospirillum brasilense*, *Azotobacter* and potassium bacteria, phosphate bacteria, etc.) can also be produced with microorganisms. In addition, microorganisms which can prevent and control plant diseases include bacteria (*Pseudomonas*, *Agrobacterium*, etc.), actinomycetes, fungi (*Trichoderma*), virus (a variety of weak virus) and so on.

7.1.4.4 Application in metallurgy

Gold mining and metal leaching recovery can be done with microorganisms. The autotrophic bacteria like *Thiobacillus ferrooxidans* have certain oxidation ability, which can oxidize ferrous iron into high-valent iron, and sulfur and low sulfide into sulfate. The metal ions in the metal sulfide ore (mainly tailings, lean ore) can form sulfate and will release it. Metals made with this method are copper, cobalt, zinc, lead, uranium, gold, and so on.

7.1.4.5 Application in environmental protection

Environmental pollution arising from waste water, waste gas and waste residue can be eliminated with microorganism. The microbial treatment technology for toxic wastes mainly includes anaerobic fermentation and aerobic fermentation. The former utilizes obligate anaerobic microorganisms (e.g. *Clostridium, Bacteroides, Ruminococcus, Butyrivibrio*, methanobacteria, etc.) and facultative anaerobic microorganisms (e.g. *E.coli, Bacillus*, etc.) to obtain biogas, fertilizer and feed. In the aerobic conditions, the latter utilized the mixture between some bacteria producing zoogloea and some protozoa to treat industrial and domestic sewage and waste gas.

7.1.4.6 Application in hightech research

Microorganisms play an important role in high-tech research. For example, they can provide plasmids, cosmids, viral vectors, restriction endonuclease, ligase, phosphatase, phosphokinase, etc. for gene engineering research; and biosensors (enzyme, microbial electrode and DNA chip) for medical diagnosis and for detection in fermentation process; and biological chip which may be used in microelectronic products.

In conclusion, fermentation engineering technology has already been used widely in all kinds of fields. It promotes development of other biotechnologies; at the same time, with research development of gene engineering, protein engineering, enzyme engineering, cell engineering, fermentation engineering will further expand the scope of its application, and it will make enormous contribution to industrial and agricultural production and human health.

7.1.5 Development Situation and Prospects of Fermentation Engineering

All Strains isolated directly from the nature cannot be immediately used for fermentation production due to their poor performance. Only through strain breeding can increase metabolite yield greatly, and product quality and types and process conditions be improved, so as to simplify the process. Strain breeding methods can be divided into natural selection and mutation breeding. The former includes the strains isolated from nature material and the ones obtained through spontaneous mutation. The latter is a simple, rapid, efficient screening method including the following steps, firstly utilizing physical mutagens, chemical mutagens and other mutagens to treat the uniform and dispersive microbial cells, thereby increasing the frequency of random gene mutation, and then selecting the few excellent mutants meeting the breeding requirements which are used for scientific experiments and production practice, with a simple, rapid and efficient screening method. Compared with natural breeding, mutation breeding has many advantages such as high mutation frequency, large variation amplitude, high speed, simple operation and so on. With development of molecular biology, birth of gene operation technology and cell fusion technology endowed biotechnology with new vitality. And this makes the fermentation industry enter the important development stage for fermentation production of new products with gene engineering bacteria, such as insulin, interferon, human growth hormone, bovine growth hormone and hepatitis B vaccine, which cannot be produced by general microorganisms,. The first new fermentation product produced with gene engineering bacteria is growth hormone release inhibiting factor which was developed in the United States in 1977. This new product can inhibit secretion of pituitary hormone. With 500, 000 pieces of sheep brain as raw material, only 5mg of the related factor can be extracted; but with only 10L fermented broth of the gene engineering strain, the same amounts of the product can be obtained. In addition, gene engineering bacteria will also improve general microorganism products, known as the old type of fermentation products, such as amino acids, vitamins, nucleotides, antibiotics, enzymes preparation, etc. Compared with general fermentation products, the yield and quality of these products produced with gene engineering stains have been improved significantly.

From the viewpoint of development of the biotechnology all over the world, it has become a hotspot of science and technology competition between developed countries. The United States, Japan, Europe and other major developed countries and regions, compete to develop biotechnology research and development work. They have established independent government agencies and a series of biotechnology research organizations, made the medium and long-term development plan by 2020, and given a strong support in both policy and funds. At the same time in these countries enterprises have invested heavily in research and development of biotechnology products, and have made many significant achievements, so that industrialization of biotechnology develops rapidly. Over the past 10 years, Chinese biotechnology industry has also made rapid development and has become the world's important competitor in the fermentation product market. The yield and export of a lot of fermentation products have increased rapidly. Citric acid production technology has entered the rank of the most advanced and the productivity ranks the first in the world. Other products such as glutamic acid and lysine have certain advantages in technology and technical level. At the same time, research and development of contemporary biotechnology products has made a great achievement. Hybrid rice first created in China has been extended to cultivate 200km^2 (2,000,000 acres), and the average yield has increased by more than 10%. Transgenic plants have been successful. Recombinant symbiotic nitrogen fixing bacteria and pest-preventing engineering bacteria have involved in large area field experiments. *In vitro* fertilization (IVF) in cattle and sheep, and transgenic fish have entered pilot-scale experiments, and animal bioreactors have made great progress. Four kinds of genetic engineering drugs have been approval to be on market. Antibody engineering has made some achievements and has got application in clinical gene therapy. Some gene therapies have reached the international level. The protein engineering of human insulin, human urokinase hormone, glucose isomerase and chymosin has reached the world level. However, from the overall viewpoint, China is still in the initial stage, not only in traditional biotechnology industry, but also in research, development and industrialization of contemporary biotechnology, compared with developed countries.

In conclusion, fermentation engineering is one of the important parts of biotechnology. It is the only way which must be followed for industrialization of cell products. Biochemical engineering, developed on the basis of combination of biotechnology with contemporary technologies, is a new engineering technology. It not only provides efficient bioreactors, novel isolation technology and media, and contemporary engineering equipment technology for the traditional fermentation industry, modification of traditional pharmaceutical industry and emerging biotechnology industry, but also provides production equipment units, optimized process, automatic on-line control, engineering concepts and technologies of integrated system design, and for the basic theory of biological process optimization and control. Biochemical technology plays an important role in the biotechnology industry and makes the application scope of biotechnology more extensive, downstream technology constantly updated, and in the meanwhile that increases yield and quality of biotechnology products significantly. Biochemical engineering technology has become the bridge and bottleneck of the biotechnology industry. Research of production process and technology has become an important aspect for promotion of biotechnology industrialization.

The rapid development of biotechnology has provided a variety of organism cells which can produce commercial products through fermentation. Therefore, fermentation engineering is the foundation to implement industrialization of biotechnology. Birth and development of biotechnology and development is involved in many disciplines; with development of genetics, microbiology, cell biology, animal science, botany, biochemistry, molecular biology, chemistry and chemical engineering, quantum biology, applied physics, electronics, computer science and other basic and applied sciences and technologies, great changes will happen to fermentation engineering, so as to create bigger economic benefits for human beings.

7.2 Elementary Knowledge on Microorganisms

Microorganisms or microbes are a kingdom of lower organisms with small size and simple structure, which can be seen only with the aid of a microscope. They consist of prokaryotic microorganisms including bacteria (eubacteria and archeobacteria), actinomycetes, blue bacterium, rickettsia, mycoplasma, chlamydia and spiral body; eukaryotic microorganisms including yeast, fungus and mushroom, mono cellular algae and protozoa; and noncellular organisms, virus (vertebrate virus, invertebrate virus, plant virus, and microorganism virus) and subvirus.

7.2.1 Structures of Microorganism Cells

Four categories of microbes including bacteria, actinomycetes, yeast and mold have very wide applications in food and fermentation industry. Microbial enzymes and metabolites are used in some industries. Microbial biomass and its inclusions are used in other industries. In the following text structures of the four categories of microbial cells have been briefly introduced.

7.2.1.1 Cellular structure of bacteria

Bacteria are unicellular prokaryotic microorganisms. In spite of small size, their structure is complex, and includes basic structure and special structure (Fig.7.3). There are many bacteria usually used in the fermentation industry, such as *Bacillus subtilis*, *Lactobacillus*, *Acetobacter aceti*, *Corynebacterium*, *Propionibacterium sherman ii*, acetone butanol *Clostridium bacillus*, *Bacillus brevis*, *Xanthomonas*, *Leuconostoc mesenteroides*, *Gluconobacter oxydans*, etc., which are used for production of lactic acid, acetic acid, amino acid, amylase, inosinic acid, propionic acid, acetone, butanol, xanthan gum, dextran (dextran), vitamin C, etc.

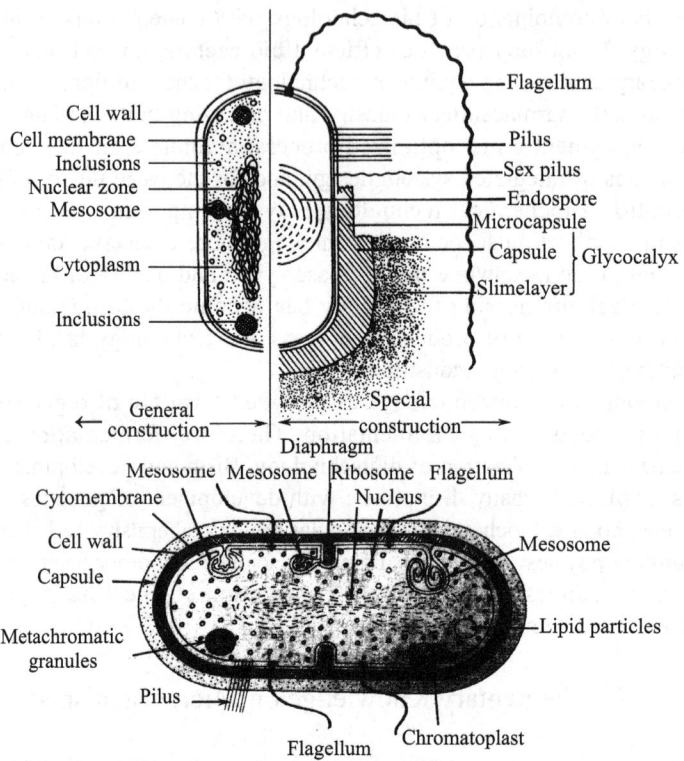

Fig.7.3 Cellular structure model of bacteria

1) Basic structure of bacteria

The basic structure refers to the cellular structure which all bacteria have, including cell wall, cell membrane, cytoplasm, nucleus, ribosome and inclusions.

The cell wall coats outside of the cell membrane, and is a colourless transparent and elastic structure with firm texture. The role of the cell wall is to provide rigidity and strength for a cell, preventing the cell from osmotic lysis when placed in diluted environments so as to maintain the cell shape. The cell wall is related with flagellum movement. The eubacteria are frequently divided into two groups, Gram-positive bacteria (G^+) and Gram-negative bacteria (G^-), on the basis of their different results of the staining test devised by Christian Gram in 1884. The differential results of the staining test are mainly because of different chemical compositions of the cell wall in these two types of bacteria. These two types of bacterial cell wall are very different (Fig.7.4) in chemical composition and structure. The cell wall of G^+ bacterial is thick (20-80nm). Its main components are peptidoglycan and teichoic acid and a small amount of surface proteins. The lipid generally cannot be found in the cell wall of G^+ bacteria. But the cell wall of G^- bacteria have only a thin layer of peptidoglycan(10-15nm). Its structure and chemical composition are more complex than those of G^+ bacteria. There are outer membrane protein (porin, non-microporous protein, and lipoprotein), phospholipid (lipid bilayer) and lipopolysaccharide in outer layer of peptidoglycan. Therefore the cell wall of G^- bacteria has a multilayer structure containing no teichoic acid.

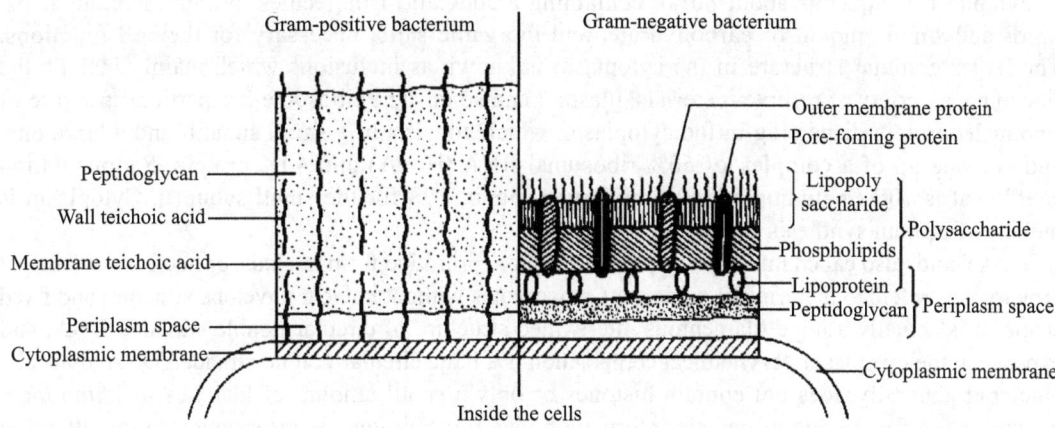

Fig.7.4 Structure comparison of cell wall between G⁺ and G⁻ bacteria

The cell membrane, also called the cytoplasm membrane, is a layer of soft and elastic semi-permeable membrane attached to the inner wall of the cell, which tightly wraps the cytoplasm. The cell membrane exhibits a "sandwich" structure observed under an electromicroscope with two dark layers and a light layer (Fig.7.5). It is composed of phospholipids, proteins and small amount of carbohydrate. The basic structure is the phospholipid bilayer which arranges symmetrically with a liquid flow. Peripherin and integral protein at different levels of integration are embedded in the phospholipid bilayer, doing "floating" or" iceberg" movement. Its main function is selectively controlling transport and exchange of nutrients and metabolic wastes cell inside and outside, so as to maintain normal intracellular osmotic pressure. The cell membrane is the site for synthesis of the cell wall and the capsule, and generation of metabolic energy, and is associated with flagellar movement. Mesosome, also called intermediates, is a tubular, layered or cystic structure by cell membrane invaginating. It is generally believed that mesosome is involved in formation of phragmoplast and cell division, cell wall synthesis, spore formation, DNA replication and separation from each other, and bacterial respiration.

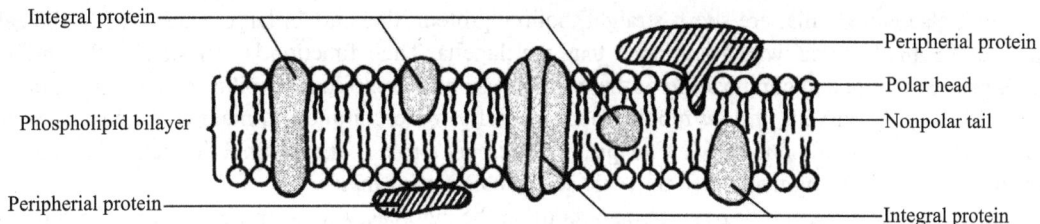

Fig.7.5 Structure model of cell membrane

The cytoplasm, also called cell plasma, is all colourless, transparent, thick gelatinous matters and particulate matters except the nucleoplasm within the cell membrane. The basic elements of cytoplasm is aqueous(about 80%), containing a cocktail of molecules, protein, nucleic acids, lipids and small amount of carbohydrate, and inorganic salts, necessary for the cell functions. The larger granular structure in the cytoplasm is known as inclusions which mainly include the ribosome, a variety of reserve materials, plasmid and so on. The ribosome is a particle structure of ribonucleoprotein dispersing in the cytoplasm, which consists of a small subunit and a large one, and is made up of a complex of 60% ribosomal RNA (rRNA) and 40% protein. Sedimentation coefficient is 70S (including 50S for the large subunit, 30S for the small subunit). Cytoplasm is the site of protein synthesis in the cell.

Nucleoid, also called nucleoplasm, nuclear region, primitive form nucleus, or nuclear genomes of prokaryotes, refers to the original nucleus of prokaryotes without nuclear envelope structure and fixed shape. It is actually a huge filamentous intertwined structure of circular double-stranded DNA, and exposed in the cytoplasm. Its chemical composition is a large circular double-stranded DNA molecule. Nucleoid generally does not contain histones or only a small amount of histones. It is the main material basis for storage of genetic information, can transmit genetic information to the offspring through self-replication, and can control cell growth and reproduction, genetic variation and other vital activities through transcription and translation.

2) Special structure of bacteria

Special structure of bacteria, common not all bacteria, found in the cell are glycocalyx, flagella, pili spores etc.

The glycocalyx is a layer of transparent mucous gelatinous substance with uncertain thickness, coating outside of the cell wall in some bacteria. According to fixed layer level and thickness, it can be divided into 4 types, the capsule, microcapsule, the mucous layer and zoogloea. The capsule is a thick layer (in thickness of 0.2μm) of mucous material secreted by some bacteria, with certain shape, fixed on the cell wall surface, which is composed of the majority of polysaccharide, a few peptides or proteins, or polysaccharide and polypeptide. The main function of the capsular is protection of the cell (avoiding the dry damage, host phagocytic leukocytes and preventing from toxicity of chemical drugs and heavy metals, etc.), and involvement in surface attachment. In addition, the capsular is the site for storage of nutrients and accumulation of certain metabolic waste, as well as certain pathogen virulence factors, and related with pathogenicity.

Flagella are one or a plurality of elongated, wavy bending filaments on the cell surface of some bacteria. Its chemical composition is mainly flagellin. The flagellum is a good antigen (H antigen) with movement power which is the most effective way for prokaryotes to realize taxis.

Pili, also called cilia, are short, straight, hollow protein filaments in large quantities, on the cell surface of some bacteria, which is thinner than the flagella. Their function is adhesion on the surface of objects. Structure and composition of the pili is the same as sexpilus. Sexpilus is longer and thicker than pili, slightly curved, hollow tubular, and generally more common in male G^+ bacteria strains. Each bacterium has only one or a few sexpili. Sexpili play a role of the channel through which female strains transfer DNA fragment when the male and female strains mate.

The spore is a circular or oval dormant body in the cytoplasm with strong resistance on adverse environmental conditions, formed due to dehydration and concentration of the cytoplasm at the late growing stage of some bacteria. Core water content of the spore is very low (40%), and it contains a special DPA-Ca (2, 6-pyridine bicarboxylate calcium) and heat-resistant enzymes. It is made up of thick keratin and dense spore coat, so that the spore has the ability of strong resistance on adverse environmental conditions such as high temperature, dryness, radiation, and toxic chemicals.

7.2.1.2 Structure of actinomycetes

Actinomycetes are prokaryotic microorganisms forming a branching hyphae and conidia. They are reproduced mainly through formation of conidium, and a few of them form sporangium. In liquid culture conditions they are reproduced mainly in a way of hyphal breakage. More than 70% of antibiotics are produced by actinomycetes, such as streptomycin, erythromycin, aureomycin, gentamycin, etc. Common actinomycetes are mainly from the following genus: *Streptomyces*, *Micromonospora* and *Nocardia* etc. The structure of *Actinomycete* mycelia is the same as bacteria, including cell wall, cell membrane, cytoplasm and nucleoplasm body. According to morphology and functions of actinomycete mycelia, it can be divided into 3 parts: the substrate mycelium, aerial hyphae and sporotrichial. Taking the genus *Streptomyces* as an example, in the following text we will describe their structure (Fig.7.6).

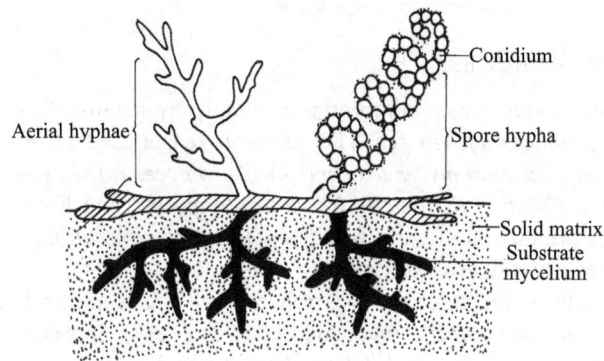

Fig.7.6 Structure model of *streptomyces*

1) Substrate mycelium

Substrate mycelium also known as vegetative hyphae or primary hyphae, prostrates on the surface of the medium or grow in the culture medium to absorb nutrients.

2) Aerial mycelium

Aerial mycelium, also called as secondary mycelium, is formed by nutritional hyphae when they develop to a certain stage and grow onto the top of culture medium. It can be used to reproduce offspring and transport nutrients.

3) Sporotrichial and conidiospore

When aerial mycelium growth to a certain stage, the majority of aerial mycelia can form spore hypha by means of differentiation called sporotrichial. Different species have different sporotrichial morphology and aerial mycelium arrangement. Various *streptomycetes* have different appearances of sporotrichials. They are straight, waved, hooked, or spiral, and have stable shape. The ways of orientation include alternation, whorled form and clustered form. Their shapes are stable and are important basis for classification and identification (Fig.7.7).

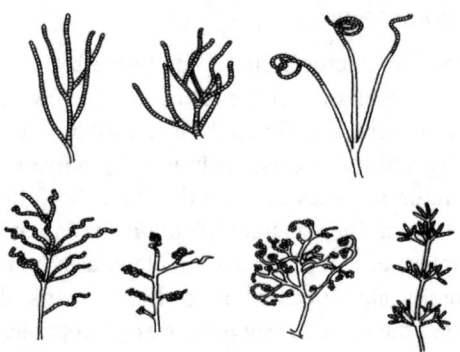

Fig.7.7 Various sporotrichial morphology of actinomycetes

7.2.1.3 Structure of yeast cells

As a unicellular eukaryote, yeast is reproduced mainly by means of asexual budding. And in sexual reproduction it produces ascospores. The common yeasts used in fermentation industry are *Saccharomyces cerevisiae, Saccharomyces uvarum, Saccharomyces rouxii, Candida* (such as *Candida tropicalis, Candida utilis, Candida lipolytica*), *Rhodotorula glutins*, and *Eremothecium ashbyii*. They are respectively used for production of wine, bread, soy sauce manufacturing, fatty, riboflavin, as well as single cell protein used as food, drug and feed.

Yeast has a typical cellular structure (Fig.7.8), including cell wall, cell membrane, nucleus, cytoplasm and its inclusions. The cytoplasm contains mitochondria, ribosomes, endoplasmic reticulum, microbody, centrosomes, Golgi bodies, spindle, vacuoles and reserve material. In addition, some kinds of yeasts have bud scar (birth mark), and some species have capsule, pili and other special structures.

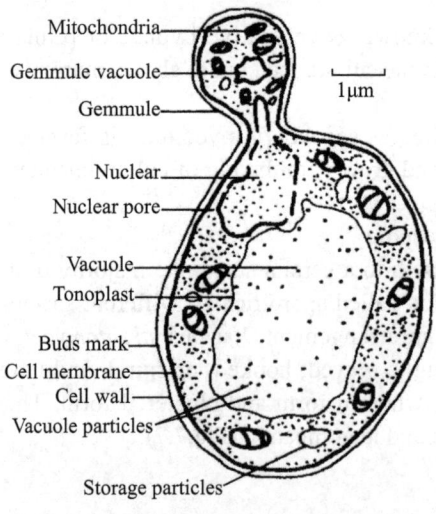

Fig.7.8 Structure model of yeast cells

1) Cell wall

The cell wall locates at the outside of a cell and surrounds the cytoplasm membrane. It is made up of glucan, mannan, proteins, lipids, chitin and inorganic salt. It gives the cell rigidity and strength and maintains cell morphology.

2) Cell membrane

Cell membrane structure of yeast is similar to that of bacteria. It consists of proteins, lipids and a small amount of sugar. The basic structure of cell membrane is two layers of phospholipid molecules, and sterol and protein are also embedded. One of the important differences between prokaryotic microorganisms and eukaryotic microorganisms is that sterols (mainly ergosterol) are contained in the cell membrane of eukaryotic microorganisms.

3) Cell nucleus

The cell nucleus of yeast, including nuclear membrane, chromatin and nucleolus, is the real integral nucleolus which is enwrapped by nuclear envelope. It is the important difference between the nucleus of eukaryotic microorganisms and the nucleoid of bacteria. The nucleus is the main site for storage, replication and transcription of genetic information. It can control cell growth, reproduction, heredity and variation.

4) Cytoplasm and inclusions

Cytoplasm is a transparent, viscous colloidal water solution. It is the site for cell metabolism and metabolite storage and transport. Cytoplasmic inclusions mainly include ribosome, mitochondria, endoplasmic reticulum, centrosomes, Golgi apparatus, spindle, microbody, plasmid, vacuoles and stock granules. The ribosome is free in the cytoplasm, and can attach to the endoplasmic reticulum. Its sedimentation coefficient is 80S (60S for the large subunit, and 40S for the small subunit). Most of the ribosomes form polyribosomes. The reticulum is formed by means of collection and parallel arrangement of the bilayer membrane systems in different shapes and sizes in the cytoplasm. The endoplasmic reticulum connects with the cell membrane and the nuclear membrane. Mitochondria are the rod-shaped or spherical organelles locating within the cytoplasm. Each cell has 1-20, up to hundreds of mitochondria. It is the site for biological oxidation and respiration of yeasts and other eukaryotes. The cytoplasm of the vigorous yeast cells is uniform and dense, the vacuole is very small; while 1 or 2 larger vacuole can be visible in the aging cell (growth in the mid-term to later period) which contains various storage granules, such as metachromatic granules, glycogen, fat particle and so on. It is a marker for judgment of maturation of yeast cell.

7.2.1.4 Structure of mold cells

Mold is a generic term for filamentous fungi and is not a taxonomic name. Generally, any fungus, which grows on solid nutrient matrix to form a villous, araneose, gossypine or carpet-like shape, is called mold. Most of molds are reproduced mainly asexual spores, and a few by means of sexual spores. Common molds used in the fermentation industry include *Rhizopus, Mucor, Absidia, Monascus, Aspergillus, Penicillium, Trichoderma, Cryytococcus neoformans, Ashbya gossypii*, etc. The hyphae of Fungi is a tubular filament germinated from fungal spores, is enwrapped by cell wall, is the basic unit of fungal vegetative body, and is divided into two types, non-septate hyphae and septate hyphae.

1) Non-septate hyphae

This type of hyphae has no septum, and the whole mycelium is a long tubular single cell containing multiple nuclei in the cytoplasm [Fig.7.9 (a)]. During the hyphal growth process, only nucleus division and increase in cytoplasm are observed, but the cell number does not increase. Common fungi with non-septate hyphae are *Rhizopus*, *Mucor* and *Absidia*.

2) Septate hyphae

The type of hyphae has some septums, and the whole mycelium is made up of many cells. The hypha between adjacent two septums is called a hypha cell. Each cell contains one or more nuclei [Fig.7.9 (b), (c)]. There is one or more pores which can make cytoplasm between cells communicate with each other on the septum so as to carry out exchange of substances. During the hyphae growth process, the number of cells increases with nuclear division. Common fungi with septate hyphae are *Penicillium*, *Aspergillus* and *Trichoderma*.

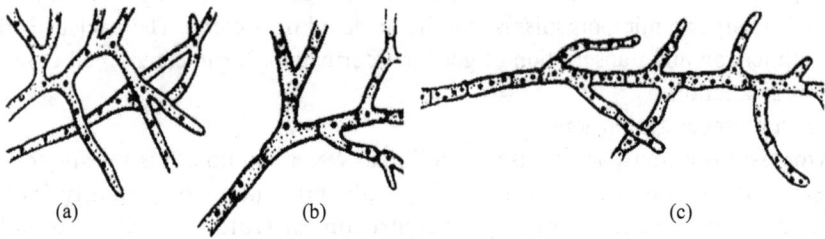

Fig.7.9   Morphology of fungal hyphae
(a) Non septate hyphae;   (b) (c) Septate hyphae

When the mold spores fall on the suitable solid nutrient matrix, they germinate to produce hyphae which continue to branch toward the two sides. Many branched hyphae interwine to form a population which is called mycelia. Different fungal mycelia have functional differentiation. Vegetative mycelia refer to those gathering on the nutritional matrix to play the function to absorb nutrients. Aerial mycelia refer to those growing out of medium. Some aerial mycelia can differentiate into reproductive mycelia which have reproductive function when they grow to a certain stage. Reproductive mycelia can produce reproductive organs and reproductive cells.

7.2.2 Nutrients Absorption Patterns of Microorganisms

Nutrients in environment can be gradually utilized by microorganisms when they are absorbed into the cells. Microbes can produce a variety of metabolites constantly during cell growth. In order to guarantee normal growth of microorganisms, these metabolites must be transported timely to the fermented broth to avoid accumulation within the cell and to prevent from causing poisonous effect to cells. All microbial cells don't have any special feeding organ or special excretory organs, therefore they can absorb nutrients and discharge metabolic waste by cell membrane. The cell membrane of microbes has a high selectivity for the nutrients which are transported in a transmembrane way. Generally, it can directly absorb some fat-soluble and water-soluble substances of small molecules. But those macromolecular nutrients such as polysaccharides, protein, nucleic acid and fat are hydrolyzed into small molecules by the corresponding extracellular enzymes, which can be absorbed by microbial cells. According to characteristics of material transportation, so far it has been generally believed that, except protozoa, the other major categories of cellular microorganisms have four nutrient absorption modes, including simple diffusion, facilitated diffusion, active transport, group translocation.

7.2.2.1 Simple diffusion

Simple diffusion is also called passive diffusion. This diffusion is driven by both the concentration difference of one substance between inside and outside of the cell membrane. Without the aid of carrier proteins, this diffusion relies solely on free movement of molecules through the cell membrane, and the molecules diffuse from the area at high concentration to that at low concentration, until both concentrations of the substance inside and outside the cell membrane are equal. Simple diffusion is not the main way of nutrient absorption of cells, and only suitable for absorption of a few of substances of small molecules, such as water, water soluble gas (O_2, CO_2) and polar small molecules (such as urea, ethanol, glycerol, fatty acid etc.), some amino acids, and ions.

7.2.2.2 Facilitated diffusion

Facilitated diffusion refers to the mode of nutrient transportation, which must be performed with the aid of specific carrier proteins (also called osmosis enzymes, similar to the role of enzymes, most of them are inducible enzyme) on the cell membrane and does not consume energy. With the aid of conformational changes of the carrier protein, under the condition without energy consumption, the solute at high concentration outside the cell membrane can be accelerated to diffuse into the intramembrane, until the concentrations of the solutes inside and outside of the cell membrane are equal. The nutrients transported into the cell in the mode of facilitated diffusion include simple sugars, amino acids, vitamins and inorganic salt. It is usually found in eukaryotic microorganisms for example, transportation of glucose in yeasts.

7.2.2.3 Active transport

Active transport is a mode of nutrient transportation which needs energy, in which the nutrients was are transported from the extracellular environment at low concentration into the intracellular one at high concentration through conformational changes of specific carrier proteins on the cell membrane. The nutrients bind with the carrier protein to form the carrier-solute complex outside the cell membrane, which go to the inside of the cell membrane to arouse carrier configuration changes under action of energy, then the affinity of the complex decreases, and finally the nutrients are released and the carrier can be used again. Some nutrients such as sugar (lactose, glucose, galactose, arabinose, melibiose, etc.), amino acid (alanine, serine, glycine, etc.), nucleoside, lactic acid and glucuronic acid, as well as certain anions (PO_4^{3-}, SO_4^{2-}) and cations (Na^+, K^+) are absorbed through active transport in microbe cells.

7.2.2.4 Group translocation

Group translocation refers to a substance transportation which needs both specific carrier proteins and energy consumption. Its characteristic is that a complex transport system exists to complete material transport and the chemical structure of solute molecules changes before and after transportation. Group translocation is mainly used to transport sugars and their derivatives (such as lactose, glucose, mannose, fructose, maltose, N-acetylglucosamine), butyric acid, purine, pyrimidine, etc. Group translocation is mainly found in the anaerobic and facultative anaerobic bacteria, such as *E.coli, Salmonella typhimurium, Staphylococcus aureus and lactobacillus*. For example, during transportation of glucose in *E.coli* and lactose in *Staphylococcus aureus*, these sugars have reaction of phosphorylation and exist in the cytoplasm in a form of phosphoric acid sugar, and the phosphate group in phosphates sugar comes from phosphoenolpyruvate (PEP). Therefore, the group translocation is also called phosphoenolpyruvate-phosphates sugar transferase transport system (PTS), in short, the phosphotransferase system.

These mentioned above are four modes of nutrient absorption in microbe cells except protozoa. They are similar to those by which metabolites are discharged from microbial cells. The four transport types of nutrients are compared in Table 7.1.

Table 7.1 Comparison between four transport modes of nutrients in microbial cells

Items compared	Simple diffusion	Facilitated diffusion	Active transport	Group translocation
Specific carrier proteins	No	Yes	Yes	Yes
Transport speed	Slow	Fast	Fast	Fast
Solute transport direction	From thick to thin	From thick to thin	From thin to thick	From thin to thick
Balance and concentration	Equal	Equal	High concentration inside cell	High concentration inside cell
Molecular transportation	Nonspecificity	Specificity	Specificity	Specificity
Energy consumption	Unwanted	Unwanted	Need	Need
Solute changes before and after transportation	No chemical changes	No chemical changes	No chemical changes	Chemical changes
Carrier saturation effect	No	Yes	Yes	Yes
Examples	H_2O, CO_2, O_2, glycerol, ethanol, a few amino acids and salts etc.	PO_4^{3-}, SO_4^{2-}, sugar (eukaryotic microorganism)	Amino acids, lactose and other sugars, K^+, Ca^{2+}, other inorganic ions	Glucose, mannose, fructose, purine nucleoside, fatty acid, etc.

7.3 Fermentation Process Control

According to culture methods and requirements for fermentation conditions, culture of industrial microbial seeds can be divided into anaerobic culture and aerobic culture during the fermentation process. Seed culture requires certain culture conditions (including components of culture medium, strain quality, temperature, humidity, culture time, inoculation quantity, age, pH value, dissolved oxygen and so on) and the suitable culture method (including the batch culture, semi continuous culture method, i.e. cyclic cultivation), so as to guarantee normal seed mass propagation. Fermentation production requires certain fermentation conditions (including components of culture medium, inoculation, temperature, pH value, dissolved oxygen, cell concentration, feeding, etc.) and fermentation method (including batch fermentation, fed-batch fermentation and continuous fermentation). By culture of microorganisms and utilizing the cell enzyme system, some components in the suitable culture medium are converted to new cells and metabolites through complex biochemical reactions.

7.3.1 Fermentation Methods and Operation Modes

7.3.1.1 Fermentation methods

1) Aerobic fermentation method

Microbial anaerobic fermentation is also called static fermentation. When target microorganisms are inoculated into the culture medium, in the absence of air in a closed fermenter, some substances are converted to metabolites, such as alcohol, acetone and butanol, lactic acid, glycerol, methane (biogas). It is a strictly anaerobic fermentation during which oxygen in fermenter should be pumped, or water is used to produce hydrogen reacting with oxygen in the vacuum environment. Anaerobic and facultative anaerobic bacteria are dominant in the strains used for this production.

Production of alcohol with starch as raw material includes the following steps, firstly utilizing glucoamylase to convert starch and dextrin into simple sugars, afterwards *Saccharomyces cerevisiae*'s performing anaerobic fermentation in a closed fermentation tank to produce alcohol and carbon dioxide through the EMP pathway as follows:

$$\text{Starch} \xrightarrow{\textit{Aspergillus niger} \text{ glucoamylase}} \text{Glucose} \xrightarrow{\text{Yeast EMP pathway}} \text{Ethanol} + CO_2$$

The fermentation enzyme system in *Saccharomyces cerevisiae* is very complex, and mainly includes: ①invertase, it can transform sucrose into glucose and fructose; ②maltase, it can decompose maltose into glucose; ③zymase, it refers to the generic term of enzymes which participates in catalytic conversion of glucose to ethanol and CO_2. Alcohol fermentation does not need free oxygen. If the oxygen exists, sugars will be completely decomposed into water and CO_2, and a great amount of biomass and energy will be obtained simultaneously. Therefore, the whole process of alcoholic fermentation is done in a sealed fermentation tank.

2) Aerobic fermentation method

Aerobic bacteria and facultative aerobic bacteria are dominant during the microbial aerobic fermentation, and their growth environment must supply sterile air to maintain a certain level of dissolved oxygen in fermentation broth, so that the cell can grow and perform fermentation. Aerobic culture is divided into shallow pan culture, solid thick layer ventilation culture and liquid submerged ventilation culture which now mainly has been adopted in fermentation production. During the culture process, fermentation technology conditions should be controlled so as to meet the requirements for normal growth of microbes and accumulation of products. Glutamate fermentation as shown in Fig.7.10 is a typical aerobic fermentation.

Glucose is converted to pyruvate via the EMP pathway and HMP pathways, and then pyruvate can be turned into acetyl coenzyme A which participates in the TCA cycle to produce α-ketoglutaric acid. With glutamate dehydrogenase and presence of NH_4^+, α-ketoglutaric acid is converted to glutamate through reduction amination. Glutamic acid fermentation is due to abnormal metabolism of microbial cells, and it can be accumulated only when normal metabolism in bacteria has disorders. And when biotin is limited, permeability of the cell membrane will change, and glutamate is ready to escape from the fermented broth. Therefore, during the actual production process, control of dissolved oxygen level and growth factor content in the medium should be noted well Glutamate biosynthesis pathway is shown in Fig.7.10.

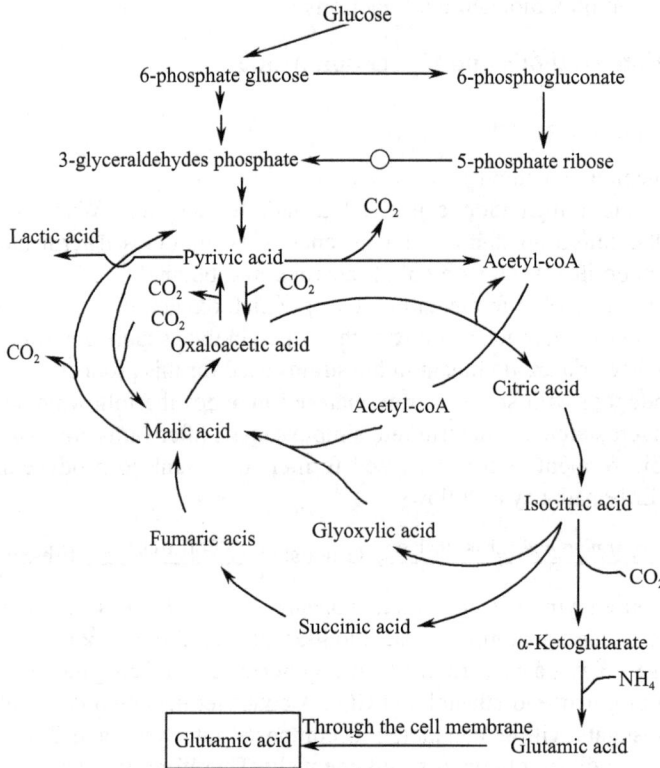

Fig.7.10 Glutamate biosynthesis pathway

7.3.1.2 Fermentation operation modes

1) Batch fermentation

Batch fermentation is defined as a completing whole fermentation process in a closed system. In addition to regulation of pH, other materials are not added during the fermentation. In batch fermentation, microbial growth can be divided into adjustment (stagnation) phase, exponential phase, balance (stability) phase and decline phase, the microbial growth environment changes constantly, and a small quantity of and several classes of products can be produced. This operation mode has advantages, when the operation condition changes or production of new products is needed, it is easy to change the processing countermeasure; when microbial contamination occurs, it is also easy to terminate operations. In addition, the requirements for composition of production raw material are not strict.

2) Fed-batch fermentation

The fed-batch fermentation is defined as a culture method to intermittently or continuously feed one or more kinds of fresh media in the batch fermentation process. It is a transition method between the batch fermentation and continuous fermentation and it is a good method for control of fermentation process and is widely used in the fermentation industry, such as fermentation production of antibiotics, amino acids, enzymes preparation, SCP, vitamin, nucleotide, organic acid, organic solvent, hormone drugs, etc.

Comparing with batch fermentation the advantages of fed-batch fermentation are as follows. ①Fed-batch fermentation may be operated to maintain very low matrix concentrations in a fermentation system, to remove substrate repression effect, product feedback inhibition and repression effect of glucose and other catabolites, i.e. to remove repression effect to rapidly utilize carbon sources. ②This method can avoid cell excessive growth due to feeding too much at one time in batch fermentation and the negative effect of oxygen supply and consumption imbalance. It can improve rheology properties of the fermentation broth and can reduce cell quantity to some extent and increase conversion rate of the target product. The appropriate cell concentration can be maintained to meet oxygen supply and demand balance. ③Microbial cells can be automatically controlled in a consecutive series of transitional stages so as to adjust cell quality; the transition state at some stage can be repeated, and it can be used as an approach for theoretical study.

3) Continuous fermentation

Continuous fermentation may be considered as an open system in which the medium is continuously added to the bioreactor and the equal volume of fermentation medium is simultaneously removed, so that microbial growth and metabolic activities remain in the strong stable state in the fermentation tank. During the continuous fermentation, the environmental conditions, such as pH value, temperature, dissolved oxygen, the concentration of nutrients, product concentration, microbial cell concentration, specific growth rate, maintain relatively stable (from beginning to end remaining basically unchanged), which can be adjusted through external control so that the cell can continuously grow at a constant growth rate during fermentation. The greatest feature of the continuous fermentation system is that the cell growth rate and metabolite synthesis rate remain in a constant state so as to achieve the purpose for stable and speeded culture of microbial cells or production of a large quantity of metabolites. So the production efficiency and the utilization rate of equipment can be improved significantly.

7.3.2 Fermentation Process Control

7.3.2.1 Fermentation media

In terms of microorganism nutrition requirements, all of microbial growth and accumulation of fermentation products need carbon source, nitrogen source, inorganic elements, H_2O, energy and growth factor. Aerobic organisms also need oxygen. Carbon source is the basic substance which will supply energy for cell life activities, composition of cells and metabolites. Nitrogen is the main source of constituting cell substances and metabolites, such as protein, amino acids and nitrogen metabolites. Microbial growth and biosynthetic process also need macroelements and microelements, such as magnesium, sulfur, phosphorus, potassium, manganese, etc. Some special micro-growth factors such as biotin, thiamine, inositol, etc. are essential for auxotrophic microorganisms. A variety of biochemical reactions in the living organisms must be performed in aqueous solution, nutritional substances can be utilized by microorganisms through the cell membrane when they are dissolved in the water. Production of other products also requires inducers, precursors and accelerants.

The medium is a mixture of various nutrients on the basis of certain proportion, which provides material composition for microbial growth and reproduction, and biosynthesis of various metabolites. Composition of the culture medium has important influence on microbial growth and breeding, product biosynthesis, separation and purification of products, quality and output. Microbial nutrition activities rely on secretion of a large number of enzymes which decompose macromolecules, such as protein, carbohydrate, fat and other nutrients in surrounding environment, into small molecular compounds absorbed by microbes with the help of permeability of the cell membrane. Different microbial growth or biosynthesis of different fermentation products requires different media which have some similarities in design, i.e., they must provide all necessary nutrients for microbial growth and reproduction and synthesis. For large scale fermentation production, in addition to the microbial needs, we must pay attention to prices and sources of raw material for the culture media.

Considering carbon source and nitrogen source, coordination between rapid utilization of carbon (nitrogen) source and slow utilization of nitrogen (carbon) source should be noted to play their advantages and avoid their shortcomings. The appropriate ratio of carbon to nitrogen is very important for medium composition. If nitrogen source is excessive, reproduction of bacterial cells will be exuberant and the pH value will be too high, which does not favour accumulation of metabolites. If nitrogen source is insufficient, the number of bacterial cells will be small and then the yield will be influenced. If carbon source is excessive, the pH of fermentation broth will be lower; but if carbon source inadequate, cell senescence and autolysis will take place in the fermentation broth. Furthermore, the improper ratio of carbon to nitrogen also affects proportional absorption of nutrients which directly affects cell growth and product formation. The optimal ratio of carbon to nitrogen is not the same at different growth stages of bacterial cells. Because carbon is used as both the carbon frame and the energy source, the amount of carbon is more than nitrogen source. The carbon-nitrogen ratio is about 100:0.2-2.0 in general fermentation industry. But in fermentation of amino acids, because molecular of amino acid contains nitrogen, the ratio of carbon to nitrogen is relatively high. We should also pay attention to physiological acidic salt, physiological alkaline salt and pH buffer addition and collocation. According to pH changes in strain growth and synthesis of product, as well as the optimal pH control range in the existing process and equipment conditions, selection of physiological acidic, alkaline substances and their amount should be considered from a comprehensive viewpoint, so as to ensure the pH can be maintained in the optimal state during the whole process of fermentation.

7.3.2.2 Effect of temperature on fermentation and its control

The temperature is one of the important influencing factor in fermentation, which can influence various enzyme reaction rates, change the synthesis direction of bacterial metabolites, and influence metabolic regulation mechanisms in microorganisms, physicochemical properties of fermentation broth, fermentation kinetics characteristics and product biosynthesis. Therefore, temperature has effects on cell growth and metabolite formation. In order to ensure normal fermentation, the optimal fermentation temperature which is suitable both for cell growth and for metabolites synthesis must be controlled during the fermentation process. The optimal temperature varies with strains, medium composition, culture conditions and cell growth phases. In theory, only one culture temperature should not be adopted, but different culture temperatures should be chosen according to different fermentation stages during the whole fermentation process. At the growth stage, the most suitable growth temperature should be chosen, but at the product secretion phase, another optimal temperature should be chosen. For example, in production of glutamic acid, the temperature is kept at about 30-34℃ to allow optimal growth of producing bacteria, and the optimal temperature for formation of glutamate is 34-36℃. Therefore, at the early stage (0-12h) of seed culture and fermentation of glutamic acid, the optimal growth temperature should be 30-32℃(Beijing *Corynebacterium* Asl.299) or 32-34℃(*Corynebacterium crenatum* Asl.542), which is useful for biosynthesis of protein and nucleic acid from nutrients which are used for reproduction of bacterial cells. The middle and later stages (after 12h) of the fermentation focus on accumulation of glutamic acid in the fermentation medium, and at the stages bacterial growth has been discontinued. But the optimal temperature of glutamate dehydrogenase is 32-34℃(Asl.299) or 34-36℃(Asl.542), therefore at the later stage appropriate increase in fermentation temperature is favorable to glutamate accumulation.

7.3.2.3 Effect of pH on fermentation and its control

The values of pH have great effect on cell growth and metabolite formation in the following aspects, enzyme activity; charge state on cell membrane, cell membrane permeability; absorption of nutrients and metabolites excrete; nutrient dissociation degree; metabolic pathways and their products.

Nutrient metabolism in the medium is the very important reason for variation of pH values which is the comprehensive result of microbial metabolism. Fermentation pH is different with different strain, even if the same strain, the optimal pH values for growth and the one for product synthesis are different. For example, in *Clostridium acetobutylicum* fermentation, cell can grow well under the neutral pH, but the product yield is very low, and in the actual fermentation the suitable pH value is 4 to 6.

In general, the optimal growth pH value can be determined according to the experimental results. Because the optimal pH value of the same fermentation product is related with strains, culture medium composition and culture conditions, when we determine the appropriate pH value for some product effect of incubation temperature should be also considered. If temperatures are different, the optimal pH values may vary.

After learning about the requirements for the optimal pH value in the fermentation process, various control methods should be adopted. Firstly, when we take into account basic formula of the fermentation media, various components should have proper proportion between them so that changes of the fermentation pH values lie in a suitable range. During the batch fermentation process, buffer salt is often added to media to control changes of pH values. Secondly, acidic or alkaline substances can be directly added to media to control pH values.

7.3.2.4 Changes and control of concentrations of dissolved oxygen

Concentration of dissolved oxygen is one of the most important parameters in aerobic fermentation. Microbes in liquid can only use the dissolved oxygen except for those in gas-liquid interface which can also use the oxygen in gas phase. The oxygen solubility in fermentation liquid is very small, only 0.22mmol/L, therefore, in order to increase concentration of dissolved oxygen, constant ventilation and stirring can meet the demand of aerobic microorganisms for dissolved oxygen. The quantity of dissolved oxygen is mainly determined by ventilation and stirring speed and also associated with the diameter-height ratio of a fermentation tank, the liquid layer thickness, types of agitator, agitating blade diameter, medium viscosity, fermentation temperature and tank pressure. In actual production, the stirring rate is constant, oxygen supply level changes through ventilation adjustment. The quantity of dissolved oxygen will have different effects on cell growth, product formation and its yield.

The dissolved oxygen for microbial respiration and metabolism in the medium must keep balance with oxygen consumption of microbes, in order to satisfy microorganisms' utilization of oxygen. The minimum oxygen concentration maintaining microbe respiration in the culture medium of aerobic fermentation is called critical dissolved oxygen concentration. Therefore, critical dissolved oxygen concentration and optimal oxygen concentration must be considered for each kind of fermentation products during fermentation, so as to make the amount of dissolved oxygen in the fermentation liquid maintain the optimal range of the dissolved oxygen concentrations. The optimal concentration of dissolved oxygen is related with product synthesis and characteristics of anabolism in microbes, and can be determined through experiments.

7.3.2.5 Effect of cell concentration and matrix on fermentation and their control

Cell concentration refers to cell content in a unit volume of culture liquid. Cell concentration has close relation with cell growth rate. Cell growth rate is related with species of microorganisms and their genetic characteristics. The growth rate varies in different microorganisms. The more complex cells are, the longer time is taken for cell division.

Cell concentration has great influence on the yield of fermentation products which is proportional to cell concentration at appropriate specific growth rate. The higher the cell concentration is, the higher the product yield is. However, if cell concentration is too high, consumption of nutrients will be too rapid and toxic metabolites will accumulate, which may change the metabolic pathways of microbial cells, especially make dissolved oxygen in the culture medium decrease significantly make it become a restrictive factor. Therefore, in order to obtain the highest product yield, the optimal cell concentration will be controlled during fermentation. In some culture conditions, growth rate is mainly influenced by nutrient concentration. Therefore, cell concentration is regulated by adjusting the medium concentration during fermentation production. Firstly, the appropriate ratio between ingredients of the basic culture medium should be determined to avoid too high or low cell concentration, and then concentration should be controlled by feeding in the middle of fermentation. Secondly, cell concentration can also be controlled by controlling sugar supplement in the production process by utilizing the amount of carbon dioxide produced in metabolism of microbial cells.

Matrix is the material basis for metabolism of microbial cells. It relates to both growth and reproduction of microbial cells and metabolite formation. Its types and concentrations have close association with cell growth and metabolism.

In batch fermentation, cell growth rate depends on the concentration of matrix. Within the upper limit of the concentration of nutritional matrix at which the specific growth rate reaches the maximum, the growth rate of microbial cells has linear correlation with the matrix concentration. When the concentration is higher than the limit of the matrix concentration, the high osmotic pressure resulting from the high concentration of matrix will lead to cellular dehydration and inhibition of growth, which causes decrease in the specific growth rate of cells. This effect is known as the substrate inhibition.

The concentration of carbon source has obvious effect on fermentation. Abnormal cell reproduction caused by high concentration of carbon source have adverse effect on cell metabolism, product synthesis and oxygen transmission. If carbon source produced repression effect is in excessive consumption, synthesis of product will be inhibited obviously; on the contrary, just providing the amount of carbon source for life maintenance, cell growth and synthesis will stop. Therefore, control over the concentration of carbon source is very important for fermentation industry.

Nitrogen source concentration also has significant effect on fermentation. Therefore, the nitrogen concentration should be controlled in production by feeding and reasonable collocation between rapid and low utilization of nitrogen source.

7.3.2.6 Effect of carbon dioxide on fermentation and its control

Carbon dioxide is a microbial metabolite, as well as a substrate of certain synthetic metabolism, and it is the important index of cell metabolism. CO_2 has stimulating effect on cell growth. However, CO_2 usually inhibits cell growth, which leads to reducing carbohydrate metabolism and respiratory rate of microbes. When CO_2 concentration in the exhausted gas is higher than 4%, microbes will decrease metabolism of sugar and the respiratory rate.

CO_2 also has impact on microorganism fermentation. Dissolved CO_2 in the fermentation broth has a stimulating effect on fermentation of some amino acids and antibiotics. Except for the effect on cell growth, morphology and product synthesis, CO_2 may affect acid-base balance in the fermentation fluid so as to reduce the pH value of fermentation broth, or react with other chemical substances, or react with metal ions essential for growth to form carbonate precipitation, or decrease dissolved oxygen concentration due to excessive consumption of oxygen, all of which can indirectly influence cell growth and synthesis of products. Because the solubility of CO_2 is greater than oxygen in a fermentation broth, as the pressure increases, the content of CO_2 increases faster than that of oxygen. Therefore, in order to eliminate the CO_2 effect, these factors, such as the solubility of CO_2 in culture media, fermentation temperature and ventilation, must be considered.

According to the effect of carbon dioxide on fermentation, CO_2 concentration must be controlled. If CO_2 has stimulating effect on fermentation production, its concentration should be increased; otherwise its concentration should be tried to reduce to remove inhibition of product. Dissolved oxygen concentration and concentration of CO_2 are controlled simultaneously by improving aeration and the stirring rate in the fermentation process. In addition, the concentration of CO_2 is also affected by the fermentation tank pressure. By increasing the tank pressure, with increase in the dissolved oxygen concentration, the concentration of CO_2 increases. Because the solubility of CO_2 is much greater than that of oxygen, in the high tank pressure the concentration of CO_2 in liquid phase is high, which is nor favorable for discharge of CO_2, cell metabolism and other parameters.

7.3.2.7 Effect of foam on fermentation and its control

In microbial aerobic fermentation process, due to effect of ventilation and stirring, microbial metabolism, medium composition and physicochemical properties, it is a normal phenomenon that a lot of foam is produced in the fermentation broth. The reasons are as follows:①ventilation and stirring; ②microbial respiration and metabolism; ③the foaming material in the fermentation medium. When during fermentation contamination of other bacteria and phages occurs, foam abnormally increases. Although some small amount of foam is a normal phenomenon in aerobic fermentation, the excessive foam brings negative effect on fermentation, such as decrease in the fermentation tank loading coefficient, decrease in the oxygen transfer coefficient, etc. If the foam is too much and is not controlled, it will lead to a great amount of "escaping liquid" causing economic losses, and at the same time the fermentation broth escapes from the exhaust pipe or the tank top seal, which will increase contamination chance. When foam formation is more serious, ventilation and stirring cannot be performed, cell respiration is impeded, resulting in metabolic abnormalities or cell autolysis in advance. Cell autolysis will lead to formation of more foam. Therefore, controlling foam is the basic elements to ensure normal fermentation. In fermentation industry, mechanical defoaming or chemical antifoam agent defoaming, or both of them at the same time are adopted usually.

The basic task of fermentation engineering is to utilize the inherent capacity of microorganisms efficiently to obtain maximum biological products at the minimum energy and material consumption. Therefore, effective control of the fermentation process must be done. Based on the monitoring of a variety of parameters, qualitative and quantitative description of the production process, the purpose of effectively controlling the process of fermentation is achieved.

Microbial growth and metabolism is a dynamic alteration process, which belongs to an open system, namely, cells exchange all kinds of materials constantly with the external environment. Fermentation parameters can correctly reflect the conditions of fermentation and metabolic changes. Especially variation of pH values in growth and metabolism of microbes is the comprehensive index of cell growth and metabolism. Various parameters can be effectively controlled through on-line or off-line detection.

7.4 Fermentation Equipments

7.4.1 Anaerobic Fermentation Equipments

Fermentation equipment is the main part in a fermentation plant, and can provide the proper place for life activities of microorganisms and their biological metabolism. According to the degree of microbial need for oxygen, microorganisms can be divided into aerobic, anaerobic and facultative anaerobic microbe. Therefore, there are different fermentation equipments to meet the oxygen needs of different microorganisms. For example, alcohol and beer are anaerobic fermentation products produced by facultative anaerobic yeast, and the fermentation equipment is anaerobic fermentation equipment without oxygen in fermentation and fermentation is relatively simple. This section focuses on alcohol and beer fermentation equipment.

7.4.1.1 Alcohol fermentation equipment

In order to convert sugar into alcohol by *Saccharomyces cerevisiae*, and increase the conversion rate, many factors must be considered in design of alcohol fermentation tanks. In addition to meeting the necessary conditions of yeast growth and metabolism, we still should consider cooling facility and discharge in fermenter, equipment cleaning, repair and manufacturing, convenient installation, and so on.

1) Structure of alcohol fermentation tanks

The fermentation tank is a common sealed type which is designed to retrieve released alcohol and CO_2 gas discharge and to utilize comprehensively CO_2 gas. The tank body is cylindrical, and the bottom cover and the top cover are dish-like or cone-shaped. As shown in Fig.7.11, there is inlet 5 of liquor and yeast, the pipe pressure gauge 3, the CO_2 waste gas recovery (EGR) pipe 4, the

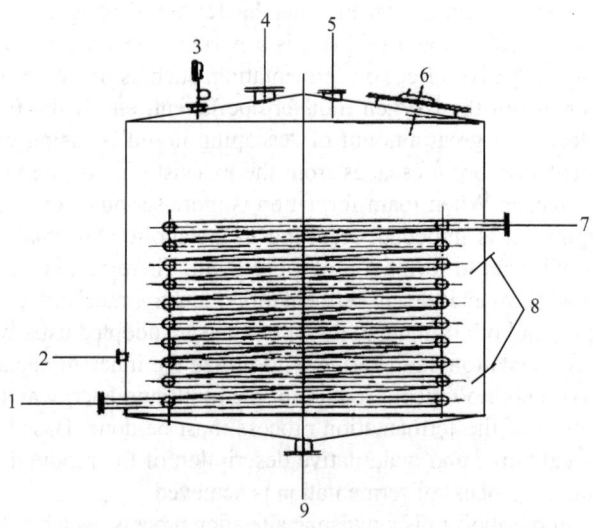

Fig.7.11 Alcohol fermentation tank
1-Cooling water entrance; 2-Sample connection; 3-Pressure gauge;
4-CO_2 Outlet; 5- Liquor and yeast entrance; 6-Manhole; 7-Cooling water outlet;
8-Thermometer; 9-Fermentation liquid and sewage outlet

manhole 6 for cleaning and observation and maintenance of the interior of the tank, as well as various interface pipes for monitoring instruments on the tank top. The bottom of the tank is equipped with the fermented broth and sewage outlet 9; the tank body is equipped with the sampling pipe 2, the thermometer connection pipe 8, and the cooling water inlet pipe 1 and the outlet pipe 7.

2) Fermentation tank cooling

A certain amount of biochemical reaction heat can be released in the alcohol fermentation process, and may make the fermentation temperature rise and directly influence yeast growth and formation of metabolites if the heat is not removed in time. Therefore, cooling device should be equipped with the fermentation tank. For small and medium fermentation tanks, the tank top with spray shower for cooling of the outer wall surface film of the tank; and for large fermentation tanks, the outer wall cooling area of the tank cannot meet the cooling requirements, so the tank is cooled with both a cooling coil pipe and a spray cooling device in the outer wall. In addition, tank outside pipe spray cooling method has some benefits, such as uniform cooling fermentation broth and high cooling efficiency. The tank outside pipe spray cooling method requires a water collecting tank installed at the tank bottom and around the tank, sewage is drained from the outlet of the water collecting tank into the sewer, so as to avoid wetness and water accumulation in the fermentation workshop, which may affect sanitation of the workshop and the operation.

7.4.1.2 Beer fermentation equipment

In recent years, large-volume open top fermenters have been used in domestic and foreign breweries, and especially in China, regardless of the old factories or newly built plants, almost all of them adopt the vertical fermentation tank with cone-shaped bottom (referred to as a conical tank).

The fermenter is a vertical cylindrical metal device with conical bottom, which is made of stainless steel sheet or plate coated with epoxy resin. As shown in Fig.7.12, the diameter-height ratio of the cylindrical part of a large tank, with the diameter 2-5m and the height 10-30m, is 1.5-6, and its volume is 40-600m^3 and the common volume 150m^3 or 200m^3. The height of the wort in fermenter should not be too high, or too high pressure will be produced and convection will be enhanced, which may influence on full reduction of diacetyl by yeasts. It is recommended that the wort height be up to 15m, and also 14m or 12m in the conical tank. The angle of the conical bottom should be 60° and this facilitates in yeast precipitation and accumulation. According to different tank volumes, several strip cooling jackets should be equipped at the upper, middle and lower part, and the conical bottom of the fermentation tank, respectively. This is called four-section cooling; the lower section and the conical bottom section integrate into one, therefore there are three sections of microcomputer-controlled temperatures. The inside of the jackets is injected with alcohol water at $-3°C$ to $-4°C$ or salt water, and liquid ammonia for cycling can also be used. The outside of the cooling jackets is wrapped with 20cm thick polyurethane or polystyrene foam plastic insulating layer, whose outside is again coated with a layer of thin metal plates used as the protective layer. The top cover and the cone bottom also should have some heat preservation devices. So the tank can be fully put outdoor to become an open top fermentation tank. The automatic washing devices 7 are installed inside the fermentation tank, and now spraying ball washing has been adopted. The tank body is equipped with the sampling hole 2 at its mid-lower section, and the platinum resistor (Pt100) interfaces for control over temperature, and the cooling agent inlet 8 and outlet 9 are set at the upper, middle and lower sections of the tank, respectively; and the manhole 12 (not shown in the figure) in the tank body is designed for observation and repair of the fermentation tank. The tank top is equipped with the pressure gauge 3, the exhaust valve 4, the safety valve 5, the manhole 6, and automatic capping device. When the tank pressure exceeds 0.1MPa, CO_2 will discharge from the exhaust valve at the tank top, so that the pressure inside the tank can be maintained less than 0.1MPa, usually 0.08MPa, otherwise it will inhibit the yeast ethanol fermentation. If the tank pressure exceeds 0.2MPa, the safety valve at the tank top automatically will open in order to control the content of CO_2 in the tank. The sterile wort and yeast inlet 11 is set up at the conical angle of the tank bottom. When the fermentation is exuberant, all the cooling jackets are used to maintain a proper fermentation temperature (called starting cooling in plants), and the yeast cells deposited at the conical bottom are discharged outside the tank by opening the conical angle valve 11. Mature beer is released from the beer outlet 10 on the side of the conical bottom. In some breweries, the yeast mud is discharged at the outlet with a movable pipe, and then is put in the yeast mud storage tank for repeated use, but the yeast cells cannot be used for more than 5 generations. Then the mature beer is separated from the movable pipe for centrifugal isolation, and removing of residual yeasts and impurities in the beer.

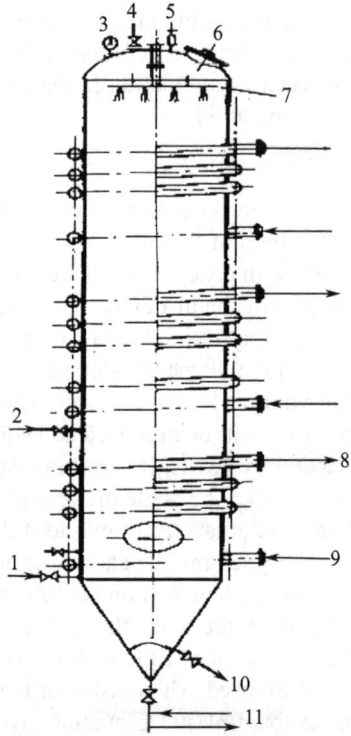

Fig.7.12 A vertical fermentation tank with conical bottom
1-Washing CO_2 entrance; 2-Sampling hole; 3-Pressure gauge;
4-Exhaust hole; 5-Safety valve; 6-Manhole; 7-Scrubber; 8-Ice brine outlet;
9-Ice brine entrance;10-Beer outlet; 11-Malt wort entrance (or beer exports)

In order to saturate CO_2 in fermention broth, the CO_2 inlet 1 for washing and purifying is installed on the side at the bottom of the fermentation tank. When CO_2 is inadequate during the fermentation process, CO_2 is injected into the fermentation liquid from the inflatable pipe hole.

The conical tank adopts the CIP (cleaning in place) washing system which can automatically wash according to a predetermined program. The conical tank can be used as a fermenter, as well as a beer storage tank, and can complete the whole prefermentation and postfermentation process. The bottom of the tank is conical, and is helpful for yeast precipitation and accumulation and makes collection of yeast simpler. In addition, it is easy to control, convenient to clean and sterilize, and can realize automatic and semi-automatic control of the temperature, the pressure, the loading height and other parameters. It also can realize mechanization of the operation process, so as to reduce labour intensity and improve working conditions. Its loading coefficient is high up to 85%-90%, so that it has high equipment utilization rate; and its production cycle is shorter 1/3 than traditional fermentation, improving the production efficiency. Furthermore, it also has some other benefits, such as less loss of fermentation extract, recyclable CO_2, saving investment and small occupied area, etc. Therefore, the conical tank has become the main fermentation equipment in contemporary beer factories.

7.4.2 Aerobic Fermentation Equipment

Fermentation equipment is one of the most important equipments of the fermentation engineering. In order to obtain the maximum production efficiency, an excellent culture device should be designed with tight structure, good liquid mixing performance, high mass transmission rate and high heat transmission rate, as well as reliable detection and control instruments.

Aerobic fermentation device (also called ventilating fermentation equipment) should increase the oxygen dissolving rate by ventilating and mixing, so as to meet the requirements for aerobic microbial growth and metabolite accumulation. For instance, aerobic fermentation tanks have been adopted in production of amino acids, citric acid, enzymes and antibiotics.

At present according to the characteristics of aerobic microbes, many types of aerobic fermentation tanks have been developed. According to energy input modes, the aerobic fermentation tank is divided into three types: internal mechanical stirring fermentation tank, external liquid stirring fermentation tank and air spray lift fermentation tank.

Internal mechanical stirring fermentation tank has internal mechanical agitator to stir the fermentation broth. Mechanical agitating fermentation tank, mechanical agitation self-suction fermentation tank and Sleeve fermentation tank belong to this kind.

External liquid stirring fermentation tank relies on an external cycling pump for mixing the fermentation broth, or a Venturi tube is set up at the liquid entrance of the fermentation tank, relying on the liquid flow at a high speed to inhale air so as to mix the gas and the liquid mixture. For example, Venturi tube fermentation tank belongs to this type.

Air spray lift fermentation tank relies on compressed air as a way of energy input, which makes fermentation broth turn up and down to homogeneous mixing. High tower fermenter and deep well aeration sewage treatment tank belong to this type.

7.4.2.1 Mechanical agitating fermentation tank

Mechanical agitating fermentation tank is also called standard or general fermentation tank which is most widely used in fermentation plants and has mechanical agitator to mix the air and the fermentation broth so as to obtain the oxygen dissolved in the mash and ensure the oxygen supply for microbial growth and fermentation.

The geometric sizes of various parts of the standard tank have certain proportion, as shown in Fig.7.13. The ratio of the tank cylinder height (H) to the internal tank diameter (D) is generally 1:1.7-1:4. In novel high level fermentation tanks, the ratio is 10 times over, which can help increase the air utilization rate. Small fermentation tanks are equipped with 1-2 sets of agitators, and large fermentation tanks have 3 or more than 3 sets. The ratio of the agitator diameter (d) and the fermentation tank diameter (D) is generally 1/2-1/3; the ratio of the spacing distance (S) between two sets of agitators to the agitator diameter (d) is 1.5-2.5; and the ratio of the spacing distance (S) between three sets of agitators to the agitator diameter (d) is 1-2. The distance between the set of agitators at the top and the liquid level should not exceed the agitator diameter. The set of agitators at the bottom approaches the air duct outlet, and the distance from the tank bottom B is generally equal to the agitator diameter D, it should not be less than 0.8D, otherwise liquid circulation will be affected.

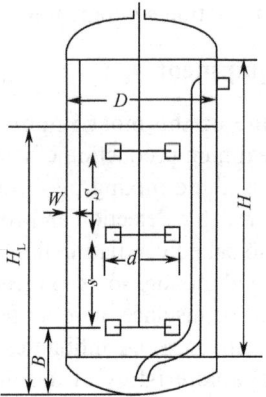

Fig 7.13 Sizes of standard fermentation tank

H-Height; D-diameter; D-Agitator diameter; W-Baffle width; B-Distance of agitator bottom;
S-Distance between two stirring paddles; H_L- Height of liquid level

  Mechanical stirring ventilating fermentation tank mainly comprises a tank body, agitator, baffle, air distribution apparatus, defoaming device, cooling device, shaft seal and others.

  The tank body is a sealed container formed by welding the cylinder to the dished or elliptical heads at the two ends. The type of fermentation tank has many characteristics, such as force uniformity, less dead corner, easy discharge of materials. It is made of carbon steel or stainless steel, and large fermentation tanks are made of 2-3mm stainless steel lining or composite of stainless steel. Tank thickness sizes depend on the diameter and the tank pressure. Because the fermentation tank needs to maintain a certain tank pressure in fermentation and sterilization, it must stand with pressure. Therefore, the tank body should be designed according to the pressure container.

  There are multiple agitators to arrange on the stirring shaft configuration according to the fermentation tank volume in order to blend fermentation liquid fully. The number of agitators of the tank depends on the liquid height in the tank, fermentation liquid properties and agitator diameter. The stirring shaft generally extends from the tank top into the in-tank, its middle section is fixed on the inner wall of the tank with a steel strip, and the lower section is fixed at the bottom of the tank. The tank also has a baffle plate, defoamer, air distribution device, shaft seal and heat exchange device. The baffle plate may change the direction of liquid flow from radial flow to axial flow, which makes the liquid to turn intensely so as to increase quantity of dissolved oxygen. Defoaming device utilizes mechanical power to break foam, so that the exhaust gas and liquid can be separated and liquid material is recovered, so as to increase the tank load. The role of air distribution device is to blow the air into fermentation liquid, and make the air distribute evenly. The shaft seal can seal the slot between the tank top or the tank bottom and the stirring shaft, preventing leakage and contamination.

7.4.2.2 Air lifted circulation fermentation tank

The tank refers to the air-lift (belt) fermentation tank (Fig.7.14), also known as circulating ventilation fermentation tank. It utilizes air power to make liquid rise in the circulating tube, and along a route for circulating. For those media containing a small amount of solid substance such as molasses, hydrolyzed sugar or sucrose, the air-lift fermentation tank is more advantageous.

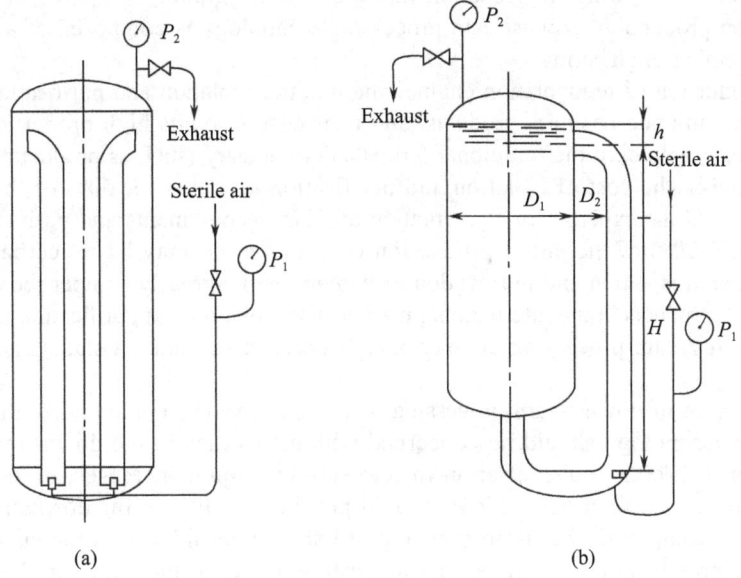

Fig.7.14 Air belt lifting circulation fermenter
(a) Internal circulation belt air-lift fermenter; (b) External circulation belt air-lift fermenter

According to installation position of the circulation pipe, the air-lifted fermentation tank can be divided into the internal circulation type (a) and the external circulation type (b) (Fig.7.14). The air-lift fermentation tank of $100m^3$ is made by welding 12-16mm thick stainless steel cylinder to the ellipsoidal head. The tank cylinder diameter is 3.8m, its hight is 11.6m, the effective volume 80-85m^3, the optimal diameter-height ratio 1:4-1:6, and limited height is 22-24m.

The working principle of external circulation air-lift fermentation tank is that the ascending pipe is set up on the outside of the tank, connects with the bottom and the upper part of the tank to form a circulating system. An air nozzle is installed in the lower part of the ascending pipe, and the air is sprayed into the ascending pipe at the speed of 250-300m/s. With the aid of the nozzle the air bubble will be segmented into fine bubbles, which contacts closely with of the fermentation liquid in the ascending pipe. Since the fermentation liquid in the ascending pipe has low specific weight and the compressed air produces spraying kinetic energy, the liquid rises in the ascending pipe, and the liquid in the tank drops and runs into the ascending pipe, forming a repeated circulation by which dissolved oxygen can be supplied for the fermentation liquid to ensure normal fermentation.

7.5 Downstream Processing Technology

7.5.1 Introduction to Downstream Processing Technology

Downstream processing engineering is one part of biological engineering. Biochemical products are made by means of microbial fermentation, enzyme reaction process or animal and plant cells culture. Downstream processing engineering is a process in which the product is isolated and purified from fermented broth, the reaction liquid or culture liquid. Compared to strain breeding and fermentation production, downstream processing technology is composed of a number of unit operations in chemical engineering.

During production of fermentation engineering product, isolation and purification are important for obtaining the ultimate business products, and their cost accounts high proportion of the whole product cost. For example, in the traditional fermentation industry (such as production of antibiotics, ethanol, citric acid), the cost of isolation and purification accounts for 60% of the entire factory investment cost, and the expense for fermentation of DNA recombinants and purification of protein accounts for 80%-90% of the entire production cost, this trend may be exacerbate unceasingly. Therefore, backward isolation and purification technology will seriously hamper the competitiveness of bioengineering products in the international market. The isolation and purification technology often is called the downstream processing engineering, in order to indicate its status and importance in bioengineering.

The importance of downstream processing technology makes people realize that development of the upstream technology should be concerned with difficulties in the downstream technology. Otherwise, even if product concentration in fermentation liquor increases, products cannot yet be obtained. So the upstream technology should provide conditions for downstream extraction of products. For example, in the strain breeding, we should consider the following questions: the strain cannot be producing or producing less impurities similar to the target products; the original intracellular products should be changed into the extracellular ones; and intracellular inclusion bodies should be formed, and they can be settled in the condition of low centrifugal force following cell disruption. These powerful measures may be solve some difficulties in downstream processing.

7.5.1.1 Characteristics of downstream processing technology of biotechnology

The fermentation broth is a complex multiphase system, and it is Newtonian liquid and has high viscosity. The solid and colloid substances which disperse in the system have compressibility, its density is similar to the liquid, it is very difficult to isolate the target product from such a complex system. Isolation and purification of bioengineering products is different from chemical purification production. Its main characteristics are as follows. ①Target product concentration is low, but the final product requirements to high purity. The content of impurities (such as microbes containing cell debris, metabolites, residual medium and short fibers) are so high that extraction of products needs several steps of operation. ②The product to extract have low stability; and the external conditions such as heat, extreme pH, organic solvent, enzyme and mechanical shear stress may lead to their inactivity or decomposition. Especially bioactivity of protein is related with some cofactors, the presence of metal ions and spatial configuration of molecules. ③Fermentation or culture is batch operation. Because of high biological variation, the fermentation liquid in different batches is not identical. This requires downstream processing technology should have certain elasticity, particularly can treat contaminated batches. ④The expense of downstream processing engineering is high, but the recovery rate is low. For example, antibiotics will lose about 20% after refining. Therefore, the downstream processing cost has become an important factor for constraining producers from raising economic benefits. Research aim of downstream processing engineering is to improve the recycling rate of products, reduce the cost of isolation and purification, otherwise, bioengineering cannot have industrialized economic benefits.

7.5.1.2 General procedure of downstream processing technology

Because of the purpose of fermentation being different, the desired microbial metabolic products including bacterial cells, primary and secondary metabolites are different. Requirements for product quality and so isolation and purification steps may have various different even combinations. But the downstream process of most of microbial products is divided into 4 stages according to the general procedure, i.e. fermentation liquid (or medium) pretreating and filtering (solid-liquid separation), extracting (purification), refining (highly purified)and finished product processing (Fig.7.15).

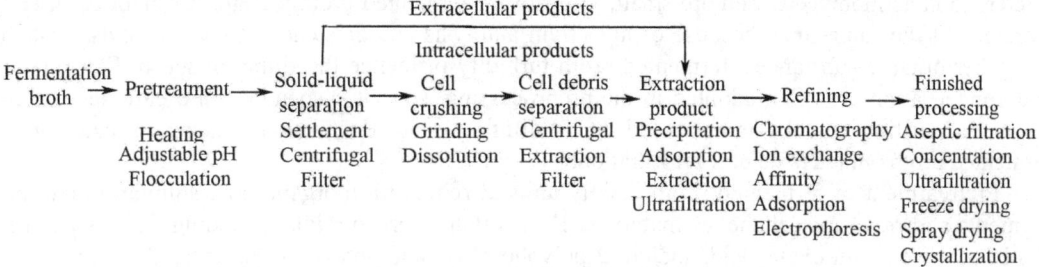

Fig.7.15 Downstream processing procedure of microbial products

The pretreating and filtering of fermentation broth can accelerate the solid-liquid phase separation, and increase the filtration speed, by adopting coagulation and flocculation technique. In order to reduce the resistance of filter medium, cross flow membrane filtration technology can be used. If the product is some intracellular metabolite, at first cells should be broken, and then cell debris should be separated. The aim of preliminary purification is to remove the impurities with the target product in properties very different. This step can make the product concentrated, and obviously improve the product quality. Common methods for isolation are precipitation, adsorption, extraction, ultrafiltration, etc. Highly purifying, also called refining, often employs highly selective isolation techniques, to remove the impurities similar to the product in chemical and physical properties. The typical purification methods are chromatography, electrophoresis, ion exchange, and so on. Finished product processing is to get qualified products. Concentrating, crystallizing and drying are important techniques.

Downstream processing engineering is composed of various chemical unit operations. Because biological products have different properties, many unit operations should be used. Some traditional unit operations including precipitation, extraction, adsorption, drying, distillation and evaporation have more mature theories, while other newly developed ones such as cell crushing, membrane filtration and chromatographic separation lack complete theories. Ion exchange technique is a status between them. In the following text we will describe various methods for isolation and purification one by one.

7.5.2 Pretreatment of Fermented Broth

When isolating and purifying the target product, fermented broth must be pretreated and filtered at first. After solid and liquid phases are separated, metabolites can be further isolated and purified with various physical and chemical techniques.

7.5.2.1 Purpose and meaning of Pretreatment

Since the target product in the fermented broth has very low concentration and is mixed with a lot of soluble and suspended impurities together, before the solid-liquid separation, pretreatment must be done. The purpose of preprocessing is to change physical properties of the fermented broth, accelerate settlement of the suspension solid substances, in order to facilitate smooth implementation of subsequent extraction and refining as much as possible to make the product move into the phase suitable for subsequent process (mostly liquid phase); to remove part of impurities (such as protein, heavy metal ion, pigment, toxic substances, etc.); and to adjust to the suitable temperature and pH value.

The speed of separating solids from the suspended depends on physical properties of the liquid. Fermented broth of different types in different culture conditions has different rheological properties. Bacteria and actinomycetes cell are small, viscosity of fermented broth is high. It cannot be directly filtrated. At the same time, because of mycelium autolysis and existence of nucleic acids, proteins and other organic substances, fermented broth turbidity influence the filtration speed. Flocculation and condensation are often adopted in the preprocessing, so that suspended solid particles volume increases, settling velocity is enhanced, or by diluting or heating which reduces viscosity of the fermented broth, which is helpful for filtration.

Pretreatment of fermented broth mainly aims at removing inorganic ion, soluble protein and pigments and other miscellaneous materials. High-valent inorganic ions including Ca^{2+}, Mg^{2+}, Fe^{2+} can be removed with oxalic acid, sodium tripolyphosphate and potassium ferrocyanide, respectively. Oxalic acid reacts with Ca^{2+} to produce calcium oxalate precipitate, which can promote protein coagulation so as to increase the filtration speed. Sodium tripolyphosphate reacts with Mg^{2+} to form soluble complexe. Potassium ferrocyanide reacts with Fe^{2+} to form Prussian blue precipitate. Removal methods of soluble complex protein are as follows. ①The fermented broth pH is adjusted to the isoelectric point of protein, in order to precipitate some protein. Since the majority of protein isoelectric points are within the range of pH4.0-5.5, most of proteins cannot be removed completely only by regulation of pH. Other methods must be used. Proteins can form precipitate with some anionic substances such as trichloroacetate, sodium salicylate, tungstate, picrate, tannate, perchlorate, etc. in the acidic solution; but in the alkaline solution, proteins may precipitate with some cations such as Ag^+, Cu^{2+}, Zn^{2+}, Fe^{3+}, Pb^{2+} to remove most of proteins. ②Protein can form precipitates when alkali metal salt is added as a dehydrating agent. ③Proteins denature and precipitate when fermented broth is treated with heat, because proteins irreversibly denature at 70-80℃, and some even at 50℃. So within the range of heat resistance for target products themselves allow, the heating method can remove soluble protein impurity, as well as reduce the viscosity of fermented broth, so as to accelerate filtration speed. ④Organic solvents (such as ethanol, acetone and others) or surface active agents are added to make proteins denature. ⑤Flocculating agents may be added to make soluble protein colloid generate aggregation or flocculation. Inorganic flocculating agents including aluminum sulfate, calcium chloride, ferric chloride, basic aluminum chloride and inorganic polymer (such as polyferrous sulfate, polyferric chloride etc.) can make the colloidal repulsion potential decrease, and then precipitation occurs, which is called agglomeration. Organic flocculants include neutral, positive and negative flocculants such as polyacrylamide, sodium polyacrylate, poly quaternary amine ester. These polymeric compounds make colloid agglomerate and precipitate through bridging action of their long chains, which is called flocculation. ⑥The impurity protein can be removed through adsorption. For example, in extraction of tetracycline antibiotics, the gelatinous precipitate of ferrocyanide potassium is formed by utilizing synergy between potassium ferrocyanide and zinc sulfate which can adsorb proteins. In fermented broth of *Bacillus subtilis*, addition of calcium chloride and sodium hydrogen phosphate often form gel, which can absorb protein, bacteria and other insoluble particles wrap them to form precipitate. The pigment substance may be secreted by microorganism, also derived from raw material (such as molasses, corn syrup, etc.) in the fermented broth. Physical and chemical properties diversity of the pigment have increased difficulty of decolorization, commonly used decolorizing methods in industrialized production including ion exchange resin method, ion exchange fiber method, activated carbon adsorption, etc.

7.5.2.2 Filtration of fermentation broth

In order to effectively isolate, extract and purify fermentation products, waste bacteria, solid impurities and suspended solid matters must first be removed to ensure liquor clarification. Centrifugal isolation and filtration are the common methods in the present fermentation industry. High speed centrifugal is often adopted to isolate bacteria and yeast cells in the fermented broth, while for the cells of larger sizes such as fungi and actinomycetes, filtering method is generally adopted. Because of high viscosity of the fermented broth, its filtering speed is slow, and it takes the heavy manual labor, which is currently the weak link in the fermentation industrial production. Microbial fermented broth is not Newtonian fluids, and is difficult to filter, but filter liquor must be clarified to benefit product refinement. Therefore, the filtration velocity of the fermented broth must be improved.

7.5.2.3 Microbial cell disruption methods

The majority of microbial metabolites such as penicillin acylase, alkaline phosphatase, etc., are present within a cell. A lot of gene engineering bacteria form intracellular inclusion bodies which are not secreted. Only a few enzymes, such as alkaline protease, pectinase and hemicellulase produced by bacteria, saccharifying enzyme and cellulase produced by mold, can be directly secreted outside cells in fermented broth. The fermented broth for extracellular products can conveniently be pretreated and filtrated to get the clear filtrate so as to carry out the next step of purification. But for intracellular products, cells should be collected firstly to crush them, then the target product is moved to the liquid phase, and finally the cell debris is isolated from the liquid phase. Cell debris isolation method is usually centrifugal isolation, but it is very difficult. A new method is binary aqueous phase extraction which can make the cell debris distribute mostly in the solid phase by selecting the appropriate conditions.

According to types of external force, cell disruption methods can be divided into two categories, mechanical method and non-mechanical method. Mechanical method is widely used in laboratory and industrial production; and ultrasonic method is mainly applied in laboratory. Non-mechanical method is almost forgotten in the laboratory application period. For example, enzymatic method and chemical penetration method are now being development actively. Others such as squeeze method and freezing melting method are often used in laboratory, but restricted by many factors, and so they do not have industrialized application. People are still looking for new methods, such as laser crushing method, high speed forward flow impacting method, freezing spray method.

With high capacity and rapid cell disruption, mechanical processing is the common method selected for microbial cell disruption. Its principle is that cells are subject to high shear stress generated by high pressure and stirring glass beads at a high speed or ultrasonic effect and then are broken. But when this method is used, in order to avoid mechanical energy consumption and generation of too much heat, measures should be taken to freeze and prevent substances with bioactivity from damage. Common mechanical methods are mechanical grinding method and high-pressure homogenization method.

Non-mechanical method for microbial cell disruption includes enzyme treatment, osmotic shock method, freezing and thawing process, heat treatment and chemical splitting method, etc.

7.5.3 Isolation and Extraction Methods of Fermentation Products

7.5.3.1 Precipitation method

Precipitation is a process where the solute in a solution changes from liquid phase to solid phase to separate out by changing the conditions or adding a reagent (acid, alkali or salt substance). It utilizes the characteristic of some fermented products which can form insoluble salts or complexes by reacting with a reagent to precipitate, in order to achieve the purpose of isolation and purification of products. This method is often used for initial isolation of products, so as to obtain biological substances from the fermented broth after pretreating and filtering to remove bacteria and other debris precipitate, and then to perform further purification. This method has advantages of low cost, simple equipment, high yield, high concentration times and simple operation. According to different precipitating agents, method of precipitation can be divided into the following categories: ①isoelectric point precipitation method; ②salting out method; ③the organic solvent precipitation; ④nonionic polymer precipitation; ⑤polyelectrolyte precipitation method; ⑥salt complex precipitation method; ⑦thermal denaturation and cid-base degeneration precipitation, and so on. Now, the precipitation methods are widely used in extraction of amino acids, enzymes and antibiotics from fermented broth. Main extraction methods of amino acids are isoelectric point method, hydrochloride method, metal salt method and organic solvent method. Salting out method and organic solvent precipitation method are used in enzyme extraction.

1) Isoelectric point precipitation method

Isoelectric point precipitation method utilizes the principle that ampholyte has the lowest solubility in electroneutrality to isolate and purify the products. At low ionic strength, pH value is adjusted to the isoelectric point (pI), in which all sorts of amphoteric electrolyte can make a net charge of zero and reduce its solubility. Different amphoteric electrolytes have different isoelectric points, and can be isolated through their isoelectric points. Antibiotics (tetracycline, oxytetracycline, etc.), amino acids, nucleotides and other small molecular substances, as well as proteins, enzymes, nucleic acid and other biological macromolecules are all amphoteric electrolyte. The method is suitable for ampholytes with strong hydrophobicity such as proteins. For example, casein can form crude aggregates at its isoelectric point, but some of hydrophilic proteins such as gelatin, when adjusting the pH to the isoelectric point do not produce precipitation in solutions with low ionic strength. Therefore, in order to make this kind of material precipitate out, other precipitating factors should be considered.

Many protein isoelectric points are in the acidic range, and many inorganic acids such as phosphoric acid, hydrochloric acid, sulfuric acid and others are permitted by many food standards, and the price is low. Therefore, the residual acid is not eliminated normally, the purification operation next step can be directly carried out. Its shortcoming is that acidified liquid easily cause inactivation of some proteins, because protein is sensitive to low pH. Isoelectric point precipitation method and salting out method, organic solvent method and other methods can be combined to improve the effect of sedimentation. In addition, isoelectric point precipitation method is often used in ion exchange method and other extracting and refining methods.

In the fermentation industry, glutamate, aspartate, cysteine, tryptophan and phenylalanine are extracted with isoelectric point method. For example, the principle of extraction of glutamic acid with isoelectric point method is that glutamic acid is amphoteric electrolyte. It can be dissociated into cations (R-NH_3^+) and anions (R-COO^-) in solution. When using hydrochloric acid to adjust pH value of the fermented broth to the glutamate isoelectric point of pI=3.23, positive and negative charges will be equal, the total net charge is zero, and glutamate solubility reaches the minimum and it exhibits oversaturation to crystallize and separate out.

It is worth noting that many proteins or amino acids have their pI changed after binding to metal. For example, pI of insulin is 5.3, but after it combines with Zn^{2+} to produce insulin zinc salt, its pI increases to 6.2. As another example, pI of glutamate is 3.22, and after it combines with Zn^{2+} to form glutamate zinc salt, the pI decreases to 2.4. Therefore, after adding metal ion, for isoelectric point precipitation, it should be noted that the pH be adjusted.

2) Salting out method

Salting out method is also called neutral salt precipitation method. It utilizes the properties of neutral salts which can damage colloidal properties of protein, enzyme; neutralize charge on particles, eliminate the hydrated membrane surrounding the colloid, and induce protein precipitation. It is widely used in enzyme extraction.

If a certain amount of salting agent (neutral salt) is added into the enzyme protein solution, because of its higher hydrophilicity than the proteins, the salt can bind to a large number of water molecules. Water film around the protein colloidal particles degrades gradually and disappears so as to make the particles dehydrate, and while neutral salt dissociation to neutralize the electrical charges of the protein particles. The protein particles will lose mutual repulsion. So the particles collide with each other due to irregular Brownion movement, the particles bind each other to form a huge conjugates under molecular affinity effects, and then the floccule precipitate separates out.

Salting out agents have many kinds, such as ammonium sulfate, sodium sulfate, magnesium sulfate, sodium dihydrogen phosphate, sodium chloride, ammonium chloride, etc. Generally the order of their salting-out effects from the higher to the lower is $MgSO_4$, Na_2SO_4, $(NH_4)_2SO_4$, NaH_2PO_4. One of the most common salting out agent is $(NH_4)_2SO_4$. Its solubility at 40℃ is 81% with high solubility, even at low temperature. During the salting-out process, it can be dissolved without heating. Other salts must be heated to dissolve at high concentrations of salt solution, but heating will affect activity enzymes, we should pay attention to this in extracting enzyme.

The amount of salting agents is associated with the types of precipitated enzymes and properties of impurities in the enzyme liquid. The highest amount should be used as the standard. Concrete amount can be determined through the contrast test and production practice. For example, the amount of BF7658 amylase salting-out agent $(NH_4)_2SO_4$ used in current production is 40%. Stirring can accelerate salt dissolution and destruction of colloidal solution, which is useful for sedimentation. But strong agitation will produce foam and induce denaturation and oxidation of enzyme protein surface, which will decrease activity of the enzyme and destruct the precipitation to form small particles, which lead to incomplete sedimentation, and then directly increase filtration difficulty.

When choosing salting-out temperature, we should consider impact of temperature on activity of enzymes, i.e. enzyme should be salted out at the temperature at which it is stable. However, salting-out effect should also be considered at the same time. The suitable temperature can be determined through experiments. In general, the lower temperature is, the more difficult precipitation is, but there are contrary cases. In fact, the temperature range for salting-out is bigger, and so salting-out is generally performed at room temperature.

For choice of pH value for salting-out, no effect on activity of enzymes should also be considered. Because proteins most easily precipitate at their isoelectric point, the pH value of the isoelectric point can be chosen as the pH value for salting-out. But in order to prevent the pH value from affecting enzyme activity, the pH value for salting-out is most within the pH range which makes enzymes stable.

3) Organic solvent precipitation

Organic solvent precipitation utilizes organic solvents which can dissolve in water and vice versa to make the product precipitate. This method is often applied to extraction of fermentation products, such as enzymes, amino acids, and antibiotics.

Many organic solvents such as acetone, ethanol and methanol can make water-soluble biomicrocules and nucleic acids, polysaccharides, proteins and other biomacromolecules precipitate. This precipitation is a result of a variety of affects, but the main effect is to reduce the dielectric constant of water solution. When the organic solvent concentration increases the hydration degree of charged groups or hydrophilic groups on the surface of protein molecules reduces, or the dielectric constant of solvents decreases, so that electrostatic attraction of the charged solutes increases, they attract each other to cause agglutination. For those molecules (such as proteins) having a surface hydration layer, action of organic solvent and water makes the hydrated layer thickness on the protein particle surface decrease constantly. Finally, because of irregular Brownian movement the protein colloidal particles without hydration layer collide with each other and aggregate under action of molecular affinity. Adopting different concentrations of organic solvents, different solutes can be precipitated, namely, with step sedimentation method, the purpose for isolation and purification is achieved.

The method has the advantages of higher resolution than the salting-out method. A solute precipitates only in a relatively narrow range of organic solvent; precipitation does not need desalination; organic solvent density is low, which is very different from that of precipitates so as to perform solid-liquid separation easily; and organic solvents evaporate easily, without residues in finished products, therefore it is suitable for food and medicine preparation. Drawback of organic solvent precipitation is that it is ready to cause protein denaturation and inactivation, and organic solvents are flammable, and explosive. It has high requirement for safety.

Precipitation efficiency of protein by different organic solvents is determined by protein type, temperature, pH values and impurities and other factors. The order of the precipitation efficiency from the higher to the lower is acetone, ethanol and methanol. Therefore, organic solvent precipitation conditions, such as type, and concentration, should be determined by experiments. Ethanol is the most common precipitating agent, during the precipitation process. After ethanol is mixed with water, a great amount of dilution heat is given off, which increases the solution temperature significantly to affects activity of thermolabile enzymes greatly, so in production the measure to add a small amount of organic solvent for several times while stirring repeatedly is taken to avoid the temperature rise. Effect of organic solvents for enzyme precipitation ability is related with temperature. Generally, the lower the temperature is, the more complete precipitation is, and so cooling must be noted during precipitation.

7.5.3.2 Solvent extraction method

Solvent extraction method is one important and efficient method for purification and for concentration of microbial metabolites. The method can be applied to solid-liquid extraction and the liquid-liquid extraction. Compared with chemical precipitation separation, the method has higher degree of separation; compared with ion exchange, it has higher selectivity and faster mass transmission speed. Compared with distillation method, it has higher production capacity, lower energy consumption, shorter production cycle, more convenient continuous and automatic operation, therefore it has been widely used. Using organic solvent extraction method not only can extract antibiotic, organic acid, vitamin, hormone and other fermentation products, but it also can develop many new extraction techniques, such as reverse micelle extraction, supercritical fluid extraction, liquid membrane extraction, etc. So, the requirements for isolation of a variety of enzymes, proteins, nucleic acids, peptides, amino acids and other genetic engineering products can be met.

7.5.3.3 Adsorption method

In the early days, adsorption method was applied to isolation and purification of various fermentation products, such as proteins, nucleic acids, enzymes, antibiotics, amino acids, and air purification and sterilization. It usually utilizes adsorbents to decolour and get rid of heat source, histamine and the other impurities. The adsorbents used earlier include kaolin, alumina, acidic clay and other inorganic adsorbents, ion exchange resin, activated carbon, molecular sieve and cellulose etc. Adsorption method has many advantages, such as simple operation, simple equipment, using no or few organic reagents, little pH change in the production process, and application to bioproducts of poor stability. Its disadvantages are poor selectivity, low yield, discontinuous operation and so on, inorganic absorbent performance is not stable, but with development and application of gel adsorbents, macroreticular polymer adsorbents, the disadvantages of adsorption method have been overcome, and it has regained importance.

Adsorption is a procedure where a substance is concentrated from the mobile phase (gas or liquid) to the solid surface, so as to be isolated. The adsorbent is known as the solid on whose surface occurs adsorption, and the adsorbed substance is called adsorbate. The essence of adsorption force between adsorbents and adsorbates is the van der Waals force. It is a generic term of a group of molecular attraction forces, including three forces, namely directional force, induction force and dispersion force. Directional force is the force that is generated between polar molecules; induction force refers to attraction between polar molecules and nonpolar molecules; dispersion force is the attraction effect between non-polar molecules. Different solid surfaces have different free energies, and so have different adsorption capacities for the other substances. The higher free energy of a surface is, the stronger its adsorption capacity is.

Solid substances are mainly divided into porous type and non-porous type. Non-porous solid has small specific surface area, in order to increase its specific surface area by crushing make its particle size to get smaller. Porous solid surface is composed of the outer surface and the inner surface. Inner surface area is hundreds of times greater than outer surface area. It has high adsorption capacity, so porous adsorbent is applied more widely.

1) Hydrophobic or non-polar adsorbents

As a typical adsorbent, activated carbon (alkaline, acidic or neutral) is applicable for adsorption of nonpolar substances in polar solvents (such as water), and is mainly used for decolorization of fermentation products and initial extraction and refining of many antibiotics.

2) Hydrophilic or polar adsorbents

Hydrophilic adsorbents such as silica and alumina are applicable for adsorption of polar substances in nonpolar or low polar solvents. Aluminum oxide and magnesium oxide are alkaline absorbents; magnesium sulfate is neutral, and acidic silica gel adsorbent and aluminum silicate (bentonite) belongs to the acidic adsorbent.

3) Ion exchange resin adsorbents

Various organic ion exchange resins also belong to the polar adsorbents, and simultaneously have the properties of ion exchange resin. Common ion exchange resins used for decolorization include 717[#] strong-basidity quaternary-amine type resin and porous weak-basidity 390 styrene amine type resin.

7.5.3.4 Ion exchange method

Ion exchange resin is a synthesized artificial polymer, which is insoluble in acids, alkalis and organic solvents. It has the capacity of ion exchange, and its chemical stability is good. These molecules can be divided into two parts, one part is higher-molecular-weight group with multivalent immobility, which constitutes the resin frame and has action to maintain insolubility and chemical stability of the resin; another part is mobile ions known as active ion, which constitutes the active groups of the resin. Active ion is cation, it is called cationic exchange resin, and vice versa. Active ion in resin can adsorb and exchange with external ions, because of ion exchange resin has acidic or basic groups, which can exchange with anions or cations. The principle of ion exchange includes complex adsorption, absorption, penetration, diffusion, ion exchange, ion affinity and other physical and chemical processes. Because of its advantages such as simple equipment, convenient operation, easy realization of automatic control and high efficiency, ion exchange is widely applied in purification, desalination, concentrate, conversion, neutralization and decolorization of bioproducts.

When the fermented broth flows through the ion exchange column, the ions in the fermented broth are adsorbed or diffused into the inside activity center of the resin through its surface. These ions can exchange with the original free ions of the resin; the exchanged ions spread from the inside activity centers onto the surface of the resin; and they diffuse from the resin surface into the solution. After elution, exchanged ions are eluted and isolated. As long as biological substances can form cationic, anionic or amphoteric ions under some conditions, substances that can be dissolved in the water, can make use of this method for isolation and purification.

7.5.3.5 Crystallization method

Crystallization is the process in which the solute is precipitated to form crystals from the solution. Most of biological substances may form crystals from the solutions under certain conditions. Therefore, it is one of the methods for isolation and purification of biological macromolecules such as protein and enzyme.

1) General properties of crystals

A crystal is made up of many unit particles (including atoms, ions and molecules) with the same properties regularly arranged, and has continuity, homogeneity from the macroscopic view. Crystals and amorphous substances in the nature have essential differences, the crystal has many direction-oriented properties (such as electrical and optical), i.e. crystal anisotropy is known as in the same direction with the same properties and in different directions with dissimilarity, while the amorphous substances do not have such properties. In addition, the crystals also have symmetry. Thus the crystal is defined as many particles with the same properties in three-dimensional space which are regularly arranged in a lattice of solid substance.

Regular structure inside the crystal requires that crystal must be formed by the same ions or molecules which can arrange periodically in a certain direction at a certain interval, therefore, only relatively pure substances can form crystals. Most of biological substances of small molecules, such as organic acids, sugars, nucleotide, amino acid, vitamin and coenzyme, generally can polymerize directionally to form molecular or ionic crystals after reaching certain purity, because of their relatively simple structure. Some biological molecules such as polysaccharides, protein, enzyme and nucleic acid, due to high molecular mass and complex structure, are not easy to assemble directionally and more difficult to form crystals. But biological macromolecules are polymerized by the same or similar small molecular monomers, thus biomacromolecules also have the ability to form crystals. But this ability has close relation with structure and shape of large molecules. For example, most of globulins, enzyme proteins, uniformly sized viruses (such as plant rotavirus) are easy to crystallize. However, to date, crystals of some nucleic acids, proteins and polysaccharides with complex structure and high asymmetry have not yet been obtained so far.

The crystallization process is highly selective, and whether a substance can crystallize depends mainly on its properties. It must also be under certain conditions that the substances with the crystallization ability can form crystals.

2) Crystal formation conditions of biological substances

The crystal forming conditions for various biological substances include sample purity, solution concentration, temperature, pH value, crystal seed, metal ion, nature of solvents and other conditions. Among these factors, the decisive factor is sample purity and solution concentration.

Macromolecular or micromolecular substances, which only reach certain purity, can form crystals. But the extent of purity has no certain standards and varies with substances. Although some substances don't have high purity, if conditions are suitable they can be separated out from the solution to crystallize. Cystine in hair hydrolysate is a typical example. But for most proteins, the general requirement for purity is above 50% under which they can form crystals. It is because that the crystallization process is highly selective, sample purity may affect on crystallization process.

Some impurities have effects on directional alignment of the crystallized solute on the crystal surface, so that the crystallization cannot be performed or is performed slowly. Some impurities can combine with crystalline solute to form complex. Some impurities are adsorbed on the crystal surface to only influence crystal colour, but not formation of crystals. The impurities in fermented broth can be adsorbed to remove by adding a small amount of adsorbents (1%-2%), such as activated carbon, activated clay, diatomite, etc.

Another condition of crystallization is changes of solubility. Proper concentration can increase the opportunity of solute molecular collision, so as to make molecules align and aggregate at a certain speed to form crystals. If the concentration is so high that it reaches the saturated state, the rate of solute separate out from solution will be higher than that of crystal formation, and so amorphous solids but not crystals, or the crystal with high impurity, can only be obtained. If the concentration is too low, crystal formation rate will be by far lower than the crystal dissolution rate, and it is impossible to get crystals, due to solution in the unsaturated state. Therefore, only solution with a supersaturated state or low saturated state, i.e., the crystalline formation rate is slightly higher than the crystal dissolution rate, crystals can be obtained. As shown in Fig.7.16, the concentration should be controlled in the stable area above the saturated zone and under the supersaturated region, in order to obtain good crystals. The crystallization mother liquor is concentrated to a suitable concentration in production, and then various precipitating agents such as a salt solution or organic solvent are slowly added into the liquor until the solution becomes milky turbid. At this time the crystallization liquor is just in a low oversaturation state. When the liquor is placed still at a suitable temperature, and then crystals can be slowly separate out.

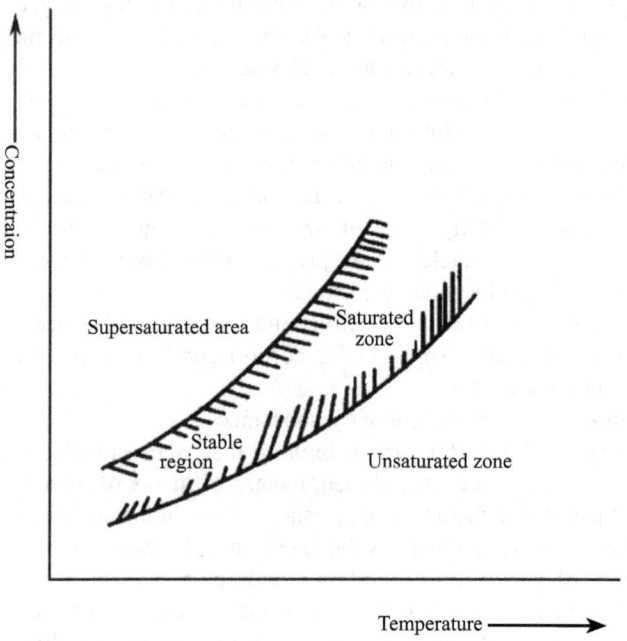

Fig.7.16 Crystallizing concentration area chart for crystal formation

Except in a few cases, crystallization is usually performed at low temperature. Solubility of biological substances can be descended, cannot be denatured. But when crystallized in a neutral salt solution, the temperature can be chosen within the range from 0℃ to room temperature. Generally, crystallization in organic solvents requires lower temperature.

The pH value of crystallization solution generally is selected, usually in the isoelectric point nearby crystallized biological substance (such as protein or enzyme), which can be beneficial for crystal precipitation.

For those proteins and enzymes which are not easy to crystallize, adding trace crystal of these substances often leads to formation of a great amount of crystals. Sometimes scraping the container wall gently with a glass rod can also achieve this goal. Among the proteins or enzymes which need to add the crystal seed for formation of crystals, most do not have high crystal yield. In order to obtain glutamic acid crystal with uniform size and consistent crystal form, the solution often should be concentrated to oversaturation state. When some certain quantity of crystal seed in a certain size is added into the saturated solution, the saturated solute will slowly diffuse around the seed and arrange on various crystal planes of the seed, so as to make the crystal grow. This process is called initial crystallization of crystal seed.

When some proteins and enzymes crystallize, it still needs to add metal ions. For example, when ferritin crystallizes in ammonium sulfate solution, rhombic crystal can be formed after adding a small amount of cadmium ions. Enolase can form crystal after adding mercury salt.

3) Crystallization methods

According to the modes for preparation of supersaturated solutions, crystallization methods can be divided roughly into concentrating crystallization, cooling crystallization, chemical reaction crystallization and salting out crystallization.

The so-called concentrating crystallization means that the product solution is concentrated by means of reduced pressure concentration, so as to make the solution reach a oversaturated state, and then solute crystallizes and separates out. This method is applicable for those substances whose solubility changes little with temperature. For example, glutamate has higher solubility in pure water, when a great amount of water is removed in solution through decompression evaporation, glutamate crystal can be formed. Another example is that, griseofulvin in the acetone extraction solution also can crystallize after removing acetone by vacuum concentration.

Cooling crystallization is applicable for the substances whose solubility changes greatly with temperature. The procedure is that first dissolving the solute with water by heating, and then separating out the crystal by cooling. For example, inosine acid purification adopts cooling crystallization. And for crystallization of citric acid monohydrate, the solution is concentrated first by increasing temperature, and then gradually cooling to reach the oversaturated state, and then natural crystal can be obtained.

Chemical reaction crystallization refers to where some metabolites of microorganisms can form new substances after adding a reaction agent or adjusting pH, then their concentrations change, and when the concentration exceeds its solubility, the crystal will be formed. For example, oxytetracycline solution, the decoloured acidic filtrate adjusts pH to 4.5, and then the free alkali crystal of oxytetracycline separate out. As another example, pH of the fermented broth of glutamic acid is adjusted to the isoelectric point of 3.22 with hydrochloric acid, so as to reduce glutamate solubility and form crystal.

Some substances, such as solvents, salts, can decrease solubility of the solute to form oversaturated solution leads, so as to crystallization. For example, kanamycin is insoluble in ethanol, adding 95% ethanol into kanamycin decoloring liquid, when the ethanol concentration reach 60%-80%, crystal of kanamycin sulfate is formed after a period of agitation. Adding a certain amount of table salt can make crystal of procaine penicillin separate out more easily.

7.5.3.6 Membrane separation technology

Membrane separation technology utilizes some filter media with characteristics of selective permeation allowing certain components to permeate, but others not, so as to accomplish isolation and purification of certain substances. It has the advantages of simple equipment, convenient operation, no phase change, no chemical changes, high treatment efficiency and energy saving, the technology is getting popularity. At present, in our country membrane separation technology, including reverse osmosis, ultrafiltration, microfiltration and electroosmotic membrane and film forming technique, has been widely applied in chemical industry, food industry, biological fermentation industry, pharmaceutical industry, electronic industry, textile industry and environmental protection. so far the fermentation industry, membrane technology has been used for isolation, concentration and purification of enzymes, proteins, and other biological products.

Membrane separation technology contains many techniques. According to membrane structure and driving force, it can be divided into various methods, i.e. dialysis (DS), microfiltration (MF), ultrafiltration (UF), reverse osmosis (RO), electrodialysis (ED) and pervaporation (PV). Ultrafiltration and reverse osmosis are used the most widely. Separation principles and application of various membrane separation techniques are summarized in Table 7.2.

MF, its mass transmission impetus is pressure difference. It utilizes microporous membrane with pore diameter of 0.01-10μm to filter the solution containing particles, thereby removing particles from the solution. It is mainly used for liquid clarification and cell collection. Traditionally, MF runs in a vertical form; and now, cross-flow MF has emerged and is, used for wine and beverage processing. In general, the pure water permeation flow rate in MF is $1m^3/(m^2 \cdot min)$. The MF membrane made of nitrocellulose was commercialized in the 30s of the 20th century. In recent years, MF membrane made of tetrafluoroethylene and polyvinylidene fluoride (PVDF) has been commercialized. Such membrane has high temperature resistance, solvent resistance, good chemical stability and other advantages. The temperature range for its use is $-100°C$ to $260°C$.

Table 7.2　Various membrane separation methods and their application

Membrane separation methods	Driving force of mass transmission	Separation principle	Application examples
MF	Differential pressure (0.05-0.5MPa)	Sieving	Sterilization, recovery of bacteria, cell collection
UF	Differential pressure (0.1-1.0MPa)	Sieving	Protein, polypeptide and polysaccharide recovery and enrichment, desalination, removal of heat source
RO	Differential pressure (1.0-10MPa)	Sieving	Concentration of salt, amino acids, glucose, freshwater manufacturing
DS	Concentration	Sieving	Desalination, removal of the denaturing agent
ED	Potential difference	Electric charges and sieving	Desalination, amino acid and organic acid separation
PV	Pressure and temperature difference	Affinity between solute and membrane	Separation of the organic solvent and water, concentration of ethanol

UF utilizes ultrafiltration membrane to selectively filter various solute molecules in the solution. When the solution with the macromolecules or fine particles pass the ultrafiltration membrane at 0.1-1.0MPa operating pressure (exogenous atmospheric or vacuum pump pressure), the solvent and the solutes of small molecule can permeate, while soluble macromolecules with molecular weight of 300-1000kDa or fine particles are retained. UF is driven by differential pressure. The aperture of UF membrane is 1-200nm (or greater), and it is asymmetric microporous. Macromolecules with different molecular weights and shapes can be separated with the membrane with different apertures. Proteins, fats, glucose, pigments, pectin, virus and other materials can be trapped. It is mainly used for macromolecule separation, such as protein concentration, plasma separation, pyrogen removal and so on. The permeation flow rate of pure water is $1m^3/(m^2 \cdot h)$. Its advantages are low cost, easy operation, mild conditions, and preservation and high recovery of bioactivity of biological macromolecules.

RO utilizes the reverse osmosis membrane to impose pressure on the solution so as to overcome the osmotic pressure of the solvent, and so the solvent permeate the reverse osmosis membrane, but organic substances and inorganic salts in the solution are all retained. Unlike ultrafiltration, in the RO process, reverse osmosis membrane selectively only allows the solvent (usually water) to permeate and prohibits the solute to permeate, and all soluble substances (including salts, sugar, ions and other substances with relative molecular weight greater than

150 Da) are retained, which is therefore also called as dehydration technology. The technology is mainly used for concentration of substances of small molecules such as ethanol, sugar and amino acid. The operation pressure for RO is up to 1.0-10MPa. Ideal reverse osmosis membrane should be non-porous. Cortical asymmetry membrane is often used, and its aperture is 0.1-1.0nm. The pure water permeation flow rate is generally $1m^3/(m^2 \cdot d)$.

DS is concentration difference between both sides of the membrane, which makes the solute from the side of the membrane at high concentration to the other side at low concentration by membrane pore diffusion, so as to achieve the purpose of separation and concentration. DS membrane has both non-porous characteristic of the RO membrane and the characteristics of very small pore diameter of the UF membrane. Generally, symmetric or asymmetric membrane can be used for separation and concentration of macromolecules, and removal of small molecular organic and inorganic salts. At present, DS membrane is mainly used for preparation of artificial kidney and fermentation process. In the fermentation process, with its permeability of the DS membrane, the membrane with the appropriate pore diameter can be selected to allow products and harmful metabolites in the fermentation liquid to permeate and retain the bacterial cells. Thereby, it can overcome harmful metabolites inhibition on cells or key enzymes of fermentation production in the fermentation system. Its drawback is slow speed, small processing capacity and great amount of dialysate.

With electric potential difference in the direct-current electric field, ED is that cations and anions permeate the ion exchange membrane respectively so as to achieve the purpose of isolation of the electrolyte from the solution. ED membrane is a dense ion exchange membrane. Cation exchange membrane has the negative ion group on its polymer frame which is usually a strong acidic group ($-SO_3H$); and anion exchange membrane has the positive ion group which is usually a strong alkaline group [$-N^+(CH_3)_3$]. It is mainly used in water treatment, such as sea water desalination, water softening desalination and industrial water purification processing. In the fermentation industry, it can be used for purification of water for beer brewing, citric acid extraction, isolation of amino acids, whey desalinization, and so on.

PV, also called as osmotic evaporation, is based on the principle that utilizes membrane's different properties of dissolution and diffusion of various components in a liquid mixture to achieve purpose. It is a process of substance transmission from the liquid phase through the uniform membrane to the vapour phase. Vapour permeated is sucked in the condition of vacuum, and condensation outside the membrane device. In this process, the membrane plays the role of changing the vapour-liquid phase balance, and this balance is just the basic principle of distillation separation. Therefore, the process using this method to isolate industrial alcohol for preparation of anhydrous alcohol has now been industrialized.

Due to wide application scope of membrane, the membrane should have a much wider range of properties and operating characteristics. The following factors should be considered when selecting membrane: separation ability (selectivity and removal rate), separation speed (permeability), and membrane material cost. At present, there are many organic polymer materials for production of the membrane, such as a variety of cellulose esters, aliphatic and aromatic polyamide, polysulfone, polyacrylonitrile, poly tetra fluoroethylen, poly vinylidene fluoride, silicon rubber, etc. According to structural and functional characteristics, the polymer membrane is divided into the following five categories: dense membrane, microporous membrane, asymmetric membrane, composite membrane and ion exchange membrane. For more information about this refer to the related monographs. No further details are given here.

Chapter 8

Contemporary Bioengineering Technologies, Agriculture and Light Chemical Industry

Some basic areas of life science and its technologies have close association with light chemical industry and food industry. It is well known that almost all the food for human beings generally originates from organisms. Before the middle of the 19th century, chemicals used for mankind activities, such as dyestuff, grease, etc, also usually came from plants and animals; while common chemicals including alcohol, citric acid and lactic acid were manufactured with biomass by means of fermentation. Since chemical industry is closely related with the field such as national defense, energy source, material, food, light industry and pharmaceutical industry, it nowadays has turned into one of the biggest and the most important industrial sectors, and at the same time provides a wide platform for development of biotechnology. At present chemical industry is undergoing serious challenge from technological competition after reformation and opening up of China. Due to continuous rise of price of feedstuff and energy, decrease in price of chemical products and increasingly rigid restriction for environmental pollution, chemical industry has to employ hi-tech to modify their traditional products and traditional manufacturing processes.

During production of fine chemical products, as biological catalysis has advantages of simple process, low cost and high benefit, biotechnology has been playing a leading role in this field, and has brought about substantial economic benefit with chemical enterprises. Combination of chemistry, biology and electronics has great potential in the field of chemical assay. As a fruit of such combination, biosensors play a significant role in rapid progress of chemical industry, and food industry, which have broad and big market prospects.

Biochemical industry has been listed as the national key of development in the field of chemical industry, and the following aspects are important. ①The technological level of fermentation process for products including organic acids should be elevated in order to meet demands in industry, agriculture and pharmaceutical industry. ②The existing dietary structure will be changed so as to make it diversify and have healthcare and nutrition function, and combination between diet and medication is simple and feasible. ③Exploration of biological nitrogen fixation, research and development of effective biological azotobacteria agent of top quality, and investigation of active nitrogen fixation of foodstuff and economic crops so as to reduce use of chemical fertilizers, it will contribute to green revolution. Meanwhile, exploration of azotobacteria in higher animals and the intestinal tract of human body and their functions may lay basis for manufacturing healthcare products of "edible azotobacteria agent". ④Development of industrialized production of SCP may meet the needs of development of feedstuff industry, and provide more raw materials for biochemical industry. ⑤Improvement of industrialized production of acrylamide and biodegradable plastic (e.g. poly-3-hydroxybutyrate, PHB) with microorganism method not only facilitates in industrialization of bioplastic, but also benefits protection of ecological environment. Acrylamide produced with microorganism method has high purity, excellent selectiveness and as high conversion rate as over 99%. However, although microorganism method for producing PHB has unique advantages, the cost must be reduced so as to show its competition by comparison with chemical plastic. ⑥Biochemical industry products, such as microorganism polysaccharide and disaccharide(e.g. fructose), bio-pigment, enzyme preparation, sweetener, surfactant, biological adhesive may be developed effectively. ⑦Increase of industrialized production level of amino acids may meet demands in food industry, pharmaceutical industry and agriculture and forestry. ⑧Research and development of biopesticide will make microorganism insecticide, microorganism microbicide, microorganism feedstuff utilized widely in agriculture, forestry and animal husbandry. ⑨Development of bioenergy, so called green energy, may use pollution-free new energy to protect ecological environment. Bioenergy includes gaseous bioenergy such as marsh gas and hydrogen which can be used in research and development of fuel battery with wide application potential, and liquid bioenergy such as ethanol produced by making use of agricultural wastes. ⑩Biochemical technologies may be used to treat wastes from chemical industry, and hence microbiological technique has great application potential and brilliant promise, and will play a significant role in environmental protection and its industrialization. Promotion of renewal resource and cellulose engineering may create new sources of raw materials. For example, cellulose synthesized by bacteria is just one example, especially, those wasted organic substances, such as organic trash, food industry scraps, and etc, may be converted into chemical products by means of contemporary microbiological techniques, so as to develop new resource from wastes. Application of biotechnology may make wastes harmless, minimized, and energy-oriented, and finally achieve the goal of industrialization of environmental protection. Development of late-model bioreactors is a crucial factor to obtain a large number of biochemical products. Upsizing, diversity and automation of bioreactors are more helpful for industrialized production. Besides, the technological level of isolation and purification of final products should also be increased.

At the end of the 20th century biotechnology substitute for 20% of chemical processes, and market share of biochemical products in the world reached 6 billion dollars and accounted for 9% of all the biological product market. And hence, application of contemporary biotechnology will certainly give great contribution to development of chemical industry including biochemical industry.

Because of development of biotechnology, production of food by traditional methods in food industry will be replaced with biological method, i.e., bioreactors containing animal or plant cells or microorganisms will be become production places. Due to soaring of world population, supply of foodstuffs and protein has become a big problem, while production capacity of bioreactors is greater than that of usual agricultural production. The trend to produce food with the aid of biotechnology is also expressed as development of some new food additives, vitamins, amino acids, citric acid and their wide marketing potential. Furthermore, people also have urgent demand for increase of microorganism protein in food.

8.1 Contemporary Bioengineering Technologies and Chemical Industry

8.1.1 Biocatalyst and Chemical Industry

8.1.1.1 Contemporary biotechnology provides a new solution

One embodiment launching into industrial application relates to production of acrylamide with nitrile hydratase through hydrolysis of acrylonitrile. Acrylamide, a monomer polymer, is very expensive (over 20 thousand yuan/t), and has market shortage. It is widely used as paper strengthening agent, oil recovery agent, adhesive, coatings, soil firming agent, flocculating agent and biochemical reagents, and is mainly imported at present in China. Although raw materials for production of acrylamide are most rich in China. For example a great amount of acrylonitrile is emitted as a side product in petrochemical industry, This raw material polluted environment seriously as it was not properly utilized. If chemical methods are adopted, due to high temperature and high pressure, big investment in equipment, special difficulties in isolation and refining process, high cost and poor quality, acrylamide produced in China cannot compete with the counterpart abroad.

In China research on biological methods for transforming acrylonitrile to acrylamide has lasted for 10 years. Chinese Academy of Oil Science and Shanghai Institute of Pesticides have newly developed microbial strains whose activity on nitrile hydratase has attained international advanced level, and they have started middle scale trial production of 50t/a. Therefore, biocatalysis method for production of acrylamide is the most advanced method in the contemporary world. It can reduce production cost much more, may create big economic benefit, and has great significance.

8.1.1.2 Biological conversion increase competition of products in the market

Biological conversion with enzyme method may simplify traditional process, and increase conversion efficiency. For example, chemical synthesis process of D,L-tryptophan needs 7 steps of reactions: it has most complex procedure and cost as high as 1800 yuan/kg, and can pollute environment greatly, and so batches production cannot conducted. Enzyme synthesis of L- tryptophan needs only one step of reaction, and has high yield of synthesis and low cost of one-tenth of that of chemical synthesis method. Those commercial products produced with one-step microorganism method have many more advantages over those with chemical method technologically and economically. Over thirty years ago synthesis process of steroid included many steps of chemical reactions such as hydrolysis, dehydroxylation, epoxidation, oxidation and reduction, dehydrogenation and esterification, ester hydrolysis and isomerization. In these about 30 years, anti-inflammatory steroid, progesterone, steroid hormones and sterol anaesthetic sold in the market are all manufactured by using various microorganism conversion methods. Biological conversion of steroid is a good example about conjunction between chemical reasoning and microorganism specificity and diversity.

8.1.1.3 High efficiency complete complex organic reaction by biotechnologies

It is impossible to synthesize various enzymes by utilizing organic synthesis methods, while due to complex chemical structure medicinal antibiotics nowadays have to be manufactured with biological methods. Alkannin produced by *lithospermum* cells is a complex anthraquinone compound, cannot be synthesized with any chemical methods, while it can be manufactured on a large scale by utilizing biotechnologies to culture *lithospermum* cells artificially. In the eighties of the 20th century in Japan *lithospermum* cells were cultured to manufacture 300kg alkannin annually, so as to meet 40% of demands in Japan. Alkannin has strong antimicrobial activity, and also plays action of detumescence and anti-cancer. In recent years, it has been reported that the main component of alkannin, acetylshikonin also has function of anti-AIDS, and has wide attention in pharmaceutical industry. Alkannin with bright colour causes no harm to the human body, and is a valuable member in the family of edible pigments and cosmetic pigments, and is sold at as high a price as about 7000 dollars/kg. In China Nanjing University of Chemical Industry cooperates with Nanjing University to carry out medium scale trial of production of alkannin with biological method.

1, 6-diphosphofructose (FDP) is very hard to synthesize with chemical method, but after adopting enzyme method its synthesis may be completed in only one step, and the rate of conversion to fructose is as high as 60%. The medium scale production trial which has been just completed in Nanjing University of Chemical Industry has indicated that cost (including equipments and labor cost) of FDP synthesized by adopting enzyme method is as low as one-twelfth of that of imported counterpart. FDP is a drug used for treatment of cardiovascular diseases ad diabetes, and it is very expensive as the price of the product imported from Italy is 12 thousand yuan /kg.

8.1.2 Production of Chemical Products with Fermentation Method

8.1.2.1 Production of ethanol and methanol

As "green energy" of the new generation, development of ethanol and methanol has become hot subject across the world, and especially, production of ethanol with fermentation method by utilizing wasted organic substances in agriculture and industry has been the key research theme on construction of late-model energy and development of environmental protection industry. According to incomplete statistics, there is 0.5-0.6 billion tons of crop straw every year in China. It may be utilized to produce 0.2-0.3 billion tons of ethanol, which can provide both clean energy and necessary chemical products, and it would be also favorable to the environment protection. According to reports abroad, productivity of fuel ethanol produced through microorganisms and algae accounts for 10% of the total liquid fuel in the world. It is also reported that *zymomonas mobilis* immobilization technique has advantages over yeast in production of ethanol. If yeast and cellobiase are co-immobilized, the efficiency of conversion of cellobiose to ethanol may be increased. In addition, mixed culture method is adopted to increase productivity of ethanol. For example, mixed fermentation of the strain utilizing glucose and that utilizing xylose (mainly referring to hemicellulose) may increase productivity of ethanol by 30%-38%. Application of genetic engineering techniques to production of ethanol has achieved breakthrough. According to reports from Taiwan, China, when the key enzyme genes (i.e. *pde* and *adh*) are transferred into *E.coli* at the same time, saccharification and fermentation can be performed concurrently. Newly constructed "engineering *E.coli* strain" may produce ethanol in concentration of 6.1% (*m/V*) by fermenting for 145h at 30℃ with 10% glucose as the substrate, and the conversion rate of sugar can reach 96%. From this we can see that application of this new technology to production of ethanol with fermentation method has great potential. As is reported, when "engineering *Saccharomyces cerevisiae*" is used to produce ethanol with xylose as the only carbon source, the yield is 1.3g/L (shaking flask test). As a liquid fuel, methanol will probably replace automobile fuel in 2030. It is expected that at that time over 60% of automobiles will use methanol fuel to drive, and in Germany this is already on trial. So far, annual productivity of methanol across the world has reached more than 22.3 million tons, but in China that is only 800 thousand tons. Methanol is popularized and used as methanol gasoline, which has increasing demand and bright market prospect.

8.1.2.2 Production of xylitol

As an artificial sweetening agent, xylitol is used in production of chewing gum, and its related products may prevent against tympanitis induced by *pneumonococcus* and has other functions. Production of xylitol with microorganism fermentation has exciting development prospect because agricultural wastes such as crop straw may be utilized fully as raw material. Scaled production of xylitol may be turned into reality by breeding high-efficiency engineering bacteria. Straw is popular stalk waste in the rural area, and may be hydrolyzed into hemicellulose containing 65.15% xylose, 24.57% glucose and 10.7% arabinose. They are mixed as substrates to produce xylitol in the concentration of 27g/L in cultured liquor by fermenting with *Candida guilliermondii* for 48h, with 15% residual xylose, and substrate use ratio of 96%. Raw materials for production of xylitol is not limited to straw, bagasse is also used as the raw material. This technology has value for commercial application.

8.1.2.3 Production of glycerol

Glycerol is one important multi-functional raw material for chemical products, and widely used in many industries, such as pharmaceuticals, coatings, textile, printing and dyeing, explosive, paper making, tanning, printing, photography, metal process, electrical materials, rubber, cosmetics, printing ink, food process, and in everyday life. So far over 1700 products contain the component of glycerol. At present annual demand for glycerol in the international market has reached over one million tons, but actually its annual total productivity only accounts for about half of the demanded quantity. Because of high cost of fermentation technology, glycerol always develops slowly, but research work is still being carried out actively and has made some progress. The research group from Jiangnan University has bred out the osmophillic yeast wl-200-5 as the strain for production of glycerol with high content of over 12% in fermented broth by fermenting in the mixed solution containing 25% sugar, and using starch as raw material. For this technology, the conversion rate of sugar is 45%-53%, the yield 30g/(L·d), and glycerol extraction efficiency over 80%, which attains international advanced level.

8.1.2.4 Production of inositol

As a raw material for pharmaceuticals, also called cyclohexanhexol, inositol is used to manufacture inositol tablets for treatment of hepatitis, hepatocirrhosis and fatty liver, and is also used as nutrient in food industry. Inositol manufactured by one company in China was sold at the price of 80-90 thousand yuan/t in domestic market, and its price is 160 thousand yuan/t in the international market. Its precursor 6-phosphate inositol (phytate phosphorus) also has high medicinal value, and has efficacy on colon cancer, liver cancer and breast cancer. As a plant fodder, bran coat may be used as raw material containing 50%-70% phytate phosphorus which may be decomposed into inositol and phosphorate with the aid of phytase produced by some microorganisms. Generally, natural strains only produce a small amount of phytase. However, Yao Bin from Chinese Academy of Agricultural Science adopted recombinant DNA techniques and constructed "engineering *Pichia pastoris*" which could produce phytase in high yield through fermentation. In order to increase yield of inositol phytase, immobilization technique may be adopted. Application of this production process will play a significant role in export of inositol-associated pharmaceuticals and earnings in hard currency, treatment of organic wastes (side products after process of grain, cotton and oil), and protection of ecological environment. In a word, using agricultural wastes as raw material, production of various alcohol products with microorganism fermentation method is a developmental trend to deserve support. Microorganism conversion of various kinds of organic wastes facilitates in formation of environmental management and industry of environmental protection. Besides, traditional fermentation industry may be modified with contemporary biotechnologies, and such modification includes construction of engineering bacteria, use of immobilized cells, design of late-model bioreactor and improvement of downstream techniques. All of these are basis and guarantee for industrialized production of alcohol products.

8.1.2.5　Production of lactic acid

L-lactic acid has been widely used in food industry, pharmaceutical industry, agriculture and chemical industry. It is one important sour agent and a preservative in food industry. Lactic acid may be used to prepare beverage which may decrease blood pressure; and poly-L-lactic acid may be used to manufacture bioplastic. The fermentation industry of lactic acid in China is developing rapidly, its annual productivity has increased to over 3000t and its exported quantity accounts for 80%. Nowadays, a production base of annual L-lactic acid productivity of 2000t in Shanghai has been founded and which laid a useful basis for industrialized production of L-lactic acid. A research group from Jiangnan University chose a *thermophilic rhizopus* strain to produce lactic acid at 36℃ with initial sugar content of about 12%, using corn powder or sweet potato powder, with the technological parameters, acid producing rate of about 11%, sugar conversion rate of approaching to 90% and obtained L-lactic acid accounting for over 96% of the total acids. This new technology will help promote industrialization of L-lactic acid in China and research and production of poly-L-lactic acid. Japanese enterprises have been making efforts to develop production of lactic acid with lactic acid bacteria, and then use obtained lactic acid as raw material to produce degradable bioplastic, polylactic acid which may used to manufacture surgery suture, artificial bone and artificial skin. Such plastic made from poly-lactic acid has high strength and transparency, and melting point as high as 185℃. In 1994 fermentation production line of annual productivity of 100t was founded in order to provide necessary condition for expanding industrialized production of lactic acid. One American company also has launched on the market the product of poly-lactic acid which is used to manufacture bottles, barrels, jars and women health products.

8.1.2.6　Production of long-chain dicarboxylic acids

This kind of organic acids are important raw material for chemical products. For example, 15-carbon dicarboxylic acid (DC15) is key raw material for synthesis of precious and famous spice such as normuscone and musk ketone. Over the years, research on production of long-chain dicarboxylic acids (C12, C13, C14, C15, C16, C17) in China by fermentation method has been carried out. Some researchers adopted *Candida tropicalis* mutant as the producing strain to produce 15-carbon dicarboxylic acid in concentration of 73.5g/L (yield of the starting strain, 35.8g/L), by using pentadecane (nC15) as the substrate, and culturing in shaking flasks for 4 days, with product purity of 94% (product purity of the starting strain, 69%). Afterwards, a 16L fermentation cylinder was used to culture mutant for 7 days, the yield of DC15 reached 130g/L; and for a 2.5t fermentation cylinder to culture mutant for 6 days, the obtained average yield, 176g/L.

8.1.2.7　Production of itaconic acid

Itaconic acid (also called methylene succinic acid) is key raw material in chemical industry. For example, glass steel made from it may substitute for traditional steel materials, and may be used in manufacture of plane and ship making, and has wide use and big demand. Productivity of itaconic acid is over ten thousand tons annual in the world, its price 45 thousand yuan/t, and production cost per ton about 15 thousand yuan. Japanese researchers screened out one strain for high yield of itaconic acid, strain TN 484, from *Aspergillus terreus* strain IF 06365. The yield of itaconic acid obtained from this strain that was reported to 82g/L in shake flask test with the fermentation medium containing glucose of 160g/L, which was higher than the starting strain by 1.3 times, this yield has been the highest one in present literature. At present, Sichuan Institute of Food and Fermentation Industry has been successful in production of itaconic acid with fermentation method, and has founded a middle-scaled factory with annual production capacity of 500t of itaconic acid. This factory may produce qualified products in batch whose quality reached the standard of products imported from Japan. China becomes the third country which can perform industrialized production of itaconic acid with fermentation method. The engineered production strain used in China is mutant of *Aspergillus terreus*, which is fermented in a 5t fermentation cylinder to produce itaconic acid whose yield is 70g/L. So far, the production scale has attained annual productivity of 400t. Nowadays, in Yunnan province, China, an advanced strain for production of itaconic acid has been bred and is used to perform test production successfully, and the annual productivity can reach 300t. Furthermore, the aim should be to make full use of sugar wastes (e.g. wood chips) as raw material to produce itaconic acid, because this can be good for both product development and environmental protection. One research group from Tianjin University of Science and Technology adopted the strain bred by the group to produce itaconic acid in the yield of 39.2g/L.

8.1.2.8 Production of cellulose and dyeing agents

Cellulose is one indispensable raw material in textile industry, paper-making industry and food industry, and is very widely used in other fields. Production of cellulose with microorganism fermentation method is a new increasing spot in biochemical industry; and production dyeing agents by microorganism will not only add new types of products in dye industry, but also provide new raw material source for textile, printing and dyeing products, food and cosmetics, and additionally help protect environment. Cellulose is also the richest natural polymer on the earth. Apart from plants, some microorganisms including "genetic engineering bacteria" can produce cellulose, too.

Acetobacteria xylinum is a typical *bacterium* producing cellulose, and structure of cellulose produced by it is similar to that of cellulose produced by higher plants. During synthesis the *bacterium* firstly makes glucose molecules bind to form glucan chains (polymerization), and then the glucan chains arrange in line to further form microfiber (crystallization). The single-stranded cellulose produced by this strain, extends out of the cell and then interwind with those extending out of the other cells to form a tight film, which is much like plant cellulose. Such cellulose produced by bacteria is insoluble in water, so firm as to exhibit gel-shaped, and may form a fiber layer in thickness of several centimeters including about 1μm fiber, which has very high water-holding capacity. Some experts predict, practical application of such bacteria cellulose will take many years, and at that time it will probably compete with some cotton textile products.

"Genetic engineering bacteria" are used to produce cellulose. American researchers at first isolated and cloned cellulose synthetase gene, and then identified the gene. The first strategy is to introduce this gene into *E.coli* or photosynthetic bacteria; and the second one is to transfer this gene to cotton in order to increase productivity and quality of cotton fiber. In the former strategy the introduced gene is expressed, i.e. "engineering *E.coli*" manufactures cellulose extracellularly. If the gene is transferred to photosynthetic bacteria successfully, the photosynthetic bacteria producing cellulose may be cultured on a large scale in a natural lake to manufacture cellulose. As for introduction of the cellulose synthetase gene into cotton crops, on the one hand, cotton productivity may increase, and more solid fiber material and cotton absorbing dyes may be obtained; on the other hand, the produced natural cellulose becomes pure. Certainly, under the lab conditions the cellulose produced by *Acetobacteria xylinum* has almost 100% purity. Furthermore, introduction of exogenous genes may make that the produced cellulose have properties of both natural cotton fiber and chemical one. All of these including bacteria, "engineering bacteria"and "engineering plants" create late-model fiber source for textile industry; at the same time *Acetobacteria xylinum* may play great roles in research on functional food. In Taiwan Province, China, such bacteria fiber polysaccharide is called "Na-Ta", which is developed as a functional food which has characteristic of high-content fiber and low-content fat, and eating "Na-Ta" frequently may protect against colorectal cancer. Of course, to increase productivity of "Na-Ta", the key lies in breeding advanced strains.

Although late-model cellulose has so many advantages, it is still under research further so as to make it turn into a practical application. And so it will really contribute to textile industry, paper-making industry and food industry. Therefore, various approaches to develop biological cellulose raw material source are important trends.

Purple bacteria pigment is one of biological dyeing agents, isolated from wastes of silkworm silk. It can be used as one dye in printing and dyeing of textile products, so as to make them possess natural colour and softness; and also used for dyeing of cosmetics and food. Such biological pigment can dye silk and cotton fiber products, as well as some chemical fiber products. It is also reported that indigotin dye may be produced with bacteria. In fact, with development of science and technology, some textile products still obtain various natural colours without dyeing, and such textile material with different natural colours has come into being. Transgene technique transferred the gene producing pigment to the genome of cotton to construct transgenetic colourful cotton plants. Exogenous genes expressed really, give cotton with various colours (e.g. red, blue, black and brown). This technique not only may save dyeing agents very much to simplify dyeing procedure, but also provide new raw material sources for textile, printing and dyeing industry. From this it can be seen that the production process combining natural biological dyeing agents with natural colorful cotton fiber will certainly give new contribution to textile, printing and dyeing industry.

8.1.2.9 Production of ethylene

Ethylene is important basic raw material in petrochemical industry, like steel and petroleum. Its production capacity is one crucial marker for judgement of economic strength of a nation. Ethylene can be used to produce material of polyester fiber, chemical fiber, pure polyester cloth; and ethylene or propylene is also used to promote maturation of fruits in agriculture and forestry. Nowadays ethylene is manufactured mainly in an approach of chemical synthesis. Application of biotechnology, especially microbiological technology, makes ethylene resource expand, and probably make organisms into the object for development of ethylene. In recent years, some microorganisms have been found to have ability to synthesize ethylene. According to reports, many microorganisms, such as fungus, yeast, bacteria and actinomycetes all can produce ethylene. In Kumamoto of Japan, one research group used one new *cyanobacterium* to fix carbon dioxide in the air to produce ethylene. Though the yield of ethylene is very low, but after modifying the process with biotechnology, it is predicted that this technology will possibly achieve the goal of practicability in a few years. Production of ethylene is influenced by many parameters, usually most suitable temperature is 25℃; optimal pH value is 4-5; and oxygen supply must be sufficient, under these optimal conditions yield of ethylene can reach the maximum, 3000mL/h. Studies have shown, adding methionine and L-glutamic acid may increase yield of ethylene.

Effective application of contemporary new biotechnology has great potential in modification of traditional production processes of biochemical products and increase of product productivity and yield. It will certainly meet demand for economic construction and domestic life and will contribute to substantial economic benefit and social benefit.

8.1.3 Application Prospects of Contemporary Biotechnologies in Chemical Industry

The chemical industry faces pressure from environmental pollution, which forces researchers of biotechnologies to relieve environmental pollution. Biotechnologies will play a significant role in treatment of industrial sewage with special toxicity, for example, degradation of petrochemical products including naphthaline, salicylate, phenol and nitrile, and clear-off of leaked petroleum on a large scale. Biotechnologies may be used to have a control over heavy metals in sewage, remove phosphorus and nitrogen in rivers. Unlike traditional treatment methods, biotechnological method for treating wastes may change wastes into precious products, and cannot lead to second pollution.

Many other key technologies in the chemical industry field cannot meet requirements for production. For example, rapid measurement of glucose, alcohol, methane, and glutamic acid is required during process control in production; and in national defense area, rapid detection of chemical warfare agents is required. Bio-probe composed of enzymes and cells may react to the above substances rapidly, and has possibility to meet these requirements.

In more than 20 years, biotechnology industry has already been proved to create great economic benefit and social benefit. Super oil-eating bacteria constructed through genetic engineering have capacity to decompose leaked oil at high speed and at high efficiency. Reports on application of genetic engineering bacteria in chemical industry are increasing. Production of chemicals through fermentation of genetic engineering bacteria has become the main topic in this field.

Development of biological isolation technique makes cost of biotechnological products more competitive then the cost of products through traditional chemical processes. Taking bio-probe as an example, demand in medical and healthcare industry, environmental monitoring field and food industry make demand for biosensors in American market increase year by year.

As for its prospects, the most important aspect for biotechnologies lies in its application to production of generic chemicals. Consumption of raw materials for chemical products will force mankind to utilize biomass to produce raw materials for basic chemical products by using biotechnologies.

Application of biotechnologies to produce chemicals at present focuses on production of expensive chemicals, especially pharmaceuticals and fine chemical products including conversion of steroids, synthesis of antibiotics, synthesis of alkaloid and organic acids, synthesis of proteins (including enzymes) and amino acids, and synthesis of nucleic acid compounds. It is estimated that the market for application of biotechnologies to chemical products will be greater than other products, and the overall product market for biotechnologies will grow greatly. Basic technologies in chemical industry will play more and more significant roles in the course of industrialization of biotechnologies, and manufacturing of biotechnology equipments based on processes in chemical industry and equipment technologies will also make great progress in the coming time.

8.2 Contemporary Bioengineering Technologies and Agriculture

In today's times population grows increasingly, and agricultural acreage decreases continually, to develop national economy and improve people's life, great effort must be taken to develop agriculture. Biotechnologies have substantial actual benefit and tremendous developmental potential, and so development of biotechnologies is an important solution to develop agriculture. Relation between biotechnologies and development of agriculture is a close one and biotechnologies are applied in agriculture, in order to promote development of agriculture at high speed and in high quality, to provide more and better raw materials for industrial production, and to provide more and better food for people.

8.2.1 Contemporary Biotechnologies and Crop Farming

One vital aspect for development of crop farming is to improve yield of crops and product quality. Traditional breeding methods have played a significant role in crop farming; but development of biotechnologies undoubtedly may promote traditional breeding, and accelerate the course of breeding and overcome various difficulties in traditional breeding; and more importantly, it broadens application area of useful substance resources dramatically.

8.2.1.1 Application of biotechnologies to improvement of breeds

The main tasks for improvement of breeds lie in increased yield, improved varieties and enhancement of stress resistance, such as resistance on plant pathogens, insects, grass plague and salts. Traditional breeding has many problems, including long breeding period, big workload, especially difficulty in balance among increase of yield, quality of improvement and enhancement of stress resistance. Life science and technology are superb craftsmanship that has made great progress, and shows promising potential.

1) Cell technology

The theoretical basis for application of cell technology to plant breeding is totipotency of plant cells, so called "totipotency" refers to that some one organ or even one single cell is isolated from one plant and then is cultured *in vitro* independently to regenerate a whole plant. Furthermore, scientists also find that, frequency of variation occurring during culture of plant cells is higher than that during natural growth of plants by almost ten thousand times, and hence there is much bigger chance to acquire useful variation. ①Culture of pollen. Among more than 200 species of plants regenerated by culturing anther, there are over 40 ones regenerated successfully in China. And due to other complementary technologies, farming area of new breeds of common crops such as wheat, rice and tobacco has reached millions of acres. ②Culture of cells and protoplast. Culture of plant cells may be performed on a large scale in a fermentation cylinder by utilizing fermentation technique, so as to manufacture key chemical feedstuffs including plant pigments, spices, pharmaceuticals and pesticides. Industrialized production of some rare species with cell culture technique may increase economic benefit tremendously, and will not destroy the ecological balance because of scaled cutting. Like a large forest, one small fermentation cylinder may provide useful products for human beings successively. In such a way, scientists have produced a large number of drugs human being need for treatment of leukemia and heart diseases with cells from vinca, foxglove and American Ginseng. Scientists are also studying on extraction of anti-cancer and anti-AIDS drug substances from some wood cells. Alkannin is not only a natural food additive and cosmetic raw material, but also has antimicrobial and anti-inflammatory effect. However, in the nature resource of lithospermum is very insufficient, therefore, it may be predicted that industrialized production of alkannin will turn into reality soon in China. ③Cell fusion technology. Cell fusion technology developed in the sixties of the twentieth century, and now geneticists are clearing off obstacle of incompatibility during distant hybridization. Cell fusion technology will probably create a miracle. Cell fusion between soybean and rice, ryegrass and wheat, and Chinese cabbage and wild cabbage may lay a basis for further breeding of new crop species. ④Removal of plant viruses. Plant viruses are one kind of important plant pathogens. Once many plants reproduced by means of asexual propagation infect some one virus, their yield is influenced dramatically. Experts at biotechnologies excise one most tiny piece of tissue with microsurgery at the stem apex on the plant infecting the virus, and after tissue culture of the excised tissue the regenerated plants becomes virus-free plantlets. Farming practices have proved that the yield of detoxicated potato, strawberry, garlic, lily and crane increases by several times. Detoxication technology has more magical action on gardening flowers. After some rare flowers are detoxicated, blossoms become bigger and more vivid, so that these flowers may increase in their competitiveness in the international market.

2) Gene technology

Significance of gene technology for crop breeding lies in that it completely breaks species' borders. As long as one gene is useful, wherever it is from any bacteria, or any one animal or other plants, or even human, may be "grafted" to the target plant and become one part of the genome of this plant. Therefore, new plant species, with important economic value and integrating various characters from different organisms useful for mankind, may be designed and created to meet the demand of people. In addition, genes may also be trimmed and some genes in plants controlling unuseful characters may be cut off or sealed, so as to let those useful characters play actions fully. Now many transgenic plants have been bred, such as transgenic plants resistant on tobacco mosaic virus, excellent spice breeds resistant on viruses, transgenic cucumber, tobacco and tomato. People wish to increase content of protein in grain seeds, and to modify amino acid composition of protein. At present with gene technology scientists have increased total protein in grain seeds by about 1%. Australian scientists have transferred the gene of protein from beans into herbage which it is used to feed for cow, obtained milk contains more essential amino acids. The genes of some drugs may also be introduced to plants in order to produce these drugs by plant. For example, American scientists have used transgenic tobacco to express Trichosanthin (TCS) which is extracted and used for clinical trials on treatment of AIDS. South Korean scientists are trying to produce insulin by utilizing transgenic plants. In some countries in Europe scientists are studying production of interferon with plants.

8.2.1.2 Application of biotechnologies to excellent species breeding

The tasks of excellent species breeding are mainly to expand population. In some cases such as those plants whose seeds are difficult to produce and so have a small quantity, rapid propagation then appears especially important.

1) Rapid propagation

Rapid propagation, also called micropropagation, refers to rapid indoor propagation of a small piece of plant tissue on a large scale with tissue culture method. Micropropagation has special significance for valuable flowers, trees and fruit trees, and endangered plants.

2) Artificial seed

Based on the basic theory that plant cells have totipotency and tissue culture technique, scientists invent artificial seed technology. As americans like eating vegetables very much, especially celery, they produce big and fresh hybridized celery having bred it for many years. Although this species of celery is delicious, it is a pity that such celery growing is difficult, because seeds are very small and germination rate is very slow. Besides, acquirement of such hybridized seeds is very difficult, and hence price of the seed is very high. In order to decrease the cost and meet demands, a new method was developed by scientists. At first cut seedling of celery into most tiny chips which are induced to form embryoid with ability of take root germination under specific conditions, and then the embryoid is packed with a polymer as artificial seed coat, and so a kind of capsule seed like fish oil pill is made. This method overcomes poor reproduction capacity, difficult production of seeds or low germination rate in some crops. It helps keep the advantage of high yield for the first generation of hybrid, and prevent degeneration of the second generation. It makes those sterile excellent species like as seedless water melon popularize rapidly. Seeds may be produced on an industrialized scale with the aid of this technology, and this will increase automatic extent of agriculture. Various accessory components, such as nitrogen-fixing bacteria, pesticides, herbicides and fertilizer, may be added to artificial seeds, which help breed strong and healthy sprouts and make crops acquire stable and high yield.

8.2.1.3 Biological nitrogen fixation

About 100 years ago, scientists found that bacteria in soybean nodule (i.e. *Rhizobium*) could convert nitrogen in the atmosphere to ammonia which was offered to leguminous plants such as soybean, peanut, clover for synthesis of amino acids and protein; at the same time, leguminous plants transported synthesized carbohydrate (sugar) through photosynthesis to root nodule, and provided it with energy necessary for nitrogen fixation. As such nitrogen fixation takes place under the condition of mutual dependence between leguminous plants and *Rhizobium*, it is called symbiotic nitrogen fixation. Besides, other bacteria, *actinomycetes* and blue algae also can fix nitrogen. Other non-leguminous trees have nodules in the root, but it is *actinomycetes* that fixed nitrogen. In farming, nitrogen fixation quantity of microorganisms is as large as twice of that of used fertilizer. From this we can see that all the nitrogen fertilizer factories founded with effort for more than half of a century are not as powerful as those bacteria in soils which appear silent. And hence biological nitrogen fixation may be considered to be fertilizer factories without smoke.

The leading role of biological nitrogen fixation is nitrogen fixation microorganisms and algae in the nature which can be used endlessly. As scientists estimate, blue algae of nitrogen fixation in the earth may fix about one million tons of pure nitrogen for the air, which is equal to nitrogen quantity contained in 50 million tons of ammonium sulphate. If this kind of blue algae of nitrogen fixation can be utilized effectively, it is equal to setting up many natural fertilizer factories. In addition, biological nitrogen fixation may be considered to be a natural phenomenon, does not nearly need any investment and does not consume energy. As long as we utilize this a little notable benefit will be created. For instance, different types of *Rhizobium* may be isolated to reproduce in batches with fermentation cylinders of large volumes, and then are used in different leguminous plants; and so root nodule formation will be enhanced markedly in these crops and their yield will increase.

8.2.2 Contemporary Biotechnologies and Livestock Breeding

The biggest problem influencing rapid development of livestock breeding is still diseases including infectious diseases and non-infectious diseases. Nucleic acid probe developed since the eighties of the 20th century has become a necessary important method in the veterinary field, and is often used for detection of minute pathogen DNA or RNA, analysis of gene maps of pathogens, and examination of epidemic diseases. Furthermore, MAb reagents which can be directly applied to practices and research of farming and animal husbandry have become a strong industry. Nowadays, several kinds of the most successful gene engineering vaccines preventing against enterotoxin from *E.coli* in poultry have been sold in the market, and other vaccines have also been started to be used. These will play a significant role in development of livestock breeding.

8.3 Contemporary Biotechnologies and Food Industry

The food industry is one of the largest industries in the world, and also one important application area of life science and technology. Contemporary biotechnology is a comprehensive technological system, which is based on life science, makes use of some characteristics and functions of organisms and their tissues to construct new species or new strains with expected characters, and conjuncts with engineering principles to process and manufacture so as to provide commodities and service. In recent years, many emerging biotechnologies have applied to production and development of food, so as to promote rapid development of the food industry. The application mainly includes: ①utilizing gene engineering and cell engineering technologies to modify and improve foods resource; ②utilizing fermentation engineering and enzyme engineering technology to process raw materials from agricultural side products into commodities such as wine, condiment, yoghourt; ③utilizing biotechnology to carry out the second development of the existing products to form new ones, such as many functional oligosaccharides, food additives; ④utilizing enzyme technologies, fermentation technologies and bioreactors to modify traditional technologies for food process, so as to reduce energy consumption, increase yield, and improve quality of food products. Furthermore, in the other fields associated with food production, such as packaging, storage, quality detection and three-waste treatment, biotechnologies are employed more widely.

Wide application of contemporary biotechnologies in the food industry makes contemporary food industry exhibit some new features. Firstly, "green displacement" arises in model of food production. Development of life science and biotechnologies makes both agriculture and industry (especially pharmaceutical, food and chemical industry) undertake important revolution. The boundary between agriculture and industry becomes increasingly vague, and "agricultural industry" and "industrial agriculture" are emerging silently. Undoubtedly, in the biological times agriculture will be the first manufacturing shop for the food industry, in which green field and leisurely cows and sheep only can be seen, noises of machines cannot be heard, but those transgenic plants and animals can be utilized to produce various industrial products, such as erythropoietin, vaccine and various biological active ingredients, i.e. "green displacement" occurs to the mode of food production. Secondly, core of food process moves upward. Application of biotechnologies such as tissue culture, gene engineering and cell engineering makes process core in food industry move from after manufacturing to before manufacture even the whole course of manufacturing. At present, the trend of core moving upward in the food industry has become increasingly notable. Furthermore, extension of such task core to the upstream is more useful for guarantee of food safety and quality. Thirdly, the meaning of "food safety" has been changed. In the new times important changes will take place in the meaning of food safety, which not only includes "non-toxicity" and "hygiene" under the traditional meaning, but also includes safety of Genetically modified food. Sense for food safety in people will be strengthened much more than ever. The latest achievements of scientific and technological will be applied to research on food safety, so as to develop new assay and detection technologies to detect Genetically modified food. Finally, the food industry fulfills comprehensive utilization and zero emission. Gene engineering, cell engineering, enzyme engineering and fermentation engineering are adopted to utilize wastes in the food industry comprehensively, elimination of environmental pollution from three-waste and fulfillment of zero emission are the goals for the food industry in the 21st century. In a word, biotechnologies have broad marker and potential prospect in the food field, and will certainly bring about new revolution in the food industry.

8.3.1 Development of New Food Resource with Contemporary Biotechnologies

The warning bell of population crisis on the earth is ringing urgently, but the earth which we rely on for living only can provide us very limited space and food resource, and this contradiction emerges increasingly. The traditional food industry is faced with severe challenge, and how to develop new food resource is a critical problem which must be solved. Biotechnologies give a promising future to the people.

8.3.1.1 Single cell protein

Protein is basic substance to maintain lives, and major part of human organs, tissues and enzymes, hormones and immunoglobin. The issue on lack of protein across the world has existed for many years, and development of SCP is just a vital solution to it.

SCP mainly refers to protein resource from microorganisms including bacteria, yeast, and other fungi. Microorganism cells contain rich protein, for example, in yeast cells protein accounts for 45%-55% of dry cell substances; in bacteria cells, 60%-80%; in mould mycelium, 30%-50%; in monodus subterraneus such as chlorella, 55%-60%; crops with the highest content of protein, soybean just only 35%-40%.

Amino acid composition and types of SCP are as good as those of animal protein. For example, yeast cell protein enriched in nutrients, contains 7 essential amino acids, and so called "artificial meat". Generally, an adult eats 10-15g dry yeast every day which can meet the requirement for protein of human body.

In additional to protein, microorganism cells contain abundant carbohydrate and lipid, vitamins, minerals, and hence SCP has high nutritional value.

For microorganisms their age is calculated as "minutes" and "seconds", while for animals and plants as "seasons" and "years". Grains have only one or two harvests every year. One cattle of 500kg can synthesize 0.5kg protein every 24 hours; while 500kg living bacteria cells can produce 1260kg protein in the event of appropriate conditions. Production of protein with microorganism cells requires less labour than agricultural farming, and it is not limited by regions, seasons and climatic conditions, and may be conducted on the equipments which take up limited soil area. And the protein produced in such way has large quantity and top quality, and is much better than that from existing grains.

SCP plays a significant role in the fodder and food industry. As fodder protein, it is used widely in the world. For example, *Candida albicans* and *Candida utilis* are used to produce yeast cells by utilizing waste solution of sulfurous acid or oil for livestock fodder. Feeding poultry and livestock with such fodder has better effects, with rapid animal growth, high production rate of cow milk, high laying production of chickens, and enhancement of animal immunity. SCP produced with yeast and *Candida* spp. can be used as human food directly. In America, SCP was produced by using ethanol as feedstock which has acted as food. Due to complete amino acid composition, SCP is often used as a nutrient-enhancing agent to add to food so as to increase biological value of protein in various food products. SCP enriched in vitamins and minerals, is used to supplement vitamins and minerals which food needs. Besides, SCP can enhance physical performance of food, and improve flavour of food. For example, concentrated yeast protein has notable fresh flavour, and has been used as flavoring agent of fodder, meat broth, etc.

SCP has so marvelous functions that development and production of it has broader prospect in China. Since our country is a big agricultural nation, food structure mainly consists of plant protein and acceptable daily intake (ADI) of animal protein has big difference from that in the countries in Europe and in the USA. In order to build up the people's health, production of SCP with biotechnologies has great significance.

By means of physical, chemical and biological mutagenesis, or adopting gene recombinant techniques, genetic characters of SCP production bacteria are modified so as to acquire excellent high-yielding mutants. These production bacteria have made actual advance in growth features, substrate availability, protein yield and essential amino acid composition, and even protein properties and functions.

8.3.1.2 Spirulina

Algae generally contain 45%-70% of digestible protein whose amino acid classes and content are the same as single cell microorganisms. Furthermore, growth velocity of algae is more rapid than higher plants, and after harvest of products its regeneration speed is also very fast. This may be formed semi-continuous production procedure. Since algae have characteristics of high content of protein and high yield of dry substances, protein yield rate in algae is higher than that in any other plants and animals.

Among algae, spirulina are regarded as one of the most important protein resources. Mankind has eaten it for hundreds of years. For the first time, spirulina were discovered on Lake Chad in central Africa and on Lake Texcoco in Mexico. The local people ate it and made cakes with it and this is useful for health. Due to thin cell wall and low content of cellulose, spirulina are easy to digest and absorb, their digestion rate can reach 95%, while that of chlorella is only 46%. After the fifties of the 20th century, in Chad spirulina were made into food whose commercial name was "Dehe". In 1964, Belgium botanist Jean Lenoard isolated pure spirulina from edible spirulina and sold the product "Dihe" and cultured them in the lab. In March of 1967 he first published the experiment results, which created the first case for artificial culture of spirulina and led to wide interest of biologists. In 1968, French Oil Institute cooperated with one compartment to found a culture pool in an area of $700m^2$, with annual yield of about 300t, which became the first spirulina production factory. Their products were tested in human body and proved to have no toxicity, and permitted for use as food and fodder. Since the seventies of the 20th century, in Japan, Thailand, Italy, USA, Israel and Taiwan Province of China across the world have established spirulina-associated companies with annual yield of over 1000t (dry powder), whose products are mainly used as nutritional and healthcare food for enhancement of immunity and promotion of human health. A large number of international journals have reported on evaluation of efficacy of spirulina. American journal "Better Nutrition" listed physiologically active ingredients which spirulina contains and medical features, and stated that spirulina had treatment and prevention effect on diabetes, anemia, liver diseases, ulcer, pancreatitis, vision disorder, leukemia, allergy, and cancer. Japanese pharmaceutical experts published the book, *Mysterious Spirulina*, which told about efficacy of spirulina. In America it was reported that a kind of glycolipid was extracted from spirulina and even had the effect to block HIV replication. Blue algae protein can be used to make fluorescence probe, and in addition, spirulina can be used for manufacturing of rare animal fodder and hairdressing cosmetic products. In 1977, United Nation Industry Development Organization (UNIDO) authorized German Dr. Chamorro carry out the toxicity test; and from the test results USA Food Protein Consulting Corporation and United Nation Food and Agriculture Organization (FAO) confirmed that spirulina protein quality had attained the ideal protein standard.

Spirulina has benefits including high content of protein and carbohydrates, and low content of fat and cholesterol; contains various vitamins and trace elements and has important nutritional, pharmacological and developmental value; and is called "food of the 21st century".

8.3.2 Improvement and Enhancement of Food Quality with Contemporary Biotechnologies

Functional properties of food refer to any property that some ingredients can influence on use value of food, except nutritional property. Most functional properties not only influence on organoleptic quality of food, but also play an important role in determination of physical properties of food ingredients during process and storage. With gene engineering technologies breeds of plants can be improved and food quality can also be enhanced.

8.3.2.1 Improvement of plant axunge

Conjunction of gene engineering technologies with traditional breeding methods provides a new solution to improve quality of plant axunge for people. It can increase proportion of unsaturated fatty acids in plant axunge so as to offer plant oil helpful for human health. For example, the antisense gene of stearoyl-acyl carrier protein(ACP) dehydrogenase were introduced into rape, the content of stearic acids in the transgenic rapeseed increases from 2% to 40%; while after introduction of gene of stearoyl coenzyme A desaturase to some crop, content of saturated fatty acid in the transgenic crop decreases, and that of unsaturated fatty acids increase significantly. Besides, American Calgene Corporation is developing soybean oil and canola oil with high content of stearic acids, such new soybean oil and canola oil will contain over 30% of stearic acids, and may substitute for hydrogenised oil and can be used to manufacture artificial cream, liquid butter and coco butter without trans-fatty acid products which hydrogenised oil contains.

8.3.2.2 Improvement of protein quality

In the aspect of improvement of protein quality there are two main goals that increase content of essential amino acids and improve process performance of protein. The main storage protein in leguminous plants, globin has very low content of methionine, which is the first restriction amino acids, but the lysine is high content, which is just contrary to protein in cereal grains. By means of genetic engineering technologies, the genes from cereal grain plants can be introduced to leguminous plants to develop transgenic soybean with high content of methionine. In addition, Chinese scholars introduced the gene of corn alcohol-soluble protein enriching in essential amino acids to the potato plant, so as to make the content of essential amino acids in potato tuber increase by over 10%. Scientists from Gainesville University in Florida of the USA introduced the exogenous gene of wheat gluten protein in high molecular mass to common wheat and acquired the wheat with high content of gluten protein in high molecular mass. Such wheat gluten protein has excellent extensibility and elasticity.

8.3.2.3 Improvement of carbohydrate quality

With gene engineering, the composition and content of starch in plant food may be changed; and transgenic potato only containing amylose may be acquired by means of inhibition of starch branching enzyme with antisense genes. American Monsanto Company developed the transgenic potato in which the content of amylose increased by average 20%-30%. The novel potato products by frying have better potato flavour and texture, and lower oil absorption value and less oil flavour.

8.3.2.4 Improvement of fruit and vegetable quality

Improvement of fruit and vegetable food quality includes enhancement of storage property and freshness keeping performance. Ethylene is the most important and direct index to regulate gene expression during maturity of fruits. With genetic engineering the antisense gene of ACC (aminocyclopropane carboxylate) synthetase and the exogenous gene of ACC deaminase are introduced to normal plants. The ethylene synthesis-deficient mutation plant is acquired so as to achieve the goal to control over maturity of fruits. This goal has been achieved in tomatoes and brought about a revolution for storage and freshness keeping of fruits. Antisense gene technology may be utilized to breed out storage-suitable fruits and vegetables.

8.3.2.5 Modification of food microorganisms

The key of the fermentation industry is to obtain excellent microorganism strains by using common mutagenesis, hybridization and cytoplasm fusion methods, as well as gene engineering methods. Biacetyl is one important substance to influence on beer flavour, when content of biacetyl in beer surpasses the threshold, a disagreeable sour smell will occur and damage flavour and quality of beer seriously. One of the effective measures for removing biacetyl in beer is to introduce the exogenous the gene of acetolactate decarboxylase to beer yeast to make it express in yeast, since yeasts themselves have no activity of acetolactate decarboxylase. This is an effective approach to decrease content of biacetyl in beer. With gene engineering technology, the gene of amylase in mold may be introduced to *E.coli*, and further to yeast cells, so as to make yeast produce alcohol by directly utilizing starch. Such method saves the step of high-pressure cooking and about 60% of energy, and production cycle shortens much.

8.3.2.6 Production of healthcare and special food

Production and development of healthcare food and special food depend on gene engineering technology. For example, tPA gene which helps solve thrombus is cloned into cows or sheep, and then its express product will be produced in cow milk or sheep milk. Production of gene engineering vaccine, food vaccine, with transgenic plants is one hot spot of research on food biotechnologies. Food vaccine means that the associated protein (antigens) genes from some pathogenic microorganisms are introduced to the receptor cells of some plants to make them express in the receptor cells, so as to the receptor plants produce vaccine to prevent against the associated diseases. At present, over 10 kinds of edible vaccine of transgenic potato, banana and tomato have succeeded and relate to surface antigens of rabies virus and HBV, and *streptococcus* mutant strain. Since these recombinant protein genes may be stored long-term in fruit or seeds of the transgenic plants, this is very useful for storage, production, transportation and popularization of vaccine.

8.3.2.7 Food additives

In the food industry, cell engineering technology may be applied to modify genetic substance and to carry out cell culture in a planned way, i.e. cell fusion technique and large-scaled controlled culturing technique, which are used to produce various healthcare food, novel food and food additives. Cell hybridization and cell culture may be utilized to produce food additives with unique scent and flavour, such as vanilla element, cocoa and pineapple, and advanced natural pigments, such as curry yellow, carotinoid, purpurin, anthocyanin, capsaicin, indigo blue, etc. For example, by adopting cell isolation method that vanilla cells are isolated to culture artificially so as to make them reproduce; and the extracted flavoring material is the same as that from the cultivated plant, but the yield increases markedly. In breeding practices, many high quality varieties with special resistance and patience were bred to adopt new technology, such as anther culture and chromosome engineering technology, combined with conventional technology of the method. In summary, in the wave of new technology revolution cell engineering has great potential for innovation of traditional process in agriculture and food industry.

Enzymes are biocatalysts with high catalytic activity and high specificity, and may be applied to conversion of substances during production of food. After application of amylase achieved significant fruits, cellulose has been applied widely to juice production, fruits and vegetables production, instant tea production, soysauce brewing and wine making in the food industry. With development of hi-tech including gene engineering and cell engineering, people have researched and developed new types, novel use and high activity more enzymes continuously; and meanwhile, enzyme immobilization technique, enzyme molecule modification technique and mimic enzyme technique are developed rapidly. With higher catalysis efficiency and more precise selection, enzymes will be applied more widely in the food industry, in order to produce novel food meeting the demand of people and promote rapid development of the food industry.

Production of food additives of different uses and different kinds with contemporary fermentation technique is a new trend. These additives include: sweetening agents such as xylitol, mannitol, arabitol, sweet polypeptide, and etc; acidifying agents such as malic acid, succinic acid, and etc; various essential amino acids among amino acids; thickening agents such as xanthan gum, thermosetting polysaccharide, etc.; flavour additives such as many nucleotide acids, biacetyl, isobutanol, etc; flavouring agents such as fatty acid ester, etc.; carotenoid, vitamins, various active healthcare bacteria, active polypeptide, active polysaccharide, superoxide dismutase inhibition factor.

8.3.3 Food Test with Contemporary Biotechnologies

The pathogens contaminating food mainly include *Listeria* spp., *Salmonella* spp. and *Campylobacter* spp.; and microorganism toxins contaminating food include bacterial endogenous toxin and fungus toxin. Traditional test methods for food components and contaminants have many shortcomings, such as complicated operation, time consumption and poor specificity. Therefore, many new test methods, antibody detection system, DNA and RNA probe technique start to replace traditional test methods. These novel techniques with highly specificity, high detection efficacy are economical and effective, can be operated simply, and improve the safety standard on food supply.

8.3.3.1 Application of PCR technology

Rapid development of modern biology, especially molecular biology, provides advanced technical approach for food detection. Polymerase chain reaction (PCR) technology is simple, rapid, specific and sensitive, and so is widely used for detection of pathogenic microorganisms in food and infection source tracking, especially suitable for detection and identification of bacteria with *in vitro* culture problem and complex antigenicity.

Salmonella spp. is one important kind of zoonosis pathogen, mainly hosts in the intestinal tract of human and animals, and may cause foods poisoning, acute gastroenteritis and diarrhea. Consequently, for either food sanitation or animal quarantine, *Salmonella* spp. is one of the necessary items to detect. Now phenol extraction method or alkaline cleavage method may be adopted to prepare samples, and PCR diagnosis kit may be used to detect the pathogen. Results reveal, for detection of *Salmonella* spp. in the blood and feces samples from artificially infected and naturally pathogenic animal blood with the kit, the positive rate is higher than that with culture method. Furthermore, PCR method only needs several hours, and detection time is very much short.

Listeria monocytogenes is a Gram-positive bacterium with characteristic of aggregation. Among all the bacteria of *Listeria* spp., *Listeria monocytogenes* is the only human pathogen, and has caused food contamination for several times, and may result in food poisoning. Usually, this microorganism can live in many food process plants, and causes contamination of food after process. PCR has become the technical basis for detection of *Listeria monocytogenes*. The target for such detection is the gene of hemolysin (hly) protein from *Listeria monocytogenes* which is necessary for its manifestation of toxicity.

Botulinum toxin may cause poisonous diseases in various animals, and identification of the toxic strain depends on detection of the toxin. PCR technology may detect the genes of type A-E toxin. Xie Zhixun and others established the PCR method for detecting and identifying Mycoplasma meleagridis, with which 100fg DNA from Mycoplasma meleagridis can be detected.

PCR technology can be used for detection of both bacteria and spirochetes, and viruses, especially those viruses hard to culture and carry out serological assay, which can be detected rapidly with PCR method.

8.3.3.2 Application of immunological technology

Interaction between antigen and antibody is basis of immunology. Some macromolecule or small molecules substance in food may become antigens in a direct or indirect way, so that immunological technology is a rapid, sensitive, specific and effective approach to research and detect food. The result of specific antigen-antibody binding may be displayed through enzyme-catalyzing colour reaction, fluorescence reaction and radiation isotope method, and hence a series of sensitive and practical immunological technologies are established. Immunological detection method may test many substances in most low content, such as trace residuals, fungus toxin, antibiotics, hormone, and bacteria toxin.

Immunological technology has been widely used in food production and scientific research. Its application scope includes: detection of food components, such as nutritional components, flavor components and some unexpected ones; detection of some factors affecting food quality during food production and process, such as quantitative or qualitative detection of spoiling microorganisms and their enzymes; detection of food safety, such as pathogenic microorganisms or their toxins, insecticide, antibiotics and food adulteration.

During storage food may be contaminated by microorganisms such as molds, and as a result, not only this can lead to decrease in aesthetic quality and nutritional value, but also it is to be noted that some molds can produce toxins. Generally, the methods for detecting molds include culture, electric conductivity measurement, detection heat-resistant substance (e.g. chitin) and microscopic observation, which are all complicated and time-consumed; while ELISA (enzyme-linked immunosorbent assay) method may be used to detect molds in food rapidly. *Aspergillus flavus* toxin is one carcinogen and mutagen produce by *Aspergillus flavus* and *Aspergillus parasiticus*. The immunoassay method for this fungus toxin includes radiative immunoassay and ELISA. New generation ELISA introduces a series of amplification mechanisms which make sensitivity of ELISA increase dramatically. For example, the amplification mechanism of substrate cycle makes alkaline phosphatase not catalyze synthesis of the coloring substance directly, but NADP dephosphorylation to produce NAD, which involves in the oxidation-reduction cycle catalyzed by alcohol dehydrogenase and flavin enzymes, finally leading to formation of the coloring substance. Such amplification mechanism amplifies the signal of alkaline phosphatase by 250 times in comparison with standard ELISA. A kind of specific trace competitive ELISA has sensitivity of 25pg for detection of *Aspergillus flavus* toxin B1. In order to reduce cross reaction between *Aspergillus flavus* toxin B1 and B2, the MAb resistance *Aspergillus flavus* toxin B2 may be used to perform indirect competitive ELISA, whose sensitivity is 50pg. The AF immunoassay kit intended to be used for detection of *Aspergillus flavus* toxin has become one routine method in assay labs across the world.

Transgenic food has gradually entered life of common people. In order to protect rights and interests of vast number of consumers, and to meet the needs for option and the right to know and international trade, detection of transgenic food is getting attention increasingly. Since transgenic substances are probably mixed into food during farming, harvesting, transportation, storage and process, so as to bring about occasional contamination. Therefore, during either labelling transgenic food or transporting transgenic raw material and non-transgenic raw material respectively, detection of both transgenic raw material and food is necessary; in addition, for distinguishing transgenic food from non-transgenic one, in order to mark the transgenic food selectively and limit content of the transgenic raw material, accurate and effective detection technology is also required.

Detection method of transgenic food is one necessary approach to determine, produce and manage the transgenic food. The essence for detection of transgenic products is to detect whether there are exogenous DNA sequences or recombinant protein component in the transgenic products. At present, more kind of transgenic crops, the quantity is large in the world, and hence it is so hard to detect. Furthermore, comparing with the huge plant genome, content of the exogenous DNA is really too small, and so high sensitivity of detection technology can be needed. Since transgenic organisms are characteristic of containing exogenous genes and manifesting characters of exogenous genes, nowadays there are two technological approaches adopted internationally for detection of transgenic plant food. The first approach is to detect the inserted exogenous genes by mainly adopting PCR method, Northern hybridization and Southern hybridization, and biochip technology; and the second one is to detect the expressed recombinant protein by mainly adopting ELISA method, Western hybridization, and biological activity assay. Detection of gene components must be rapid, accurate, sensitive and reliable. However, due to many kinds, big quantity, especially complicated composition, the components (nucleic acid or protein) in agricultural products containing transgenic component which are to detect always are degraded and broken, and so most difficult to detect.

8.3.4 Genetically Modified Food

8.3.4.1 Concept and current situation of genetically modified food

Genetically modified food (GMF), also called biotech food, refers to the food with novel characteristics which made from animals or plants whose genes are modified by scientists in their labs. In other words, GMF means that some genes from organisms are transferred to other species to make them express the characters or products which they do not possess by means of molecular biology technique, and the transgenic organisms are used as feedstock to process and produce food which is just GMF. Through such technology mankind may obtain more qualified food, which has benefits of high yield, rich nutrients and strong disease-resistance, but has obvious shortcomings of risk to cause genetic pollution.

For contemporary people, it has already been an undoubted thing to take drug or undergo injection. Nevertheless, in the coming future, you may take other ways to treat your own diseases. When you have a cold, eat a tomato; when you have loose bowels, eat an apple; and then the disease will be eliminated as long as you eat the GMF. This is neither just a dream, nor a plot in a science fiction. With development of gene engineering technology, this GMF which treat some diseases will turn into reality. Although the existing GMF has not yet achieved the goal of treatment of diseases, due to high yield, rich nutrients and unique flavour it has already entered the market abroad. And meanwhile, this brings about much dispute; some people boost it strongly, and other people oppose it. In China, the GMF has entered people's family. In order to learn about the GMF really, and avoid dispute and terror caused when it becomes the food products in our mouth, Dr. Luo Yunbo from China

Agricultural University has always been engaged in research on gene engineering, won "National Outstanding Youth Fund" in 1998, and has profound knowledge and opinion on the GMF. Luo Yunbo once gave the definition of the GMF refers to that genes from some organisms are transferred to other species to modify genetic substance of the organisms by utilizing molecule biological technology, so as to change their characters, nutritional quality and consumption quality to meet requires for mankind. Application of the GMF in agriculture aims at solving the problem of serious environmental pollution resulting from traditional agriculture, increasing organism diversity and reducing production cost. For example, the transgenic crops resistant on insect and pathogenic microorganisms are bred with gene engineering technology by itself may kill or suppress diseases and pests, and it is not necessary to only rely on pesticides to solve the problem on diseases and pests. This not only increases yield and reduces production cost, but also protects environment. At present the genetically modified crops popularized all over the world are mainly corn, soybean, rape, tomato, etc.

According to incomplete statistics, the farming area of genetically modified soybean and corn accounts for about 80% of their total farming area. In the food industry, soybean, corn and their processed products are all necessary feedstock, and the food made from the transgenic feedstock is also the GMF which has substantial scale in quantity and variety. For over 0.8 billion hungry population, the GMF has so great development potential. With progress of our scientific technology, transgenic technology is certainly to be applied to production of food in China, hence, we should prepare early to explore how to cope with import test systems on the GMF in other countries, especially developed ones. At the same time, the test systems on the GMF in China should be established according to our national situation, so as to sanitation and safety of food products.

8.3.4.2 Classification of the genetically modified food

Production of the GMF has two ways. Firstly, the existing genes are modified to make some characters not express. Taking refreshing tomato as an example, the gene of synthesis of ethylene to promote maturity of tomato is blocked to express so as to reduce or block reduction of ethylene and prolong storage period of tomato. Secondly, other genes are introduced so as to produce new characters. For example, the gene of synthesis of oligosaccharide is introduced to the lactate acid bacteria used for production yogurt, and then content of oligosaccharide in the finished yogurt, which has special action of prevention against human cardiovascular diseases, will increase much. However, so far, this technology has still remained in the initial stage. Though to the date nobody has classified the GMF, according to practice, based on functions of transferred genes it can be classified into the following types. ①Yield-increasing type, yield increase of crops has close association with their growth and differentiation, fertilizer, stress resistance, and diseases and pests resistance, and hence the associated genes may be transferred or modified in order to increase yield. Experts think, as transgenic technology may not only reduce agricultural production cost much, but also increase the yield per unit area, it will be adopted widely in agricultural production in the future. ②Mature-controlling type, by transferring or modifying and controlling the genes associated with the fruit maturity phase, the fruit maturity phase of the transgenic organisms prolongs or comes in advance, so as to adapt to demand of the market. The most typical example is storable tomato. ③High nutrition value type, many crops lack essential amino acids. In order to change the status, the gene of storage protein in seeds may be modified to make it express protein with appropriate amino acid composition. The crops which have been bred successfully are transgenic corn, potato and kidney bean. ④Healthcare type, the gene of one antigen or toxin from the associated pathogen is transferred to grain crops or fruit trees. These grains and fruits are eaten by people, and so they supplement nutrients, as well as take vaccine which may prevent against diseases. Some GMF may prevent against atherosclerosis and osteoporosis, some preventive factors may be acquired through transgenic cow milk and sheep milk. ⑤New variety type, by gene recombination between different varieties new varieties are formed, and the GMF that made from them may have new characteristics in quality, color and flavor. ⑥Environmental protection type, the transgenic crops with function of diseases and pests resistance may reduce the risk of environmental pollution caused by use of pesticides. ⑦Process type, this type of GMF is made by using transgenic products as raw material, and has many varieties. Xinjiang Institute of Natural Colorful Cotton has obtained fibrous seeds of new cotton breeds of red, blue and black cotton whose new characters express in the present generation, with transgenic and cloning technology. In spinning and weaving with them, many and complicated printing and dyeing procedure may be spared in order to reduce production cost and decrease harm of chemicals to the human body. Since 2002, China Color Cotton Co., Ltd. has cooperated with Chinese Academy of Sciences and China Agricultural University to carry out transduction of the gene of pigment protein from coloured cotton in Hainan islands, China. Now the three breeds still have problems of uneven expression of pigment protein and failure in inheritance. If the problems are solved our coloured cotton breeds will be more various. At present the coloured cotton breeds planting in a large area in the world are only brown cotton and green cotton. Based on colour depth, the brown cotton may be divided into brown red, pink, deep brown, yellow, light yellow types; and the green cotton into blue, light blue, light gray, buff types.

8.3.4.3　Development prospect of the genetically modified food

Transgenic technology has been regarded as one advanced technology to improve agricultural technology and increase benefit of agricultural economy in science and industry area, and is called "green revolution". Professor Guo Lihe from Biochemistry and Cellular Biology Institute of Chinese Academy of Sciences thinks that the GMF is not only a science problem, but also an economic and political one. For example, the price of transgenic soybean in the USA is equal to half of that of common soybean in China, while the price of transgenic corn one-third of that of common corn in China. And so this leads to a trade problem. As he says, since gene engineering technology may increase yield of food feedstock, improve nutritional value, flavor of food, and remove disagreeable features of food, as a big agricultural country, it is so necessary for China to promote research on the GMF. But now some problems exist, in China transgenic technology and knowledge are on the way to innovation, and so technological level is lower than that in the developed countries, while introduction of abroad technologies may involve the problem of intellectual property, which has too high cost. On the other hand, added value of agriculture is intrinsically low in China, and meanwhile research capital mainly depends rather on investment from the government than civil investment, which is too small and nearly zero, and is hard to match with the big civil capital investing in research and development. In the long run, improvement of crops with gene engineering is a thing which must be done.

8.4　Contemporary Bioengineering Technologies and Paper Industry

Application of biotechnologies in paper making industry has a long history. In the ancient times of China most of paper was made by means of natural fermentation and fiber dispersing and pulping, utilizing plant feedstock such as straw, bamboo, bark, etc., which is processed properly. Essence of that technology is dispersing fiber fairly completely by means of microorganism fermentation. But due to long period, small scale, poor adaptability of raw material, the technology cannot adapt to demand of modernized production, and have always been at the original primary stage. In recent years, with rapid development of biotechnology, microorganism enzymes and gene engineering bacteria have widely been applied in paper pulping and making industry, and shown promising potential.

Lignin is the biggest barrier for effective utilization of cellulose. During chemical pulping procedure, most of lignin may be removed from fiber of raw material, but 3%-12% of it residue. This part of residual lignin may cause paper pulp to brown and decrease strength of paper. By utilizing modern biotechnologies raw material for paper making may be modified to reduce content of lignin in paper pulp, improve pulping process, increase quality of paper and reduce environmental pollution.

8.4.1 Modification of Raw Material for Paper Making with DNA Recombinant Technology

DNA recombinant technology, one effective solution to improve characters of organisms, is a procedure, in which genes are cut and spliced to form recombinant genes, and then the recombinant genes are introduced to host cells to reproduce and express. With effort work of 12 years, Jiang Liquan Lab of Michigan Technical University finally discovered a method for reducing content of lignin in woods by means of gene recombination, i.e. the gene *pt4CL1* controlling synthesis of lignin was kept to be in suppression state by using antisense RNA technology to acquire transgenic woods. In the transgenic poplar content of lignin was lower than that in the control poplar by 45%, while content of cellulose increased by 15%. It was still found that this tree grew fast and was taller than the control tree by 30%. One British company has isolated an inhibitor of the gene of cinnamyl alcohol dehydrogenase, which can inhibit synthesis of lignin. The special gene of enzyme inhibitor is introduced to excellent breeds of fast growing type poplar and eucalyptus, in order to breed out new generation feedstock for paper making. As during stewing such raw material lignin is easy to dissolve out, it can save chemical quantity and reduce energy consumption, increase yield rate of paper pulp, decrease sewage pollution load and reduce sewage treatment expense. French Organism Cell Research Center modified lignin in poplar through gene recombination to make poplar more suitable for paper making. Taiwan Province Forestry Institute in China had researched and developed transgenic eucalyptus saplings successfully, which have characteristics of high content of cellulose and low content of lignin. Apart from decreasing content of lignin, gene engineering technology can change the proportion between three lignin structure units of this tree, which aims at making lignin easier to dissolve out during pulping or bleaching.

8.4.2 Synthesis of Cellulose with Contemporary Biotechnologies

Green plants can synthesize cellulose; however, some bacteria have ability to synthesize cellulose, and such ability is the strongest in *Aacetobacter xylinum* and has potential for large-scale production. Such cellulose originating from microbes as called "microbial cellulose" or "bacterial cellulose". Comparing with cellulose originating from plants, microbial cellulose has its own obvious benefits: ①Cellulose produced by *Acetobacter xylinum* is purer, and contains 100% of cellulose and no semi-cellulose, lignin and other cell wall components, its extraction and purification process is very simple. ②Microbial cellulose has excellent physical properties and mechanical performance. Since there are many "pores" inside the cellulose, it has good water-permeability and air-permeability, and strong hydrophilism.

During manufacturing top-quality paper or special paper products, supplementation of bacterial cellulose may increase quality of the finished paper and make it have high strength, durability and water-resistance. Besides, during manufacturing carbon cellulose paperboard able to absorb toxic gas, adding microbial cellulose may increase absorption volume of carbon cellulose paperboard and reduce loss of filling in paper.

8.4.3 Biopulping

Chemical method for pulping can overcome defect of ancient paper making technology, has features of short production period, large scale and strong adaptability of raw material, and make yield and quality of paper increase much due to modernized production process. However, chemical method for pulping has still many problems including big consumption of chemical reagents and high energy consumption, big investment, high COD and BOD load in industrial sewage, and production of toxic and carcinogenic substances which may cause serious environmental pollution. Since microorganisms can effectively convert three main components in plant fiber feedstock into CO_2, H_2O and humus, the procedure is environment-friendly, and may reduce energy consumption and chemical dosage. Biopulping may help overcome those shortcomings of chemical method.

Swedish Pulp and Paper Research Institute (STFI) at first carried out such work. They pretreated bits of wood with *Phanerochaete chrysosporium* and found that pulping with the treated bits of wood may reduce energy consumption very much. Meanwhile, they bred the fungi through mutagenesis to obtain a variation strain without activity of cellulase, which reduced damage of cellulose in pulping. Biopulping has also been researched by Americans, adopting bio-mechanical method, i.e. before conducting mechanical pulping, bits of wood were pretreated with fungus, as a result of such treatment, comparing with traditional method, over 24%-30% of electricity was saved, and the finished paper product had higher strength including tensile strength, bursting strength and tear strength, and at the same time reduced production cost considerably.

Chinese researchers have carried out biopulping with non-wood feedstock, and have bred out useful *Phanerochaete chrysosporium* which may degrade lignin selectively. These fungi can degrade cell wall components of straw preferably, further degrade intercellular substances, so as to make fiber separate completely, and then degrade lignin in secondary fibrous cell walls so as to make fine fiber in secondary cell walls expose. These features are very useful for development of biopulping.

8.4.4 Biobleaching

Paper pulp bleaching with biotechnologies is an effective way to reduce sewage pollution and fulfill clean production. Biobleaching refers to a procedure of treatment of paper pulp with microbes or enzymes secreted by microbes, so as to remove lignin or help remove lignin, and increase whiteness and other performance of paper pulp.

8.4.4.1 Biobleaching with enzymes

In the traditional method for bleaching paper pulp with chlorination and alkaline treatment techniques are adopted widely. In recent years it has been found that the paper pulp bleached with traditional method contains a great amount of lignin and carcinogens, which may harm the human being through the biological chain. In 1986, Finnish scholars for the first time reported treatment of paper pulp with xylanase might make chloride at the chlorination stage decrease by 25% at the international academic conference. Reported in North America and Europe, treatment of paper pulp with xylanase, cost reduced by 20% and COD value decrease by over 85%. In recent years China also has performed research on xylanase treatment of paper pulp, and achieved excellent outcome. Use of xylanase makes whiteness of paper pulp increase by 5%-6% of units, increases strength of paper pulp, and also reduces pulp return yellow value of paper pulp. However, generally xylanase always contains a small amount of cellulose which may lead to decrease in fiber strength and paper pulp viscosity. Therefore, selection of enzyme source microorganisms and extraction of xylanase are very important, some problem has been almost solved, with selection of enzyme source microorganisms, mutagenesis breeding of strain, construction of gene engineered stain and modification of extraction and purification process. The bacteria with high yield xylanase are bred out, and it can synthesize xylanase adaptable to high pH values and meet with the actual requirements for a plant.

However, xylanase used for biobleaching experience a development history of three generations. The first generation is acidic enzyme, the second one is neutral enzyme, and the third one is alkaline enzyme. Currently, research and application of the third generation enzyme are in the hottest time. Obtaining xylanase with excellent properties including thermophilic and basophilic properties has become a hot spot in the research area. It is expected that in the coming time recombinant enzyme will be applied effectively in the bleaching process.

Since in the xylanase treatment process, xylanase can play action of auxiliary bleaching by degrading and removing deposited semi-cellulose on the surface of paper pulp, it cannot substitute for chemical bleacher. And hence, biobleaching cannot completely substitute for chemobleaching, and can reduce pollution but cannot eliminate it. In order to eliminate pollution from bleaching solution containing toxic chlorine fundamentally, it is necessary to fulfill biobleaching at last, i.e. residual lignin in paper pulp is removed with biological method completely. People expect that biobleaching can be fulfilled through direct action of ligninase on lignin. Many researchers are investigating on non-xylanase bleaching enzymes preparation, where involved enzymes include peroxydase, manganese peroxydase, laccase and cellobiase, among which laccase is the hot spot in present studies. Japan reported that biobleaching with laccase could remove 50%-60% of lignin, but this is a little far from really meant biobleaching. Structure of lignin is very complicated, and the complex between lignin and xylan is formed in paper pulp and attaches with cellulose, and is very hard to remove. It is not enough to depend on only one enzyme. Combination between xylanase and ligninase is expected to degrade lignin remaining in paper pulp completely, so as to ensure good biobleaching. In the future breakthrough for biopulping and biobleaching technology will make the paper making industry keep off pollution and fulfill clean production.

8.4.4.2 Bleaching with microorganisms

Since microorganisms have ability to degrade or selectively degrade lignin, lignin in paper pulp may be removed by means of microorganism treatment, in order to achieve bleaching efficacy or reduce usage of chemical reagents, and relieve toxic organic substance pollution on environment. In 1979, American scientists treated sulphite paper pulp directly with *Phane-rochaete chrysosporium*, then carried out alkaline extraction, and finally found that some lignin was degraded in paper pulp. Later, further studies revealed that combination with microorganism treatment and chemobleaching achieved the same effect as the traditional five-stage bleaching process, and could reduce usage of chlorine and pollution from bleaching sewage and organic chloride.

In summary, life science and biotechnologies have permeated into the whole procedure of the paper pulping and making industry. With further development of life science and biotechnologies, we believe that application of hi-biotech in the paper pulping and making industry will bring the paper making industry with a more promising prospect.

8.5 Contemporary Biotechnologies and Leather Textile Industry

8.5.1 Application of Contemporary Biotechnologies in Leather Industry

During leather making it is to remove hair on the skin of pigs, cows and sheep, and then further process can be performed to make leather. In the past dehairing process used immersion method by lime and sodium sulfide, which needed long time, many steps and big labour strength, and led to serious pollution. Utilizing enzyme method, proteins are decomposed by proteinase at the junction between hair, epidermis and dermis, so as to lose the junction between hair and skin and to reach dehairing. So far pig skin leather, napped leather and cow sole leather have adopted enzyme dehairing technology. In addition, softening raw hide is a key step in the leather industry. Leather softening with acidic proteinase and a handful of lipidase may remove smudge effectively, make leather soft and permeable, and increase quality of leather.

During process and production of cod liver oil and spice, some proteins mingle with products and cannot beremoved, and so this always affects on purity and quality of products. Proteinase may be used to remove proteins mingled in the products. Silver powder of kinetoscope film is stuck to the film with glue as a bounding adhesive. In order to recover silver on the waste film, the film may be soaked in proteinase solution to make gelatine decompose, and finally silver powder falls off to retrieve.

8.5.2 Application of Contemporary Biotechnologies in Textile Industry

8.5.2.1 Treatment of textile feedstock

In order to enhance strength and smoothness of fiber and facilitate in weaving, before manufacturing textile products, it is necessary to starch up fiber, i.e. starch is treated with α-amylase for some time to make viscosity attain a certain extent, and then it can be used as sizing agent. Before bleaching, printing and dyeing during manufacturing textile products, the attached sizing agent must be removed. In this step α-amylase is utilized to hydrolyze the sizing agent to make it remove, which is called desizing. For some textile products, during starching animal glue is used as a sizing agent, and so proteinase may be used to desize.

At present, enzyme technology has been adopted to treat fiber products in many countries. For example, Demark Novo Corporation firstly applied cellulase to treat cotton fiber of jean cloth; and jean cloth after treating shows primitive simplicity and soft feeling. With neutral cellulose, tinctorial yield of jean cloth increase during dyeing. Additionally it is still found that treatment of fiber cotton and bastose with acidic cellulase makes textile products become more natural and supple, as well as may remove hair on the surface of fiber to make fiber surface shine. With enzyme method to treat fiber surface, the contact of fiber with the skin become more comfortable. Treatment of woolen cloth with proteinase makes quality of man suit cloth improve tremendously. Under certain conditions, cellulase hydrolyze β-1,4-glucosidic bonds with cellulose. Partial hydrolysis of cellulose molecules makes that crystalline regions become smaller; and meanwhile, partial hydrolysis of non-crystalline regions makes the space between it and crystalline region bigger. Under action of external force, relative motion is easy to occur between crystalline regions. And so, control over enzyme usage and action conditions may make that bending strength of fiber decrease and rigidity decrease, so as to reduce skin irritation and make touch quality of textile products improve significantly.

8.5.2.2 Treatment of raw silk

Natural silk is mainly composed of silk fibroin and sericin. The surface of fibroin is wrapped with a layer of sericin. During manufacturing advanced silk products, degumming treatment must be done, i.e. proteinase treatment may be adopted to remove sericin on the surface of natural silk so as to increase quality of silk products. ①Proteinase treatment may replace traditional soap alkali method to remove sericin, and make degumming efficiently. After treating silk with proteinase, the finished silk products have supple touch feeling and rich texture. ②Proteinase treatment may be used for sand consolidation of real silk textile, which makes textile products have elastic and supple touch feeling. During sand consolidation, proteinase firstly removes sericin by dissolving sericin in raw silk, the fibroin stuck together at the beginning separates each other, and then some part of fibroin is further expanded and hydrolyzed. The reaction procedure may be expressed as: water molecule immersion and enzyme molecule absorption→ fibroin moisturizing and expanding → enzyme hydrolysis on fiber surface →further immersion with proteinase → cleavage of fibroin peptide bonds →decomposing protein into amino acids. From this we can see, control over enzyme action may make fibroin of real silk moisturized and expanded with little damage, and the moisturized and expanded silk forms a sand wash style of "peach tart" by means of further cold grinding. That is to say, the enzyme can be used as a sand washing agent.

8.5.2.3 Treatment of wool smudge

On the surface of wool there are some squamous substances, which are some protein polymers. Currently after treating wool surface with proteinase from *Bacillus subtilis*, these squamous substances may be removed, and the resulted wool textile products have resistance of shrinkage and anti-pilling performance. The treated wool is easy to form felt, and becomes very soft and is easy to dye.

In a word, application of enzymes in the textile industry is broadening increasingly. With further development of biotechnologies, safer enzymes with various special functions, high activity and efficiency, will emerge one by one. This not only turns over a new leaf for the traditional spinning process, but also helps protect environment. Enzymes have a promising application prospect in the textile industry.

Chapter 9

Contemporary Bioengineering Technologies and Biomaterials

9.1 Introduction

9.1.1 Biomaterials and Their Development

Biomaterials are an interdiscipline among life science, material science and medicine, and applied widely in engineering science and other disciplines. According to the definition accepted on international conferences since 1972, biomaterials refer to the materials associated with living organisms, or the ones which are implanted into living organisms to play some biological functions, and the ones simulating biological functions.

Development of biomaterials undergoes three different stages. In the 1960s, the first generation biomaterials which could be used in the human body came into being. It has characteristic of "inert biological materials", i.e. under the condition of induced minimum toxic reaction, physical junction between injured tissues may be reconstructed. These inert biomaterials have suitable physical properties and minimum rejection reaction to hosts, and make life span of millions of patients prolong 5-25 years.

In the 1980s, bioactive glass was developed that marked emerging of the second generation of biomaterials. At this stage people had already found the mechanism of junction of the biomaterials with tissues. In the middle of the eighties, the second generation biomaterials started to mature, and bioactive materials and corresponding composite materials had been used in orthopedic and stomatology field. Synthesized porous hydroxyapatite (HA), filling materials and surface coating melt implant could all provide bioactivity, and bioactive glass ceramics had been used for spinal repair. In the nineties, emerging of absorptive materials indicated the second generation biomaterials advanced to a new level. These materials may be degraded by organisms, and so be replaced with regenerated tissues finally after implanting into the human body. For example, in 1984, polylactic acid (PLA, also called polylactide) and poly-glycolic acid (PGA, also called polyglycollide) suture line for surgery application may be hydrolyzed into the final product of carbon dioxide and water. Later, such materials were used to absorptive fracture fixation plate and bone screw, and also used as carriers for drug release.

At present development of biomaterials is in the third stage. With development of cell engineering, molecular biology and gene engineering, interaction between biomaterials and cells may be controlled at the molecular level, so as to induce specific cellular response, and fulfill intended cellular adherence, propagation, differentiation, apoptosis and reconstruction of extracellular matrix. Since reaction between a large number of cells and biomaterials depends on chemical structure and topological structure of material surface, further research on interaction between biomimetic material surface and cells is the key for modification and design of the third generation biomaterials

Development of biomaterials, especially development of the third generation biomaterials is closed associated with tissue engineering. The so called tissue engineering refers to a procedure that utilizes principle and method of life science and engineering science to research and develop new generation substitute for clinical application, with functions of repair and improvement of tissues and organs in the human body, which is used for substitution for partial or whole functions of tissues and organs. Tissue engineering requires that the third generation biomaterials should have biodegradability, bioactivity and three-dimensional space structure. It is expected that such three-dimensional space structure can prove mechanical and chemical signals for cells needing, control cellular adherence, propagation and differentiation, and finally assemble into three-dimensional tissues. It is worth mentioning that the procedure for assembling cells into three-dimensional tissue includes many steps, and their combination should be harmonious like "symphonic music". The procedure lasts from several seconds to 10 weeks, and spatial size spans from nanometers, microns to centimeters. Therefore, it is faced with many technological challenges that transformation of research fruits in tissue engineering field into products to treat diseases of millions of patients. The challenge against bioengineering focuses on production of suitable and healthful seed cells which can amplify, and meet needs for on a large scale engineered tissues, preparation of three-dimensional support with excellent performance and manufacturing bioreactors which can simulate natural environment and can deal with a large scale culture cells, and establishment of the methods for preventing against tissue rejection.

9.1.2 Conditions Required for Biomaterials

At first, biomaterials should meet requirements for biological safety, i.e. they should not be toxic, allergic, irritative, genetically toxic or carcinogenic to the human body. Secondly, biomaterials should have good biocompatibility, i.e. they should have no adverse reaction to tissues, blood and the immune system of the human body. Besides, implanted biomaterials and devices should not be rejected by the immune system and can induce expected host responses.

After implanting into the human body, surface of biomaterials induces a series of host responses, including preferential absorption of protein, complement activation, cellular recruitment or adherence, etc. These responses are one part of inflammation and fibrosis responses in a broader sense. Based on various responses which biomaterial surface, such as protein absorption, immune responses, release of cytokines and grow factors, responses of target cells, inducing the expected healing approach to make tissues reconstruct, these are the requirements for the third generation biomaterials. Bioactive glass and its three-dimensional porous structure can meet with such requirement or condition.

9.1.3 Classification of Biomaterials

According to their origin, biomaterials can be classified into natural biomaterials and synthesized biomaterials; according to their use, medicinal biomaterials and non-medicinal biomaterials; according to degradability, biodegradable materials and non-biodegradable materials; and according to chemical composition, inorganic biomaterials and polymer biomaterials. If degradability and chemical composition are considered at the same time, biomaterials include inorganic degradable materials, inorganic non-degradable materials, organic degradable materials and organic non-degradable materials. If according to functions biomaterials, there are many classes of biomaterials. Therefore, due to so many preparation methods of biomaterials, we will give a simple description to some parts of inorganic phosphate biomaterials, polymer biomaterials and composite materials in the following text, and concentrate on preparation and functions of these biomaterials and their close relationship with bioengineering.

9.2 Natural Biomaterials

Natural biomaterials refer to natural macromolecules originating from animals and plants or human. Natural polymers that are medicinal biomaterials earlier used by mankind, have good biocompatibility, and almost all can be degraded; and the degraded products are not toxic. As medicinal materials, natural materials should also possess the following features: rich feedstock source with simple access; possibility to process with routine methods; physical and mechanical performance equal to that of synthesized biomaterials; no induced immune responses. To date, natural materials which can completely meet these requirements are few, but with appropriate modification or combination with synthesized materials some natural materials may be widely used in clinical medicine. Such typical natural materials include collagen, fibrous protein, chitin, chitosan, and cellulose derivative. They are mainly used for production of absorbable suture line, controlled release drug carriers and artificial skin, tissue repair and alternative, tissue isolation film, etc. It is also reported that glucan, gelatin, alginate and starch are used as drug carriers.

9.2.1 Structural Features of Natural Biomaterials

Natural biomaterials are composed of several macroelements with low atomic serial number, mainly H, O, C, N, Ca, P, Cl, K, and some trace elements, such as Fe, Cu, Mn, Zn, Co, Mo, Se, Ni, I and Mg, which are the key for function of specific enzymes in organisms.

According to strength of interaction between atoms and molecules, natural biomaterials may be divided into two major classes, strong interaction and weak interaction. The former refers to ion bond or covalent bond and hybrid bond, and it should be noted that metal bond does not exist in natural biomaterials. The latter includes electrostatic interaction between dipolar ion, dispersion force, hydrogen bond and hydrophobic bond. In biological macromolecules, it is synergism of various interactions that determine stability of their structure and variability of their conformation. Complicated structure of biomaterials to a great extent depends on interaction between their atoms and molecules.

Natural biomaterials can form complex internal structure and overall diversity under the conditions of their surrounding environment. Their complexity surpasses that of traditional materials such as metal, ceramics, etc. However, it is found that their complicated structure consists of several basic compounds, which are water, nucleotide acids (4 types), amino acids (20 types), saccharide and biological minerals (4 types). Structural complexity of biomaterials mainly focuses on self-assembling modes of the several compounds. One of the core tasks for this discipline is to research on hierarchy and self-assembling modes of biomaterials so as to discover law on their particular functions.

9.2.2 Activity of Natural Biomaterials

Cells are basic structural and functional units of organisms. In some sense, active biomaterials are all composite materials which composed of cells and cellular matrix. For example, the hard tissue bone may be regarded as the composite material to composed bone cells, extracellular matrix and inorganic minerals distributing inside and outside of matrix. Molecules constituting biomaterials are essentially almost the same as those constructing non-life materials, but biomaterials have by far more complex self-assembling hierarchy configuration and excellent functions than non-life materials. These have all close association with regulation of motion, reproduction, differentiation, renewal and reconstruction of cells. Law and mechanism of adherence, motion and phagocytosis of cells on the biomaterial surface are both very important basic knowledge. Understanding interaction between cells and cellular matrix from the viewpoint of biomaterials and developing cell-modulated growth technology will provide a great amount of information for life science and material science, which may give important guidance to research and development of medicinal materials, even tissue engineering materials, and design of biomimetic materials.

9.3 Medicinal Biomaterials

Medicinal biomaterials refer to materials used for diagnosis, treatment, and replacement of damaged tissues and organs or promotion of their functions. According to their composition and properties, medicinal biomaterials may be classified into medicinal inorganic material, medicinal polymer material and biological medicinal composite material, among which medicinal inorganic material includes metal material and bioceramics material. According to uses, medicinal biomaterial may be classified into musculoskeletal system repair material and alternative material. The former includes bone, teeth, joints, tendons. Alternative material includes soft tissue material such as skin, breast, esophagus, respiratory tract, bladder; cardiovascular system material such as artificial cardiac valves, blood vessel, intra cardiovascular tube; medicinal film material such as blood purification film, selective gas permeable film, corneal contact lens; tissue adhesive and suture line material; controlled release drug carrier material; clinical diagnosis and biosensor material; and dental material. According to biochemical reaction level in the physiological environment, biomaterials can also be classified into medicinal biomaterial, bioactive material, biodegradable material and absorbable material.

Inorganic medicinal biomaterials, polymer biomaterials and composite materials are most widely used in medicinal field.

9.3.1 Inorganic Biomaterials

Hydroxyapatite (HA or HAP), $Ca_{10}(PO_4)_6(OH)_2$, its chemical composition, crystal and structure similar to hydroxyapatite of the human skeletal system, and so is suitable substitute for the bone. The ratio of calcium to phosphorus in HA is 1:67, which is similar to the natural bone. Due to high bone conduction and bioactivity, HA has been widely used in the orthopedic surgery.

Typical HA composite material is organic-inorganic nanometer composite material. It is made of organic matter of collagen as the main component and inorganic matter of hydroxyapatite, where in mass ratio of organic matter and inorganic matter is approximately 3/7, diameter size of protein is 300nm, and diameter size of inorganic crystal is 50nm, it arranges along the axis of collagen fiber. Formation and regeneration of bone are related with extracellular matrix secreted by osteoblasts, while absorption of bone results from acidic substance secreted by osteoclasts to dissolve HAP and enzymes synthesized by them which make collagen absorbed. Osteoblasts interact with osteoclasts through signaling factors such as cytokines. During bone formation, cell membranes of osteoblasts adhere to each other to form cell population under mediation of adhesion factors. Osteoblasts synthesize collagen and release organic matter as framework of extracellular matrix, and hydroxyapatite crystal releases from matrix vesicles to form organic-inorganic composite material.

Cells are involved in formation composite material. Apatite size in the bone is smaller than cell size by two orders of magnitude. And therefore, during composite between inorganic matter and organic matter outside the cells, the procedure is self-assembling composite between the two materials and conducts spontaneously in local space out of cells. From the viewpoint of material science, implementation of self-assemble of organic-inorganic material composite by simulating local space *in vivo* is an ideal method for forming composite biomaterials. With this method, strain properties of organic material can be combined with compression resistance of apatite effectively, so as to accomplish obdurability of material. With development of tissue engineering, people come up with new requirements for composite material for tissue regeneration. Research and development of support material promoting cellular adherence, propagation and differentiation is expected, and mechanical performance should adapt to the implanted site. Therefore, various organic-inorganic composite materials are concerned more than any others.

Common HA composite material include: ①apatite-collagen composite; ②hydroxyapatite-collagen-hyaluronic acid composite; ③apatite-chitosan composite; ④porous hydroxyapatite/chitosan-collagen composite; ⑤hydroxyapatite-poly α-hydroxy acid composites. Besides the above hydroxyapatite composite materials, there are following types:

9.3.1.1 Porous hydroxyapatite composite material

This composite material may neutralize the acidic product of polycaprolactone (PCL) through calcium ion released during degradation of apatite, and so inhibit inflammatory reaction.

9.3.1.2 β-tricalcium phosphate/chitosan-gelatin composite material

Due to release of calcium ion, this composite material may promote propagation of myoblasts. The support made of this composite material may cause mild inflammatory reaction at the beginning of implantation, and gradually degrades with extension of post-implantation time. Correspondingly, inflammatory reaction of the material allays and almost disappears till 12 weeks after implantation. This composite material may be used as the support of bone tissue engineering.

9.3.1.3 β-tricalcium phosphate/poly(propylene fumarate) (β-TCP/PPF) composite material

Cell culture experiment has proved that cells may adhere, propagate and differentiate into osteocytes on the β-TCP/PPF matrix. Its bone conduction is similar to or higher than polystyrene-TCP, and this shows that this composite material has great application prospect in tissue engineering.

9.3.1.4 β-tricalcium phosphate/co-poly lactic acid (β-TCP/CPLA) composite material

This composite material integrates biodegradability of CPLA and bone conduction of β-TCP, and so may be made into guided bone regeneration (GBR) film. It can fulfill reconstruction of bone defect of over 10mm. It has another advantage of non-antigenicity, and so is expected to be applied in the oral surgery and orthopedic surgery fields.

9.3.2 Polymer Biomaterials

Inorganic biomaterials are mainly used for repair and substitution for human bone tissue, and also used as tissue engineering support and controlled release drug carrier. Repair and treatment of other tissues in the human body mainly employ organic biomaterials, i.e. polymer biomaterials, which can be classified into gel material and non-gel material.

9.3.2.1 Gel biomaterial

Gel biomaterial is a kind of aquogel substance. As cell release carrier, aquogel is implanted into the human body as minimally invasive form after mixing cells and polymer. Structure of aquogel is similar to that of macromolecule composite, and it has good biocompatibility.

According to its origin, aquogel may be classified into natural material and synthesized material. Natural polymer aquogel has been widely used in tissue engineering. Due to different origin, it has inter-run variation in structure and performance, and so has its own limitation. So people modify natural polymers or use various synthesized polymers to control accurately, so as to eliminate inter-run variation.

As a biomaterial, aquogel must have its biocompatibility at first, otherwise especially pay attention to its inflammatory reaction for organism. Implanted tissue possibly displays immune response to aquogel. Natural derivate polymers have generally good biocompatibility, while synthesized polymers may cause rejection reaction. Secondly, it is necessary to determine mechanism of ion or covalence crosslinking and phase transition behaviour. Ion crosslinking of polyvalence counter ions is a simple method for forming the gel, but such ions may change with soluble ones in the body fluid, which inevitably causes loss of original performance of aquogel. Covalence crosslinking is a common method for controlling crosslinking density of aquogel precisely, but toxicity of the crosslinking agent and degradation of crosslinking bonds should be considered. Furthermore, aquogel may be also formed through phase transition of polymers. For example, when approaching the lowest critical solution temperature (LCST), very little temperature change can make polymers transition from collosol to gel. How to control LCST to make it approach to the body temperature is the key to the procedure. The aquogel which is formed by in situ polymerization and its volume extent very small with temperature increase is good tissue engineering material.

Once gel forms their various properties, it must stabilize for some time to maintain space for tissue survival, and hence, mechanical performance is a crucial design parameter of aquogel used for tissue engineering. Meanwhile, cellular adherence and gene expression have close association with mechanical performance of support materials. Mechanical performance of aquogel mainly depends on rigidity of polymerize chain, class of crosslinking agent and crosslinking density. In addition, it has relation with phase transition of collosol and gel resulting from balance between of hydrophilism and hydrophobicity.

Control over degradable performance of aquogel is also crucial for engineered tissues. Usually, it is expected that degradation rate can match with growth rate of tissues. The requirement for degradation time varies with types of engineered tissues. Degradation mechanisms of aquogel include hydrolysis, enzymolysis and dissolution. Degradation rate of linked gel correlates with its mechanical performance, and interaction between cells and aquogel may have vital effect on cellular adherence, migration and differentiation. Inappropriate interaction will induce formation of unexpected tissues.

There are many types of gel biomaterial. Here, we will describe the following several types:

1) Collagen

Collagen may be used as effective framework material of many cells, and chemical modification of collagen may obtain better engineering framework material. American Collagen Corporation has developed effective growth frameworks such as fibroblasts, epithelial cells, keratinocytes, osteoblasts, hepatocytes and islet cells, by utilizing type I collagen.

Generally, for collagen extracted from animal tissues or obtained by digesting with proteinase, its performance cannot meet with requirements, and so further crosslinking is needed.

Crosslinking methods of collagen are as follows, crosslinking with glutaraldehyde (II) as a crosslinking agent; crosslinking with 4-butyrate thiolactone under catalysis of imidazole; crosslinking with polyphenols such as o-dihydroxybenzene and nordihydroguaiaretic acid (NDGA) as a crosslinking agent.

2) Gelatin

Gelatin is a apartially-denatured derivative of collagen, and formed by decomposing triple helix structure of collagen into single-chained molecules. Gelatin has two types, i.e. gelatin A and gelatin B. The former is formed by treating with acid before heat denaturation, while the latter by treating with alkali so as to convert amide residue of glutamine to glutamate and aspartic acid. Therefore, carboxyl content in gelatin B is higher than that in gelatin A by at least 25%.

Gelatin solution may form reversible gel when temperature decreases to below 35℃. At the moment, the large molecule chain changes from coil conformation into helix one, and in this procedure triple helix conformation of collagen is easy to restore. Temperature for physical gel gelatin transformation has association with gel concentration, assembly time and pH value.

Physical network of gelatin may decompose at higher temperature, and is not suitable for long-term using at 37℃. At the same time heat stability and mechanical stability of gelatin aquogel are not high, and so are to improve through chemical crosslinkage. N, N (3-dimethyl aminopropyl)-N'-ethylcarbodii mide (EDC) and N-hydroxysuccinimide (NHS) may be used as chemical crosslinking agents to form gelatin.

3) Hyaluronic acid

Hyaluronic acid, as an important component of glycosaminoglycan (GAGs) in natural extracellular matrix (ECMs), plays a significant role in trauma healing, and may be crosslinked with various hydrazide derivatives. Hyaluronic acid can be degraded by hyaluronidase in cells and the serum. Impurities and endotoxins in hyaluronic acid, that must be removed through purification Mechanical strength of hyaluronic acid is low, and hence its application scope can be restricted.

4) Fibrin

Fibrin is usually used as a surgical sealing agent or adhesive, and may promote wound healing. Fibrin gel may be prepared with autologous blood of patients, and so it is autologous tissue engineering framework. Fibrin may form gel at the presence of thrombin through enzymatic polymerization at room temperature. Fibrin is a natural component in the human body, and so its degraded product neither has toxicity, nor induces immune response. Feature of fibrin is that its degradation and reconstruction of wound are associated with cellular migration and activity of related enzymes during trauma healing, and its degradation rate may be regulated by controlling agents of proteinase.

5) Alginate salts

Alginate salts are polysaccharide extracted from seaweed, and often used as hemostasis material and wound dressing. Since alginate biomaterial cannot be degraded, its application is limited. Sodium periodate may be utilized to make alginate oxidized lightly, and the oxidized alginate salts may be degraded through hydrolysis. Degradable alginate gel can promote formation of chondroid tissue in animals significantly, compared with common alginate aquogel. Another problem on application of alginate salt gel in tissue engineering lies in lack of the site binding to cells. Due to strong hydrophilism, alginate salts interfere with absorption of protein, and cannot interact with cells of mammal animals. Modification of alginate salts with sugar-specific binding protein agglutinin may enhance their ability to bind to cells.

6) Agarose

Agarose is another seaweed polysaccharide, and can be dissolved in hot water over 65℃. Unlike alginate salts, double-helixes of agarose molecule strands bind into bundles, in the contact area multiple strands gathered, and therefore agarose molecules can form thermally reversible gel. Physical structure of this gel may be controlled by agarose concentration, so as to form various pore sizes. At low agarose concentration, the sponge gel with large pores and low rigidity may be formed, and it can promote migration and propagation of cells.

7) Other gels

Apart from the above gels, there are chitosan, polyacrylic acid and its derivative, polyethylene glycol and its copolymer, polyvinyl alcohol and polyphosphazenes. They can all be used as biomaterials in the tissue engineering area, and we do not describe them here any more. For further study please see related books.

9.3.2.2 Synthesized degradable biomaterials

Degradable synthesized biomaterials have many types, and in summary include polylactide, polylactone, Polyhydroxyalkanoates, unsaturated polyester, polycarbonate, and biodegradable elastomer. In the following we will give a simple account for the most typical PLA and polyhydroxyalkanoates (PHAs).

1) Polylactic acid

PLA belongs to polylactide substance. Polylactides include poly-D, L-lactic acid (PDLLA), poly L-lactic-co-D, L-lactic acid, poly-D, L-lactic-co-glycolic acid (PLGA) and PGA, among which PLA is a common one. PLA and its copolymers have wide application in the medial field, and may be decomposed into non-toxic compound through ester bond hydrolysis in the human body. The degraded products are excreted through the kidney or through biochemical pathway in a form of CO_2 and H_2O.

Lactic acid is a kind of simple chiral molecules, and has two optimal isomers, D-type and L-type, respectively. Lactic acid, which could be made with petroleum chemistry method, is racemic compounds of D, L-lactic acid, and lactic acid made with fermentation method is completely L-lactic acid. The common methods for preparing PLA in high molecular mass from lactic acid include lactide method and direct condensation method. With the two methods linear polymers are obtained. Melts of the polymers have low elasticity and poor process performance. Branch chains may be introduced to PLA by adding oxidized natural oil, which not only improves elasticity of PLA melts, but also reduces viscosity of the melts, and is useful for subsequent process.

PLA is a biomaterial whose degradation takes the longest time and it possibly can exist five years after implanting into the human body. Degradation and absorption of PLA mainly depend on molecular mass, crystallinity, orientation of chains and implant sizes. Besides, it should be considered that environment in the body may have effect on degradation. Degradation of PLA lasts long, at the later stage allogeneic reaction may occur, and so its application is limited. Allogeneic reaction is associated with crystalline residue produced during PLA degradation and decrease in local pH value. Although degradation of PDLLA cannot produce crystalline residue, self-catalytic reaction of internal acidic degraded products leads to rapid internal degradation and low external degradation, and pH value decrease will induce inflammatory tissue reaction.

2) PHAs

Poly-3-hydroxybutyrate (PHB) is the most common polymer of poly-hydroxyalkanoates (PHAs), and is often used as degradable implanting material. PHB is natural thermoplastic polyester, and is synthesized through bacterial or gene engineered strain. Many mechanical properties of PHB are equal to those of degradable synthesized polyester such as PLA, but strong fragility of crystallized PHB influences its application. Flexibility and breaking elongation of PHB may be increased by plastifying or mixing with other degradable polymers such as poly-ε-caprolactone (PCL) and poly-3-hydroxyhexanoate.

PHB is nontoxic, and has good compatibility with tissues and blood. PHB in low molecular weight exists in the blood of the human body. The degraded product 3-hydroxybutyrate is metabolite by itself, and hence implanted PHB usually has no toxicity. *In vitro* experiment of PHB film degradation indicated that the polymer had no loss within 180 days, but after induction phase for 80 days the molecular weight of the sample started to decrease. Adding some polymer or plastifying may make PHB degradation rate increase. Non-crystalline or hydrophilic additives may increase water absorption capacity of PHB so as to promote its hydrolysis, and so water absorption capacity of PHB/PDLLA mixture is stronger than PHB/PCL mixture. Degradation rate of non-crystalline synthesized PHB is higher than that of PHB synthesized by bacteria, and therefore the two can be mixed for use so as to promote degradation. Specific surface area of the polymer may be increased by mixing PHB with soluble additive (glycerol derivatives) and then dissolving out the additive, consequently promoting degradation of the polymer.

For patch materials of the gastrointestinal tract, in addition to closed tissue defect, it acts as support for tissue regeneration, endures with enzymolysis in the gastrointestinal tract, doesn't adhere to any surrounding tissue, and is flexible enough to suture. In a certain time, material should also be resistant to the gastrointestinal tract secretion of erosion, but eventually can be degraded and absorbed. PHB material may meet with all the above requirements. For some application such as repair of intestinal defect, special patch surface is required. In details, one side of material surface should be flat and smooth, so as to avoid cellular adherence; while the plane of the material contacting with the intestinal tissue should be porous, so as to promote cell adhesion and formation of scar tissue.

9.3.3 Drug Controlled Release Materials

9.3.3.1 Principle of drug controlled release

Controlled release refers to the procedure that some drugs are released from carrier materials at a constant speed and within particular time span, and common carrier materials include natural and synthesized polymer materials. Controlled release may make drug keep at the minimum concentration in the blood for disease treatment, in order to avoid the drawback of suitable drug concentration for the routine drug delivery system, and overcome the problem of drug intoxication at too high drug concentration and ineffective efficacy at too low drug concentration.

Drug controlled release may be classified into diffusion release, permeation release, dissolution release, biological adhesive release, chemical reaction and expansion controlled release. According to mechanisms for controlled release, various preparations, such as tablet, cylinder, pill, granule and capsule, may be conducted, targeting preparation may deliver drug to the target site directly, which can increase drug quantity at the site, and reduce drug dose and toxicity or side effects. Guide mechanism of targeting drugs is that guide can be done by utilizing drug recognition on some specific groups, drug particle sizes and drug magnetism, in order to deliver the drug to the target site.

9.3.3.2 Controlled drug delivery system and qualified conditions

Controlled drug delivery system should meet the following conditions: ①function of controlled drug release, in order to make blood drug concentration at the target site kept within the required range; ②function of targeting drug release, in order to make drug delivered only to the treatment target site; ③to deliver drug to the disease site directly; ④under the premise to achieve the required efficacy, to reduce drug dose as possible as we can; ⑤to minimize toxicity or side effect of drug and to ensure safety and reliability of the drug; ⑥easy to take, and ready to be accepted by patients; ⑦to have certain physical and chemical stability under common circumstances.

Controlled drug delivery system includes systemic treatment of drug system by means of oral and injection dosage that can be absorbed by the body, the intelligent drug delivery system which may change drug release behaviour because it is sensitive to acidity or alkalinity, osmotic pressure, drug concentration, blood pressure, body temperature, etc.; the drug delivery system can control drug lifetime in the body indirectly by utilizing ultrasonic or microwave radiation.

9.3.3.3 Development of drug delivery system

Before 1950s, traditional drug preparation were used for a long time; from the fifties on sustained release preparation emerged (SRP), from the seventies on, controlled release preparation (CRP); and from the eighties on, targeted drug delivery system (TDDS), and integrated dynamic driving system (IDDS).

Population and health are two major topics in the medical field. Population control and disease prevention also need a novel drug delivery system. People need reliable contraceptive preparation with long-term efficacy which has no side effect, in order to achieve effective and convenient contraceptive effect. Furthermore, long-term efficacy vaccines are needed to prevent against diseases. As a result, development of an ideal drug delivery system is not only the common wish of patients and doctors, but also the goal of scientists and drug producers.

9.3.3.4 Materials associated controlled release drug delivery system

Drug carrier materials used in a controlled release drug delivery system include natural organic and inorganic material, and synthesized organic or inorganic material, polymer material and non-polymer material, biodegradable material and non-biodegradable material. At present most drug carrier materials are polymer materials which include natural polymer material, modified natural polymer materials and synthesized polymer materials. Most traditional drug carriers are inorganic substances, with development of drug preparation, novel preparation emerge, and polymer material has performed well enough to make it become more and more important as drug carriers.

Performance of polymer material in a controlled release drug delivery system should include biocompatibility and biodegradability, i.e. polymer carriers can be degraded into micromolecule compounds which are metabolized, absorbed or excreted by the human body. If drug carriers cannot be degraded, surgical operations should be needed to remove them after drug release; degradation of polymer material must occur in a proper period; and the degraded products of polymer material must be nontoxic and do not induce any inflammatory reaction.

9.4 Tissue Engineering Materials

With the development of cell engineering, molecular biology and gene engineering, the design idea of the third generation biomaterials comes into being, i.e. interaction between biomaterials and cells is controlled at molecular level, so as to induce specific cellular responses, finally to fulfill intended cell adhesion, propagation, differentiation, apoptosis and reconstruction of extracellular matrix (ECM). Since many reactions between cells and biomaterials depend on chemical and topological structure of material surface, further research on interaction between biomimetic engineered material surface and cells is crucial for improvement and design of the third generation biomaterials.

9.4.1 Surface Engineering

So called surface engineering just refers to modification of biomaterial surface so as to optimize free energy, protein absorption capacity, hydrophilic/hydrophobic balance, electrical performance, topological structure and bioactivity of the surface, finally inducing cellular physiological responses, and creating a good artificial ECM environment. Common surface engineering of biomaterials includes immobilization and interception of bioactive macromolecules on material surface, bioactivation of bioceramics surface and formation of topological structure of biomaterial surface. Acting as tissue space for nutrients and metabolites to diffuse and permeate, ECM is also a storage place for cytokines and growth factors, and can release them to adjacent cells. Biomaterials adopted in tissue engineering include natural ECM and its derivatives such as collagen, gelatin, hyaluronic acid, chondroitin sulfate and chitosan, degradable synthesized material such as polyester including PLA, PGA and their copolymer. Advantage of natural biomaterials is that though they contain rich biological information, they are not easy to prepare on a large scale and inter-run variation of their performance is significant. Comparing with natural biomaterials, as artificial ECM material, polyhydroxyl ester may be produced on a large scale, its fine structure may be adjusted, and its mechanical properties and degradation time may be controlled. But its biggest drawback is lack of cell recognition signals, and this may interfere with specific cellular adhesion and activation of specific groups. Cell adhesion has association only with properties of material surface. When biomaterials enter the human body, firstly their surface will contact with the body environment, and therefore, properties of material surface determine the body's responses to the implanted material. If the response is very strong, the material cannot take its action. So, material surface must be modified in a biomimetic way, so as to reduce inflammation. From a viewpoint of material science, surface modification of implantable biomaterial refers to a procedure to make biomaterials adapt and admitted by tissues, so as to endow them dynamic integration capacity; from the angle of clinical application, it means that engineered material surface binds with surrounding tissues and the implanted material, acting as "artificial ECM". In the body environment, interaction between cells and ECM is performed through molecular recognition between cell membrane receptors and the corresponding ligands ECM provided. When the implant enters the human body, receptors on the cell membrane surface are actively looking for the signals through which the surface contacts with cells, so as to judge whether they are autogeneic or allogeneic. Therefore, modification of material surface may be conceived from the view of biomimetic ECM.

9.4.1.1 Effect of material surface for cells

Interaction between cells and biomaterials determines fate of the implanted material by regulating immune responses and strength of junction between biomaterials and surrounding tissues.

The affecting factors mainly include free energy, protein absorption capacity, hydrophilic/hydrophobic balance, electrical performance, topological structure and bioactivity of the material surface. These factors are independent from each other but still coordinate with each other, which leads to complexity of action of biomaterial surface for cells.

1) Surface free energy of biomaterial

Surface free energy of biomaterial has important effect on cell adhesion, elongation and growth. Generally, the surface with high free energy is more helpful for cell adhesion and cell elongation than that with low free energy. For example, studies on fibroblast indicate, only if surface energy reaches $57 \times 10^{-7} J/cm^2$, cells can elongate well.

2) Protein absorption capacity of biomaterial surface

Physiological fluid contains a large number of soluble proteins, and they form a protein absorption layer on biomaterial surface in a way of dynamic balance. Among the absorbed proteins most of them are albumin, which cannot promote cell adhesion; they still contain a small number of ECM proteins including laminin, fibronectin and vitronectin which may regulate cell contact and extension. In these proteins, particular oligopeptide region (cell bonding region) has action similar to ligand and recognizes specifically and binds with intergration receptors on the cell surface.

3) Charging performance of biomaterial surface

Charge distribution and charging capacity of biomaterial surface may affect adhesion of protein and cell on material surface. On the biomaterial surface with a positive charge, the adhesive layer of cells exhibits continuity; while on the one with a negative charge, the adhesive layer exhibits non-continuity.

4) Hydrophilic/hydrophobic balance of biomaterial surface

Hydrophilic/hydrophobic balance of biomaterial surface is an important factor to affect and regulate protein absorption. Usually, hydrophobic surface has stronger protein absorption capacity, and contact between protein and biomaterial surface often accompanies with water/protein absorption exchange; and adsorption exchange between heterologous proteins often results in Vroman effect. Meanwhile, hydrophilic/hydrophobic balance also influences on cell behaviours, and only suitable balance can promote cell growth.

5) Topological structure of biomaterial surface

One century ago, effect of topological structure of biomaterial surface was found. The cells grow along the orientation of topological structure of lower layer material surface, which is now called contact guidance. For example, specific parallel groove on biomaterial surface may enhance cell adhesion on material surface.

6) Bioactivity of biomaterial surface

Cell recognization function is self-protection for organisms, and also fundamental cause to induce transplantation rejection. When biomaterials are implanted into the human body, cells actively explore and recognize whether the implant is autogeneic or allogeneic. If the implant is found to be allogeneic, cells will at once send a signal to mobile macrophages and neutrophile cells in the immune system for attack at the foreign substance, finally leading to inflammatory reactions. In the nature, various cells all employ protein as a medium for recognization. Since cell recognization is based on interaction between intergration receptors and corresponding ligands in ECM, it is crucial for cell recognization and acceptance of biomaterials as autogeneic ECM, that biomaterial surface has intergration receptors such as polypeptide sequence of fibronectin, collagen, vitronectin and RGD.

9.4.1.2 Biomimetic engineering of biomaterial surface

Interaction between biomaterial surface and cells are influenced by many aspects. Biomimetic engineering of biomaterial surface aims at making engineered biomaterial surface induce specific physiological cell responses, in order to create a good artificial ECM environment for cell growth. In order to meet demand for normal cell growth, biomaterial surface must have some features as follows: ①good biocompatibility; ②suitable hydrophilic-hydrophobic balance of material surface; ③strong ability of specific cell recognition; ④suitable topological surface structure; ⑤ability to eliminate non-specific recognition; ⑥easy to process and give characterization. Common biomaterial surface engineering includes immobilization and interception of bioactive macromolecules on material surface, biological activation of bioceramic surface, and formation of topological structure of biomaterial surface.

Immobilization and interception of bioactive macromolecules is the basic method for material surface with biological specificity and recognition and also the method is also used widely. Just as its name implies, this method is just to immobilize bioactive macromolecules on material surface with physical absorption, embedding or chemical bonds. Common bioactive macromolecules include ECM adhesion proteins (fibronectin, collagen, laminin, vitronectin, etc.), ECM polysaccharide and its analogues (e.g. hyaluronic acid, chondroitin sulfate, chitosan, alginate salts), cellular adhesive peptides (RGD, etc.) and cellular activators (e.g. various growth factors).

There are many methods for immobilizing bioactive macromolecules on biomaterial surface, such as chemical oxidation etching technique, plasma deposition, ion beam implantation technique, radiation and UV grafting technique. With these methods the needed functional groups may be formed on material surface, so as to change topological structure of material surface and enhance interaction between materials and biological environment.

In studies on hard tissue engineering, bioceramics with calcium phosphate modified has good biocompatibility and bone integrity. When bioceramics material with bioactive is implanted into the body, a series of biochemical and biophysical reactions will take place on its contact place. These reactions at last lead to chemical bond between interfaces with strong mechanical performance, and such adhesion mode is called "bioactive fixation".

Among various performances of material surface to affect cell behaviours, effect of surface morphology is always neglected. Due to lack of particular surface with consistent topological structure, studies on effect of surface morphology on cell behaviours do not nearly make great progress. With generation and development of optical etching technique, this problem has been solved. In the past optical etching technique was applied in microelectronic technology area, and used for preparation of material surface with exact shape. With this etching technique to an accuracy of millimeters can be done on the surface of several base materials such as silicon, plastic and glass. Since diameters of typical animal cells are 10-20μm, the material surface prepared with optical etching technique is suitable for study on responses of cells and tissues to different material surface. For preparation of material surface with different structures and sizes, optical etching technique has enough flexibility. The material surface with multiple parallel grooves is simple and easy to explain, and so applied to cell studies.

Biomimetic surface engineering of biomaterials is a complex systematic engineering, which integrates requirements of material science and biology in order to optimize them. Biomimetic surface engineering makes biomaterials have a clear target orientation, and plays action of a medium to connect materials and cells and tissues. It aims at inhibiting non-specific interaction, and meanwhile introducing specific interaction sites, so as to make cells play their functions in similar ECM.

Apart from requirements for physical and mechanical performance, biomaterials must also be accepted by organisms. Organisms may try to reject, damage and shield all the things that they regard as foreigners; and integrate all the things they regard as autogeneic substances. Whether biomaterials are judged to be autogeneic or allogeneic depends on their surface performance to a large extent. ECM has great effect on cell behaviours. When cells contact with biomaterial surface, cells will look for specific adhesion ligands which represent ECM characteristics that cells are acquainted with autogeneic tissues. If ligands have particular density on material surface, and maintain certain conformation to expose the binding ends of ligand and cell membrane receptors, the responses between cells and materials are good, and can be considered to be a successful recognition. As for inert biomaterials, they have only non-specific responses to environment in the human body. In the human body, these biomaterials may be rapidly covered with deformations proteins and with various conformations and while those with bioactivity cannot induce specific protein absorption. As a result, this non-specific protein layer makes the organism to confirm that inert biomaterials will be foreign matter. Therefore, biomimetic engineering of biomaterials is to control specific physiological bonds on materials surface and to restrain non-specific protein absorption.

9.4.2 Immobilization Methods of Biomaterial Surface

In the last 30 years, research on biomaterials has made astonishing success, but most of these research fruits have not been applied to clinical practice. On one hand, biomaterials, which contact with the human body directly, are used for reconstruction of biological functions and structures, have been used in artificial cardiac valves and blood vessels, joint replacement, pacemakers, dental implants and contact lenses, to save millions of lives and improve life quality of more people. On the other hand, physiological responses to implants in organisms may cause allogeneic rejection, which will promote formation of collagen fiber outside the implanted materials, finally leading to failure in implantation. In the past research on biomaterials only stressed on development of mechanical performance, physical performance and process performance of materials from the angle of material science, and did not increase bioactivity of materials fundamentally. On account of lack of necessary biological features, it will lead to failure in repair of natural organs and simulation of their structures and functions. Consequently, needs of organisms should be considered, so as to make implants achieve optimal integration effect and reconstruction of tissues.

Immobilization of biological macromolecules on material surface is one of the methods for synthesizing materials surface with specificity and recognition. Inhibition of non-specific responses is one principle which must be followed when designing material surface to induce specific physiological responses; and the material surface to reject protein absorption is applicable for this objective, active functional groups (e.g. amino, carboxyl, hydroxyl, etc.) are suitable to immobilize on the non-absorptive film. Another method is to form perfect crystalline surface with molecular assembly technique, so as to avoid non-specific absorption, such as the self-assembly system constructed on metal film through alkyl hydrosulfide. Both of the two methods are based on physical action. And hence it must be noted that immobilization not only refers to short-term immobilization of biological macromolecules on material surface or inside of it, but also includes long-term immobilization. For a drug release system (including controlled release of growth factors), the immobilized drug should be released from the base material to the environment at certain speed; while for biomaterials used for artificial tissues and organs, the immobilized ligand proteins on the material surface should maintain their constant status for some time. Because biological macromolecules have very big molecular weight, either physical method or chemical method can lead to long-term immobilization of biological macromolecules on material surface. If biomaterials are degradable, the immobilized biological macromolecules will be released continually with degradation of the base material.

One important feature of chemical method for immobilizing biological macromolecules is that immobilization of macromolecules should be done through one spatial group often called "hanging arm". The spatial group may provide bigger space freedom, which helps enhance specificity of biological macromolecules, and is especially applicable for active molecules with small molecular weight. If the spatial group is degradable, the active molecules are released when degrading.

For immobilization of biological macromolecules on material surface with chemical inertness, the biomaterial surface must be modified so as to generate functional reaction groups (e.g. amino, carboxyl, hydroxyl, etc.) for subsequent immobilization. The methods for modification of material surface include ion radiation grafting copolymerization, gas discharge plasma technique, photochemical grafting, chemical modification (oxidation grafting) and chemical derivatization.

9.4.2.1 Ion beam technique for material surface modification

Modification of biomaterials with ion beam technique (including plasma technique) is a hot topic in contemporary modification field of material surface. The main causes arising in such a trend include three aspects: ①People have realized significance of surface engineering of implanting biomaterials; ②Biomaterials extend from metal, ceramics and polymer composite materials to materials with bioactivity; ③Comparing with other modification techniques, covalence bonding, laser technique, etc, ion beam technique is especially suitable for modification of biomaterials, its advantages lie in precise process control, low temperature reaction, ion diversity and non-balance procedure, and its drawback is relatively higher cost.

During ion implantation, ions are stimulated to a certain energy level, generally 20-200keV, and then contact with the target material surface. The ions with so large energy may penetrate surface of solid materials (e.g. metal, ceramics and polymers) to reach the depth of several hundred nanometers. For metal biomaterial, physical process of atom or nuclear collision may form highly disordered even amorphous structure in the area close to surface. Chemical process, such as formation of new chemical bonds between atoms of the base material and the implanted ions, may form the surface layer on the material surface completely different from internal mechanical and chemical performance. For polymer material, the same process may occur, but is more complicated. There are two competition processes, crosslinking and cleavage of molecular chains. Crosslinking of molecular chains can form three-dimensional crosslinking surface with high strength and wear-resistance on material surface, and cleavage of molecular chains finally leads to degradation of long-chain macromolecules. Another use of ion implantation is to prepare specific functional groups on material surface, so as to increase wettability and biocompatibility of materials. Advantages of ion implantation are modification of material surface, but no change of intrinsic properties; controllable and repeatable process, and clean process for medicinal materials; and low temperature which has no effect on product sizes.

9.4.2.2 Photochemical immobilization for modification of material surface

Immobilization of biological macromolecules on materials surface with photochemical method may avoid drawbacks of thermochemical method, and meanwhile position the activity space. Its advantages lie in covalence bonding induced by light, no need for functionalization of the target molecule, and immobilization different types of biological macromolecules including proteins (enzyme, antibody and polypeptide) and nucleic acid on surface of inert materials. In some processes, light is needed to link polymers or photosensitizers with double functional groups; and in other cases coupling agents with low molecular weight are adopted to functionalization of material surface.

9.4.3 Formation of Topological Structure of Biomaterial Surface

Among many properties of biomaterial surface affecting on cell behaviours, effect of material surface morphology is often neglected. Effect of material surface morphology on cell behaviours was discovered one century ago, i.e. cells grew in an orientation of topological structure of lower layer biomaterial surface, which is now usually called contact guidance. Due to lack of particular surface with consistent topological structure, research on effect of material surface morphology on cell behaviours has not made any great progress. With development and application of optical etching technique material surface with consistent topological structure may be prepared. Application of such material surface makes it easier to do systematic research on effect of topological surface structure on various physiological properties of cells (e.g. cell adhesion, growth and differentiation). Studies have revealed, material surface morphology may regulate cell extension and cell shape. And hence, material surface morphology becomes a main factor to affect cell growth and function expression. Such effect obviously is regulated through physiological relation between cell morphology and functions, and further research and discussion on this phenomenon will optimize design of biomaterials.

During cell culture, micropatterning technique on biomaterial surface may regulate space of cells. Common solution is to regulate cell functions (e.g. cell adhesion) by changing charging property, hydrophilism/hydrophobia and topological structure on surface of material. In addition, with micropatterning technique, biological macromolecules regulating cell functions may be immobilized on material surface, so as to fulfill visualization of effect of material surface performance on cell functions and spatial regulation of microcosmic tissues.

9.4.3.1 Early studies

For design of implant surface or matrix for cell culture *in vitro*, it is very important to learn about interaction between biomaterials surface and cells. As described above, physical and chemical properties of biomaterial surface may affect cell adhesion and other behaviours, and these properties include surface composition, surface charges, surface energy, surface oxidation layer, hardness, curvity and surface morphology. Cell adhesion on biomaterial surface conducts mainly through extracellular matrix absorbed on material surface. ECM proteins exist in the serum for cell culture, and material surface properties affect cell adhesion by influencing on ECM protein absorption or by changing conformation of absorbed proteins. When cell adhesion does not depend on ECM proteins, material surface properties may affect on cell adhesion by means of electrostatic action between cell membrane with negative charges and the base material with positive charges.

Although effect of material surface properties (e.g. surface charges and energy) on cell adhesion is very important, a few studies on effect of surface morphology on cell behaviours are reported. The known first study on effect of rough surface on cell behaviours is that Harrison cultured cells on a spider network and proved that the spider network influenced on direction in which cell moved. Weiss cultured cells on the complex network of blood protein fiber in plasma clot and also proved the phenomenon. The phenomenon that cells grew according to topological orientation of lower layer matrix surface was described as "contact surface guidance" by Weiss. Early studies on effect of topological structure of material surface on cell behaviours were based on uncertain matrix surface, such as spider network, plasma clot, fiber fabrics, etc. Though qualitative studies on them proved how cells interacted with the rough material surface. The main cause to hinder quantitative studies was lack of consistent topological structure of material surface with precise definition. In 1958 Weiss for the first time tried to prepare relatively consistent morphology of material surface, and he used a precision lathe to etch parallel grooves similar to helix lines on a glass plate. Curits and Varde used the 25μm diffraction grating replica of silicon-coating polystyrene to investigate cells' orientation. Rovensky adopted polyvinyl chloride recording disk to study, for these material surfaces, grooves are V-shaped, and line ridges are similar to convex cylinders. Obviously, due to co-action of cylinders and grooves, the structure of material surface used by Curits and Rovensky was complicated. Ohara and Buck used a drilling bit to cut out grooves at an interval space of 5-30μm mechanically on the surface of a culture dish. All these methods for pattern of material surface lack flexibility in sizes, and may lead to unevenness of material surface by mechanically cutting, so as to make studies more complicated. Therefore, early studies were only limited to contact guidance of cells, and in most cases the study results were not certain.

9.4.3.2 Topological structure of material surface with optical etching technique

Optical etching technique is one technique applied in microelectronic technology area to prepare material surface with definite morphology. For preparation of integrated circuit, the technique may be used to generate patterns or pores precisely on silicon chip surface. At first, photosensitive material is used to determine a certain pattern, and then is transferred to matrix surface. Re-dissolving procedure of prepared final structure determines kinds of illumination. For example, for groove in the size of about 1μm, UV-light may be used; while for those in the size of less than 100nm, electronic beam etching should be adopted. Since typical animal cells have diameters of 10-20μm, the material surface prepared with optical etching technique is suitable for research on responses of cells and tissues to different material surface structures. Multiple parallel grooves are applied widely to research on cell behaviours due to the simple structure which is easy to study.

Many materials are used to prepare groove surface, including titanium coated silicon, glass, quartz, silicon dioxide, organic glass. Based on different study objectives, groove size may be selected; and different solutions such as optical etching or plasma etching may be chosen. Clark adopted more advanced technique to study effect of fine groovesoncell arrangement, and employed modified laser holography technique to determine ultra-fine grooves prepared with an X-ray imprinted mould (groove width 130nm, ridge width 130nm, groove depth 100nm, 210nm and 400nm).

It is worthwhile to note that any solution to prepare structured material surface with optical etching or plasma etching technique must stress on chemical homogeneity of etched area surface and unetched area surface. For example plasma etching may change surface properties, after plasma etching polystyrene surface becomes hydrophilic. Clark adopted organic glass to prove that surface energy in etched areas and unetched areas was inhomogeneous, and proposed that this problem could be solved by conducting oxygen plasma treatment at the end of process. Brunette adopted optical etching and plasma etching technique to prepare silicon surface with grooves, which was used as a mould to replicate groove surface of epoxy resin, and finally titanium was used to coat on the replica. It is obvious that homologous groove titanium surface with such process can be prepared.

Through preparation of definite material surface with optical etching technique, it is possible to do systematic studies on effect of material surface morphology on cell behaviours, such as cell elongation, adhesion, growth and differentiation. Such technique may be used to study on basic topics in biology, such as mechanisms of contact guidance and cell migration.

The phenomenon of contact guidance refers to cell ego orientation in accordance with effect of the lower layer surface, which is proposed as mechanism for invasion of tumor cells, cell migration and adhesion is the core to understand cancer cell metastasis, topological structure is one of the factors to determine cell migration. In order to explain the phenomenon of contact guidance, many assumptions were once proposed. For example, Weiss thought that gelatinous exudates made cells orientated along the long axis of fiber. Rovensky concluded that topological structure on the guidance of cells resulted from difference between cell adhesion behaviours on material surfaces with different geometrical structures, by studying on effect of groove surface on cell migration and orientation. With an experiment about fibers in different diameters, Dunn and Heath put forward that cell contact guidance was to avoid interruption of movement. They adopted the base material in a triangular shape to show that morphology of the matrix would force limit formation of microfiber beams which were crucial for cell adhesion and migration. Ohara and Buck adopted grooved matrix in a size less than cells to prove that cells had hardness enough to penetrate irregular matrix surface.

With studies on orientation of fibroblasts and epithelial cells on V-shaped groove surface, Brunette found that although the surface parameters of adopted grooves were greater than cell sizes, groove depth and interval distance had relatively small effect on cell orientation. Their further study showed that cell orientation had association with groove depth, and cells could change their shape to keep consistent with the lower layer surface. Dunn and Brown used grooved material surfaces in different groove widths, ridge widths and groove depths to prove through regression analysis, that column formation of fibroblasts from the chicken heart was in direct proportion to ridge width. Clark adopted grooved surfaces with different parameters (depth range 0.2-1.9μm, interval distance range 4-24μm) and three kinds of different cells to prove, that when groove depth determined row formation of cells, effect of interval distance was relatively small. They proved at the same time that sensitivity of cells to topological structure of material surface had association with kinds of cells, as well as with whether interaction between cells existed.

In subsequent studies, Clark and others adopted material surface of ultra-dense grooves (interval distance 260nm, groove depth 100-400nm) to simulate fibrous ECM arranged in order. At that moment, they found column formation of cells depended on groove depth, and effect of topological structure on cell behaviours had close association with interaction between cells. This conclusion notably has contradiction with hypothesis of Ohara and Buck, because for groove surface with parameters smaller than the cell size, cells can simply stride over grooves and groove depth has no effect on cell behaviours. In a word, the above experiments demonstrate, column formation of cells on grooved surface has close association with groove depth and weak relation with interval distance of grooves. Determination of groove parameters is associated with cell sizes.

9.4.3.3 Patterning of biomaterial surface

Histology of tissue structure in the human body is a vital factor to determine tissue growth and cell movement, i.e. contact guidance. Many studies have indicated, design of surface structure of implants may increase material performance. Comparing with smooth material surface, slightly rough surface may promote bone integration, reduce formation of fibrous capsule and enhance integrity of implants. This is because rough surface enhances cell adhesion in the connective tissue, so as to make the tissue bind closely with implants.

Biocompatibility of soft tissue implant often causes function loss of implant due to formation of Cysticas. Experiments proved that micron-scale patterning surface may reduce formation of Cysticas significantly. Van Kooten studied on interaction between human fibroblasts an organic silicon surface adopted cell cycle method. There were three kinds of surfaces with grooves in width of 2μm, 5μm and 10μm respectively(represented as 2MU, 5MU and 10MU, respectively), and the grooves were smooth (SMT). The results showed, cells on SMT run into the S phase more rapidly than those on patterning surface, and cells on the 10MU surface propagated more slowly than those on the 2MU surface and the 5MU surface.

Cheroudi and Brunette proved that micropattern material surface could increase performance of the implant greatly by subcutaneous implantation experiment. Due to different micropattern sizes, responses of fibroblasts to material surface varied significantly. At the same time, they also demonstrated that micropattern material surface might promote bone integration and induce bone mineralization.

9.5 Intelligent Biomimetic Materials

Intelligent biomimetic materials have particular requirements for both artificial extracellular matrix and biomaterial surface.

9.5.1 Artificial Extracellular Matrix

Artificial ECM has pivotal role on formation of functional new tissues, to positioning cells or transfer cells to the specific site in the human body; to define and maintain particular shapes in order to provide three-dimensional space for formation of new tissues; and to guide development of functional tissues.

Artificial ECM provides adhesive matrix for implanted cells, and to deliver cells to the specific site in the human body acting as a release carrier. As a result, artificial ECM must have high specific surface area and high pore rate, so as to guarantee delivery of cells in high density. Besides, people often expect that formed new tissues have pre-determined appearance (e.g. blood vessels), which depends on correct selection of mechanical and degradable performances of artificial ECM. Artificial ECM should provide temporary mechanical support enough to stand with mechanical action in the human body, so as to maintain certain space for tissue formation. Such mechanical support requires engineered tissues that should support their own weight, so as to keep integrity of their structure. The cells for construction of tissues grow well on the degradable framework with bioactivity, which is expected to provide temporary mechanical and chemical signals for guidance of cell differentiation, and finally to form three-dimensional tissues. Cells for construction of tissues also must express suitable genes, so as to maintain tissue-specific functions of engineered tissues. Functions of the implanted cells depends on specific receptors (e.g. integrin) on the cell surface interacting with biomaterials, and soluble grow factors. In order to reconstruct functions of the defected tissue and construct bigger new tissues, engineered tissues should integrate with the blood supply system of hosts, and the blood vessels should grow into the engineered tissue to guarantee delivery of nutrients and metabolites.

9.5.2 Phenotype Control of Engineered Tissues

During tissue development, microenvironment of engineered tissues must be regulated so as to induce express pattern of cellular matrix and form new functional tissues. This relates to regulation of multiple interactions in microenvironment, which include interactions between cells and adhesion matrix, between cells, between cells and grow factors, and mechanical stimulation on cells.

9.5.2.1 Three-dimensional adhesion

As for roles of adhesion structure on cell adhesion, migration, signal and the cytoskeleton, most of them have been so far obtained from cultivation of two-dimensional plane tissues. However, three-dimensional ECM is crucial for tissues and cells, and three-dimensional environment can promote polarity and differentiation of the normal epithelium. At present, people know little about the adhesion structure of cellular matrix, which migrating cells form by interacting with three-dimensional matrix in the internal environment in tissues, especially during embryonic development. Yamada and others results revealed that validity of cell-derived three-dimensional matrix mediated cell adhesion was 6 times to two-dimensional matrix or three-dimensional collagen by experiment on 10-minute cell adhesion. Migration rate of cells in cell-derived three-dimensional matrix was faster than that in two-dimensional matrix of adhesive protein by over 1.5 times. Increase of migration rate made cells have fusiform morphology as they were in the human body.

Comparing with two-dimensional matrix adhesion, three-dimensional matrix adhesion may enhance functional activity of cells, it is necessary to study on molecular composition of different types of cellular matrix adhesion. Adhesion plaque mediated strong adhesion of rigid matrix on two-dimensional matrix synergizes with raw fiber adhesion to form raw fiber on pliable fibronection. Fibroblasts should be cultured in high density on two-dimensional matrix for several days to produce three-dimensional matrix and form three-dimensional adhesion, which certainly will make preparation and functionalization of biomaterials face with a big challenge.

9.5.2.2 Cell motility

Research on cell migration relates to disciplinary cross and development of molecular biology, chemistry, physics and computer science, and so novel technology and methods should be adopted to deal with the problem of cell motility. In 2001 the American National Sophisticated Medicine Research Institute invested 8 million dollars to establish Interdisciplinary Cell Migration Joint Research Group, whose outlay will increase to 38 million dollars in the coming five years.

Cell migration includes many complex processes to be expounded. Some researchers are employing modernized approaches to determine proteins associated with cell migration and the genes encoding the proteins, so as to further determine their roles in cell migration. In the 1970s, Abercrombie studied cell migration on solid surface with telecamera and electron microscopy technology, and observed one whole cell motion cycle. At first, new adhesion structure of eukaryotic cell formed through interaction between the extending protrusion of the cell and the matrix surface, afterwards the cell was pulled to another new adhesion structure through contraction of action filament; the cell formed adhesion structure through trans-membrane integrin, recruiting signals and structural proteins (including actin fiber); then with the formed adhesion structure as a towing point, myosin connecting with the actin fiber pulled the cell forward; and finally the adhesion structure in the rear of the cell was removed, a cell motion cycle was completed. Ingber Research Group prepared successfully the round and quadrate island-shaped surfaces in a size equal to one single cell utilizing optical etching technique, and ECM was coated on these island-shaped surfaces to culture cells, the results showed, the cultured cells were spread in a shape of island, the cells on the round island exhibited ball-shaped, and the ones on the quadrate island quadrate-shaped; and lamellipodia of the ball-shaped cells extended irregularly, but the quadrate-shaped cells mainly extended from the four edges. From this we can see that shape of cells determines their direction of stress and migration.

For design of bioactive materials, the cell migration problems in material or through material should be considered. Cell migration has close association with concentration of adhesive ligands, cell motility rate is often a quadratic function of ligand concentrations, but not a linear function, and so design of bioactive materials should consider optimization of ligand concentration. Dimilla and others put forward the models that cell surface migration rate was confirmed by receptor-ligand binding in cell adhesion and cellular mechanical property. Subsequently, Palecek and others demonstrated that cell migration rate depended on three parameters, matrix ligand level, integrin expression level and integrin–ligand binding affinity. In addition, spatial distribution of ligands in material surface also affects cell migration.

9.5.2.3 Signals associated with growth factors

Action of growth factor has two kinds of fluctuation during cell propagation by studies reveal. The first kind is acute attack, i.e. the signals are generated immediately after stimulating by the receptor growth factor, and last for 30-60min, the second kind is that signals are generated at different stages (8-12h) stimulating by growth factors, and the procedure includes activation of cell cycle depended proteins. The theory of "two kinds of fluctuation" may explain correlation of signal stimulation with the cell cycle. Sending of the signals associated with growth factors includes three stages, bind between growth factors and transmembrane receptors, kinase activity of activation receptors, and recruitment and activation of signal enzymes.

People are wondering how growth factors associated events transform into cell responses? When cells are exposed in the environment containing growth factors such as platelet derived growth factor, the final result is cell motion, propagation, differentiation or inhibition of cell apoptosis. Then, how do cells select among these possible responses, and change signal cascade so as to mediate the suitable response? Little is known about these problems. So far it has been determined that cell has responses to many signals necessary for cell division. Many researchers have proved with many methods, P13 K or inositol phospholipid-specific phospholipase Cγ/ protein kinase C (PLCγ/PKC) has association with PDGF depended cell division. Cell cycle is controlled by cell shape and the cytoskeleton tension. When growth factors contact with the whole cell, ECM, and keep stable structure, cell shape determines whether a cell progressed into the DNA synthesis phase.

Tissue engineering is a complicated procedure, which combines many cell functions in the existed injured site, and related events include a series of interaction between growth factors, cytokines, cell and cell matrix. Repair procedure of the tissue starts from several minutes after injury, includes regeneration of blood vessels, inflammation and matrix precipitation, and may be endothelialization or minimization based on different injured sites. By adding different growth factors to the corresponding injured site, at the various stages the healing procedure is guided, improved or regulated.

9.5.2.4 Mechanical stimulation

Cell shape and structure change induced by mechanical stress plays an important role in control over many functions of cell (e.g. cell growth, motility, contraction and mechanical transmission). These cell functions are mediated through the cytoskeleton (CSK). The CSK is a composite space truss system which is composed of microfilaments that connect the nucleus and cell surface receptors, microtubule and intermediate filament to cross each other in eukaryotic cells. The articular cartilage is composed of chondrocytes and highly hydrated ECM (mainly type II collagen and aggregated glycoprotein). Some studies have indicated, chondrocytes in the cartilage implant/chondrocytes in the support material will both change synthesis capacity of matrix when mechanical stimulation. Static compression inhibits cell propagation and matrix synthesis in the alginate salt support. During culturing the cartilage implant, static charges also inhibit expression of the type I and type II collagen genes, but do not affect long-term expression of the aggregated glyconucleoprotein mRNA. On the contrary, dynamic compression stimulates chondrocyte matrix synthesis and gene expression in tissues, agarose and alginate support.

Microenvironments relate to fluid content, solid matrix rigidity and permeability of the system. It is very different in different tissues and gels. The same macroscopic deformation possibly induces different mechanical signals at the cellular level, so as to represent different cell behaviours. And with the change of biochemical composition, cell binding sites and cell signals are different, cells do not bind with the support material in agarose and alginate salt matrix, while cells bind with the support through integrin in protein matrix. The difference in cell signals will change cell behaviours and cell responses to external stimulation, i.e. local mechanical signals induce different cell responses, as a result, the same macroscopic stress responses will transform into obviously different cell metabolism events in different culture system. Chemical composition and physical properties of support materials both affect cell behaviours; and hence chemical components and physical properties of support biomaterials will have great effect on biocompatibility of biomaterials. This requires that design and preparation of biomaterials correlate with cell behaviours, so as to make a success.

9.5.3 New Generation of Tissue Engineering Materials

9.5.3.1 Biomaterials with three-dimensional surface structure

Many kinds of cells can respond to topological structure of material surface in micron size. This point has been confirmed by studies, cells also can respond to nanometer scale signal. Cells recognize characteristics of material surface that act each other simultaneity, so as to induce contact guidance. Fibroblasts detect matrix through filopodia on the cell plate to determine suitable adhesion sites so as to form adhesion structure and mature actin fiber, and recruit microtubular protein to guarantee stability of contact structure.

The matrix material has important effect on formation of cell adhesion structure and cytoskeleton development. The integrin in adhesion structure links with the actin in cells, both of them have association with signal transmission, and signal transmission pathway will influence on cell differentiation for long time. Fibroblasts contact with material surface by adhering at first, otherwise they will be apoptosis, cells must spread so as to perform cell division after adhering, and unless spreading they are not easy to enter into the DNA synthesis phase. Fibroblasts are sensitive to three-dimensional topological structure of biomaterial surface. Dalby and others utilized the principle of spontaneous phase separation of hydrophobic polystyrene and hydrophilic poly bromostyrene on silicon chips during spin coating process, to construct three-dimensional structure of silicon chip surface, and the results proved the above conclusion.

Biocompatibility of artificial ECM is related with topological structure of its surface. Cell orientation, morphology, growth and differentiation are influenced by topological structure of artificial ECM. Jansen and others utilized optical etching technique to etch out grooves in micron-scale size on silicon chips, which were treated and used for culture of bone marrow cells. The results revealed, micropatterning structure could promote bone regeneration, and suitable topological structure of material surface would facilitate deposition of mineralized matrix and differentiation of osteoblasts. Cell behaviours including cell adherence, propagation and differentiation play a significant role in formation and healing of tissues and organs. Studies have indicated that material surface in a geometric shape of streak fits for cell orientation and motion, so as to form particular structure. Micron and nanometer structure domains on biomaterial surface may regulate two kinds of basic external signals, interactions between cell and matrix, cell and cell, in order to construct highly orientated cell micropatterning, finally making them arrange into the tissue structure. At the moment, the size of structure domains should be the same as the real organism structure. The average size of cells is

10-30μm, while that of functional structure domains on cell surface is 50-100μm. Cell guidance is influenced by different chemical and topological structure domains on material surface, and the study by Barbucci have proved this fact.

Topological structure of material surface is vital for cell-material interaction. Different types of cells affect topological structure of material surface in different ways. Cellular phenotype and bone contact extent of osteoblasts response to topological structure of material surface, for example, bone at first forms in grooves and crevices. Cells recognize surface characteristics by changing structure of actin fiber in pseudopodia, and at the same time act with the surface. Different types of cells respond to different topological structures (e.g. grooves/protrusion, column, point and fiber). Topological structure may affect cell adhesion, motion, orientation, morphology, cytoskeleton, contact inhibition, phagocytosis and gene expression. Studies have shown, cells have phenomenon of affinity with rough material surface, i.e. rough material surface can enhance cell response ability. Dalby implanted primary human osteoblasts on hydroxyapatite composite material surface with different rough extents, after culturing for 4 hours, the cells on the composite material surface undergoing optimization of surface roughness had thicker stress fiber and mature adhesion structure, while those on the untreated composite material surface had neither microfilament actin nor adhesion structure.

9.5.3.2 Materials for mediating cell adhesion

Materials for mediating in cell adhesion may guide cell growth, but during regeneration of cell, such as the nerve, the bone, the blood vessel and the cornea in the human body, it is required that cells should be guided with more specific signals, which originate from the traumatic site or the surrounding tissue. During tissue repair, ideal biomaterials should involve in the interaction between specific adhesive proteins expressed in the target cell and growth factor receptors selectively. Support material should guide the target cells to migrate to the traumatic site, so as to stimulate cell growth and differentiation. With progression of tissue repair, the support material is degraded by enzymes for matrix reconstruction released by cells, and then disappears.

There are specific amino acid sequences such as RGD, which exist in the adhesive structural domain of fibronectin and other ECM glycoproteins, the material surface modified with RGD may mediate cell adhesion, but this process is much complicated. Immobilizing specific amino acid sequences on the glass and titanium surface, such as arginine-glycine-aspartate-serine (RGDS), and lysine-arginine-serine-arginine (KRSR), can effectively enhance cell adhesion, propagation and deposition of calcium-containing inorganic substance. But after implanting the material into the human body, bioactive molecules in the peptide sequences immobilized on the biomaterial surface are influenced due to their interaction. For this reason, Bizios and others studied effect of nanometer phase ceramics with a crystal particle size less than 100nm on compatibility of osteoblasts, and explored a new solution for surface engineering of biomaterials. The slab material of hydroxyapatite (HA) was prepared with wet process, and was used for culture of osteoblasts after sterilization. The results revealed that alkaline phosphatase in nanometer phase HA had higher activity, and more inorganic calcium deposited. Nanometer crystal structure is expected to improve integrity of HA with the bone tissue.

Cell adhesion plays a significant role in tissue regeneration and implants repair. Cell adhesion may be controlled by changing the balance between oxidants and anti-oxidants, i.e. by modifying the cytoskeleton. Malorni and others added N-acetyl cysteine (NAC) to Dulbecco's modified Eagle's medium (DMEM), to explore adhesion behaviour of epithelium-like cells in different ECM. The results indicated that NAC anti-oxidants may enhance biocompatibility of biomaterials effectively.

It is found that understanding interaction of cell-protein-material surface has big significance for acceleration of engineering of the base material through studies. ECM provides signals about survival, propagation, differentiation, phenotype expression for cells. Adhesion between cells and ECM is mainly mediated by integrin, and the mediation process can regulate cell migration, formation, propagation and differentiation. Integrin-mediated signaling pathway is related with ECM regulation of cell development. It is also found, the biomaterial surface, which is modified by utilizing interaction between ECM components and their structural domains, may induce expected cell-material reaction.

9.5.4 Intelligence of Artificial Extracellular Matrix

9.5.4.1 Temperature-sensitive two-dimensional cell sheet conveyor system

According to histology and developmental science, the tissue is composed of cell sheets that are the smallest unit of the tissue; and the tissue is formed by cell with three-dimensional overlapping. Intelligence of artificial ECM controls over interaction between biomaterial surface and cells through cell sheets. At a certain temperature, some particular material is grafted on the culture surface to form temperature-sensitive culture plate in which cells are cultured; and then, hydrophilism of the culture plate surface changes and cell morphology changes with the temperature shift, and meanwhile cells fall off the adhesive surface. At the moment the cell sheets are recovered in a non-enzymolysis way, and so ECM and cell growth receptors are retained. With such a method, various kinds of cells, including fibroblasts, vascular endothelial cells, hepatocytes, macrophages, retinal pigment epithelial cells, are cultured and recovered. Many methods may be adopted for intelligence of artificial ECM.

9.5.4.2 Temperature-sensitive injectable aquogel

Support materials should simulate the *in vivo* environment as accurately as possible, so as to guarantee cell propagation, differentiation and maintain its natural phenotype, finally playing functions of tissues or organs. Injectable tenacious support may form *in situ* in the human body, so as to minimize immersion of implanting process into the body tissues as possible. Poly-N-isopropylacrylamide (PNIPAAm) is a kind of temperature-sensitive injectable aquogel, and manifests the phenomenon of low critical solution temperature(LCST) phase separation to form insoluble polymer. And hence, a small amount of cross-linking agent is used to make hydrophilic PNIPAAm polymerized into aquogel. At a temperature higher than LCST, this aquogel contracts through volume phase transition network to drain a great amount of water in pores and transforms into rigid matrix; and when the temperature decreases to below LCST, PNIPAAm aquogel once again swells in water. However, since "dense cortex" forms on the aquogel surface at the temperature higher than LCST, water release from aquogel is influenced, and contraction rate of the network is limited. PNIPAAm aquogel contraction and swelling process is reversible. Besides, there are other temperature-sensitive injectable aquogel materials, such as Poly-N-isopropylacrylamide-co-acrylic acid aquogel, which has no cytotoxicity. In a study, the nude mouse model was used to evaluate *in vivo* biocompatibility of this aquogel for 7 months, and the results showed that few fibrous tissue capsules were found in the aquogel matrix.

9.5.4.3 Intelligence of release carriers

Intelligence of release carriers means that carriers have the ability to retain or release carried substances according to needs. Intelligent carriers include the following types.

1) Growth factor releasing carrier

Uludag and others copolymerized PNIPAAm and protein-reactive N-acrylamide succinimide (NASI) and hydrophobic alkyl methacrylate (AMAs) to prepare temperature-sensitive PNIPAAm copolymer containing NASI, which is just a growth factor releasing carrier. LCST of this copolymer may be regulated by ester group sizes of hydrophobic alkyl succinate methacrylate.

2) DNA carriers responding for pH value

In recent years, many studies have reported, after exogenous DNA is bound with the corresponding carrier to form a complex which is introduced to the target cell and further runs into the cell nuclei to transform into the genetic information on the host DNA, the exogenous DNA is transcribed together with the host DNA into mRNA. By reading the genetic information on mRNA, the corresponding protein is synthesized on the ribosome, which is called gene therapy. Gene therapy technology is now combining with tissue engineering, which indicates a promising application prospect. Transfection of the host cell with plasmid DNA expressing a growth factor may make the growth factor release continuously in an endogenous way, so as to promote repair of the injured tissue. At present viral vectors are often adopted, because they can respond to changes of external environment including pH values, so as to transport DNA skillfully. They have higher transfection efficiency, but safety problem still exists. The key issue for design non-viral vectors lies in how to endow the vectors with virus-like intelligence, i.e. the vectors can respond to pH value changes, and so discontinuous physical property changes (e.g. hydrophilism/hydrophobia, aggregation, condensed state), so as to regulate interaction between the vectors and the cell membrane and even the nuclear membrane.

Common pH value-responsive DNA vectors are a kind of proton-sponge polymers. These polymers contain amino groups, and so can buffer the environment of the acidic lysosome within the physiological pH range. Such buffering ability may promote osmotic swelling, which helps lead to physical breakage of the endosome and also helps release DNA to the cytoplast before fusing with the lysosome. Polyethyleneimine (PEI) is just a commercialized "proton-sponge" polymer; and comparing with poly-lysine (PLL), PEI may mediate gene transfection at a higher level. Commercialized cationic liposome Lipodetamine Plus (Invitrogen, Carlsbad, CA) has stronger buffering ability.

9.6 Nanoscale Biomaterials

Nanoscale biomaterials usually refer to biomedical materials with special functions and size ranges within nanometers. Nanometer microparticles refer to the ultra-fine microparticles, and generally are 1-100nm in size which is generally much smaller than that of cells and red blood cells, so nanoscale microparticles are applied in the medical field, to manufacture special drug or novel antibodies for local orientated therapy, to perform cell separation and cell staining; to give an early diagnosis of tumor tissues even cancer cells, and to carry some drug to treat genetic diseases and cancers. When the size of microparticles is nano-scaled, the microparticles have many characteristics, such as quantum size effect, small size effect, surface effect and macroscopic quantum tunneling effect by themselves, and hence they manifest many unique properties and have wide application potential in catalysis, light-filtering, pharmaceuticals, magnetic media and novel materials.

9.6.1 Basic Characteristics of Nanoscale Materials

9.6.1.1 Surface effect

The special effect resulting from chemical environment difference between surface atoms and internal ones is generally called as surface effect. It actually refers to changes of properties of nanoscale particles, resulting from the percentage of surface atoms against the total atoms varying with particle radiuses. The surrounding of surface atoms is short of adjacent atoms, and hence their chemical bonds are in saturation. At the same time, the electronic cloud of surface atoms has strong orientation, and so exhibits strong chemical activity. This permits nanoscale particles to have advantages over any other routine chemical substance in the aspects of catalysis, absorption, etc.

9.6.1.2 Volume effect

Volume effect is also called small size effect. Since nanometer particles are so tiny in volume, they contain fewer atoms. Therefore, many phenomena cannot be explained with property of slab-shaped substance containing indefinite atoms, and these special phenomena are usually called volume effect. The theory associated volume effect is "Kubo theory".

9.6.1.3 Quantum size effect

Quantum size effect refers to a phenomenon that when the size of nanometer particles is close to or smaller than the Bohr quantum radius, the electronic energy level adjacent to Fermi energy level of particles changes from continuous energy level distribution to discrete energy level, and nanoscale semi-conductor and microparticles have discontinuous energy levels of the highest occupied molecular orbit and the lowest occupied molecular orbit, which makes energy gaps become broader. Among nanometer particles, the wave behaviour of the electrons lying in the discrete quantized energy level brings nanometer particles with a series of unique properties, such as very high optical non-linearity, specific optical and catalytic properties.

9.6.1.4 Theory of quantum tunneling

The ability for macroscopic quantity of nanometer particles such as magnetic flux to pass through potential barrier of the macroscopic system is called macroscopic quantum tunneling (MQT) effect. Macroscopic quantum tunneling effect provides the basis for micromation of electronic pieces and macromation of disks and tape storage information.

9.6.2 Requirements for Basic Properties of Nanoscale Biomaterials

Nanometer targeting drug has strict requirements for nanometer microparticle carriers, which mainly include: ①monodispersion nanometer microparticles; ②the optimal particle shape should be ball-shaped; ③average particle diameter of targeting drug carriers should be less than or equal to 30nm, optimal less than or equal to 15nm, even particle diameter distribution (narrow particle diameter distribution), $d=\pm 30\%$, i.e. the optimal particle diameter range of carriers is 10-20nm; ④suspension stability in water solvent is good, not ready to cause flocculation; ⑤nanometer carriers should have no biotoxicity; ⑥carries should be colorable, especially when necessary to conduct targeting contrast and indication; ⑦the optimal magnetic targeting nanometer drug carrier should be super paramagnetism, i.e. the carrier has high magnetic strength at the external magnetic field, and no magnetic induction remains after closing the external magnetic field.

9.6.3 Preparation and Application of Nanoscale Biomaterials

There are many methods for preparing nanoscale biomaterials: ①mechanical grinding method; ②solid phase reaction method; ③liquid phase method, including co-precipitation method, even phase precipitation method, metal alkoxide hydrolysis method, sol-gel method, microemulsion method; ④gaseous phase method, including freeze-drying method, spraying thermal decomposition method and gaseous-phase chemical reaction method.

Nanoscale biomaterials can be applied mainly in the following areas: nanometer sensor; nanometer drug/gene vector technology; nano-drug carrier targeting technique, including physical and chemical targeting and biological targeting; and nanometer biomaterial-controlled release technique. Nanometer drug carriers may be used as antitumor drug carrier, antibiotic and anti-parasite drug carrier, oral drug carrier, ophthalmological drug carrier, brain-targeted drug carrier, bone marrow-targeted drug carrier and gene drug carrier.

Chapter 10

Contemporary Bioengineering and Biological Medicine

With the birth and development of molecular biology, traditional medicine industry and pharmaceutical technology have experienced revolutionary changes, especially for biological medicine industry. The consistent development of contemporary biological and pharmaceutical technologies provides a lot of new methods and novel techniques for production of variety of new drugs, and improves the technical levels of the whole medical industry. Biological medicine industry has become one of the leading industries of the national economy in the 21st century.

Biopharmaceutical refers to the processing and preparation of products that can be employed for prevention, treatment and diagnosis of diseases, which integrate principles and methods in chemical and biological technologies, separation and purification engineering and pharmaceutical. The material such as organism and tissues, body fluids, or its metabolites (primary metabolic products and secondary metabolites) have been used for research in biology, medicine and biochemistry. In terms of the sources of raw materials, biological drugs include all kinds of natural biological active substances and its synthetic or semi-synthetic analogues prepared from animals, plants, halobios, microorganisms, and other biological materials. Thus, antibiotics, biochemical drugs and biological products belong to the category of biological drugs.

Biopharmaceutical is a technology to manufacture a variety of biological medicine using organisms or biological process under the artificial conditions. Contemporary biotechnology provides the main technological platform for biopharmaceuticals. The penetration and development of biotechnology in pharmaceutical industry has pull new power to prompt the gradual fusion between the theory and technology of drug preparation, mediated by modern biotechnology. For instance, the functions of antibiotics are no longer limited to sterilization or bacteriostatic; the production of insulin and growth hormone will no longer depend on animal organs as raw materials; and the production of Hepatitis B vaccine will no longer rely on human blood, etc. The application of genetic engineering and the development of protein engineering have not only reconstructed the old areas but also opened many new fields in biopharmaceutical. The production of human growth hormone (HGH), for example, is no longer subjected to the limitation of raw materials sources which provide effective security for clinical use. The modified human insulin transformed by using protein engineering has become more stable and effective. It's possible to produce antibody from plants, nucleic acid vaccine from yeast, natural active drugs from the culture of plant endophytes and plants hairy root systems, etc.

Contemporary biotechnology has been widely applied in drug production, gene diagnosis and gene therapy, etc. With the rapid development of industrial construction, many biopharmaceutical products entering into a wide range of large-scale industrialization period. The market for the types of biotech drugs has increased significantly, including vaccines, monoclonal antibodies, cytokines, hormones, antithrombotic factors, gene therapeutic products and antisense drugs, etc. The statistic data indicate that the total number of new drugs approved in recent 4 years is the sum of those in the past 15 years. It shows that biopharmaceutical technology plays an important role in the future economy of medical industry. The research of biotech drugs will be carried out more rapidly. The following shows a list of areas that will be expected to undergo more rapid development.

1) Development of novel vaccine

No matter in the past or at present, vaccine plays an irreplaceably important role in the prevention of many diseases. However, with the change and development of human disease spectrum, there are still many diseases (such as obesity, cancer and AIDS, etc.) which are hard to conquer. More research work needs to be done for disease prevention and treatment. As a report showed, 98 species of vaccines are currently in the phases of research and development, 61 of which will be used for tumor treatment, 6 for respiratory diseases, 4 for HIV, and the majority of them will be used for infectious disease treatment. 59 species of monoclonal antibody are in the progress of clinical research, including 31 species used in tumor, others would be used for organ transplantation, respiratory diseases, skin diseases, nerve disorders and autoimmune diseases.

2) Production of active peptides using genetic engineering

The biologically active peptides prepared by using genetic engineering technology are called gene engineering active peptides. The application and development of gene engineering technology, on one hand, make it possible to produce these active peptides, and on the other hand, more new active peptides have been discovered. For example, there are more than 50 varieties which have been found only in the category of neural active peptides, more than 10 active peptides and growth factors have also been found in the category of cardiovascular diseases. There may be extremely rich active substances such as peptides to maintain normal physiological regulation and control mechanism and to defend disease existing in human body; however, we know little about them. There may be more than 90% active peptides yet to be found in human body. Therefore, the prospects for development of genetic engineering active peptides as drugs are very bright.

3) Exploitation of protein engineering drug

The utilization of protein engineering can improve the stability of recombinant protein products, increase product activity and prolong the half-life of products in the body, elevate the bioavailability, and reduce product immunogenicity, etc. The natural insulin preparation, taken as an example, can easily form dimers and six polymers during storage, which will delay the process that the insulin enters from injection site into blood, therefore, slow down its hypoglycemic effect and increase its antigenicity. This attributes to the structure of B_{23}-B_{28} amino acid residues in insulin. Changing these residues can reduce the extent of polymerization.

4) Research and application of novel efficient expression system

So far, most of biotech drugs in the market (Recombinant DNA products), are produced in expression system of *E.coli*.(34 varieties). Some products are involved in Chinese hamster ovary cells (CHO, 14 varieties), young hamster cells (2 varieties) and *Saccharomyces cerevisiae* (11 varieties). The recombinant expression systems using the cells from fungi, insects, transgenic plants and animals are being further improved. Transgenic animals, as a new expression system, have been aroused great attention, since they can produce highly complicated active products in a cheaper way. Some products are already under clinical trials, such as $α_1$-antitrypsin, α-glucosidase and antithrombin, etc. There are 20 kinds of products under early development using transgenic sheep or cows. The production of biological drugs using transgenic animals for medical treatment may come true soon.

5) Gene therapy and gene diagnosis

The application of biotechnology has prompted the greater development of medical technology. Its research objects have been already spread from the original genetic disease to tumors, infectious diseases and cardiovascular diseases, etc. The thoughts for gene therapy are also continuously expanded, not only the addition and replacement of normal genes, also the positive regulation or negative control of genes *in vivo*, and even introduction of originally not existed genes into body. In addition, for patients with the genetic diseases, immune deficiency or tumors in incubation period caused by genetic defects, the preventive gene therapy could be performed based on the gene diagnosis results by surveying in their families. Therefore, gene therapy is a new way to prevent and treat diseases. Especially, for the treatment of those serious diseases harmful to human health, gene therapy has exhibited potential application value and prospect. The application of proteomics research will help reveal the physiological and pathological process and update the technology for early disease diagnosis.

6) Guide drug

Guide drug is also called "bio-missile". All kinds of monoclonal antibodies are used as the "guidance department" for this missile, and all the drugs for disease treatment (such as anticancer drugs, toxin protein, etc.) are employed as bullets of the missile, both of which compose of guide drug. Due to the high affinity of antibody binding with antigen, it can directly introduce the effective drugs to the pathological site and allow drugs to fully function.

7) Investigation of traditional Chinese medicine modernization

For quite a while, traditional Chinese medicine (TCM) had been in a leading position in the field of medicine in the world, and has made great contributions to the reproduction and prosperity of human being and to the prevention and cure for diseases for thousands of years. The reasons for survival and development of TCM till today attribute to its efficacy in prevention and control of disease. However, influenced by traditional concepts, the development of TCM is greatly restrained and has gradually lost its leading position in the field of international medicine. Confronted with various challenges, TCM itself should realize its modification and improvement towards internationalization and acceptance by the world, particularly developed countries. Thus, the utilization of contemporary biology in the exploration of TCM is the key for the modernization of TCM and the basis for its internationalization. Both molecular biology and TCM have common characteristic of life science in philosophy which named "nonlinear problem", and "nonlinear problem" is the key to life science. Therefore, the application of molecular biology in the investigation of TCM can not only sublime TCM theory system, enhance curative effect, but also bring TCM to the advanced levels in world medical field.

The short-term goals for the scientific research of TCM mainly focus on excavating, sorting, validation and analysis of traditional Chinese medicine information. To establish TCM in a solid position as a medical science, actions should be taken up by comprehensively exploring the TCM theory and systematizing through sorting and literature research, confirming using experimental methods, scientifically verifying by clinical experience, and exploring its therapeutic mechanism, etc. While the long-term goals mainly include transplantation and penetration; that is, transformation and penetration involved in large-scale and multi-scientific subjects, will make TCM become not only a complete contemporary academic system including theory, clinical, experimental methods and drug etc., but also a unique medical field that will meet the requirements of people to prevent and cure diseases in modern society.

In recent years, TCM has started to carry on molecular biology, absorb new research methods, expand theoretical framework, and broaden research thoughts, which opens up a new field in TCM modernization. Currently, the application of molecular biology in the investigation of TCM has made achievement in the fields of basic theory, clinical, investigation, and acupuncture research in TCM. Especially, the clarification of the molecular mechanism of "Like cures like", studies on acupuncture analgesia, and acupuncture and moxibustion for treatment of cancer have attracted great interests. ①Theory of TCM, in recent years, the prevention and control of biological method are employed to study the essence of kidney, the theory for promoting blood circulation to remove blood stasis, and anti-aging mechanism for TCM. For example, Zhong Liyong and others explained the main control point for drug to mediate kidney yang deficiency syndrome from molecular level, his research work involving in the three types of compounds for spleen-strengthening, kidney-tonifying and blood-activating on hypothalamus-pituitary-adrenal-thymus (HPAT) axis and influence on adrenocorticotropic hormone releasing factor (CRF) gene expression has been published. Results indicated that kidney-tonifying medicine can protect HPAT shaft from the inhibition of exogenous cortex ketone by improving the hypothalamus CRF mRNA expression; spleen-strengthening drugs play direct roles in promoting the immune system, while blood-activating medicine has no influence on HPAT shaft. A conclusion can be drawn that only kidney-tonifying medicine among the three kinds of compound could improve the expression of hypothalamic CRF mRNA and the function of HPAT axis. The main control points for drug to mediate kidney yang deficiency syndrome are positioned in hypothalamus; the theory for promoting blood circulation to remove blood stasis is an important content in the etiology and pathology of TCM. In recent years, with the advances in the molecular mechanism research regarding "five endogenous pathogens" including gore and phlegm turbidity, and abnormal gene expression, protooncogene expression, and gene regulation disorder stimulated by pathological metabolites, the theory for promoting blood circulation to remove blood stasis has reached a new stage. The molecular mechanism of pharmacological effects in this kind of TCM was revealed by the investigation of gene regulation intervened by this TCM and inhibition of protooncogene expression. Internationally, studies regarding the molecular mechanisms of aging have gradually gone deep into it. Studies show that part of anti-aging mechanisms of TCM are through restoring DNA repair function for old people, in addition to gene regulation to enhance the expression level of mRNAs which code antioxidant enzymes. Shen Xiaoheng and others studied the SOD and catalase (CAT) activities in aging rat liver tissue and their gene expression level, and results indicated that the soups have the anti-aging efficacy through increasing antioxidant enzyme activity and reducing free radical products by regulating and enhancing gene expression level of encoded enzyme proteins. ②Clinical in TCM, the curative effect of arsenic to treat leukemia has been accepted by medical field, which confirmed the "Like cures like" therapy in TCM, when Chinese scientists studied the pathogenesis mechanism of promyelocytic leukemia at molecular level, they found out that the main ingredient in a folk formula for treatment of Scrofula is arsenic trioxide. Their studies confirmed that arsenic trioxide could promote the apoptosis of NB4 cells in acute promyelocytic leukemia (APC) patients, shown as nuclear fragmentation. The number of apoptosis cells depends on the dose and time of treatment. ③In TCM research, the applications of biotechnology in herbal medicine identification, medicinal resources protection and natural active ingredients acquisition, etc, have made remarkable progress. Biotechnology is applied to search for natural active ingredients in TCM and produce them in great amounts, change the effective components in existing traditional medicinal materials, thus turn original medicinal plants into "genetically modified medicinal materials". For example, the exogenous genes such as enkephalin, epidermal growth factor, erythropoietin, growth hormone, human albumin, hemoglobin and interferon are successfully expressed in transgenic plants. New hairy root culture systems infected by *Agrobacterium rhizogenes* have been established with good properties in more than 40 kinds of traditional medicines, including *Ginseng, Radix arnebiae seu lithospermi, Salvia miltiorrhiza*, etc.

The applications of biotechnology have completely changed the process and production of human biological medicine for disease diagnosis and treatment, thus, to meet the requirements of new era. If the transgenic plants tobaccos containing hemoglobin genes can be used to produce a mass of artificial plasma, current status of blood supply will be changed completely. The development and application of transgenic animals, also will promote further development of medical career. The development of medical biotechnology closely related to human survival in 21st century will certainly make more contribution to protect human health.

The discovery of each new human gene has commercial development potential, and may produce new drugs for human disease detection, treatment, and prevention. It is expected that more new genes related to diseases could be found and more new drugs of medical efficacy will be developed when human genome is totally completed. Varieties of gene drugs based on nucleic acids have been increased rapidly, and more than 30 kinds of antisense drugs are undergoing clinical trials. It can be deduced that the rapid development of contemporary biological drugs will make greater contribution to human health. Therefore, contemporary biological medicine has become an important field explored by some foreign companies.

10.1 Application of Contemporary Biotechnology in Antibiotic Industry

With the number of known antibiotics rising, the chance of screening new antibiotics with traditional conventional methods becomes less and less. In order to get more new antibiotics and excellent antibiotic-producing bacteria, recombinant DNA technique was started to apply to the biosynthesis of secondary metabolites with more complicated structures in 1980s, and thus, the recombinant technology was employed in screening new microbial medicine resources and modifying drugs of microbial metabolism. With further development of molecular biology research in *Streptomycete*, recombinant DNA technology used in above areas has made great progress; moreover, the application of gene engineering technology has made it possible to produce new type of microbial antibiotics and new metabolic products.

The transformation of antibiotics using biotechnology is mainly embodied in the aspects of improving the production ability of stains and modifying the current strains to produce new metabolites by using gene recombinant technology. With the more deeply understanding of structure, function, expression and regulation of biosynthesis genes and resistance genes in some antibiotics, the research and application of recombinant microorganisms which are used to improve the yield of metabolites and find new products have attracted more and more attention, currently there are 23 species of cloned synthetic genes for antibiotics.

10.1.1 Improve the Output of Antibiotics

For a long time, the high yield strains for industrial antibiotics production are screened by mutation breeding of physical or chemical methods. Although the current mutation breeding technology is still the main means to improve microbial strains for industrial production, it has been successfully reported that gene engineering technology are used to modify genes directionally and improve the gene expression levels to reconstruct production capacity of strains. Improving antibiotic production by gene engineering could be considered from the following several aspects.

10.1.1.1 Direct screening of high-yield producing bacteria by gene clone technology

The increased dose of a certain gene related to production in cloned strains (gene at rate-limiting stage or regulatory gene) could improve the output. Despite its random selection and heavy workload, it is worthy of trying if the output measuring method is simple and convenient.

10.1.1.2 Increase the copy numbers of genes involved in rate-limiting steps of biosynthesis

Some stages in the antibiotic biosynthesis may be the rate-limiting steps for the whole synthesis pathway. By recognizing the rate-limiting bottleneck in synthesis pathway, and trying to guide gene copy number to improve the enzyme systems during this stage, it is likely to increase the output of final antibiotics, if the increased intermediate products don't produce feedback inhibition on some steps of synthesis pathways.

10.1.1.3 Strengthen positive regulation

Key regulatory genes are often embedded in gene clusters to control synthesis of antibiotics in many *Streptomycete*, which are often the components of gene clusters contributing to antibiotic biosynthesis and their own resistance. Positive regulatory genes regulate structural genes through some positive control mechanisms to accelerate the production of antibiotics. Negative regulator genes regulate structural genes through some negative control mechanisms to reduce the output of antibiotics. Therefore, to increase the action of positive regulatory genes or decrease the function of negative genes is a feasible way to increase the yield of antibiotics. Introduction of additional regulatory genes into wild type strains provides the simplest way to obtain high yield of products.

10.1.1.4 Increase resistance gene

Resistance genes often link with biosynthesis genes, and probably their transcriptions are also closely linked, which is the essential component to activate the transcription of biosynthesis genes. Therefore, resistance gene must be firstly transcribed, and once the resistance is established, the transcription of biosynthesis gene can proceed. The generation of antibiotics is closely related to the resistance of strains to its own antibiotic. Antibiotic production level is determined by both antibiotic biological synthetase and its own resistance enzyme, which provides basis for the improvement of antibiotics production by increasing their self-resistant level to improve strains.

In addition, one of conventional breeding methods is to use protoplast fusion to enhance unit production of microbial metabolites. In order to improve the unit yield of certain antibiotics, the protoplast of one producer can be fused with another high yield producing strain with similar biosynthesis pathway to antibiotics bacteria to obtain a new fusion strain. For example, the interspecific protoplast fusion between daunomycin and tetracycline producing bacteria, because both the antibiotic biosyntheses come from polyketone pathways, the unit production of daunorubicin in fusion stain has been obviously improved. The interspecific protoplast fusion of paromomycin producing bacteria and high yield mutant strain of neomycin producing bacteria can produce a recombinant with 5 or 6 times more unit yields of paromomycin. The difference of chemical structures between two antibiotics is only one hydroxyl and one amino, and their synthesis pathways are also very similar.

10.1.2 Ameliorate Antibiotic Component

Many antibiotic-producing bacteria can produce multi-components antibiotics, although these components posses similar chemical structures and properties, their biological activities are quite different, which provides big challenge for fermentation, extraction and refining of effective components. With the more deeply understanding of all kinds of antibiotics synthetic pathways and the development of gene recombination technology constantly, the application of gene engineering method can directionally (or selectively) modify antibiotics producing bacteria to obtain high yield strains of producing effective components only and simplify downstream separation process.

The application of gene engineering technology in modifying strains and creating new hybrid antibiotics has provided a new source for microbial. Hybrid antibiotics are new compounds with antibacterial activity produced via genetic recombination technology.

10.1.3 Improvement of Antibiotic Production Process

The biosynthesis of antibiotics is generally sensitive to oxygen supply and the lack of dissolved oxygen is often the limiting factors for highly productive fermentation. In order to maintain cells to stay in aerobic condition, what traditionally can be done is to change the operating conditions and reduce cell growth rate or culture density. At present, the way of increasing oxygen supply level is usually focused on how to improve dissolved oxygen level or gas-liquid mass transfer coefficient, increase the input of sterile air in fermentation tank, as well as disperse air using all kinds of stirring devices to meet the requirements of oxygen by bacteria.

Oxygen molecules entering into the liquid cross several layers of boundary membranes and then enter into cells to reach breathing organelles which consume oxygen and generate energy through physical diffusion. If hemoglobin with high affinity to oxygen is introduced into cells, breathing organelles can easily obtain adequate oxygen, reducing the sensitivity of cells to oxygen, and improving the control strength of dissolved oxygen during the fermentation. Therefore, cloning hemoglobin gene into antibiotic producing bacteria using recombinant technology and expressing hemoglobin in cells, will likely resolve the conflicts between supply and demand of dissolved oxygen and improve the utilization rate of oxygen by improving their own metabolism function.

For example, cloning the hemoglobin gene of one kind of filamentous bacteria into *Actinomyces* can promote aerobic metabolism, cell growth and synthesis of antibiotics. With limited oxygen, Vitreoscilla hemoglohin(VHb) is induced and its synthesis can be amplified for several times. This vitreoscilla globin gene, (*vgb*) has been cloned in *E.coli*. The intracellular localization study shows that a large number of VHbs exist in intercellular area, and its function is ensuring more oxygen for cell. The maximum of induced *vgb* expression occurs under the minimal oxygen conditions (the dissolved oxygen levels in air are below 20% saturation) and regulation in transcription level. The transcription levels are lower under completely anaerobic conditions and may achieve maximum inducing effect in low oxygen but incomplete anaerobic environment. The condition of oxygen-depleted could promote cell growth and protein synthesis. The research and application of hemoglobin gene engineering will greatly reduce energy consumption in antibiotic industry and other fermentation industry.

Introducing a high temperature-resistant regulation genes or heat-resistant biosynthesis gene into antibiotic-producing bacteria could increase the fermentation temperature, thus, to reduce the cost of temperature controlling during production.

10.2 Gene Diagnosis and Gene Therapy

Traditionally there are mainly three ways for disease diagnosis: clinical diagnosis, serology diagnosis and biochemistry diagnosis, which are all based on the changes of disease phenotype. However, the phenotypic changes are not specific in many cases and occur in a later time, resulting in the consequences of confused diagnosis and delayed treatment. Genetic diagnosis, a new method for clinical diagnosis, is to diagnose disease or human condition in genetic level. It is an important achievement of molecular biology, molecular genetics and gene engineering technology in medicine in recent 10 years. These new disciplines and technologies have many characteristics, such as advanced, accurate and quickly updated, and prompt the continuous development and mature of gene diagnosis technology. And the scope of its application is expanding continually.

The occurrence of disease is nothing but the combined effects of internal and external factors. The disease pathogenesis and early pathological changes are due to gene changes, namely endogenous gene mutations or exogenous gene invasion. Endogenous gene mutations include amplification (such as fragile X syndrome), transposition (such as the transposition of chronic myelogenous leukemia cancer gene *C-abl* and *BCR*) and point mutation (such as point mutation in *p53* gene), etc. When various pathogens infect human body, they can cause the invasion of specific exogenous genes and proliferation *in vivo*, resulting in all sorts of infectious diseases. Some viral nucleic acids can integrate with host genome chromosome or activate oncogenes, block tumor-suppressor genes and lead to tumorigenesis, such as HBV and liver carcinoma, EBV and nasopharyngeal carcinoma, etc. Genetic diagnosis refers to the diagnosis method used to search for the abnormal changes of endogenous genes as well as discover and identify the existence of pathogenicity of exogenous genes, and then achieve more specific, sensitive and quicker diagnosis for diseases

10.2.1 Gene Diagnosis

10.2.1.1 The principle of gene diagnosis

The method of contemporary molecular biology is used to detect whether gene structures and expression functions are normal and how they are distributed. DNA probe and the single strand of target genes form hybrid molecules, with DNA (gene) or mRNA as target genes, which is the basic principle of gene diagnosis. Detection of mRNA provides evidence for whether gene function is normal or not at gene expression level, and comparing with direct DNA detection, has the following advantages: ①gene transcripts are equivalent to genetic template amplification, so the sensitivity of mRNA detection is higher than DNA detection; ②there are no intron sequence in mRNA, which is more convenient for the treatment of large genes such as DMD genes, especially when using PCR method.

Based on the principle of complementary bases, denaturation and renaturation of DNA double strands, nucleic acid molecule hybridization is to use the fragment of known sequence of mononuclear nucleic acid as probe and to test the presence of complementary cognate nuclear acid sequence in samples. Probes are labelled with appropriate markers (such as radioactive isotope) and detected using appropriate methods (such as autoradiography) after hybridized with target genes. There are mainly three types of technologies involved: DNA probe hybridization, PCR and others.

10.2.1.2 The significance of gene diagnosis

Of different levels of diseases shown in biological individual, genes are the fundamental determinants. Gene diagnosis aim directly at pathogenic genes, not only can accurately make diagnosis to all kinds of clinical disease status, but also can further understand the causes and types of diseases. Accordingly, it's possible to substantially classify and genotype some difficult-and-rare-cases, and provide evidences for treatment especially for emerging gene therapy. Gene diagnosis has become the supplements and updating technology of traditional disease diagnostic methods, including clinical diagnosis, biochemical diagnosis and immunological diagnosis.

The significance of gene diagnosis is not only limited to disease diagnosis, but also plays a positive role in judging individual's susceptibility to certain diseases, predicting disease outbreak and providing preventive care for the diseases. For instance, human leucocyte antigen (HLA) complex consist of closely linked gene clusters and play an important role in the process of immune regulation. It is well-known that HLA gene polymorphism is related to the genetic susceptibility of some diseases, for example, the proportion of HLA-DR4 carriers in rheumatoid arthritis patients is as high as 70% in white people, however, the positive ratio in normal patients is only 28%.

In addition, gene diagnosis is very promising in promoting prenatal and postnatal care, having solved the problems of many genetic diseases that are difficult to be diagnosed in early stages and become an effective means for prenatal diagnosis of genetic diseases. Genetic diagnosis can tell in the early weeks of pregnancy whether the fetus is sick, and prevent the occurrence of diseases from seriously damaging the health by performing abortion if necessary, which can greatly improve the quality of population. As the most populous country in the world, China will be faced with the grim situation of population quality. Of about 5 million disabled people, a considerable number of them have congenital malformations or carry hereditary diseases, bringing huge pain and heavy burden to individuals, families and society. Therefore, widely performing prenatal gene diagnosis on these diseases can reduce the birth of children in disease, lower disease incidence, and have great social and economic significance.

The problems with bone marrow and organ transplantations are rejection response. The use of immunity inhibitors will evoke many adverse reactions, therefore, the ideal way is to conduct tissue-typing before the operation. The application of gene diagnosis technology to analyze and display genotype is able to better match tissue-typing and improve the success ratio of transplantation. In many countries, HLA gene grouping technology has gradually become the conventional technology chosen by organ transplantation recipients.

The fact that certain proportion and degree of mutations exist in some infectious epidemic pathogens is the main reason why epidemic diseases are difficult to control effectively. Using classical biological and serological methods can only make sure the serum type of pathogens, but not understand the genetic differences between isolation strains in the same serotype. The analysis of the homology in isolation strains from different regions and different years and variants through gene diagnosis is beneficial for the clarification of variation trend of pathogens, evolutionary rules, proliferation pathway and dominant popular strains, etc., providing evidence for predicting the outbreak of epidemic diseases. Therefore, the application of gene diagnosis has great potentials in preventive medicine.

Since the detection objective of gene diagnosis is based on genes, so the features of gene diagnosis include high specificity, accuracy, sensitivity, and strong adaptability as well as unlimited probe sequence and sources. Gene diagnosis provides early rapid diagnosis for some important diseases that are difficult to be diagnosed by traditional medicine. For instance, gene diagnosis in infectious disease can detect the presence of not only growing pathogens, but also the latent pathogens. Taking another example, the early diagnosis of tuberculous meningitis is very important for treatment and prognosis. The process of conventional culture and identification of *Tuberculosis Bacili* needs a few weeks, however, applying PCR in diagnosis takes only one working day.

10.2.2 Gene Therapy

From the end of 1980s to early 1990s, with the development of gene recombinant technology, people began to try to use gene therapy to treat genetic disease, malignant tumor, cardiovascular disease, diabetes and some occupational diseases. All of those diseases result from gene defect, mutation or abnormal expression, and have not been treated satisfactorily. In June 1990, the National Institutes of Health (NIH) of United States used double copy of retrovirus as carrier and inserted adenosine deaminase(ADA) gene as mini gene into them. The transfected cells were administrated into one 4-year-old child patient with ADA defects, and the patient felt much better. The first success of gene therapy greatly inspired the confidence of scientists, therefore, scientists around the world launched their research on this new treatment-gene therapy. There is another patient in Canada with family hypercholesterolemia caused by defected LDL gene that is not able to efficiently remove LDL from blood, resulting in the overaccumulation of LDL in blood and heart damage. After receiving gene therapy for 5 months, the patient's blood cholesterol levels dropped 30% to 50%, and finally maintained at the level of 17%-20% lower than those in serious illness. The breathtaking effect of gene therapy greatly promotes its research progress. In 1991, there were more than 20 programs regarding gene therapy running in the United States and Europe. In Asia, China has carried out the related research earlier and made good progress. Changhai hospital, the affiliation hospital of Second Military Medical University, and Genetic Institute of Fudan University have performed genetic treatment for two β-hemophilia patients and stay in a leading position.

Gene therapy is a new research field of contemporary medicine and biology. Originally, the concept of gene therapy was brought up to eradicate hereditary disease, therefore, strictly the meaning of gene therapy refers to in situ repair of gene defects or replacement by normal gene. Due to the technical difficulties, it was just the goal of gene therapy. Currently, what gene therapy can do is to transform the gene with normal function into a patient's body, exert its function, and correct the lack of protein in a patient or give the body new resistance for disease. The implementation of this gene therapy must meet two requirements: ①Mature DNA cloning technology, that is, the gene fragments of related disease are cloned, performed for DNA recombination and elements are constructed with target genes. ②Effective gene transfer method, namely, target gene is transferred into a patient's body and highly expressed. More generalized gene therapy includes the regulation of gene expression on genetic levels. At the same time, the research object of gene therapy has been spread from original genetic disease to tumour, infectious disease and cardiovascular disease, etc. The research focus has shifted to the aspects of gene therapy in tumour. Gene therapy is not only a new treatment method, but also a new pharmacology. Different from traditional pharmacology, gene therapy will transfer a special kind of active substances into human body, and express in specific space at a specific time, so as to achieve the purpose of disease treatment.

Germ cell gene therapy refers to the genetic correction of reproductive cells or early embryonic cells, since the genetic change of reproductive cells or early embryonic cells will certainly have an influence on next generation, ethical barriers and technical difficulties make germ cell gene therapy difficult to proceed. Somatic gene therapy is to use somatic cells as receptor cells, only involved in genetic change of somatic cells without affecting next generation. After years of research and discussion, it has currently been widely accepted as one of treatments for serious diseases.

10.2.2.1 Forms of Gene Therapy

1) Rectified gene therapy

Rectified gene therapy includes gene replacement, gene correction and genetic augmentation. Gene replacement is achieved through homologous recombination to replace mutated genes with normal genes. Gene correction is *in situ* repair specific for sequences of mutated genes by site-directed recombination. Genetic augmentation is to transfer the gene with normal functions into cells with genetic defect or gene lost so that the normal product will be expressed to make up the function of gene defects. Theoretically, genetic displacement and gene correction are the ideal methods in gene therapy, however, because of technical limitation, genetic augmentation is still the most commonly used method. It is most suitable for the treatment of recessive monogenic disease, while for dominant genetic disease, the application of this method has been limited to some extent because abnormal gene product may affect cell function. The same issues exist in gene therapy for tumour and infectious disease, thus at the same time as the function of defect gene is restored, actions must be taken to try to reduce the expression or overexpression of defect genes.

2) Regulated gene therapy

Regulated gene therapy belongs to more generalized gene therapy. The improved symptoms are achieved through the regulation of expression of certain genes using antisense technology and the technology to remedy the function of defective genes by the compensatory role of other genes.

There are mainly three types of antisense technologies used in gene therapy. The first one is to restructure specific antisense genes to expression vectors that are guided into target cells to transcribe antisense RNA, which binds with target RNA and form double strands to block the translation of mRNA. The second type is to artificially synthesize antisense oligodeoxyribonnucleotides that are introduced into human body after chemical modification and enter into cells by phagocytosis to combine with DNA and form nucleotide trimers, which can affect the combination of transcription factors and block the initiation of transcription, or bind with mRNA to form RNA-DNA heterochain to affect gene translation. The third one is nuclear enzyme, namely, catalytic RNA molecules that can catalyze, cut, and degrade the RNAs of abnormally expressed genes. Specific ribozymes with the structures of hammerhead and hairpin have been designed. The most important feature of ribozyme-mediated gene expression and inhibition is its specificity at cutting site. One point mutation at cutting sites is powerful enough to make the enzyme lose its cutting ability.

Antisense technology is often used in the treatment of tumours and viral diseases. The result of treatment depends to a large extent on whether it can effectively guide antisense nucleic acids, ribozymes or oligomeric DNAs into target cells.

Functional remedy of gene defects through the compensatory effect of other related genes. For example, to treat Mediterranean anemia caused by the mutation of β-globin chains, people try to increase the synthesis of γ-globin chain by activating the methylation-inactivated γ-globin gene and form fetal hemoglobin (HbF,$\alpha_2\beta_2$) with excess of α chains, hoping to make up the function or compensate the synthesis defects or deficiencies of β-globin chains of patients to finally relieve their clinical symptoms.

10.2.2.2 Gene therapy pathway

There are two ways to implement gene therapy. One is *ex vivo*, the other is *in vivo* method. The *ex vivo* one refers to choosing appropriate target cells to modify genes and culture *in vitro*. The cells with target genes transferred and expressed are screened, and then transported back to patients' body. The principles of selecting target cells are mainly based on the following aspects: ①the target cells have to be easily removed from the body and conveniently and effectively returne to body; ②the target cells can proliferate in a large amount *in vitro* by using conventional cell culture method; ③exogenous genes can be effectively transferred by gene transfer technology that is currently available; ④after returning to body, the target cells can express exogenous genes effectively and continuously. Of course, the target cells should be selected by considering patient's condition, exogenous gene and the way of transfection, etc. Currently, the commonly used target cells are lymph cells, hematopoietic stem cells and fibroblasts, liver cell, muscle cells and tumour cells, etc. The *in vivo* method, called in vivo direct transference. It refers to directly transferring the target gene into target cells or tissues through carriers, and making it expressed in the body. The *ex vivo* method is more traditional, safer, and more easily controlled, requiring less for the efficiency of gene transference. It has disadvantages of involving in numerous steps and complicated and difficult technologies. The *in vivo* method is easy to operate and, generalize, but requires advanced gene transfer technology. At present, *in vivo* method is not matured yet and has the problems such as short effect, immune rejection and safety, but it is the research direction for gene therapy. The development of gene therapy toward clinical depends on the maturation and perfection of *in vivo* method. Therefore, to find ideal gene transfer vector is the key for gene therapy to be employed in clinical. The gene transfer vectors used in gene therapy mainly include viral vectors and non-viral vectors. In most cases, viral vector has higher transfer efficiency, but encounters some problems such as immunogenicity of viral product, the targeting of gene transfer, and time and level of gene expression. Generally, non-viral vector has no toxicity, but the efficiency for *in vivo* gene transfer is low. In recent years, the viral vector has become the main ways of gene transfer, especially for *in vivo* study.

One of most important aspects of the efficacy and safety for gene therapy is the *in vivo* regulation of gene expression. The so-called targeting issue in gene therapy is whether the exogenous genes can be accurately and effectively guided into specific cells and tissues and effectively expressed in the body, namely, the accurate locations of genes *in vivo* in terms of space and time and the regulation of gene expression levels. That is the key issue in the application of gene therapy, and thus has become a hot topic in the research field of gene therapy.

Some viral infections seriously harmful to human health, such as viral hepatitis, AIDS and T cell tumors caused by infection of type I T lymphocyte virus (HLV-1) are the research object of gene therapy. Currently, great efforts have been made into the gene therapy of AIDS, mainly by transferring the appropriate genes to target cells to interfere with the regulation of HIV gene expression.

10.2.2.3 Prospects of gene therapy and existing problems

The success of clinical trials in gene therapy has brought out a revolution in human medicine. Thereafter, gene therapy has been developing rapidly and become one of popular research topics in life science. The study object of gene therapy has been spreaded from the original genetic diseases to tumours, infectious diseases and cardiovascular diseases, etc. The thought of gene therapy is continuously developed, not only involving in the addition and replacement of normal genes, but also involving in the positive or negative regulation of genes in the body, even guiding the genes that do not exist into the body. In addition, for patients with genetic diseases, immune deficiency or in tumor incubation period caused by genetic defects, preventive gene therapy can be conducted based on gene diagnosis by understanding patients' family history. Therefore, it can be concluded that gene therapy is a new way of prevention and treatment of diseases, which has potential application value and prospect for treating diseases that are seriously harmful to human health.

Although gene therapy advances very rapidly, investigators still face many problems. Especially, gene transfer efficiency is low, the vector itself also has some disadvantages, and the expression of guided gene is difficult to control, etc. Since 1995, the United States Recombinant DNA Advisory Committee (RAC) has strictly examined more than 100 gene therapy clinical trials. The results have shown that only ADA gene therapy and TK/GCV treatment for brain tumour exhibited partial effectiveness, the effect of gene therapy for others are difficult to be judged due to mild symptoms with selected cases and lack of appropriate comparisons. In 1999, the failure of a gene therapy clinical trial conducted in the Institute of Human Gene Therapy of Pennsylvania University cast a shadow over prospect of gene therapy. Most gene therapy researchers have agreed that, currently, gene therapy must return to the basic research: ①study the pathogenesis of diseases, and further understand the critical genes causing diseases and their mechanism of action; ②make better gene transfer vectors, improving efficiency of gene transfer, targeting and safety; ③study thoroughly gene expression and regulation, making efforts to accurately control the gene expression of exogenous genes in space, time and levels; ④establish optimal experimental animal models, providing more reliable data for clinical applications. As a new treatment, gene therapy will bring good fortune to human beings.

10.3 Genetic Engineering Vaccine

Before the emergence of recombinant DNA technology, there are two types of vaccines available to human, one is the pathogen processed chemically, namely the inactivated vaccine, the other is the attenuated virus or bacteria, which no longer makes people sick, namely the attenuated vaccine. Both vaccines play an important role in disease prevention through the effect of existing surface antigens on B lymphocytes. Many vaccines have been developed for a long time, however, the vaccines that truly work well are limited due to the following reasons: ①there are too many types of pathogens which mutate continuously and appear as new subtypes with poor cross immunity effect, such as influenza vaccine; ②the antigenicity of pathogens is too weak, and immunogenicity is even worse after being inactivated, such as *vibrio cholera* inactivated vaccine; ③pathogens are relatively complicated with different life cycles and antigenicities, making it hard for preparation of vaccines, such as malaria vaccine; ④so far some pathogens have not been able to be artificially cultured, so that the related research work is restricted, such as hepatitis B-virus; ⑤attenuated vaccines have the potential risks of restoring infection activity. Every year there are a few children who get infected with polio due to the inoculation of poliomyelitis vaccine.

The emergence and development of gene engineering technology make vaccine research enter into a new stage. It has solved above problems and difficulties to a certain extent.

The general methods and procedures to develop genetic engineering vaccines can be divided into two parts: screening and verification of protective epitopes, and cloning and expression of a particularly coded gene.

At present, one problem that we need to pay attention to is that, the immunities induced by several promising antigen candidates screened can only partially function. How to use different immune functions of different antigen epitopes from different molecules to improve protection is possibly the problem that still need to be solved. It is envisaged that by combining the amino acid sequences on functional epitopes of different antigen molecules and designing synthetic joint epitope peptides, polyvalent composite vaccines can be produced.

The research progress in genetic engineering vaccines of hepatitis β, tumours and schistosomiasis are presented below. Nucleic acid vaccines generally belong to the category of genetic engineering vaccines and are also introduced here.

10.3.1 Nucleic Acid Vaccines

Since the 20th century, the vaccine immunizations have undergone two major innovations. The first one is the attenuated or inactivated vaccines developed by Pasteur and others, the second one is to use natural ingredients in intact organism or subunit vaccines. Although they improve the immunities to certain extent, but still have issues of lack of safety and greater risks, especially for patients with lower immune function. Since genetic engineering vaccines could not truly reflect more complicated space structures of antigen proteins, thereby, only induce humoral immune response and may be hard to induce cellular immune response which is important to prevent some microbial infections. Therefore, nucleic acid vaccines emerge with the development of gene therapy and transgenic technology.

Nucleic acid vaccines include DNA vaccines and RNA vaccines. They are made by gene fragments encoding pathogen antigens that can cause protective immune response and vectors, and mainly brought into the body by direct injection. The nucleic acid vaccine injected into body does not integrate with host chromosome, but can express proteins, and then induce various immune responses including humoral immune response and cellular immune response.

Currently, the name of nucleic acid vaccine has not been unified. It was originally named as genetic immunization by some scholars. In 1994, the World Health Organization (WHO) suggested using nucleic acid immunization to replace genetic immunization. At present most studies use plasmid DNA as immunization molecules, so Whalen and others consider that DNA-mediated vaccine or nucleic acid vaccine is more accurate. In addition, some scholars use the name of naked DNA vaccine or naked RNA vaccine.

Nucleic acid vaccine can be inoculated through a variety of means and pathways to appropriate location of the body. The different inoculation methods or ways may have an impact on immune effect. Fynan and others compared systematically the nucleic acid immune effects using different inoculation methods, the flu virus nucleic acid vaccines were injected into the epidermis of mice by gene gun, and simultaneously nucleic acid vaccines were directly injected into control mice. The results showed that the immune effect is 600-6000 times better when the mice were vaccinated using gene gun instead of direct injection of nucleic acid vaccine. Obvious immune effect can be obtained at the dose of 100-200μg nucleic acid vaccine when using regular intramuscular injection, while only 0.04-0.4μg purified DNAs were needed to achieve the same results when using gene gun to immunize mice. The high immune effect of gene gun may relate to its highly efficient transformation.

In 1986, Benvenisty and others studied the absorption, distribution and expression of DNA in different tissues by directly injecting radioactive isotope labeled DNAs to newborn rats. After intraperitoneal injection of DNA 40 hours, the radioactivity was firstly found in the intestine and liver, and then in all organs, of which the liver has higher radioactivity, 48 hours after injection, the degradation products of DNA were found in urinary bladder. In addition, some scientists found that when PRs-CAT plasmids containing chloramphenicol acetyltransferase (CAT) were injected into mice quadriceps femoris, CAT activity was detected 48 hours after the DNA injection or 18 hours later after RNA injection. Based on above observations they speculated that directly bringing the genes encoding antigenicity proteins into human muscle tissue may obtain the same immune effect as inoculation of antigenicity material. In 1992, Tang and others further proved that direct injection of human growth hormone gene can express protein in animal and induce humoral immune response, which drew wide attention in the academic community. The report by Ulmer and others was considered to be classic in nucleic acid vaccine research. Through direct intramuscular injection of plasmid containing gene sequences encoding influenza virus core protein into mice, the mice received protective immune response. Their research work started a new era of gene immunotherapy and brought out the third vaccine revolution.

In 1992, Rabinovich and others called the recombinant plasmid consisting of gene fragments encoding some antigens and vector plasmid as nucleic acid vaccine. When nucleic acid vaccine was injected into the body, exogenous antigen may persistently be expressed and stimulate the body to produce immune response. Since 1993, some laboratories have constructed and studied nucleic acid vaccines of many pathogens, such as hepatitis B virus surface antigen (HBsAg) nucleic acid vaccine, Hlv-gP160 nucleic acid vaccine, mycoplasma nucleic acid vaccine, DNA rabies vaccine, mycobacteria DNA vaccines, *Plasmodium yoelii* circumsporozoite protein (pyCSP) nucleic acid vaccine, etc. Currently, the nucleic acid vaccine research has entered the pre-clinical testing phase.

Nucleic acid vaccine can provide immunity for host and has the following advantages. ①After a host is immunized with nucleic acid vaccine, the target gene can continually express exogenous antigens in the body and present to lymphocytes through antigen presenting cells (APCs); then stimulate the body to produce humoral immune response and cellular immune response, so that the body can obtain immunity, The effectiveness of vaccine are the same as natural infection of pathogen. ②Many subunit vaccines can not induce cytotoxic T lymphocyte (CTL), while nucleic acid vaccine can induce CTL, so nucleic acid vaccine can prevent intracellular infectious diseases. ③Nucleic acid vaccine can not integrate with host chromosome DNA, but can be expressed in host cells, that is to say, nucleic acid vaccine can play the role of vaccine with no risk. ④Single inoculation can induce long-term even lifelong immunity, and don't need to strengthen the inoculation. ⑤Mutation or throwback does not occur in nucleic acid vaccine, causing no pathogenic harm to host. However, attenuated live vaccines may have atavism or mutation any times, therefore, it's not safe to use them in some patients, especially in patients with immunodeficiency diseases. ⑥Research has shown that taking up DNA plasmid doesn't make host produce anti-DNA antibodies, and won't cause autoimmune diseases. ⑦The nucleic acid vaccines don't need to be stored at low temperature, and can be prepared in powder, which is convenient to store and use. However, many subunit vaccines need to be kept at low temperature, limiting their extensive use.

Although there are a lot of advantages for nucleic acid vaccine, there are also some disadvantages and safety issues, for example: ①the ability to produce immune response stimulated by nucleic acid vaccine is usually weaker than the immune reaction caused by natural pathogens infection; ②the expression level of nucleic acid vaccine target gene is not satisfactory; ③long-term low level expression of exogenous antigen has the possibility of causing immune tolerance. Further research is needed to find out whether or not exogenous DNA can produce anti-DNA antibody and lead to autoimmune diseases; ④potentially, DNA has risks of integrating into host chromosome; ⑤how long nucleic acid vaccine will last and express proteins in the body remains to be studied. What is described above presents new challenge for nucleic acid vaccine research.

10.3.2 Cancer Vaccines

In the early 20th century, Ehrich firstly suggested developing cancer vaccine to be used for active-specific immunotherapy for cancers. The development of molecular biology and emergence of genetic recombinant cells provide new means for the change of cancer cell types. The use of gene transfer technology has successfully brought a variety of exogenous genes into cancer cells, and to a certain extent, improving the immunogenicity of cancer cells. The availability of cancer genetic engineering vaccines has brought new hope for human cancer treatment. Some vaccines have already entered into the stage of clinical treatment. The rapid development of DNA vaccine research makes it possible to open up new way to cancer treatment.

According to the different mechanisms of producing immunogenicity, the study of cancer vaccine can be divided into the following three categories:

1) Naked DNA vaccine

Since Wolff and others reported naked DNA technology in 1990, gene immune method with plasmid as carrier to directly bring coded sequences and essential control elements of DNA into tissue has been aggressively studied in experimental research of cancer treatment.

2) Carrier-mediated cancer vaccine

Since virus itself can induce strong immune response, early in 1977, Wallack made cancer vaccines for active immunotherapy by infecting cancer cells with pox virus. In recent years, along with the development of cytokines and genetic engineering technology, the viral carriers that can express a variety of cytokines and cancer antigens have been developed successfully. Especially, the preparation of cancer vaccines using recombinant poxvirus carrier to express exogenous genes has shown remarkable effectiveness of cancer treatment.

3) Gene-modified cancer vaccine

At present, little is known regarding cancer specific antigen molecules, and the preparation of antigen-specific cancer vaccines faces numerous challenges. Currently, cancer cells themselves are normally used as cancer vaccine. By changing its genetic background through genetic modification, tumorigenicity is reduced, and accordingly, the immunogenicity is improved. Gene-modified cancer vaccines include MHC molecules, co-stimulus signal B7 molecules and various cytokines gene-modified cancer vaccines.

10.3.3 AIDS Vaccine

In 1981, the first case of acquired immune deficiency syndrome (AIDS) was reported in the United States. Thereafter, AIDS spreads rapidly throughout the world. In 1985 the first HIV/AIDS patient was found in Zhejiang, China. AIDS has become another great challenge following cancer that current medicine and biology must face. The pathogenic agent of AIDS is HIV virus, which belongs to *Lentivirus, Retroviridae*. The research and deveoplment of HIV vaccines have been going on for more than 20 years similar to other viruses, because HIV virus has the characteristics of many morphologies and constant shifting, making it hard for development of AIDS vaccine. But this doesn't mean that there is no future for HIV vaccines. Instead, many clinical cases show that the prospects for development of HIV vaccine are promising. For example, for many years people have realized that some high-risk groups have not got infected although long-term exposure to HIV; some patients infected with HIV-2 can produce cross protection for HIV-l; certain newborns can automatically scavenge HIV virus in the body through an unknown mechanism. What's more noticeable is that some AIDS patients' immune system can effectively restrain HIV virus for as long as 10 years. Currently, many believe that among HIV vaccines, DNA vaccine has better future. The U.S. Food and Drug Administration (FDA) officially approved to carry out a large-scale human clinical trial of AIDS DNA vaccine in June 1998, and several other therapeutic vaccines were also under the trials. For example, the injection of the complex of gp120 and gp41 isolated from HIV virus to AIDS patients can produce specific antibody, thus increasing the amount of antibodies to attack viruses. The AIDS vaccine developed by China was officially launched for human trials in March 2005.

10.3.4 Hepatitis B Vaccine

Hepatitis B (HB) is one of the severest viral hepatitis. The severe Hepatitis B and some chronic Hepatitis B diseases can aggravate to liver cirrhosis or liver cancer, and endangering patients' life. The ultrastructure of Dane particles and the molecular structure of hepatitis B virus DNA have been clearly understood (Fig.10.1). The surface of Dane particles is covered by one type of protein, called surface antigen (HBsAg). HBsAgs consist of three types of proteins encoded by three different genes, respectively. ①Small surface protein (S protein). The polypeptides composed of 226 amino acid residues are encoded by S gene, which is the main component of HBsAg and HBV envelope, and is also the main protein of HBV. ②Medium surface protein (M protein). M protein is composed of S protein and former S2 protein (polypeptide containing 55 amino acid residues encoded by former S2 gene), which has one human serum albumin (pH value SA) poly-receptor sites. ③Large surface protein (L protein). L protein is composed of M protein and former S1 protein (polypeptide containing 108-109 amino acid residues encoded by former *s1* gene). HBsAg antigen has strong immunogenicity and can be used to prepare for vaccines; however, blood-borne antigen (PDV) is difficult to meet the needs of mass vaccination. After HBsAg genes were fully understood, they were first expressed in *E.coli*, and proteins were extracted. But the expression yield was low, and the immunogenicity of the product was weak, and it was later turned to yeast and Chinese hamster ovary (CHO) cells expression system.

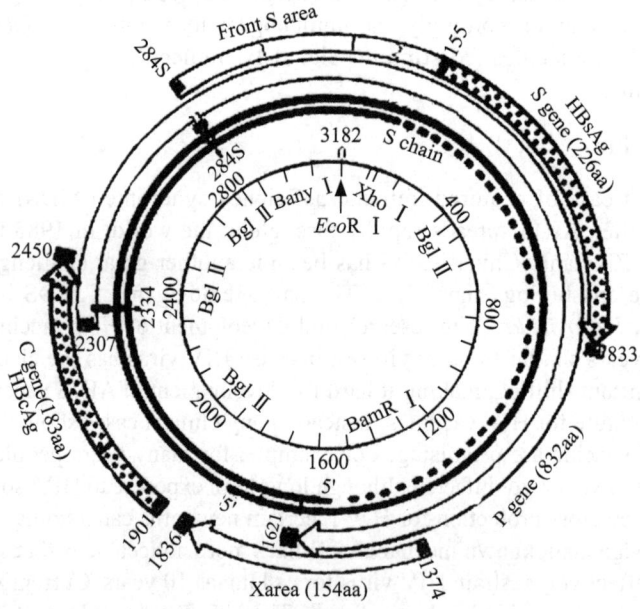

Fig.10.1 The structure of HBA DNA

10.3.4.1 Yeast expression system to prepare for Hepatitis B vaccine

After yeast expression system of genetic engineering vaccine was established, researchers inserted 31kb of HBsAg genome DNA into yeast expression vector, and transformed the plasmid into yeast *Saccharomyces cerevisiae*. The transcription is promoted by strong promoter encoded for glycerin aldehyde dehydrogenase (GAPDH) *I* gene. The vector contains DNA replication start points with both bacteria and yeast (shuttle plasmid), combining with ADH-1 as end point to generate gene cassette, which can be transformed into *E.coli* or yeast for replication and expression. When above expression vectors entered into yeast, yeast cells in fermentation tank reached high density, producing a lot of viral proteins similar to natural polypeptides, which was almost 1%-2% of total protein amount in yeast. Recombinant protein can form protein polymers (20nm in diameter) with the same properties as the immunogenic accumulation body in patients with hepatitis B. Through the SDS-PAGE analysis, the polypeptides presented the band with relative molecular weight of 23,000Da, but lacked the saccharification band with relative molecular weight of 28,000Da (Fig.10. 2).

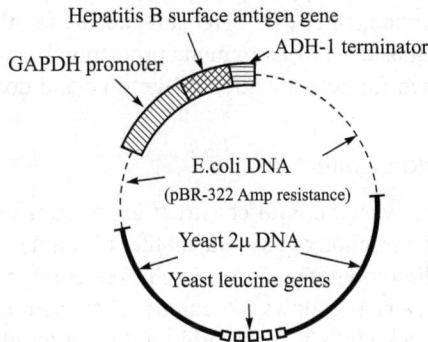

Fig.10.2   Hepatitis b antigen express in recombinant yeast cell

The production process of Hepatitis B vaccine by yeast is roughly listed as follows:

Yeast fermentation in large tank→cell disruption, antigen recovery→ adsorption of antigen by silica gel→hydrophobic chromatography→gel filtration→formalin treatment→$Al(OH)_3$ adsorption→ antisepsis by thiomersal→ dispensing.

10.3.4.2 Chinese hamster ovary (CHO) cells expression system

The establishment of CHO expression system is to construct plasmid by inserting cDNA of dihydrofolate reductase (DHFR) from mice into HBV sequence, transfecting DHFR-deficient CHO cells-to screen for DHFR positive cells. The recombinant CHO cells containing DHFR expression vector are cultured and screened through methotrexate (MTX), and can be cloned to efficiently express HBsAg. Those cells grow fast and can be cultured in monolayer or suspension, and the secreted HBsAg vaccine has the advantages of strong immunogenicity, simple process in purification of antigen, which is suitable for large-scale industrial and continuous production.

The efficacy testing has shown that the immunogenicity of CHO cell vaccine is stronger than that of PDV, and the antibody titer produced by chimpanzees vaccinated with CHO cell vaccine is higher than that of yeast recombinant HB vaccine. The CHO vaccines (10μg/dose) or PDV (20μg/dose) were inoculated to healthy medical staff by conventional method. No significant differences in anti-HBs response between the two groups were shown one month after inoculation of three doses; six months later, the anti-HBsGMT in CHO cell vaccine group was 7 times higher than that in PDV group; seven months later, 3-6 times higher in CHO cell vaccine group; and 24 months later, positive turning rates of anti-

HBs in CHO cell vaccine group and PDV group were 95% and 71%, respectively. Obviously, the protection efficacy of CHO cell vaccine group lasted longer. At the same time, the ratio of no response or low response to anti-HBs in CHO cell vaccine group was remarkably less than that in PDV group, which showed that the immunogenicity of CHO cell vaccine is stronger than that of PDV; however, there is no difference in side effects between two vaccines.

For many years, the research in CHO cell vaccine has made great progress. At present, CHO cell vaccine has already been used in France to replace PDV and promoted for a wide range of use. In Israel, Singapore and other countries, CHO cell vaccine has been widely used for humans. It can be predicted that CHO cell vaccine will gradually replace PDV and be accepted in a wide population. However, although a lot of researches have been done on CHO cell vaccine production, immune sequence, dose and immune response, etc., and plenty of resources have been accumulated except for chronic uremia patients, the immunogenicity of CHO cell vaccine to other types of diseases remain to be further discussed. And the persistence of its immune protection requires more in-depth research, in order to provide reliable evidence for determination of the time and dose of inoculation to strengthen immunization.

10.3.4.3 Poxvirus expression system

Recombinant HB pox virus vaccine is to construct expression vector by the recombination of plasmid containing P7.5 poxvirus promoter gene and another plasmid containing HBsAg gene. HBsAg are secreted out of cells via proliferation of pox virus in chicken embryo cells at the expression level of 1-2μg/mL. Its production process is as follows: co-culture of chicken embryo cells with recombinant poxvirus; harvesting culture supernatant after viral proliferation; after ultrafiltration and concentration, performing KBr and sucrose centrifugation; pepsin digestion and formaldehyde inactivation; addition of adjuvant, and final dispensing.

Poxvirus has played an important role in the prevention of smallpox. Although smallpox has disappeared, poxvirus, as an effective carrier which can express a variety of exogenous genes in eukaryotic cytoplasm, has drawn great interest for nearly 10 years. The application of recombinant DNA technology has made it possible to construct recombinant virus that can express one or more genes in exogenous microbes and maintain its infectious properties to a variety of host cells. There are many advantages for poxvirus expression system, with similar infectious property to parent strain, it can stimulate not only humoral immune response but also cellular immune response. It possesses high capacity of inserting an exogenous gene, thus to develop combined vaccines simple preparation without causing carcinogenicity, easy for mass production; with good thermal stability, it's convenient for storage and transportation. Although most of recombinant viruses have been proved to be effective by animal experiments, and be able to induce optimal immune protection, there are still some problems in using smallpox virus as a carrier for genetic engineering inactivated vaccine, such as nerve toxicity for subjects and other side effects, the repression effect of preexisting human anti- poxvirus antibodies to vaccines, etc.

10.3.5 Other Vaccines

Due to the change of environment that humans live in and the wide use of drugs, the diseases caused by the resistance of pathogenic microbes become difficult to treat with existing drugs, so scientists are trying to treat the diseases in ways of manipulating immune systems.

10.3.5.1 Anti-bacterial vaccine

Since the immune activities of antigens outside the bacterial cell surface is stronger than that of intracellular antigens, people come up with the idea that antibacterial vaccines can be obtained by inserting pathogenic surface antigens into the cell surface of non-pathogenic bacteria, for example, bacterial flagellum. Flagellum consists of flagellum protein, which is the common outer surface structure of many bacterial surfaces. If the flagella of non-pathogenic bacteria carry some specific antigenic determinants of pathogenic bacteria, then vaccine recipients can easily get protective immunity. Cholera vaccines are made using this method.

10.3.5.2 Falciparum vaccine

Malaria is one of major parasitic diseases threatening human health, especially in the developing countries where the disease incidence is high. *Plasmodium falciparum* is the most harmful pathogen. Because of its high infection, rapid proliferation and high fatality rate, there are no effective preventive measures available so far.

Plasmodium is well adapted to the parasitic mode of life in human body, and has many ways of escaping the attack from human immune system. For example, in most of its life cycle, plasmodium lives inside the cells of human body, staying extracellularly in very short time and entering into the cells quickly, thus can avoid the attack of antibodies and specific killer cells. Moreover, plasmodium at each liberation period (sporozoite, merozoite, and gametophyte) has different surface antigens with weak immunogenicities; even if having induced the host's immune response, due to the late-occurred antibody, the attack of antibodies falls behind the changes of antigens. In addition, the antigens of the same Plasmodium usually have high degrees of polymorphism. In the study, according to the life cycle of plasmodium and its ways to avoid immune attack, researchers obtained plasmodium vaccines from the following three aspects: ①inducing the production of antibody against circumsporozo-ite protein (CSP) by vaccine receptors; ②inducing the production of antibodies against merozoite surface antigen by vaccine receptors, and antigen related to erythrocyte invasion as well as surface antigens of infectious red blood cells; ③surface antigens of gametophyte or surface antigens of zygocyte are used as vaccine components to stop the gametophyte integration or zygote development so as to block the spread of plasmodium falciparum.

10.3.5.3 Leprosy vaccine

The current leprosy vaccine is prepared by adding BCG to inactivated *Mycobacterium leprae*. Clinical experiments have shown that most patients could produce cellular immunity to *Mycobacterium leprae*, causing *Mycobacterium leprae* to greatly decrease in numbers in the body or even disappear. However, because of its poor repeatability of the results, the vaccine is still in the research stage.

10.4 Application of Contemporary Biotechnology in New Drug Development

10.4.1 Screening Model for Drug Innovation

The process of new drug development is roughly shown in Fig.10.3. The first step in new drug discovery is to look for lead compounds through all sorts of optimal screening models, thereby, the establishment and application of screening models are very important. The traditional screening model is often established at intact cell levels, for example, direct screening for anti-infection drugs using a variety of pathogenic microorganisms as targets, or screening for anti-tumor drugs in a variety of *in vitro* tumor cell lines, but it is getting more and more difficult to obtain new lead compounds. With the purpose of discovering new drugs more efficiently, the application of gene engineering technology to screening new drugs can make drug screening fast and reasonable. Along with the advances in microbial genomic research program, and the discovery of disease related genes, a batch of new drug screening targets have been established. With the progress of research in proteomics, and the perfection of cloning techniques, the higher-level three dimensional structures and active center of enzymes and receptors related to many important diseases have been continuously revealed and clarified. Therefore, the use of gene engineering technology can produce a great quantity of targeted enzymes and effective parts of receptors to be used for screening, and providing a solid foundation for the development of high throughput screening (HTS). Currently, it is very common to screen tens of thousands of samples in a week using HTS technique. The practical results indicate that the application of this method makes new drugs screened at the molecular levels with high positive incidence, which greatly accelerates the pace of new drug development. Molecular biology and gene engineering are playing more and more important roles in new drug research and development.

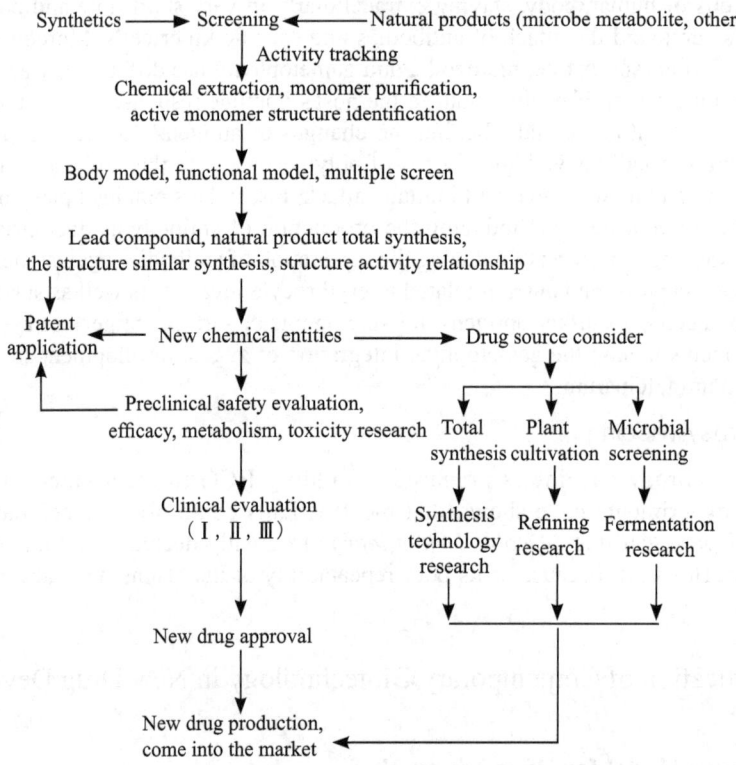

Fig.10.3 The process of research and development of new drugs

10.4.1.1 Targeted enzymes

Many diseases are caused by functional disorder of enzyme, therefore, screening of new drugs targeting at enzymes is a very important means. The eventual treatment can be achieved by inhibiting enzymatic activity to repair the disorders of regulation mechanisms, or promoting enzymatic activity to restore the normal physiological process. In research and development of enzyme preparations, gene engineering technology can be used to clone and express the genes encoded for target enzyme in order to eliminate the interferences of tissue homogenate on other enzymes and produce enough target enzymes. For instance, the eukaryote rat DNA polymerase β has been successfully expressed in recombinant engineering bacteria, which provides plenty of target enzymes for large-scale screening of DNA polymerase inhibitors. It is well-known that many anticancer drugs can cause DNA damage in cells, but this damage can be repaired under the effects of 3', 5'-endonuclease, DNA polymerase, 5',3'-exonuclease and ligase, respectively. Therefore, if the inhibitors for DNA repair system can be screened, it is possible to develop low toxicity anti-cancer drugs.

10.4.1.2 Receptor

Receptor refers to biological macromolecules existing in cell nucleus, such as glycoprotein, lipoprotein and nucleic acid, etc. The specific sites in its structure can accurately recognize and selectively bind to certain specific ligands. Ligand refers to biologically active material which can selectively bind to receptors, including endogenous substances (such as neurotransmitters and hormones) and exogenous active substances (such as drug). With the development of molecular biology and more understanding about receptor, such as its molecular structure, functional mechanism, signal transduction mechanism, and receptor interactions, the drug screening by using receptors can be more close to the physiological approach of drug function. For example, human immunodeficiency virus (HIV) is the pathogen of human AIDS. The diameter of this RNA virus is about 100μm, and the viral envelop consists of bilayer lipid molecule and two glycoproteins gp41 and gp120. In addition to two types of multiple copies of protein subunits p17/18 and p24/25 in the center of virus, there are two worm-like RNA genomes and molecular reverse transcriptase. The infection of HIV virus with human body firstly starts by specifically binding of its surface glycoprotein gp120 to surface protein CD4 on helper T cell (namely T4 cells), subsequently leaving gp120 at the cell while the virus enters into T4 cells. Under the effect of reverse transcriptase, the viral DNAs catalized by viral DNA polymerase are produced using its RNA as templates, and then integrated into host genome DNA to replicate and proliferate. Since gp120 stays on the surface of CD4 T cell, the infected CD4 T cell can act as HIV virus to infect another CD4 T cell to form a large polynuclear syncytial mass. The people with infected CD4 T cells that have lost their immune functions suffer from to AIDS. Therefore, using the combination of gp120/CD4 as target to screen for antagonists provides the possibility of obtaining drugs for AIDS treatment.

Gene engineering technology is applied to the expression of gp120 genes on the surface of Burkitts lymphatic sarcoma cells. Those cells are mixed with human T cells as well as tested samples, if Burkitts lymphatic sarcoma cells bind to CD4 on the surface of T4 cells, polynuclear syncytial mass can be observed, showing that there is no binding antagonist existed in tested samples; otherwise, it demonstrates that there is gp120/CD4 binding antagonist in tested samples. The recombinant gp120 obtained by genetic engineering techniques can also be used for binding with CD4. The antagonism of test samples are detected using ELISA to screen for gp120/CD4 binding antagonist.

The receptors of β-adrenaline and 5-hydroxytryptamine (5-HT1A, 5-HT1D) have been successfully expressed in *E.coli* or yeast, having confirmed that the functions of these receptors are exactly the same as those from mammalian tissues.

10.4.2 Development of New Drug Resources

Since British scientist Fleming discovered penicillin in 1928, tens of thousands of antibiotics and various physiological active substances produced by microbes have been publicly reported. But in the future, it will be very difficult to find new microorganism drugs by only relying on the traditional screening models and measures. Therefore, opening up new sources and exploring new species of various organisms have already become the consensus of medical researchers from many countries. In recent years, several world-famous pharmaceutical companies have made huge investments in this aspect financially and physically to expect for new findings.

Different microbial populations live in different environments. In order to adapt to the surrounding environment during evolution, microbes are likely to form different *in vivo* metabolic pathways from those living in ordinary environment. Therefore, the chance of producing new microbiological drugs has greatly increased. At present, medical scientists in various countries of the world are paying more and more attention to microbial populations living in extreme environment such as microorganisms in Antarctic and Arctic, under high temperature in crater, in deep sea and in space as well as living symbiosis or parasitically with plants or animals.

With continuous development of contemporary biotechnologies such as gene cloning technology, cell culture, and DNA amplification, some research institutes begin to clone the total DNA of soil microorganisms that are hard to isolate into pre-set host system, which can be used to produce various "industrial strains" carrying unknown microbes DNAs. Following the fermentation, primary and secondary screenings of "industrial strains", new compounds are expected to be discovered. There is no doubt that the approach to using cloning technology to explore the potential of unknown soil microbes is encouraging for the benefits of humans, and also providing an alternative way for the advent of new lead compounds produced by new microbes. The premise of conducting the research in this area is to make sure that there are no gene contaminations involved. Especially for those "industrial strains" with unknown genes, appropriate management measures must be strengthened to prevent unpredictable contaminations and risks from bringing to human society.

In addition, isolation and investigation of endophytes are new effective ways to expand medical microbial sources, which has attracted great attention of scientists in many countries. The well-known antineoplastic drug paclitaxel can't be widely used due to the lack of resource. American scientists isolated several strains of microorganism from the bark of taxus, one of which can produce trace amounts of paclitaxel as confirmed by screening results, paving a solid foundation for expansion of taxol sources. The isolation of endophytic microorganisms from plant with medical values will inevitably draw more and more attention of researchers.

The 60.9% of the Earth's surface that human beings live on is occupied by ocean, more than 2/3 of global biological species belong to marine organisms. The large variations of marine environment result in tens of thousands of marine biological species, which provides abundant resources for the research and development of new drugs. Taken marine microbes as an example, they are remarkably different from terrestrial microbes in terms of living environment, not only tolerant to salt but living in the extreme environment such as high pressure, lack of (no) light and oxygen, and cold climate. How to successfully isolate, culture and perform drug screening in the laboratory closely depends on the application of biotechnology.

10.4.3 Developing New Sources of Lead Compounds

In order to constantly obtain more new lead compounds through various screening technologies, it's not only necessary to make efforts on targets, but also to provide a large number of compounds available for screening. The establishment of compound library through various technologies has already become one important task for researchers in many countries.

10.4.3.1 Combinatorial biosynthesis technology

With the advance of microbial genome research and the application of combinatorial chemistry principle, combinatorial biosynthesis technology has achieved rapid development in recent years. The so-called combinatorial biosynthesis refers to directional synthesis of a series of new compounds based on the understanding of biosynthetic pathway of microorganism and cloning the genes related to biosynthesis and regulation, as well as subsequent *in vitro* deletion, addition, replacement and recombination of genes from different sources (interspecies or outerspecies), which are then put into an appropriate microbial host. With the continuous increase in numbers of biosynthesis gene blocks discovered, all kinds of unnatural or natural compounds can be designed through combinatorial biosynthesis technology. At present, combinatorial biosynthesis technology is mainly used in some important antibiotics with polyketide biosynthesis pathway.

Among the microbial medicines we have ever found, there are many drugs produced through polyketide biosynthetic pathway. The known polyketide compounds have reached more than 10,000, of which the compounds used as drugs for treatment have exceeded 10 billion dollars in global annual sales, including those important antibiotics in clinical applications, such as erythromycin, tetracycline, rifamycin, amphotericin, daunomycin, lovastatin, pravastatin, tacrolimus FK506, and rapamycin as well as avermectin and tylosin for agricultural use. The genes for most of polyketide synthases (PKS) have been successfully cloned with 10 years efforts. These research results have revealed the models of polyketide biosynthesis in details, including several enzymatically synthetic steps, such as the selection of initial synthetic unit, the extension of carbon chains (or the condensation of salts of acetic acid, propionic acid or butyric acid), the reduction of ketone group, aromatization and cyclization. These reactions are completed through a series of polyketone synthases and regulated by related genes. The most outstanding research on PKS was performed by the team of the most famous British *Actinomycetes* geneticist professor Hopwood. Their research on the PKS genes of actinorhodin and the studies performed by Dr. Khosla and others at Stanford University laid the foundation for the combinatorial biosynthesis.

The current study has shown that the structures of genes of polyketide biosynthesis gene can be divided into two categories, i.e., PKS I type and PKS II type. The former is mainly macrolides compounds, such as erythromycin, etc., and the latter primarily refers to rubidomycin and tetracycline containing aromatic rings. In addition, polyketide compounds themselves can be modified by hydroxylation, glycosylation or methylation. A recent study found that the gene structure of nonribosomal peptide synthetase (NRPS) biosynthesized through peptide antibiotics is very similar to that of PKS; moreover, the pattern of chain coexistence of NRPS and PKS was found in epothilone biosynthesis gene clusters. Accordingly, it expands the categories for generating new compounds by gene combinatorial biosynthesis, most likely introducing the gene units of peptide chain synthase into PKS gene cluster to broaden the physiological activity of polyketide compound and probably enhance its solubility in water, etc. Because of advances in researches on microbial diversity, controllability of microbial metabolic pathways and feasibility of genetic manipulation, combinatorial biosynthesis has rapidly become another new research focus after combinatorial chemistry in China and aboard.

10.4.3.2 Surface display technology

In recent years a new gene technology called surface display technology has extensively been utilized in the research and development of polypeptide libraries and screening for new antibodies, receptors, vaccines and other new biotech drugs. The so-called surface display technology refers to cloning of oligonucleotides or some gene fragments in genome synthesized from cDNA following combinatorial synthetic method into the specific expression vector, making exogenous peptides (or protein structural domains) displayed on the surface of phages, microbial cells or ribosomes using gene fusion expression systems. Those displayed polypeptides can keep their relatively independent spatial structures and biological activities. Surface display technology can not only be used for the construction of antibody libraries, development of subunit vaccines, etc., but also for screening and development of polypeptide as lead compounds.

Since the advent of surface display technology, it has attracted great attention of scientists due to its convenience, low cost, and large capacity of database. By using surface display technology, people can easily prepare in the laboratory for a large number of polypeptides with similar structures but ordered changes. New lead compounds with specific biological activities can be obtained from above polypeptides selected via HTS using yeast two hybrid technology and gene chip technology. It is expected that new types of antibody drugs screened through surface display technology will be officially available on the market within the next 10 years.

Chapter 11

Infection and Immunity

11.1 Infection of Pathogenic Microorganisms

Pathogenic microorganisms are defined as those microorganisms which cause infectious diseases in human beings or animals. By means of interaction between pathogenic microorganisms and the human body or animals (hosts), they alter physiological activities and functions with each other. On one hand, pathogens invade into the human body to damage cells and tissues in the host; on the other hand, utilizing various functions of immunity and defense hosts make every effort to kill, neutralize, and clear off pathogens and their metabolic toxins. Rivalry between pathogens and hosts determines progression and outcome of the whole infection course. Besides, environmental factors have great effects on infection. Whether infectious diseases can break out, therefore, depends on both pathogenic ability, or pathogenicity of microorganisms or their virulence, and resistance, or immunity of the human body.

Infection refers to growth and proliferation of pathogenic microorganisms in the specific sites after invasion into the human body so as to generate virulence. During this course, sometimes the body has no corresponding clinical symptoms, which is called as dormant infection or carrier state; and sometimes, the corresponding clinical symptoms, which is called as dominant infection. And infectious diseases specifically refer to the latter condition, i.e., infection manifesting the corresponding clinical symptoms. The human body can react as immune response for stimulation of pathogenic microorganisms. In normal conditions, immune response has the following functions: ①repelling and eliminating invaded pathogenic microorganisms, i.e., anti-infection immunity; ②excluding wastes and useless cells produced during self-metabolism, which is called self-stabilization function; ③disposing autologous cells mutated resulting from some biological or physical and chemical agents, immune surveillance function. These effects are useful for the human body. However, in some abnormal conditions, ①the human body has strong response to invade pathogenic microorganisms or other foreign substances (e.g. penicillin), which is called allergic reaction; ②self-stabilization function disorders and then autoimmune diseases occur; ③immune surveillance function disorders and then tumour occurs. The immune responses in the abnormal conditions are harmful to the human body.

11.1.1 Infection Modes and Routines of Pathogenic Microorganisms

Infection is classified as exogenous and endogenous infection. The infection originating from those agents outside the hosts is called exogenous infection, whose infection source is patients, healthy bacteria-carriers, or bacteria-carrying plants or animals. When normal bacteria flora imbalance resulting from abuse of antibiotics leads to decrease in immune functions of the body, the normal flora in hosts may cause infection which is called endogenous infection.

Different pathogens invade the human body in different routines. Most pathogens cannot permeate through the intact skin, but enter the body through the natural opening of the body and traumatic wound, or through iatrogenic routines such as guide tube, intravenous infusion and surgical incision. A very few pathogens (e.g. schistosome and hookworm) can permeate through the skin; and some (e.g. poliovirus, measles virus) can permeate through the mucosa and then arrive at the specific sites by means of blood circulation to cause pathogenic changes; and some (e.g. diphtheria bacillus) can adhere to the mucosa to grow and reproduce, in order to form one local locus, producing toxins and resulting in various symptoms. Many pathogens have characteristic of organotropy, i.e. pathogens have high selection for their infected tissues or organs, for example the hepatitis viruses invade only the liver cells while the pneumococcus species only the mucosa of the respiratory tract. In general, defense strategies for exogenous infection in the human body are most to prevent further spreading or re-infection of pathogens, while for endogenous infection it is important to destroy infection sources.

Pathogens generally infect the body in the following modes:

1) Respiratory and digestive tract infection

Many pathogens, such as *tubercle bacillus*, *diphtheria bacillus*, respiratory tract viruses, *chlamydia pneumoniae*, may be transmitted through saliva and sputum of patients or bacteria-carriers and dust with them. The food contaminated by patient's excrete is also one of the major transmission modes, and contaminated water sources, furniture and insects such as fly, cockroach, and so on are main transmission media of some intestinal tract pathogens such as typhoid bacillus and Shigella dysenteriae and hepatitis viruses.

2) Trauma and contact infection

Some pathogens, such as pathogenic *staphylococcus*, *streptococcus*, *Clostridium tetani*, leptospira and viruses (e.g. human immunodeficiency virus, HIV), may lead to infection by invading into the human body through injured skin mucosa. Some pathogens utilize blood sucking insects as media to transmit infectious diseases, which is also one mode of traumatic infection. For example, common rat flea may transmit plague; malaria mosquito, malaria; and culex, Japanese encephalitis virus. Some pathogens such as *Brucella* spp. may invade into the intact skin; *gonococcus* and *Chlamydia trachomatis* may invade into the normal mucosa; and leprosy bacillus may be transmitted by means of close mankind-mankind or mankind-animal contact or utensil contamination.

3) Vertical transmission

The procedure of pathogens transmitted through the placenta or the birth canal directly of parents to offspring is called vertical transmission of infectious diseases. Such pathogens are most common in viruses such as herpes virus, type B hepatitis virus, human immunodeficiency virus (Fig.11.1).

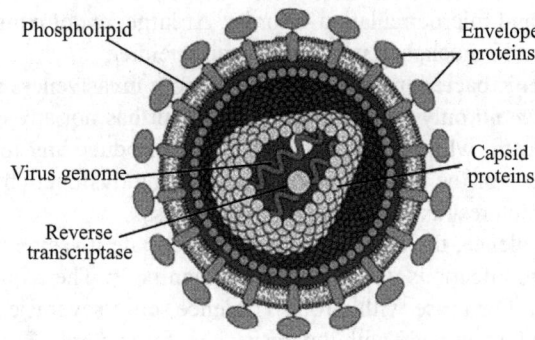

Fig.11.1 Human immunodeficiency virus (HIV)

11.1.2 Mechanism of Infection of Pathogenic Microorganisms

Pathogenicity of pathogenic microorganisms depends on their virulence, quantity and invasion routine. And virulence of pathogens depends on their invasiveness and toxin. For the specific host, bacteria causing diseases in the host is called pathogenic bacteria, and otherwise non-pathogenic bacteria. However the two have no apparent border. Some bacteria do not cause diseases in general conditions, but cause diseases in some special conditions with some changes, and then are called as conditional pathogenic bacteria. Strength of pathogenicity of pathogenic bacteria is called virulence, based on invasiveness and toxin.

11.1.2.1 Invasiveness

Pathogens have the ability to break through defensive barriers of the human body, inhabit, proliferate and spread in their hosts to damage the body's functions, which is called as invasiveness and mainly includes bacteria capsule and enzymes. The capsule provides bacteria protection against phagocytosis of phagocytes from the host. For example, type S pneumococcus with capsule has strong virulence, while its virulence decreases after losing the capsule.

Pathogenic bacteria can produce enzymes associated with their invasion into the human body during growth and reproduction, and their pathogenicity mainly depends on these enzymes, which are also called invasive enzymes. These enzymes include hyaluronidase, collagenase, coagulase, deoxy ribonuclease, lecithinase, hemolysin, streptokinase, etc.

11.1.2.2 Toxins

Based on origin, nature and functions, bacteria toxins can be classified as exotoxins and endotoxins. Exotoxins refer to one kind of metabolites, mainly protein, secreted into the environment during growth and proliferation of pathogenic bacteria, which have strong antigenicity and toxicity, and can selectively act on the specific target organs. For example, *Corynebacterium dipheriae* can produce diphtheria; *Clostridium tetani*, tetanotoxin; and *Clostridium botulinum*, botulinus toxin.

Endotoxins are lipopolysaccharides produced by Gram-negative bacteria, which exists in the outer layer of cell wall of bacteria and belong to component of the cell wall. They usually bind to the cell, and can release only if cytolysis occurs. Endotoxins act on multiple cells and systems including leukocytes, platelet, complement system, blood coagulation system and so on, so as to result in fever, leukocytosis, hypotension and microcirculation disorder. At large, endotoxins have multiple complex actions, which are not toxic too much, and not organ specific, either.

For some one pathogenic bacterium, it does not need both invasiveness and toxins for virulence. For example, *Clostridium tetani* only produces tetanotoxin but has not any invasiveness, it still can cause diseases and even death; while pneumococcus cannot produce any toxins, but can reproduce very much in the lung tissues of the human body to lead to lung dysfunction and even hosts death in case of severe infection, which results from so high invasiveness.

Apart from certain virulence, the quantity and the appropriate invasion routine are necessary for pathogenic bacteria to cause infectious diseases in the human body. The required quantity of bacteria depends on their virulence. For those with strong virulence, only several cells are enough to cause infection, especially for the human body with low resistance, for example *Bacillus pestis* requires only several cells for invasion into the body with low resistance to cause plague, However, those bacteria with low virulence require a large quantity to cause infection. For example, several hundreds of millions of salmonella cells are always required to cause food poison, maybe because a large number of cells can release a great quantity of toxins so as to overwhelm defense of the human body. Some particular pathogens need particular invasion routine, for example *Clostridium tetani* must invade into the deep wound to cause infection which is called traumatic infection, but cannot cause the infection by invading into the gastrointestinal tract orally. On the contrary, *typhoid bacillus* and *dysentery bacillus* must invade into the digestive tract orally so as to cause infection.

11.2 Immune Response in Human Body

Immunity is one important strategy which organisms develop gradually during the long evaluation course to defend infection and maintain completeness of the human body. Immune defense function of the host is divided into two major classes, nonspecific immunity and specific immunity which coordinate each other to fulfill the role of resistance on infection and protection of the body together, but sometimes can also bring about pathological damage.

11.2.1 Nonspecific Immunity in Hosts

Nonspecific immunity is the general physiological defense function of the body, and also called innate immunity; it is formed during germ development, originates from congenital inheritance, and can defend against invasion of exogenous foreigns without special stimulation or induction. It mainly consists of surface barriers, cellular barriers and humoral barriers.

11.2.1.1 Surface barriers

1) Physiological barriers

The outer surface of the healthy body is covered with continuous and intact skin structure, whose external corneous layer is firm and impermeable and forms one effective barrier for blocking microorganism invasion. Meanwhile, lactic acid in secrete of the sweat gland and long-chain unsaturated fatty acids in secrete of the skin gland both have ability to kill and inhibit bacteria. The respiratory tract, the digestive tract and the urogenital tract are all covered with mucosa, which has not strong function as one surface barrier, but various appendixes and secretes. The mucus secreted by the mucosa functions as one chemical barrier, and can prevent virus from invading into cells by competing with receptors of the cell surface for neuraminidase. When microorganism and other foreign particles fall into the mucus of the mucosa surface, the body can repel them by mechanical means such as cilia movement, cough and sneezing, and at the same time can clear them off through tears, saliva and urine. Many types of humoral secretary fluid contain antimicrobial components, such as lysozyme in saliva, tears, latex, nasal mucus and sputum, gastric acid in gastric fluid, spermine in sperm. Those persons with too much smoking and wine drinking are subjected to have trachitis, bronchitis and pneumonia due to their attenuated cilia movement.

2) Local barriers

There are special structures in some sites in the body, which can form the local barriers to prevent microoranisms and macromolecules from invading in the body, and play an important role in protection of the local organ and maintenance of constant local physiological environment. For example, the blood brain barrier, composed of the wall of the brain capillary vessel and astrocytes on the outside of it, can prevent substances in the blood including pathogenic microorganisms and its metabolic products from dispersing into the brain, so as to keep stability of the central nerve system; the placenta barrier, composed of the endometrium of one pregnant female and the chorion of the fetus together, can prevent pathogenic microorganisms from infecting the fetus through the placenta.

3) Symbiotic bacteria flora

A large number of normal bacteria floras exist in the cavity tract connecting the human body surface with its external environment. They can inhibit growth of most bacteria or fungi with pathogenic potential, by competing for necessary nutrients on the surface, or by producing inhibitors such as colicin, acids and lipid. Long-term clinical application of a great amount of one broad-spectrum antibiotic may suppress those bacteria in the intestinal tract which are sensitive to the antibiotic, so as to break down antagonism between these normal bacteria floras, which always leads to dysbacteriosis, such as methiciltin-resistant *Staphylococcus aureus* enteritis.

11.2.1.2 Humoral barriers

1) Complement system

As normal components in the serum of the human body and animals, complements include more than 20 kinds of proteins, which are synthesized in hepatocytes and macrophagocytes. They have the function to complement action of antibodies in the antigen-antibody reaction. Antibodies act to discern invaded heterogenous cells and activated complements, i.e.at first antibodies bind to antigens to form the antigen-antibody complex which then activate complements, to attack invaded cells.

After activation complements may lead to irreversible damage of the cell membrane which results in lysis of invaded cells which include Gram-negative bacteria, viral particles with the lipoprotein envelope, red blood cells and so on. This function of complements is most significant for resistance against pathogenic microorganisms and elimination of aged cells which have pathological damages. Various fragments produced during complement activation have multiple physiological functions mainly including chemotaxis, promoting activation and phagocytosis of phagocytes and clearing off the immune complex (antigen-antibody complex),which is a crucial part of innate immunity of the body. At the same time, complement components have compound functions of immune regulation to involve in specific immunity of the human body. Each fragment of complements has its own specific receptor, and these receptors distribute in many kinds of cells. Fragments of complements play their roles through their receptors.

2) Interferon (IFN)

Interferon is a kind of inducible proteins produced in the host cells with stimulation of various stimulators such as virus, with low relative molecular weight, high activity and multiple functions, which can be classified as α, β and γ classes based on their chemical structures and as type I, II and III based on their own specific receptor on the cell surface. Alpha and beta interferon are produced by leukocytes and fibroblasts respectively, and belong to type I interferon; γ interferon is mainly produced by T lymphocytes, and also called type II interferon or immune interferon. After produced and released from cells, this kind of inducible proteins acts on the same kind of cells to make them acquire multiple "immune ability" of antivirus and antitumor. Interferon acts on host cells to make them synthesize antiviral proteins which can control over systhesis of viral proteins, influences on assembly and release of viruses and has broad-spectrum antiviral functions; and meanwhile it has multiple action of immune regulation. Type I interferon has high activity of antivirus, while for type II interferon such activity is low but it has stronger action of immune regulation.

3) Lysozyme

Lysozyme is a kind of alkaline proteins which are not heat-resistant, mainly originate from phagocytes and can be secreted into serum and various secretary fluids, and can hydrolyze peptidoglycan in the cell wall of Gram-positive bacteria so as to lead to cytolysis.

11.2.1.3 Cellular barriers

1) Phagocytes

Phagocytes have ability to phagocyte invaded pathogen particles, which can be enhanced much more by stimulating phagocytes to activate when the corresponding ligands bind to the various receptors on the surface of them such as complement receptors and antibody receptors. Phagocytes enrich in lysosomes, which contain many enzymes such as hydrolases, lysoenzyme, and other antimicrobial substances. When pathogens invade phagocytes can permeate through the wall of the capillary vessel with action of chemokines and adhesion molecules to arrive at the infection locus, then phagocytize the pathogen to form phagosomes, further fuse with lysosomes to form phagolysosomes, and then kill the phagosized pathogen by means of oxygen consumption and anaerobic pathway. After phagocytosis, in most cases pathogens are killed by phagocytes, then digested and decomposed by hydrolytic proteases, polysacchridases, nucleases and lipases, and finally excreted out of the body with undigested residue, which is called complete phagosis. Some pathogens of endocellular parasites such as *Mycobacterium tuberculosis*, *Leprosy bacilli* and *Brucella melitensis* have avoidance mechanism by which they can resist on phagosis of lysosomes or lysoenzyme; and hence in the body with low immunity these pathogens are phagosized but cannot be killed, and can be moved with phagocytes to bring about spreading over the body, which is called incomplete phagosis.

Besides, phagocytes can secrete soluble cytokines, which cannot only enhance bacteria-killing action to promote inflammation, but also play function of immune regulation. At the same time, as antigen-presenting cells, they are crucial part of specific immunity.

2) Natural killer cells

Natural killer cells (NK) belong to lymphocytes, and mainly distributed in the peripheral blood and the spleen. They have functions to kill directly target cells without sensitization in advance and without helper cells or molecules. NK cells can kill target cells directly by releasing perform and granule enzymes, and can kill target cells by releasing tumor necrosis factors (TNF). Some tumor cells and those infected by microorganisms may become target cells of NK cells. In addition, NK cells present the activity earlier than other killer cells, and so play an important role in antitumour and anti-infection, especially in anti-infection of viruses.

11.2.1.4 Inflammation

Inflammation is a series of local and general defense responses when the human body undergoes harmful stimulation, can be regarded as a result of comprehensive action of nonspecific immunity, and has functions to clear off harmful foreign, to repair injured tissues and to maintain self-stability. Harmful stimulation includes various physical and chemical agents, but focuses on infection of pathogenic microorganisms.

With infection of pathogens, stimulation and injury of tissues and microvessels may rapidly result in release of many soluble media. For example, platelets activated by LPS of bacteria attach to local collagen and basement membrane of vascular endothelium, and release many active components such as 5-hydroxytryptamine and coagulation factors, which cause coagulation, cascade reaction of kinin and the fibrinolytic system. Therefore, accompanying with inflammation, redness, swelling, pain, fever and dysfunction may occur. At the advanced stage of inflammation, there is a repair procedure in which fibroblasts, epithelium and macrophages and cytokines involve are involved.

11.2.2 Specific Immunity

Specific immunity is one immune function which is produced after undergoing stimulation by foreign antigen in the body's life course, for example microbial infection or vaccination of some vaccine, and also called acquired immunity with acquisitiveness, high specificity and memory.

According to its acquiring modes, specific immunity may be classified into active immunity and passive immunity. Active immunity is produced as the body's undergoing stimulation, and lasts long. Passive immunity refers to reception or infusion of immune cells and molecules from other immunized individual to acquire immunity, and lasts short. Active immunity and passive immunity can also be classified into naturally acquired and artificially acquired patterns, respectively.

Based on effectors in the effector phase of specific immune responses to specific foreign stimulators, specific immune responses can be classified into humoral immune response and cellular immune response. Humoral immune response mainly relates to humoral effectors such as antibody, while cellular immune response mainly cellular effectors such as T cells.

11.2.2.1 Immune system

The material basis of acquired immunity is immune system in the human body. The immune system is made up of immune organs, immune cells and immune molecules.

1) Immune organs

Based on their functions immune organs are classified into central immune organs and peripheral immune organs. The central immune organs are the sites for development and differentiation of immune cells, including bone marrow, thymus and bursa of Fabricius. The bone marrow is the site for development of hemapoietic stem cells (hematopoietic progenitor cells); the thymus, T lymphocytes; and the bursa of Fabricius, B lymphocytes. Mammal animals have the bursa-equivalent organ, which is the bone marrow. The peripheral immune organs are the sites where immune cells reside and immune responses occur, including the lymph node, the spleen and the mucosa-associated lymphoid tissues which are distributed in the respiratory tract, the digestive tract and the urogenital tract, and mainly consist of the tonsil, the appendix and the mesenteric lymph nodes.

2) Immune cells

Immune cells include lymphocytes, granulocytes and mast cells, monocytes/macrophages, dendritic cells; broadly speaking, also include red blood cells and platelets, and progenitor cells of various cells. All of these cells originate from multipotential hemapoietic stem cells in the bone marrow.

Lymphocytes are highly specific, and can be classified into T cells, B cells and class III (neither T nor B) lymphocytes.

Stem cells are the most primitive blood cells, mainly distributed in the bone marrow, and a little quantity of them exists in the spleen. They can perform cell division continuously to keep a particular quantity. Various blood cells are derived from cell division and differentiation of the stem cells, and so are called hematopoietic stem cells. They are shaped like one small lymphocyte, have a nucleus containing nucleolus and a little cytoplasm, are poorly basophilic, and contain no granules.

Macrophages are cells which have strong ability to kill and phagocytize microorganisms. They still play an important role in specific immunity as in nonspecific immunity. Macrophages are necessary for both cellular immune response and humoral immune response. They can present the antigen information to T cells or B cells after recognizing and processing antigens.

The red blood cells may carry pathogenic particles to the liver and the spleen through blood circulation in which the pathogens are phagosized and cleared off. Because in blood circulation the number of the red blood cells is much bigger than the white cells, immune adhesion of the red blood cells is one of the main pathways for the body to clear off pathogens.

11.2.2.2 Immune molecules

Immune molecules include membrane surface molecules and humoral molecules.

Membrane surface immune molecules mainly consist of membrane surface antigen receptors, major histocompatibility antigens, leukocyte differentiation antigens and adhesion molecules. B cells and T cells have their own specific surface membrane antigen receptor BCR and TCR, respectively, which can recognize different antigens to bind with them, so as to start immune response. Major histocompatibility antigens are marker molecules of the body, involve in T cell recognition of specific antigens and interaction between various immune cells during immune response, and also many constrain the NK cells from damaging self-tissues for mistake. And they are the key molecular basis for the immune system to distinguish between self and non-self antigens.

Humoral molecules mainly include antibodies, complements and cytokines. Cytokines (CKs) are peptides of low relative molecular weight, including interleukin (IL), colony stimulating factors (CSF), interferon, tumor necrosis factors (TNF), transforming growth factors (TGF), and etc. CKs have multiple functions of immune regulation, and some of them are cytotoxic (e.g. TNF) and antiviral (e.g. IFN) and directly involve in the effector phase of immune response. Complements and antibodies are the major humoral components in nonspecific and specific immune responses.

11.2.3 Allergic Reaction

Allergic reaction, also called hypersensitive reaction, is abnormal humoral or cellular immune responses which occur after the body receives stimulation of the same antigen or hapten, to lead to tissue damage or physiological function disorder.

The antigen resulting in allergic reaction is called allergen, such as heterologous serum proteins, allogeneic histoproteins, microorganisms, and some drugs such as penicillin. Common allergic reactions are penicillin-anaphylactic shock, and serum diseases, bronchial asthma, food hypersensitivity, contact dermatitis, urticaria, allogeneic skin grafting, and so on. Allergic reactions include rapid allergy and delayed allergy. The former is associated with antibody and may be transferred passively with serum containing antibody; while the latter has no relation with antibody, but with sensitized lymphocytes, and may be transferred passively with sensitized lymphocytes.

11.3 Antigens and Antibodies

11.3.1 Antigens and Their Features

11.3.1.1 Concept of antigens

Antigens (Ag) are a kind of substances which can stimulate the human body and animals to produce antibodies or sensitized lymphocytes, and can have specific reaction with these products *in vivo* or *in vitro*. Therefore antigens possess two kinds of abilities: one is to stimulate the body to form specific antibodies or sensitized lymphocytes, which is called immunogenicity, also called antigenicity; the other one is to have specific reaction with their corresponding antibodies or sensitized lymphocytes produced due to stimulation of the antigens, which is called reactogenicity. Those antigens with both immunogenicity and reactogenicity are called complete antigen. Some substances cannot stimulate the human body and animals to produce antibodies or sensitized lymphocytes singlely, but can have specific reaction with the corresponding antibodies or sensitized lymphocytes. These substances are called haptens, also called incomplete antigen. Most of oligosaccharides, all lipids and some simple chemical drugs are all hapten. Structurally, the chemical groups of antigens can stimulate lymphocytes to produce immune response and have specific reaction with the products of the immune response are called antigenic determinants. The number of antigenic determinants which antigens carry is called antigenic valence, and generally antigens are multivalent.

11.3.1.2 Features of antigens

1) Foreignity

Foreignity refers to that the antigens invading into the human body are different from components of the tissues and the cells in the body. Such antigens include: ①heterologous substances, such as heterologous serum; ②allogeneic components, such as red blood cells from different blood types, other organs from different individuals; ③autologous insulated components, such as lens proteins in the eye, sperm cells and thyroglobulin, which are fixed in some specific sites in normal circumstances, insulated from the cells producing antibodies, and so cannot cause production of self-antibodies, but can also promote production of self-antibodies which are called auto-antibodies when entering blood flow. Reaction between auto-antibodies and auto-antigens results in auto-immune diseases. Allergic ophthalmia and thyroiditis are typical examples of autoimmune diseases; other autologous tissue proteins may denature to become autoantigens due to physical and chemical agents and biological agents such as ionizing radiation, burn, contamination of some chemicals and microorganisms, so as to cause autoimmune diseases, such as lupus sebaceus, leucopenia, chronic hepatitis, and so on. In general, autologous tissue proteins have no antigenicity to the autologous body.

2) Antigens must be macromolecules with certain chemical composition and structure

Antigens have large molecular weight, and so most of proteins are good antigens and some polysaccharide, nucleic acid and teichoic acidic ester with complex structure also have antigenicity. Some substances with small molecular weight such as penicilin, have no antigenicity in nature, but can act as hapten, i.e. when entering the body with atopy they can bind to proteins in the body to acquire antigenicity, finally resulting in allergy. On the contrary, proteins with large molecular weight will lose antigenicity when it is hydrolyzed into substances with small molecular weight, and hence, vaccines usually enter the body through injection but not oral routine, so as to avoid loss of their antigenicity due to digestion in the gastrointestinal tract. Furthermore, antigens also need certain chemical composition and structure.

3) Specificity

Specificity of antigens is determined by particular chemical groups on the surface of their molecules. Antigens may bind to their corresponding antibodies specifically (Fig.11.2). In recent years structures of some natural proteins have been discovered. It has been found that there are many different determinants on the surface of protein molecule, and their specificity as antigens attributes to different amino acids and their sequences.

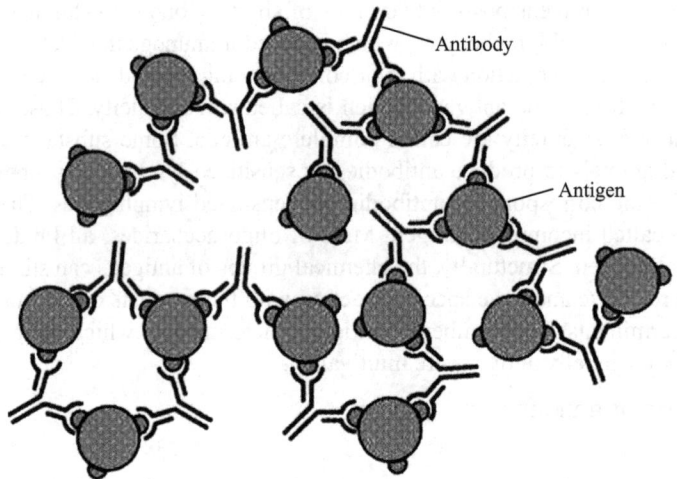

Fig.11.2 Antigen antibody binding

11.3.2 Structure and Functions of Antibodies

11.3.2.1 Structure and classes of antibodies

Antibodies (Ab) are immunoglobulin (Ig) with specificity which is produced by B cells after B cells are transformed into plasma cells due to antigen stimulation of the human body or animals. There are many kinds of antibodies, but based on their physical and chemical properties and immunological traits, they are divided into 5 basic classes, which are denoted accordingly as IgG, IgM, IgA, IgD and IgE.

The five classes of immunoglobulin have almost the same structure, and are all composed of 4 peptide chains, among which the same long chains are called heavy chain (H chain) and the same short chains as light chain (L chain). There are disulfide bonds between H chain and L chain, H chain and H chain, and the Ig monomer molecule composed of the 4 chains has the generic molecular formula, H_2L_2. The basic structure each Ig chain is a ring configuration which is formed by the peptide of about 100 amino acids with β-pleated sheets agglomerated through the disulfide bond, and is called functional region. The L chain consists of two functional regions, in its N-terminus amino acid sequence varies and so is called variable region (VL), while in its C-terminus amino acid sequence is conserved relatively and so called constant region (CL). The H chain is composed of one N-terminus V region and 3-4 C-terminus constant regions. The V region of the L chain and the V region of the H chain together make up the antigen binding site, and the Ig monomer molecule contains two antigen-binding sites and is so called two-valence antibody. The structure of antibody is shown in Fig.11.3.

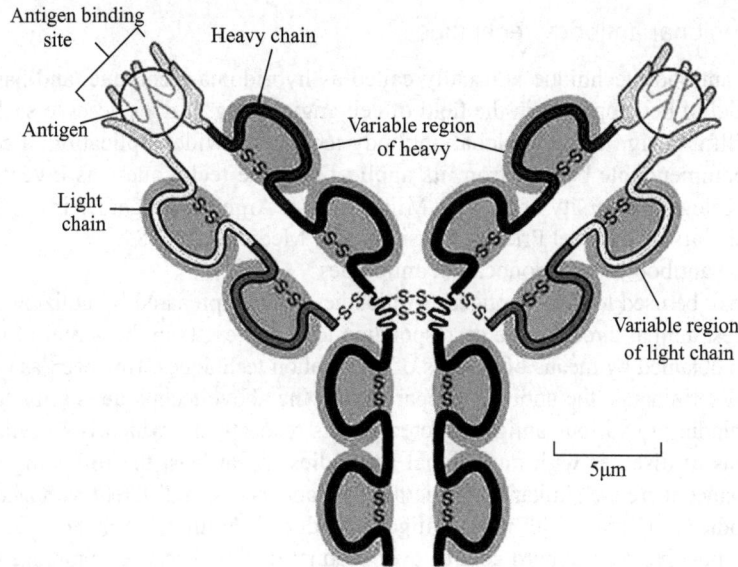

Fig.11.3 Structure of antibody

11.3.2.2 Distribution and physiological functions of Ig

1) IgG

Approximately 40%-50% of IgG exists in the serum of the human being, and the remaining are distributed in various tissue fluid. In the serum it is the major Ig, and accounts for about 80% of the total Ig. IgG is the key anti-infection antibody and the only antibody which can permeate through the placenta. Most antibodies with functions of antibacteria, antivirus and antitoxin belong to IgG.

2) IgA

IgA accounts for about 13% of the total human serum Ig, and exists in the blood mainly as monomer IgA which accounts for 85% of total IgA, while IgA in the external secreted fluid is almost dimmer. Secretary IgA can play function of anti-infection immunity in the local mucosa of the body.

3) IgM

Due to large molecular weight, IgM exists only in the blood vessel, and accounts for about 6% of the total Ig. when the animal is immunized by the antigen, in its blood IgM is detected firstly and then IgG. IgM may promote phagocytosis of particulate antigens (e.g. bacteria) by phagocytes, which is called opsonization. IgM has the strongest coagulation action and cytolysis ability among the 5 classes of Ig.

4) IgD

IgD mainly exist in the serum, and account for about 1% of the total human serum Ig. Its function is not clear, and may be associated with some allergies diseases.

5) IgE

The content of IgE in the serum is the Least, and it accounts for only 0.002% of the total human serum Ig. It can bind to the mast cells in the tissues and basophilic leukocytes in the blood, so as to cause some allergies.

11.3.2.3 Monoclonal antibody technique

Monoclonal antibody technique is usually called as hybridoma technique, and has always been regarded as a model of development in the field of cell engineering. It does deserve such high remark on account of skillful design of monoclonal antibody technique, wide application after coming into being and most commendable benefit from its application. The technique was invented by Georga Kohler from Cambridge University and Cesar Milstein from Argentina together in 1975. As a result, both of the two scholars won Nobel Prize in Physiology or Medicine in 1984.

1) Multiclonal antibodies and monoclonal antibodies

Antibodies may be used to detect antigens. Antibodies may be prepared by utilizing certain antigen to immunize one test animal directly. The corresponding antibody exists in the serum of the immunized animal, and can be obtained by means of a series of purification techniques. However, as one antigen has several antigenic determinants, the antibody prepared with the above technique is a mixture containing many antibodies binding to various antigenic determinants respectively, which is so called multiclonal antibody. Diagnosis of diseases with multiclonal antibodies has at least the following shortcomings: ①low specificity, since there are similar antigenic determinants between different pathogens and so such multiclonal antibodies may react with those antigens produced by different pathogens, the technique exhibits high false positive rate; ②hard control over product quality, because immunized animals vary individually, so after immunization with the same antigen, contents of produced antibodies recognizing different antigenic determinants are different, and also the antibodies differ from each other; ③taking long time and more steps during preparation and high cost.

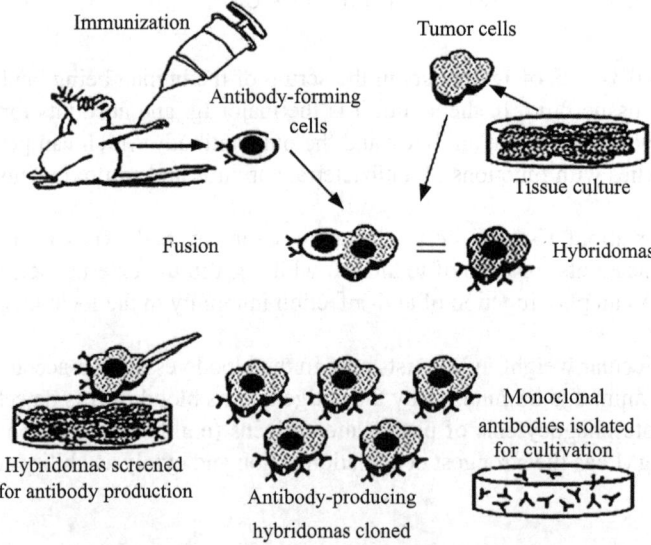

Fig.11.4 Preparation procedure of monoclonal antibodies

However, monoclonal antibody technique is a procedure to cultivate fused cells *in vitro* at large scale which can produce a large number of antibodies by utilizing cell fusion technique (Fig.11.4). As monoclonal antibodies recognize only one particular antigenic determinant, it has many apparent advantages over multiclonal antibodies such as high specificity, homologous component, high sensitivity, high productivity, and easy standardized production process. So far ten thousands of monoclonal antibody products have been developed in the world, among which many have been in the market.

2) Application of monoclonal antibodies

Although monoclonal antibodies are used mainly for *in vitro* diagnosis of pathogen infection, their applications are not limited to this field, and the application scope is fairly wide and includes:

(1) Identification of pathogenic microorganisms: including clinical diagnosis of bacteria, virus and parasite infection diseases, and detection of pathogens possibly contaminated in food or environment.

(2) Determination of hormone level: use for evaluation on endocrine function and pregnancy test, especially early pregnancy test.

(3) Detection of tumor-associated proteins: early diagnosis and efficacy evaluation after treatment of tumor by detecting tumor-associated proteins such as carcino-embryonic antigen, α-fetoprotein.

(4) Determination of content of one drug in the blood: determination of prohibited drugs, and determination of contents of treatment drugs such as gentamicin so as to choose the appropriate dose.

(5) Tumour treatment: conjugating those drugs for treating tumour with specific monoclonal antibodies with anti-tumour activity to prepare so-called "biolistic particle", which can make anti-tumour drug gather in the tumour locus by utilizing specific binding between monoclonal antibodies with tumour, so as to reduce side effect of those anti-tumour drugs.

(6) Application in other fields: detection of pathogens in plants and animals, and isolation of rare and precious substances with biological activity.

Chapter 12

Biodiversity and Environmental Management

12.1 Loss of Biodiversity and Its Reasons

According to article 2 in "The Convention on Biodiversity", biodiversity is defined as various organisms from land, ocean, other aqua-ecosystems, and the ecological integrity composed of them, including diversity of intra-species, inter-species and ecosystems. In total, biodiversity refers to all the living organisms and the integrity composed of them on the earth, including gene diversity, species diversity, ecosystem diversity and landscape diversity. Protection of biodiversity refers to the gene, species and ecosystem levels protection.

At present, scientists are paying more and more attention to protection of ecosystems. An ecosystem refers to the environment in which one species lives. The tropical rain forest in South America is just an ecosystem, the central park in the city, the forest or the beach for camping, and the wet land such as mangrove, mash and coastal estuary are all also ecosystems. Ecosystems in good conditions can supply habitats for various species including plants, insects, birds, giant higher animals, and so on.

All the species are crucial for maintaining survival of the whole ecosystem. The ecosystems all over the world correlate closely to each other, and effect on an ecosystem may spread far away to another one. If one link is lost in an ecosystem network, the whole network will start to collapse. Unfortunately, the network of biodiversity has been damaged a lot so far.

12.1.1 Contents of Biodiversity

A variety of organisms are living in the earth and variants are happening in different species, even in the same species. Various species get together to form a community. The communities and their living environment influence each other to form an ecosystem.

12.1.1.1 Gene diversity

Generally speaking, gene diversity refers to the total hereditary information in all the living organisms in the earth. While narrowly speaking, it mainly refers to the total genetic variations occurring in individuals within one species or in different individuals within a community. For example, rich varieties of domestic animals and crops have rich genetic diversity. The gene can be transmitted from the parental generation and determines the character of the phenotype among the filial generation. The majority of genes in the same species are the same, but some genes undergo some mutations which may result in observable or invisible phenotypic and physiological changes. Nowadays, the potential significance of genes is still not well known for the limited cognitive ability, and therefore protection of genetic diversity has not been concerned much, which is an urgent problem.

12.1.1.2 Species diversity

According to the records of taxology, the species number of the living animals, plants, and microorganisms in the earth is approximately 2,000,000. However, the data may be corrected according to the change of new species annually. From 1974 to 1977, the American scientists found sulfur bacteria, which can survive at high temperature (100℃) and pressure (1.15×10^5 kPa) in the submarine hot spring at the depth of 10,000m locating in the eastern Pacific Galapagos Islands. In the late 1970s, living microorganisms were proved to exist in the sky at the height of 85km. Moreover, the scientists from former Soviet Union discovered *Coccus*, *Bacillus*, and *Fungi* many times when drilling in the Antarctic glacier at different depths (4.5-295m). In one word, the organism species in the biosphere are to be discovered in the future. According to scientists' estimation, the species number of living organisms in the earth approximately reaches more than 5,000,000.

12.1.1.3 Ecosystem biodiversity

In certain geographical regions, the species of various animals, plants, and microorganisms living in the same environment constitute one community with specific composition, structure, and function. Then the communities of different species and their non-organism environment then constitute one ecosystem. The complicated interaction exists within one community, between different communities and between living environments, and the major ecological processes consist of energy transition, water circulation, soil formation, and the interactions between organisms such as competition, predation, symbiosis, and parasitism.

12.1.1.4 Landscape biodiversity

Landscape refers to the heterogeneous land region which is made up of a group of repeated and dependent ecosystems, and has three characteristics of composition, function, and dynamics. Herein, as a structural property, the landscape heterogeneity just attributes to its composition which takes effect on their function and dynamics. The landscape diversity in the earth surface is caused by the interaction between human beings and the nature, for instance, the landscape of agriculture, forest, grassland, desert, city, fruit garden, and so on.

12.1.2 Threats to Biodiversity

At present, the major five threats to biodiversity are population, food, resource, environment, and energy crisis.

It's estimated that the rate of species extinction caused by human beings is 1000-fold of the natural one. Currently, there is one species to extinct on the earth every day and the rate of intra-species mutations and the entire natural ecosystem disappearance is higher than that of species extinction. Therefore, the loss rate of global biodiversity has increased alarmingly in recent years. At the beginning of the 1980s, the tropical rainforests were destructed at the rate of 1140 million hm^2/a, yet the rate reached 1700-2000 million hm^2/a at the end of 1980s, resulting in 50% reduction in the tropical rainforests area in the earth. Meanwhile, 1/3 of the temperate forests across the world have been cut down and nowadays this is at risk. Wetland, the ecosystem with abundant biodiversity and high productivity, plays important roles in modulating the water flow, removing the sediments and pollutants, and supplying habitats for waterfowl, fish, and other species. Now many wetlands have been drained into farmland or pools for aquaculture, or eroded by the cities. But for the wetlands with forest they are being destroyed by cutting damage. Coral reefs are able to endure the wave shock, but they are also sensitive to nutrients, water temperature, light changes. Hence, as the clean tropical water is polluted by the sludge, fertilizers, and sewage, these slow-growing animals will be ruined and replaced by the fast-growing algae. In addition, coral reefs are in excessive excavation and even in use for burning lime, thus severely threatening survival of the coral reef ecosystem. It's estimated that 5% of plant species have been extinct in recent decades in China. In 2000, the forest coverage rate in China reached 50% but now only about 13%, the data also includes the later artificial plantations at present. The area of mongolica forest is declining, and broad-leaved Korean pine is only present in defined protected areas. The aging deciduous broad-leaved forests have disappeared, and the evergreen broad-leaved forests have been also in serious damages and only disperse discontinuously in different mountains. Besides, the area of soil erosion increases in the tropical and subtropical zones with loss of forests, and the available pasture area can be reduced due to overstocking in grassland. For the worse, alkalization has taken place in specific area.

At present, many ecosystems are suffering grievous damages all over the world, and thus various living organisms lose their habitats.

12.1.3 Reasons for Biodiversity Loss

Biodiversity loss attributes to many factors. Population growth and endless demand for organism resource are the most important one. The world's population has risen dramatically and now it has been more than 5 billion. In order to survive, human beings have to consume 39% of land productivity yielded through photosynthesis of plants, algae, and bacteria. Yet the global population is still increasing and it is expected that the number will be over 8 billion by 2020.

12.1.3.1 Reasons for biodiversity loss

The human activities result in biodiversity loss, mainly including the following aspects: ①habitats loss and fragmentation; ②predatory overexploitation; ③environmental pollution; ④variety simplification in agriculture and forestry; ⑤invasion from introduced species. Moreover, global climate change, construction of dams and reservoirs, reclamation of land from the lake, development of the new mines, and a variety of natural disasters are all threats to terrestrial and aquatic biodiversity. Therefore, it is imperative to find the right way to coordinate the conflict between biodiversity protection and economic development needs.

12.1.3.2 Grading of threats to species

For biodiversity protection, the extent to which species are threatened must be determined at first.
Extinct species: One species that has not been discovered in the wild nature for 50 years affirmatively.
Endangered species: The group (species or subspecies) which are faced with the danger of extinction. They will be impossible to survive if the endangering factors continue to exist. This group includes those species whose numbers are reducing to the critical level or whose habitat area are shrinking so dramatically that they are deemed to be in danger of extinction at any time.
Susceptible species: The groups which are believed to turn into "endangered species" soon as long as the causal factors exist continuously. It includes those species whose majority or entire group are declining due to overexploitation, severe habitat destruction, or other environmental disturbance; and meanwhile, it still includes those species which are threatened severely within the distribution area

in spite of abundance in populations.

Rare species: The groups that are in small populations all over the world but now not "endangered species". These group usually live within restricted geographical areas or habitats, or are thinly scattered in a wide area.

Indeterminate species: The group which cannot be assigned to an appropriate grading due to no enough information.

Grading of threatened-species still needs to be studied further, and the current one is employed mainly in quality and is subjective.

12.2　Significance of Biodiversity Protection

Biodiversity is diversity of organism resources. Some organisms have been utilized as resources, yet much more ones are potential resources whose value is not yet known. Generally, when utilizing organism resource, people don't always pay more attention to the value of biodiversity, and just take it only for use. Thus, only if the huge contributions of biodiversity to domestic economy and environment are well known, more concern about biodiversity protection can be aroused.

12.2.1　Values of Biodiversity

Biodiversity satisfies the basic food demand of human beings. It's estimated that there are more than 80,000 species of terrestrial plants all over the world, but only 150 species are planted for food in large area. In the world 90% of food derives from 20 species throughout the world. At present, 75% of food comes from seven kinds of crops, including wheat, rice, corn, potato, barley, sweet potato, and cassava; and the former three kinds account for over 70% of the total yield. Moreover, various poultry, livestock, fish, and seafood supply proteins essential for humans, and diverse vegetables, fruits and fungus to meet the basic necessities of daily life. With population growth and life improvement, there are great needs for developing food crops and cereal crops. Hence, development of new food and improvement of varieties of poultry, livestock, and aquatic products are imperative; and both of them are involved in the hybridization to wild species or wild relatives with strong resistance to enhance their stress resistance.

12.2.1.1　Direct value of biodiversity

The majority of drugs closely related to human survival are obtained from plants, animals, microorganisms, or their metabolites. So far, scientists are working hard to screen active ingredients of drugs for treatment of human diseases. According to the World Health Organization, 40% of drugs derive from the natural resources in the developed countries, while in the developing countries 80% of people rely on traditional medicines or the compounds acquired from the nature or through chemical synthesis for disease treatment. In China, there are approximately 5000 species of medicinal plants in record, about 1000 in which are in normal usage. A considerable number of animals are utilized as major drugs by humans, for example, hirudin is a precious anticoagulant, bee venom is used for arthritis treatment, and some kinds of snake venom are helpful for hypertension treatment.

Experimental animals for medical research play an important role in development of the pharmaceutical industry, for example monkeys for polio vaccine and armadillos for leprosy vaccine. China is the big country of macrofungi plantation and has utilized tuckahoe, hedgehog hydnum, Ganoderma lucidum and Cordyceps sinensis as medicine ingredients for long time. Although microbial species are abundant and have close association with human life and health, only a small part of microorganisms are well developed and utilized, and much more are to be studied. Now the vast majority of antibiotics and vaccines are obtained from microorganisms, and they have made great contributions to prevention and treatment of infectious diseases. In fact, they are the meritorious statesman in increasing the average human life expectancy. Besides, microorganisms can be applied to large-scale production of enzymes, organic solvents, alcohol, amino acids, vitamins, and so on. What's more, biodiversity also provides mankind with all kinds of industrial raw materials, like wood, fiber, rubber, papermaking raw materials, natural starch, grease, etc. Yet more and updated organism resources need to be developed in order to provide the necessary raw materials and new energy for various industries.

12.2.1.2　Indirect value of biodiversity

The indirect values of biodiversity are related to functions of an ecosystem and may be much higher than the direct values, which often stems from the indirect ones, as the harvested animals and plants can't live without their living conditions, which are an integral part of the ecosystem. Species without consumption and production value may play effect in supporting the valuable ones in the ecosystem. The indirect value of biodiversity can be seen as the value of environmental resources and its significance can be summarized into the following aspects. ①Energy fixation. The photosynthetic plants and algae can fix the solar energy, so that the initial energy resource for all the biological survival enters into the food chain from these organisms and energize all species including human beings, with the food chain flow. ②Climate regulation. Ecosystems affect both macroclimate and local climate, including temperature, moisture and air. ③Hydrology stabilization and soil protection. The plant roots embed into the soil in depth to make it more permeable for rainwater. Runoff in the vegetation-covered place is slower and well-distributed than that in the bared place. In the forest-covered areas, the flood can be weakened during the rainy season and water still remains in river during the dry season. In the areas with well-grown vegetation, covered with vegetation and litter layer, direct rain wash of soil will be reduced so as to protect soil, reduce erosion, maintain land productivity and prevent from collapse; and protect seashore and river bank, and prevent lake, river, and reservoir from siltation. ④Storage of necessary nutritional elements and promotion of element circulation. Various nutritional elements are stored in all the organisms in the ecosystem and element exchange and circulation between the organisms and their living environment are promoted through life activities to maintain the ecological processes. ⑤Maintenance of the evolutionary process. Ecosystem functions, including pollination, gene flow, cross-pollination and the organism-organism or organism-environment interactions, are important for maintaining the evolutionary process and environmental benefits. ⑥Absorption and degradation of pollutants. Some organisms are resistant to pollutants owing to their genetic characteristics and therefore able to absorb and degrade pollutants. Some others can degrade organic wastes, pesticides, and the pollutants in the air and water. While some organisms are just sensitive to pollutants, thus can be indicative significance for environmental pollution.

12.2.2　Objective and Strategies of Biodiversity Protection

12.2.2.1　Objective of biodiversity protection

Protection of biodiversity is closely related to sustainable economic development. The objective of biodiversity protection is to protect and utilize the organism resource by protecting diversity of gene and species, without destructing crucial habitats and ecosystems to guarantee sustainable development of biodiversity. The process to achieve this object can be divided into three basic parts: ①saving biodiversity; ②studying on biodiversity; ③utilizing biodiversity sustainably and wisely. Saving biodiversity refers to take effective measures to protect genes, species, ecology, and ecosystem, so as to make efforts to prevent degeneration of the critical natural ecosystems, effective management and maintain biodiversity of land and water which have been disturbed by humans, finally making the damaged species return to the original ecological environment. Studying on biodiversity refers to illustration of composition, distribution, structure, and function of biodiversity. It aims at understanding effect and role of genes, species, ecological environment, and ecosystems, and finding out the complicated relationship between the disturbed ecosystems and the natural ones so as to maintain and develop biodiversity on the above basis. Utilizing biodiversity sustainably and wisely make it become inexhaustible, finally maintaining and improving human living conditions. The best way to utilize biodiversity is to place it in a natural state to keep its ecological and cultural value. In order to realize the above objective, biodiversity protection must be regarded as part of the overall planning of countries and regions. Protection of biodiversity can be realized only under the strong leadership of decision-makers and active participation of people.

12.2.2.2 Strategies for biodiversity protection

The book of Global Biodiversity Strategies offers five key strategies, making effective implementation of them much more possible. First, the guidelines of national and international policies must be developed to maintain sustainable utilization of organism resource and biodiversity protection. Second, it's necessary to create conditions for effective protection work of local communities and encourage it. Besides, biodiversity protection activities must be carried out deeply in the working and living places. Third, the facilities for biodiversity protection need to be strengthened and applied widely. The natural protection zones in the world are just the important facilities that can protect biodiversity. In combination with other protection facilities like zoos, botanical gardens, aquariums, seed banks, and gene banks, they can protect most of biodiversity in the world and are helpful for effect of biodiversity. However, if lack of expense and labour, they cannot play the action. Fourth, the ability to sustainable utilization and protection of biodiversity must be enhanced all over the world, especially in the developing countries. Biodiversity protection will not be achieved unless people know well about distribution and value of biodiversity, the mechanism how it affects their own life and pursuit, and the approach to manage biodiversity to meet the needs without reducing biodiversity. Finally, protective action must be promoted through international cooperation and national planning.

Biotechnology provides reliable technical gurantee for biodiversity, especially genetic resource preservation. Organs, tissues, cells or protoplasts of excellent plant varieties can be stored at ultra low temperature ($-196°C$), and regenerate the whole plant as long as necessary. For instance, animal embryos or sperms may be stored with this technology. In addition, DNA fingerprinting technique can be used for newborn offspring identification and reproductive behavior research. Molecular marker technology may be used to research on competition of pollens with different genotypes and for more accurate estimation of genetic diversity of one population. Random amplified polymorphic DNA (RAPD) technology which is used routinely in the biotechnology field has also been applied in biodiversity protection, particularly in the genetic diversity research.

Furthermore, biotechnology is also helpful for sustainable utilization of biodiversity. In a broad sense, plant tissue culture is a pretty mature technology in the biotechnology field. A small number of organs, tissues, cells or protoplasts isolated from rare, endangered and valuable plants are used to culture and regenerate a whole plant which can be multiplied with the rapid propagation technology.

Human survival and development are closely related to the environment all the time.

Australopithecus learned to walk upright and manufacture tools, and eventually evolved into human beings through struggles against the environment. Today human beings have become the masters of the earth, and their activities can take huge effect on the environment, which will act on humans themselves. Human beings have suffered from their intentional or unintentional activities, like environmental pollution, resources destruction, and ecological imbalance. The devastating flood in 1998 in China is just an example. We have only one earth, the only place where human are living. Therefore, to protect living conditions and biodiversity is the responsibility of all the mankind. Only if human beings realize that we must properly handle relationship between long-term and immediate interests to guarantee sustainable economic development, the global environmental problems can be settled step by step. Just the Chinese ecologist Lu Shugang said, every organism in the earth is like a small inappreciable screw on an aircraft, disappearance of several species like loss of several screws will not influence normal flight. However, the earth, our common homeland, will face a tough predicament one day with disappearance of such more "screws".

12.3 Biomonitoring and Evaluation of Environment

Environment refers to the human-centered external influence and power that act on human and also its scope or realm, and that is the surrounding space where human beings are living. The relationship between human and environment is manifested by human production and consumption activities, which means exchange of substances, energy and information between human and environment. Human can obtain substances, energy and information in the form of resources from the environment through production activities, which are then emitted into the environment in the form of three-waste through consumption. Hence, both production and consumption activities are affected by the environment and affect the environment in turn, and they develop with development of the human society.

12.3.1 Significance of Biomonitoring and Evaluation

The natural environment is one of the material conditions of human survival. It realizes its occurrence and development in accordance with its own intrinsic law. Thus, there must be inevitable contradictions between the objective properties of the natural environment and the human's subjective requirements, and between the objective developing process of the natural environment and intentional activities of human beings. Hence the natural environment is not only utilized, but also changed, and it gradually turns into the environment more suitable for human survival during the aimed and planned utilization and modification process. The new living environment will act on human beings subsequently. Human change themselves while changing the objective world. The living environment of human beings has differed more and more from the original natural one.

In recent one hundred years, the world's population growth rate has reached the peak in the history of mankind. Among the new population, 90% are born in the developing countries, and some of them are suffering from forest destruction, soil erosion, wood lack, desert spread, and so on. It's well known that human are both producers and consumers. As producers, all the human production activities demand for the natural resources (land, energy, mineral resources, and biological resources), and this will result in aggravation of environmental destruction with population growth and production scale improvement, which will need more resources and bring more wastes. As consumers, there will be more wastes with population growth, intensifying environmental pollution.

Resource issue is another major problem that people are facing today. The natural resources as the fundamental materials are important for human production and development. However, the needs and consumption for resources are increasing with population growth and economic development, leading to resources decrease, degradation, and exhaustion.

Land resource loss, especially arable land loss, soil degradation, deterioration, salinization, and desertification, has become a global issue. The global spare resources for development and utilization have become fewer and nearly exhausted in many regions. In addition, abuse of pesticides and fertilizers causes soil pollution. This series of problems pose a serious threat to human survival.

With one-third of the global land covered by forest, the forest is a source of lumber and plays important roles in water storage, climate regulation, soil conservation, livelihood supply, biodiversity conservation, and so on. The serious disorder between deforestation and afforestation pace brings about land barreness, soil erosion, weather change in small regions, river flow reduce, endangered species increase, and so on.

Water is the source of life, no water, no life. With rapid growth of population and development of industrial and agricultural production, water consumption is increasing day by day. In addition to the natural conditions, destruction of water resource resulting from water pollution is also one of the main reasons causing the water crisis. Water shortage has become a widespread concern in the world.

Environmental pollution, as a significant social problem, has been begun by the industrial revolution. At that time, people just paid attention to production but not to the serious consequences of environmental pollution. In the 20th century, especially after the World War II, industry, transport and science made great development, environmental pollution gradually expanded from local pollution to regional pollution, from single air pollution to pollution of water, soil, food and many others, brewing a lot of world shocking nuisance events. With the aggravation of environmental pollution, people realized severity of environmental pollution. Since the 1950s, two upsurges have occurred to environmental pollution. One was in the 1950s and the 1960s when the environmental pollution was so serious in industrially developed countries that it directly threatened the safety of people's lives, became a major social problem, aroused strong public dissatisfactions, and affected smooth development of the economy. As a result, the United Nations Conference on Human Environment was held in Stockholm in adoption of the "Declaration on the Human Environment" to arouse the attention of the world. The other one happened at the beginning of the 1980s, accompanied by environmental pollution and a wide range of ecological damage. The environmental problems that have effect in wide range and result in serious destruction mainly include acid rain, ozone layer depletion, "greenhouse effect", and so on. Global Climate Change Conference held in Hamburg, Germany in 1988 pointed out that the earth would be in doom unless the greenhouse effect was blocked, so that all the governments were fully aware of severity of the global environmental pollution and the necessity of pollution prevention and control. It's necessary to strengthen management of the environment in order to improve the polluted environment and prevent new pollution.

Environmental pollution comes from both nature and human activities. Natural disasters, also known as environmental issues, such as volcanic eruptions, earthquakes, forest fires, typhoons, floods, landslides, epidemics, and other such problems, are regional or local ones. Environmental damage or pollution caused by human production or living activities is the second environmental issues. We usually refer to the problem of environmental pollution as the man-made environmental problem caused by effect of human activities on the surrounding, environment but not the natural disasters.

Environmental pollution is composed of water pollution, air pollution, soil pollution, and solid waste pollution. Water pollution refers to the water quality deterioration and use value decrease caused by over-loaded discharge of man-made harmful and toxic substances. Air pollution is the phenomenon that the harmful and toxic gases and suspended solids are involved in the atmosphere, and their concentration exceeds the allowable amount of the atmospheric environment, resulting in deterioration of air quality and direct or indirect harm on human, animals, and plants. Soil pollution refers to the soil quality deterioration as the accumulated concentration of the pollutants in the soil exceeds the environmental capacity of the soil. Solid waste pollution is the phenomenon caused by the industrial solid waste and garbage piled up in the environment.

From the development trend of environmental science and protection work in certain countries, it can be divided into three stages. The first stage is the emergency stage referring to the emergent management of serious environmental pollution. The second one is combination of prevention and management that is mainly involved prevention against pollution and implementation of comprehensive management. The third one is improving and beautifying the environment, which emphasizes on comprehensive characteristics of the environment.

Biological monitoring is an approach to utilize manifestation of organisms (animals, plants, and microorganisms), to environmental pollutants, i.e. distribution, growth, development, physiological and biochemical indicators of organisms, and change of the ecosystem in the polluted environment, to determine the situation of environmental pollution. And the organisms used for monitoring and evaluating quality, change, and pollution degree of the environment is called indicator organism. Different species have different responses and changes to the toxins, pollutants, and their contents in the environment. Compared to the instrumental and chemical monitoring, biological monitoring and evaluation can reflect both the combined effect of environment and materials but also the historic conditions of environmental pollution, and thus has the advantages of continuous monitoring, high sensitivity and simplicity.

12.3.2 Classification of Environmental Biomonitoring

Biological monitoring can be divided into the following three types according to the purpose of monitoring:

12.3.2.1 Surveillance monitoring

Most of such monitoring belongs to routine monitoring that collects long-term continuous data, evaluates the environmental quality, judges and determines the pollution trend in the wider regional context, and especially assesses the operating results of the conventional biological treatment processes, like the treatment method using activated sludge, biological membrane and so on, through establishment of water quality, air and soil monitoring network to provide effective basis for better treatment.

12.3.2.2 Research monitoring

Because of diversity and complexity of environmental pollutants, it's difficult to determine the types of pollutants and the mechanism just from the overall organism manifestation results. So this requires studying the pollution mechanism, pollution degree, patterns of pollutant migration and variation, and the nature of pollution on the basis of surveillance. And the results serve the surveillance monitoring in turn.

12.3.2.3 Intended monitoring

Such monitoring, usually determining the range and the degree of environmental pollution in emergency by means of mobile monitoring with monitoring vehicles and ships, air monitoring, remote control, remote sensing, includes the monitoring of emergency events, such as emergent epidemic diseases and sudden fish death.

Biological monitoring can also be divided into population and community ecological monitoring, based on ecological level.

Biological monitoring can be classified into biological monitoring of air pollution, water pollution, soil pollution, water pollution control engineering, in accordance with the monitoring objects.

12.3.3 Biomonitoring and Evaluation Methods

12.3.3.1 Atmospheric biological monitoring

Atmospheric monitoring methods utilizing indicator organisms mainly include the following:

1) Evaluation and judge of air quality with plant damage symptoms

Although different types of plants demonstrate different resistance to the same pollutants, the plants show almost the same damage symptoms as long as the concentration of the pollutants exceeds the damage threshold of various plants. For example, when sulfur dioxide acts on broad-leaved plants, there will be irregular necrotic spots between the leave vein and the leave margin, and the color will turn yellow or white. And the old leaves will fade under the long-term effect of the low pollutant concentration. When polluting the conifer trees, sulfur dioxide will bring about banded necrosis on the coniferous top. For fluoride pollution, the broad-leaved plants present the symptoms of light brown or brown-red necrosis at the leaf margin and tip, and occasional small spots between the veins; and the conifers present the symptoms of brown to reddish-brown necrosis and the entire leaf necrosis in severe cases. It's just the plant damage symptoms that support determination of pollutant types.

At present, biological monitoring of air pollution is guided in quantitative and standardized direction gradually. As the responses of some plants to the air pollutants are often more sensitive and obvious than animals and humans, atmospheric biological monitoring mainly means monitoring of plants. It can be used to examine the extent of atmospheric pollution through study on the changes of plants enzyme systems and the germination rate, and it can be used to estimate the local air quality though changes of different species type and number in the plant communities.

2) Biological monitoring with lichens and mosses

Both lichens and mosses belonging to cryptogam are primitive lower plants, and the symbionts of algae and fungi. They are sensitive to sulfur dioxide and hydrogen sulfide. Lichens will disappear as long as the average concentration of sulfur dioxide reaches 0.015-0.105mg/L Mosses are the second indicator plant only next to lichens. The majority of mosses can't survive when the concentration of sulfur dioxide in the atmosphere exceeds 0.017mg/L. And the local pollution state can be determined according to species changes of lichens and mosses in quantity, frequency, coverage, and their internal and external damage symptoms. The species are few or even absent in the heavily polluted regions and the observable species are increasing with the mitigation of pollution degree.

3) Microbial monitoring

Microorganisms can't grow and propagate independently in the air because of lack of nutrients that can be used directly. They are brought into the air mainly through dust, water drops, dry cast from human and animal bodies, respiratory excretions and other ways, and then spread with the air flow. The number of the microorganisms in the air is related with the human and animal density and activities, plant numbers, soil and land cover, temperature, sunlight, air flow, and other factors, the groups vary with the environment. The majority of microorganisms in the air are non-pathogenic and many microorganisms are sensitive to air pollution. These microorganisms can be used as indicator organisms or for research of cytological damage. For example, *E.coli* is highly sensitive to the smoke generated by the light reaction of O_3 and hydrocarbons and the mixed pollutants can cause death of *E.coli* even at the very small concentration (μg). The pure O_3 is toxic to *E.coli* and can bring about the cell surface oxidation and the inclusion exudates cell death consequently. Luminescent bacteria are also excellent tools for determining the cytological damage caused by air pollutants as they can grow in dark and their bioluminescence can be easily determined. Moreover, microorganisms can be indicator for carcinogen pollution based on the common polycyclic hydrocarbons which can stimulate the bacteria to produce distortion in the air. For example, the carcinogenic contaminants, 3,4-benzopyrene can increase the metabolic activity and cause abnormal cell growth of *Bacillus cereus*. The phenomenon that benzopyrene can affect the growth of *Bacillus megaterium* result in the formation of large particles cells, etc. can be utilized to study the cell injury caused pollutants levels and the characteristics of the impaired cell. For now, the microorganisms on the instruction of air pollution are still in the research stage without practical use and demand for the development and application of the practical technology.

12.3.3.2 Biological monitoring of water bodies

Biocommunity monitoring of water body pollution is ecological monitoring of water pollution, which mainly refers to monitoring and evaluation of the extent of water pollution by utilizing changes of aquatic biological communities, species types, individual number, damage degree, and concentration of the poisons and pollutants enriched in aquatic organisms. The monitoring methods are the following:

1) PFU method

PFU refers to the approach that utilizes polyurethane foam unit (PFU) as an artificial machine to collect the microbial communities in water which are used to test the diverse parameters of community structure and function to evaluate the water quality. Furthermore, indoor toxicity test can also be used to forecast the toxicity of the industrial wastewater and chemicals on the microbial communities, and the degree of water body pollution.

2) Biological index method

Biological index method utilizes mathematical formulas to reflect changes in the structure of organism populations and communities, in order to evaluate the value of environmental quality. This method was first proposed to evaluate the water pollution extent by Beck in 1955. He discovered that amphibians could be divided into A and B two categories. A represents the sensitive one that has never been found under polluted situation, and B represents pollution-resistant one that is present under polluted situation. On this basis, the biotic index is calculated as the following formula:

The Biotic Index (BI) = $2n_A + n_B$

wherein, n_A and n_B mean the number of category A and B respectively.

When BI is equal to zero, the area is seriously polluted. When BI ranges from 1 to 6, the area is moderately polluted by the organic substances. When BI ranges from 10 to 40, the area is clean water without contamination.

Besides, there are still many other biotic index methods, like Tsuda Matsunae index method, Trent biological index method, Shannon-Wiener method, and so on.

3) Residue monitoring method

Biological residue monitoring method, also called aquatic biological method, means monitoring and evaluation of the water quality by determining the sewage content of organisms. The test can be carried out in fish, algae, and so on. And fish toxicity test is applied widely. Fish respond very sensitively to the water environment, a series of toxic responses, such as abnormal behaviour, physiological dysfunction, tissue lesion, and even death, will occur in fish as the water pollutants reach a certain concentration or intensity.

To use aquatic plants and animals for biological evaluation, the evaluation criteria must be determined at first, and then test spot deployment, sampling, and test are conducted, consequently, classification is made, based on statistical evaluation.

4) Bacteriological monitoring method

Bacteria can grow in various natural environments. When the natural water is polluted, some chemicals contained in the water will have negative influence directly or indirectly on organisms. In addition, organic substances will affect microorganisms at certain temperature and dissolve oxygen, and thus threaten other organisms. Therefore, biological test of water is very important, and in particular, intestinal bacteria test plays significant roles in health science. In practical work, the total number of bacteria, especially the indicator ones for fecal pollution, is tested to determine water quality. However, difficulty in bacterial identification and technology hampers its application in environmental monitoring to some extent.

5) Biological test

Biological test, one of the most important methods for biological monitoring of environmental pollution, refers to evaluation of pollution according to changes of physiological functions in organisms caused by pollutant toxicity. This method is in wide use to determine both single factor pollution and multiple pollutions in the environment, because it is simple, requires no special equipment, can reflect the comprehensive condition of toxicity and pollution.

12.3.3.3 Biological monitoring and evaluation of soil pollution

Biological monitoring of soil pollution is realized through characteristics of microorganism, animal and plant variation and their tolerance. Similar to air monitoring, the sewage contents of plants can be an indicator for monitoring and evaluation of soil quality. Some indicator animals can be used for soil monitoring and evaluation. There are a variety of microorganisms in the soil and they play pivotal roles in substance cycling, soil fertility and plant nutrition. As the soil is polluted, microbial species and number of microbial flora will change. The aim of microorganism monitoring of soil pollution is to determine properties and degree of soil pollution to provide basis for improving environmental health.

With continuous enhancement of people's awareness, more attention is paid to the simplicity and efficiency of environmental test methods. New monitoring and detection methods, like biosensors, gene chips, nucleic acid probes, and so on, can conduct fast continuous online assay, and therefore have broad application prospect in the field of environmental monitoring.

12.4 Bioremediation of Polluted Environmental

Bioremediation, a man-made engineering behaviour to remove or eliminate harmful pollutants through bioconversion or degradation, is the most effective way to improve quality of environment. According to different types of organisms, bioremediation can be divided into microbial remediation, plant remediation, enzyme remediation and biological surfactant remediation. According to types of the pollutants, it can be classified as bioremediation of oil pollution, heavy metal pollution and pesticide pollution. According to different polluted areas, it can be divided into remediation of soil pollution, groundwater pollution, natural water pollution, etc. This unit focuses on microbial remediation of environmental pollution.

The basic principle of microbial remediation is that degradation abilities of original microorganisms (indigenous microorganisms) or inoculated domesticated microorganisms are enhanced by adjusting the environmental conditions (soil pH, humidity, temperature, ventilation and nutrient additive) in the polluted areas, so that chemical pollutants can be degraded rapidly and completely. There is a dynamic equilibrium among microbial populations in the natural environment to control the quantity and the group effectively through changing the environmental conditions (nutrition). Generally, microorganisms acting on pollution molecules are not one single strain but a class of related strains. Metabolic reactions of pollutants in microorganisms are diverse, and beneficial or harmful to an ecosystem (Table 12.1).

Table 12.1 Microbial effects on chemical pollutants

Action types	Chemical changes
Degradation	Conversion complex compounds to simple products, sometimes mineralized
Combination	Conversion of compounds to more complex structure
Deoxygenation	Conversion of compounds to non-toxic compounds
Activation	Conversion of compounds to more toxic compounds

Phytoremediation refers to utilizing absorption, enrichment, or degradation of the specific types of plants to reduce or remove the pollutants in the contaminated environment. Phytoremediation is more suitable for environmental protection and has got more and more attention all over the world. It takes low investment and maintenance cost, simple operation, high efficiency, and potential and significant economic benefits, without secondary pollution. Since its introduction in the 1980s, phytoremediation has been one of the hotspots in the international academic circles and has begun to enter the primary stage of industrialization.

12.4.1 Bioremediation Methods

Bioremediation technology is mainly classified into *in situ*, *ex situ*, and *in situ-ex situ* combined bioremediation.

12.4.1.1 *In situ* bioremediation

This method means directly adopting treatment technology in the polluted area, for this method, the polluted objects need no moving, and the cost is low. But the treatment process is difficult to control and time-consuming. Therefore, it is applicable to treatment in the areas that have been polluted for long time or on a large scale. The stable engineering technologies for in situ bioremediation that have been formed and can be learned are as follows:

1) Bacteria inoculation method

This method means that the appropriate bacteria (super bacteria, mixed bacteria) are inoculated directly to the polluted area, and supply the necessary nutrients for growth of these bacteria to degrade the pollutants.

2) Biological culture

This method means adding hydrogen peroxide and nutrients regularly into the soil in order to mineralize the pollutants completely into carbon oxide and water through metabolism of original microorganisms.

3) Biological ventilation method

This method means providing nutrients or oxygen for the environment by air sparing also called biosparing, bioventing, bioslurping, so as to promote microbial degradation ability, and improve or modify the environment in consequence.

4) Phytoremediation

This method means assimilating or degrading pollutants by planting appropriate plants which are natural or genetically modified.

12.4.1.2 *Ex situ* bioremediation

This technology has advantages of accomplishment of limiting disperse and migration of pollutants at the beginning of pollution, reduction of pollution scope and easy control of the process. However, it costs more than *in situ* treatment method and requires various bioreactors.

12.4.2 Main Factors Affecting Bioremediation Technology

Many factors take effect on bioremediation, and mainly include properties of pollutants, microbial ecological structures, environmental factors, and so on.

1) Properties of pollutants

There are various pollutants in the environment, they are different in structure and properties and thus different in biodegradability.

2) Microbial ecological structure

The major contributor to pollutant degradation is the microorganisms with biodegradability.

3) Environmental factors

Improvement of the environmental conditions can promote biodegradation of organic pollutants. Improvement of the environmental conditions is usually achieved by means of ventilation and soil tillage to ensure the oxygen supply and maintain the proper temperature. Other approaches include appropriate supplement of nutrients (N, P) to promote the growth of microorganisms and increase the rate of degradation metabolism, water activity regulation (water content), and environmental pH adjustment to the optimum range of microbial metabolic activities.

12.4.3 Application of Bioremediation Technologies

The bioremediation technologies are applicable to many types of pollutants, among which oil and oil products, polycyclic aromatic hydrocarbons and hydrocarbons, heavy metal, water pollution cause people's attention. During management of some environmental pollution, bioremediation has been successfully applied.

12.4.3.1 Bioremediation of oil-polluting area

Oil is an important energy resource and may cause environmental pollution during exploitation, transport, storage, treatment and utilization. It's estimated that annually more than ten million tons of oil enter the ocean and does harm to organisms and ecological environment due to wars, disasters, and other accidents. In China, a plenty of farmland surrounding oil field in North China cannot be cultivated because of oil pollution and much money as compensation has to be paid to the farmers every year.

Bioremediation of oil-polluting area usually includes two methods. One is promotion of growth of the microorganisms by adding nutrients (easy to be used). As the microbial populations are exposed to the specific contaminated environment for long time, some subspecies can form limited metabolic capacity to use or degrade the pollutants. Growth of the special microorganisms is similarly affected by the nutritional status. After adding the necessary nutrients like N, P, and so on, growth of these microorganisms will be accelerated so as to promote degradation of the pollutants. This principle was applied to clear away the contaminated oil pollution in Alaska sea area caused by Cruise Exxon Valdez successfully in the 1990s, which has been field bioremediation on the largest scale so far.

The other method is to increase the quantity of the beneficial microorganisms by culturing with bioreactors, and then inoculation of the mixed microbial group into the polluted area for propagation and fermentation. And this method has been applied successfully in some cases.

12.4.3.2 Bioremediation of groundwater pollution

Currently, the groundwater pollution is so serious that it does obvious harm to human health. A large number of organic compounds, heavy metals and landfill percolate in the soil transfer partially into the groundwater with the aid of percolation, so as to make hardness, mineralization degree of water, the total number of bacteria and *E.coli* exceed the standards much, consequently resulting in groundwater pollution.

Since the 1990s, in situ bioremediation has been applied and replaced the traditional ground treatment technologies gradually. Actually, the method is related to enhancement of natural microbial degradation of organic substances in the groundwater in different ways. Because of deficiency of dissolved oxygen and lack of nutrients, microbial growth slows down. Therefore, it's necessary to add various nutrients, and form groundwater flow in the layer through water pumping toward the ground surface so as to disperse microorganisms evenly, which can promote absorption of nutrients and ensure oxygen supply, finally promoting microbial degradation ability. The chart flow of management of groundwater pollution with in situ bioremediation is presented as follows (Fig.12.1):

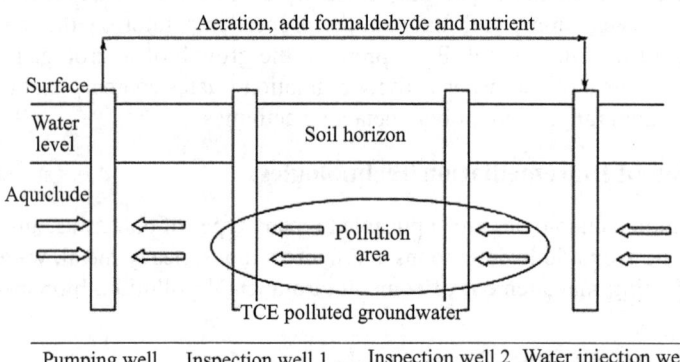

Fig.12.1   *In situ* bioremediation of groundwater pollution

In addition to the method of combination of groundwater pumping with reinjection system illustrated in Fig.12.1, there are other approaches for *in situ* bioremediation.

1) Biological injection method

For this method, the pressured air is injected under the groundwater to promote volatilization and degradation of organic matters in the water due to the air flowing.

2) Microbubble method

For this method, the bubbles containing 125mg/L surfactant are infused into the polluted groundwater and can supply abundant oxygen for bacteria and improve microbial metabolism. In a word, it is a highly effective and economical method.

3) Organic clay method

For this method, organic modifiers with positively charges and cationic surfactants bind to the surface of the negatively charged clay through chemical bonds to form the organic clay, and the toxic compounds are absorbed through the surfactants on the clay to conduct biological degradation.

12.4.3.3 Bioremediation of heavy metal pollution

Heavy metal pollution is harmful to environment and toxic to human. Mercury pollutants enter into the human body and penetrate the blood-brain barrier through blood circulation to impair the brain tissues. Chromium pollutants can form chromium-sulfur protein in the blood and accumulate in the organs, like the liver, the kidney, and so on to impair visceral organ.

Unlike organic matters, metals can't be biodegraded and only can be removed from the environment by means of biological adsorption. The current common bioremediation methods can be classified into the following types.

1) Biological adsorption

It's a method for separating toxic heavy metals with cheap inactivated cells, and applicable to industrial wastewater treatment and groundwater purification. Biological adsorption agents include abundant organism resource in the nature, including algae, lichens, fungi and bacteria. Some groups with chelating metal and coordination ability, such as sulfhydryl, carboxyl, hydroxyl, etc., exist on the surface of the cell wall of these microorganisms. These groups form ionic or covalent bonds with the adsorbed metal ions to complete metal adsorption. Furthermore, the metals can aggregate on the cell surface through precipitation or crystallization. Some insoluble metals can be captured and aggregate through extracellular secretion or the pores on the cell wall. So the particles formed via washing and drying these cells with acid or base are stable bioadsorbant which are usually filled in the adsorption column so that the heavy metals are captured when the metal-containing waste liquid flows through the column.

2) Biological accumulation

Metal ions are adsorbed and transported into cells through metabolisms, and accumulate so as to remove the metal pollutants from the environment.

12.5 Treatment of Water Pollution

Water, an important part of the environment, is one of the common substances in the nature and indispensable for human life.

With rapid development of industry, discharge of sewage has resulted in serious pollution of the natural headwaters recently. Because of limited fresh water for direct use and decrease of the available resources caused by water pollutions, the water crisis is presented in many regions on the earth. The polluted water is unsuitable for use and destroys the water resource and threatens human health. Thereby, to control water pollution and protect the water resource comprehensive treatment sewages of production and living is needed, and treatment of the discharged sewages with bioengineering technology can be carried out to meet the emission requirements.

Free discharge of the industrial and living sewages has serious impact on the water quality of rivers, lakes, and oceans, and leads to severe water pollution. In order to solve this problem effectively and maintain the quality of water resource, it's necessary to control the water quality index to determine the pollution degrees. As a result, it can provide scientific basis for treatment solution selection, treatment depth and comprehensive management.

12.5.1 Sewage Index and Emission Standards

12.5.1.1 Sewage index

Water quality refers to the physical, chemical, and biological integrated characteristics of the water and its contained impurities. The quality index of water represents types, compositions and quantity of the impurities in the water, and is the specific measuring standards for water quality assessment. The important indices for sewage pollution extent include physical index, chemical index, and biological index. The physical index includes total solid, suspended substances, turbidity, odour and taste, colour, chroma, and so on. The chemical index mainly includes biological oxygen demand (BOD), chemical oxygen demand (COD), total organic nitrogen, total organic carbon, dissolved oxygen, pH, and so on. The biological index includes total bacterial count, coliform group, pathogens, and so on.

1) BOD

It refers to the amount of dissolved oxygen consumption needed by aerobic microorganisms to conduct oxidative decomposition of organic pollutants in 1L sewage at 20℃ for 5 days, and is expressed as mg/L. It can reflect the content of organic substances in the water indirectly. The higher BOD value indicates, the higher organic content in water. In practice, it is usually presented as BOD_5.

2) COD

It is the amount of oxygen consumption for oxidation of organic substances in the water with chemical oxidants in the unit of mg/L, and reflects the content of organic substances in the water indirectly. Higher COD value shows higher content of organic substances.

Commonly oxidants include potassium permanganate ($KMnO_4$) and potassium dichromate ($K_2Cr_2O_7$). $KMnO_4$ has weak oxidation capacity that only part of organic substances can be oxidized. Yet $K_2Cr_2O_7$ has strong oxidizing capacity so that the majority of the organic pollutants can be oxidized. Thus, the amount of the oxygen consumption determined with $K_2Cr_2O_7$ is usually used to represent the content of total organic substances in the sewages.

3) Solid

Total solid (TS) refers to the weight of the residue of unit volume of water sample after evaporation and drying at 103-105℃. Suspended solid (SS) refers to the residual solid left in the water sample which is filtrated at first, and then evaporated and dried. SS is an important indicator for solid content of water pollution. Dissolved solid (DS) is the solid staying in the filtrate of the water sample after filtration.

4) pH value

Water power of hydrogen is expressed as pH value which has wide effect on the forms of pollutants in water and various water treatment processes.

5) Phenolic substances

The phenol content is one of the most common indicators of water quality at present.

6) Content of toxic substances

Toxic substances are toxic to microorganisms, other organisms and human beings, and include some heavy metals such as mercury, arsenic, chromium, plumbum, and etc., and organic carcinogens such as 4-nitrobiphenyl, N-subnitro dimethylamine, and so on.

12.5.1.2 State standards for waste discharge

For human's health and survival, every country makes detailed requirements for sewage discharge according to their own actual situation. The specific provisions for industrial wastewater discharge in China are as follows:

BOD, 30mg/L; COD, 100mg/L; SS, 70mg/L; pH, 6-9; chroma (dilution ratio), 50; and the other indices can be seen in the state standards GB8978-96 for detail.

12.5.1.3 Treatment methods for wastewater

The aim of wastewater treatment is to separate pollutants from the wastewater in some ways or convert them into nontoxic and stable substances through decomposition in order to purify the wastewater. Selection of treatment methods depends on properties, compositions and conditions of pollutants in the wastewater and the requirements for water quality. Contemporary sewage treatment methods can be generally divided into four categories, including physical treatment method, physicochemical treatment method, bioengineering treatment method and chemical treatment method.

1) Physical treatment method

Physical treatment method means treatment, separation or recovery of pollutants in the wastewater with physical method like interception, sedimentation, grease trap, sieving, filtration, evaporation and centrifugation. For example, filtration can remove suspending particles in the water, and evaporation is applied to concentrate the non-volatile soluble substances in the wastewater.

2) Physicochemical treatment method

This method refers to utilization of physicochemical processes to treat and recover pollutants that can't be eliminated with physical method. It includes adsorption, floatation, extraction, air stripping, steam abstraction and membrane separation. For instance, floatation can remove emulsion droplets or suspending particles in the relative density of 1; adsorption is applied to remove trace hazard substances such as heavy metals, pesticides and detergents; and extraction is used to recover phenol and heavy metals according to different allocation of soluble pollutants in two phases.

3) Chemical treatment method

Chemical treatment method means treatment and recycling of soluble pollutants or colloids with chemical reagents or other chemical methods including coagulation, neutralization, oxidation-reduction, electrolysis, ion exchange and so on. For example, neutralization can be used to neutralize acidic and alkaline wastewater, and oxidation-reduction is utilized to eliminate reductive or oxidative pollutants.

4) Biological treatment method

Biological treatment method refers to treatment of colloid and organic substance pollution through biological effect of microorganisms to achieve the purpose of purification of water. It includes activated sludge, biofilm, aerated pool, and so on, and is mainly applicable for organic pollution. It can be divided into aerobic and anaerobic biological treatment, based on the respiration characteristics of microorganisms. In accordance with growth types, it can be classified as suspension growth type and immobilization growth type. Besides, it can be divided into continuous and batch biological treatment; and it can be categorized into plug-flow and completely mixed flow treatment. Based on action principles, it can be generally classified as follows (Fig.12.2):

Aerobic biological treatment refers to degradation of organic substances to carbon dioxide eventually through a series of oxidative decomposition reactions in the presence of oxygen. Anaerobic biological treatment makes use of facultative and anaerobic bacteria to decompose organic matters under anaerobic conditions in order to convert organic matters to methane at last.

For better biological treatment of wastewater, generally the quality of wastewater needs to meet the following requirements:

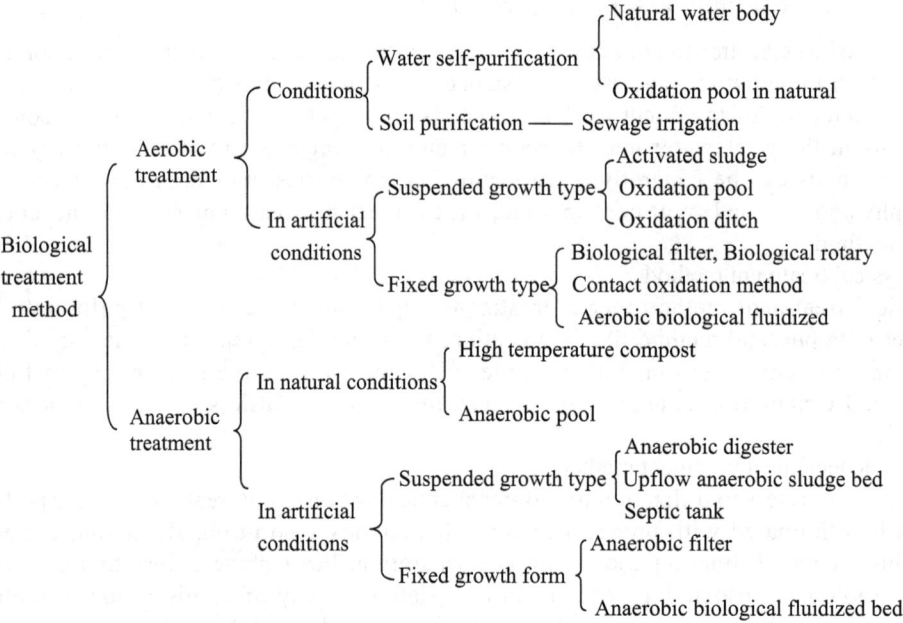

Fig.12.2   Classification of biological wastewater treatment

1) Power of hydrogen

The practical results of biological treatment of wastewater demonstrate that the suitable pH of wastewater is 6.0-9.0. However, the suitable pH values of some wastewater treatment system are in a small range. To reduce cost, activated sludge is usually domesticated to reduce or avoid addition of acid and alkaline during the process of wastewater treatment. In general, if the pH value of industrial wastewater is not suitable for microbial action, pH value of wastewater must be adjusted by adding waste acid or waste alkaline of adjacent factory, but the drag-in of the heavy metal pollutants that are difficult to biodegrade should be avoided to prevent now pollution.

2) Temperature

The optimum temperature is 25-30℃ for aerobic or anaerobic microorganisms. Usually, the water temperature at 20-40℃ is suitable for propagation of microorganisms and the treatment result will be ideal. The lower water temperature is helpful for microbial effects under the aerobic conditions. The higher the water temperature is, the lower the dissolved oxygen content in the water is.

3) Concentration of toxic and harmful substances

The toxic and harmful substances contained in the wastewater, like heavy metals, should be eliminated and controlled in the permitted concentration range as shown in Table 12.2. Some nonmetallic substances, such as hydroxybenzene, formaldehyde, phenol and so on, are toxic, inhibit the effect of microbial enzymes, and influence biological oxidation of organic matters in the wastewater. Therefore, the concentration of these substances should be limited. Due to the strong adaptability of microorganisms to the environment, during wastewater treatment gradual increase in the concentration of toxic substances makes microorganisms adapt to the new environment, which can treat the wastewater containing toxic substances effectively. This is the very method for treating toxic and harmful wastewater through microorganism domestication.

Table 12.2 Permitted concentration of toxic and harmful substances in wastewater

Toxic substance	Allowable concentration (mg/L)	Toxic substance	Allowable concentration (mg/L)
Cr	10	benzene	100
Cu	1	glycerin	5
Zn	5	resorcinol	100
Pb	1	Caprolactam	100
As	0.2	Benzoic acid	150
Fe	100	methanol	200
Potassium cyanide	8-9	toluene	7
Sulfide	40	phenol	100
Cyanide (CN)	2	formaldehyde	160

4) Nutrients

Microorganisms are the vital new force for biological treatment of wastewater and their growth, propagation and metabolism are related with the nutrients in wastewater. The nutrients mainly include carbon, nitrogen, phosphorus, sulfur, trace potassium, calcium, magnesium, iron and vitamins. During the wastewater treatment process, the missed nutrients should be complemented according to the sources of the wastewater. The domestic wastewater has all the above nutrients and the industrial fermentation wastewater can basically meet the nutrition need of microorganisms and the requirements of biological treatment. But amount and proportion of nutrients in different types of industrial fermentation wastewater are different.

5) Amount of microorganisms

Microorganisms play important roles in wastewater treatment which needs the microorganisms in high activity and great amount. Hence, it's necessary to screen and enrich the effective strains with excellent ability of wastewater treatment in order to increase the microbial population and number. In some cases, the domestic wastewater addition, sludge recycling, feces supplement and other measures are needed for inoculation and culture.

6) Dissolved oxygen

According to the test results, the respiration rate of aerobic microorganisms can't be affected only if the dissolved oxygen content in the wastewater is more than 0.3mg/L. So the oxygen should be sufficient in aerobic wastewater treatment but absent in anaerobic wastewater treatment.

12.5.2 Methods for Wastewater Treatment

12.5.2.1 Aerobic biological treatment

1) Activated sludge

Biological treatment methods for purifying wastewater or sewage are primarily based on activated sludge which is flocculent mud granules, and considered as a group of biota, protozoa fauna, algae, and the absorbed organic and inorganic matters, with strong capacity of adsorption and degradation of organic matters.

(1) Principle of activated sludge process

The activated sludge treatment process for wastewater is a procedure to place wastewater into an aeration pool to make the organic substances as a skeleton (mainly clay and also can be flocculent precipitate) in the wastewater with the sludge containing a large number of microorganisms. And so the organic substances can be decomposed into carbon dioxide, water and other simple matters by the microbes. Besides, the sludge with the aid of its strong bio-adsorption ability can adsorb suspending substances, colloidal matters, pigment and toxic substances. Thereby, the organic substance content in the wastewater can be reduced obviously and suspending substances, coloured substances and toxic substances can also be removed to different extents after the activated sludge treatment.

The activated sludge process can be divided into the following three phases according to the action principle: ①removing BOD of the wastewater through biological adsorption; ②removing BOD through aerobic microbial oxidation and degradation; ③removing BOD through the intracellular respiration.

The activated sludge has strong biological adsorption capacity. Biological adsorption includes both physicochemical adsorption and biochemical reactions of microorganisms to take the organic substances as the storage materials. When biological adsorption reaches the physicchemical adsorption equilibrium, the uptake capacity for the bio-storage substances reaches the saturation state, then aerobic microbes' oxidation and degradation in the second phase are started to remove BOD. During this process, the dissolved substances in the wastewater penetrate the cell wall to complete rapidly a series of biochemical reactions such as oxidation, synthesis and so on under the action of intracellular enzymes. The solid in the wastewater surrounding the cell walls is adsorbed and converted to soluble substances through extracellular enzymes secreted by microorganisms, then penetrates into the cell and participates in biochemical reactions in the presence of intracellular enzymes. Some parts of the organic substances (C, H, O, N, P, S, etc.) are oxidized and degraded into CO_2, H_2O, NH_3, PO_4^{3-}, SO_4^{2-} and so on respectively, and the other parts into necessary nutrients for microorganisms to form the new protoplast. Consequently, the microorganisms in the sludge are propagated rapidly and secret viscous substances to form bacterial colloids. With removal of BOD in the second phase, the amount of BOD in the wastewater decreases and the corresponding BOD in the microorganisms is reduced. Then the process enters the third phase in which the substances are oxidized to carbon dioxide, water, and ammonia eventually to maintain microbial cells. Oxidation and degradation in the activated sludge process proceed according to the following chemical reaction formulas:

Organic oxidation:
$$C_xH_yO_2 + O_2 \longrightarrow CO_2 + H_2O + \Delta E$$

Cell material synthesis:
$$C_xH_yO_2 + NH_3 + O_2 \longrightarrow (\text{Cell material}) + CO_2 + H_2O - \Delta E$$

Cell material oxidation:

$$(\text{Cell material}) + O_2 \longrightarrow CO_2 + H_2O + NH_3 + \Delta E$$

The activated sludge treatment process for wastewater aims at removing BOD of wastewater, precipitating and separating suspending substances to purify wastewater through the above oxidation and degradation reactions.

The key of activated sludge wastewater treatment is sedimentation ability of the sludge related to adsorption and oxidation. As oxidation lags behind adsorption, the sludge is light and its sedimentation ability is weak. Oppositely, as oxidation runs too fast, the sludge will break up and the sedimentation and adsorption abilities will decrease. As a result, the clear processed water is unavailable. In conclusion, the activated sludge treatment of wastewater demands for the significant sedimentation capacity of the sludge first of all.

(2) Microorganisms in the activated sludge

The activated sludge is composed of various microorganisms among which bacteria and protozoa are playing the primary roles. In addition to microbes' oxidation and degradation and cellular synthesis, the wastewater treatment can be completed through the zoogloea formed by microbial population. The zoogloea wraps the wastewater to conduct biochemical adsorption. Bacteria in the form of mono-cells are difficult to settle down but easy by forming zoogloea, and the suspending substances in the wastewater will settle down together. So the treatment of wastewater can be finished in a short time.

The zoogloea formed in the activated sludge is generally bacteria, including *Pseudomonas*, *Flavobacterium*, *Brevibacterium*, *Achromo-bacter*, *Aerobacter*, *E.coli*, and so on, based on identification of the bacteria isolated from the sludge. Sometimes there may be yeasts and molds in the activated sludge in treatment of certain types of wastewater.

(3) Culture and domestication of the activated sludge

During the activated sludge treatment process for wastewater, there must be sufficient amount and concentration of sludge, so culture and domestication of the sludge are necessary. The sludge culture is to supply proper conditions for growth and propagation of the microorganisms, which then form the sludge and reproduce in great numbers, so as to reach the concentration required for wastewater treatment.

The bacteria culture in domestic sewage plants is simple, but complicated in the industrial wastewater treatment system. The methods for bacteria culture include the following ones: ①bacteria amplification culture by multi-level; ②bacteria culture by dry sludge; ③bacteria culture by industrial wastewater; ④domestic sewage bacteria culture and then domestication with industrial wastewater which is toxic and difficult to be degraded.

To shorten the time, the bacteria culture and domestication processes are usually combined in the practical work. Domestication of sludge means decrease of the amount of domestic sewage and adding nutrients in the late phase of bacteria culture in the industrial wastewater, and gradual increase of the industrial wastewater proportion to 100% at last. For industrial wastewater treatment, especially toxic industrial wastewater, domestication of sludge is necessary in particular. The aims of sludge domestication include: ①reproducing the bacteria that can degrade the special substances and removing the bacteria that are not accustomed to the conditions; ②promoting and improving production of the enzymes produced by the bacteria with degradation activities. During the sludge domestication, microorganisms are adapted to the increased concentrations of toxic substances, and nutrients are supplemented if necessary. By increasing the concentration of toxic substances and the amount of water, the rate of load increases is 10%-20% everyday, and the COD value should be controlled at about 500 during the domestication process.

(4) Performance index of activated sludge

In order to increase the contact area between sludge and wastewater, and improve the treatment effect, the sludge should be present in the form of loose particles, and easy to adsorb and oxidize the organic substances. The index for measuring performance of the sludge are as follows. ①Concentration of activated sludge. It refers to the mass of suspending solid (*MLSS*) or volatile suspending solid (*MLVSS*) per unit volume of mixed liquor in the aeration pool, and is expressed as g/L or mg/L. Apparently the concentration of the sludge can directly reflect the microbial content in wastewater. The *MLSS* value of the sludge in the aeration pool usually keeps at the range of 2-6g/L, mostly at 2-4g/L. ②Sludge volume ratio (*SV*). It refers to the volume ratio of the settled solid to the wastewater after settling for 30 minutes. The *SV* value can indicate the normal amount of the sludge in the aeration pool. The higher *SV* is better for rapid separation of the sludge from the water. Generally the *SV* value keeps at 15%-30%. ③Sludge volume index (*SVI*). It's also called sludge index (*SI*) for short, and refers to the volume of 1g dried sludge after settling the wastewater in the aeration pool for 30 minutes, in the unit of mL/g. The *SVI* actually reflect the loose degree of the sludge. The higher the *SVI* is, the looser the sludge is. And this can improve the wastewater treatment effect for the greater surface area. However, the sludge will be too loose to maintain strong sedimentation ability with too high *SVI*. Generally, a healthy sludge aeration pool should have the *SVI* value controlled within 50-150mL/g.

The relationship among the three above ones can be expressed as the following equation:

$$SVI = \frac{SV \text{percentage} \times 10}{MLSS(\text{g/L})}$$

For example, if the sludge sedimentation ratio of an aeration pool is 24% and the settlement concentration is 3.0 g/L, the sludge volume index will be:

$$SVI = \frac{24 \times 10}{3.0(\text{g/L})} = 80(\text{mg/L})$$

(5) The technological process of activated sludge treatment

After processed in the primary precipitation pool, raw wastewater goes into an aeration pool containing a great amount of sludge, and stay for some time, most of the organic substances in the wastewater are absorbed, oxidized and degraded by the activated sludge, and subsequently the wastewater runs into a precipitation pool in which, the sludge settles down under gravity, and the supernatant can be discharged. The treatment process of activated sludge includes plug-flow, biological adsorption and completely mixing aeration treatment. The flowchart of completely mixing aeration is shown in Fig.12.3.

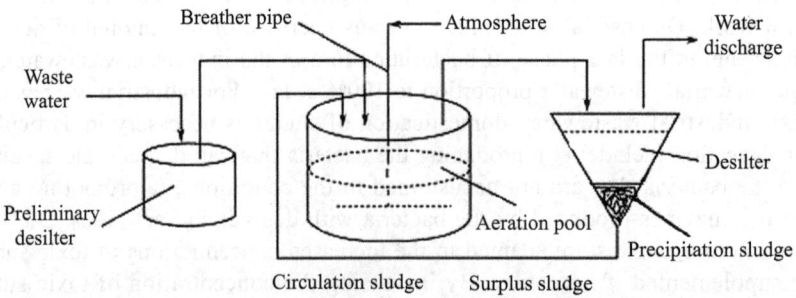

Fig.12.3 The Process of completely mixing aeration treatment

In order to ensure the best performance of the aeration pool, there must be sufficient activated sludge. Therefore, the settled sludge is pumped to the front of the aeration pool to repeat adsorption, oxidation and degradation of the organic substances in the wastewater.

In the normal continuous production conditions, the microorganisms in the activated sludge perform metabolisms utilizing the organic substances in the wastewater; and the amount of the sludge is increasing all the time. As the concentration reaches the threshold, emissions of partial sludge may be needed. This part of discharged sludge is called surplus sludge which can be cycled into aeration pool.

This process has characteristics of sludge reflux, which ensures the certain concentration of sludge in the aeration pool and is helpful for continuous run of the system.

(6) New progress in activated sludge treatment

Many process studies on the conventional activated sludge method have been carried out for decades. Especially in the recent decade, people have been making efforts to improve the bearing capacity for the concentration of flooding wastewater organic substances, increase the treatment performance and efficiency, strengthen and enlarge the functions of activated sludge. So the activated sludge process is developing toward the aspects of high speed, high efficiency, low consumption, and so on. The main progresses include the following new methods: deep aeration, pure oxygen aeration, addition of chemical coagulants and activated carbon, deep well aeration, biological contact oxidation, and so on.

2) Biofilm method

For the microorganisms on the surface of the biofilm are all aerobic, it's very necessary to keep ventilation. The filter materials can't be too thin, or they will be blocked more easily and influence the performance in consequence. Moreover, the filter shouldn't be too thick. The ordinary filter layer (upper layer) is in depth of 30-40mm, and the supporting layer (lower layer) 50-70mm. The filter layer in the high load filter is in depth of 40-70mm, and the lower layer 70-100mm. The biofilm thickness is appropriate at about 2mm. The structure model for biofilter is shown in Fig.12.4.

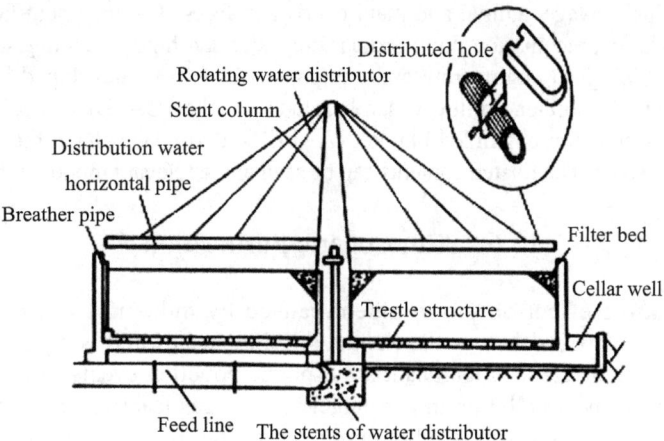

Fig.12.4 The structure of biofilter

The basic flowchart of a biofilter is the same as that of activated sludge, and consists of primary precipitation pool, biofilter and secondary precipitation pool. In order to prevent the filter layer from being blocked in the biofilter, it's necessary to set up a primary precipitation pool to remove the suspending and colloidal particles in the wastewater. The secondary precipitation pool used for separating the fallen biofilm is in small volume, because the moisture content in the biofilm is lower than that in activated sludge and its settling rate is faster.

The industrial wastewater containing organic substances is fed from the filter top, penetrates the filter layer, distributed on the filter material surface evenly by the water distributors, and flows down along the filter gap. The microorganisms and suspending substances are entrapped by the filter materials and form the biofilm. The microorganisms can adsorb the organic substances on the filter surface as nutrients, propagate rapidly, further adsorb and degrade the organic substances in wastewater, and finally enter the collecting drain at the bottom to discharged out of the pool.

3) Photosynthetic bacteria treatment

Photosynthetic bacteria gain energy through photosynthesis, respiration and fermentation or denitrification. Because of photosynthetic bacteria having the ability to endure and decompose the wastewater with high concentration of organic substances, they have been mainly used for industrial wastewater treatment in food fermentation industry, leather industry, and printing and paper industry in the foreign countries in the recent decades. Photosynthetic bacteria are so widespread in nature that they can be found in any place in the presence of sunlight. Photosynthetic bacteria can only utilize H_2S, NaS_2O_3, and organic substances as the hydrogen donor for carbon dioxide reduction, as they conduct photosynthesis with light energy. And this is an important characteristic in photosynthetic bacteria classification.

Oxidation pool process means completing wastewater treatment by means of microbial purification in the natural or man-made/man-improved pools and pothole pools where the organic sewage can stay for long time. In the recent decade, the oxidation pool for industrial wastewater treatment is in rapid development and mainly applied in the USA and Canada. In China, it is mainly used for secondary treatment, as well as full treatment of industrial wastewater.

12.5.2.2 Anaerobic biological treatment

The processed objects in anaerobic biological treatment are organic industrial wastewater in high concentration, municipal sewage, animal and plant debris and feces. The treatment facilities at the early stage include double-layer precipitation pool, ordinary digester and high-speed digester. In recent years, many new anaerobic biological engineering and equipments have been developed including anaerobic contact digestion, anaerobic biofilter, up-flow anaerobic sludge blanket (UASB), anaerobic fluidized blanket (AFB), anaerobic attached film expanded blanket (AAFEB), two-phase anaerobic process, combined anaerobic process, and so on. The related contents can be seen in the relevant monographs.

12.6 Atmospheric Purification

The other negative effect on environment caused by industrialized mass production is atmospheric pollution, which destroys the human's living spaces and is significantly harmful to human physical and mental health. Significant atmospheric pollution events, such as the smog event in London, the photochemical pollution in Los angeles, toxic gas leakage accident in Bhopal, India, the minamata disease, bone pain, and rice bran pollution caused by water and atmospheric pollution in Japan, took place and shocked the world in the history. Atmospheric pollutants are mainly toxic and harmful gases emitted through human activities, industry, agriculture and livestock, factory and mine, transportation, volcanic eruption, nuclear explosion, and so on. Carbon dioxide, carbon oxide, sulfur dioxide, and other toxic gases emitted through industrial activities are responsible for formation of global greenhouse effect and acid rain.

Among many kinds of atmospheric pollutants, about 100 types of them are harmful to environment and human. They can be classified as primary and secondary pollutants according to the formation process. The primary pollutants are the pollutants discharged directly from pollution sources. The secondary pollutants, more toxic than the primary pollutants, are the new pollutants produced through reactions between unstable primary pollutants and atmospheric materials or interactions between primary pollutants. The common pollutants include sulfuric acid, sulfate aerosol, nitric acid, nitrate aerosol, and so on.

Atmospheric pollutants can affect water, soil, and crops through a variety of channels and also enter the human bodies through the respiratory tract, diet (like drinking water), and skin contact to result in direct, short-term, or long-term harm to human health and ecological environment. Atmospheric pollution is one of the most important environmental issues in the current world.

The atmospheric pollutants can be divided into solid, liquid and gas according to their physical state; and they can also be divided into particulate matter, sulfur oxide, carbon oxide, nitrogen oxide, hydrocarbon and halogen, according to their chemical properties.

The key of preventing atmospheric pollution is to have control over pollution sources. The air pollutants produced in various production processes can be formed, one is the aerosol pollutants like dust, smoke, droplets, and many other particulate pollutants, and the other is gaseous pollutants, such as SO_x, CO, CO_2, NO_x, H_2S, NH_3, organic gases, and so on. The aerosol pollutants can be separated out through external forces like gravity, inertial force and centrifugal force for their large mass; and this process is usually named dedusting. The gaseous pollutants are treated through adsorption, condensation, combustion, catalysis and other approaches. This section briefly describes the methods of atmospheric purification.

12.6.1 Biopurification Method

Biological purification of waste gases refers to conversion of gaseous pollutants to less toxic and even non-toxic substances utilizing metabolism of microorganisms. There are so many kinds of microorganisms present in the nature that almost all the inorganic and organic pollutants can be converted. It has been used for waste gas treatment and control. In particular, it has gained great progresses in microbial treatment of volatile organic gases, stench treatment in metal factories and solid waste composting chemical plants, and coal desulfurization technology to control production of SO_2. Biological purification doesn't need regeneration and other treatment processes. Compared with other methods, it has benefits of simple instruments, low investment and operating cost, excellent performance, straightforward management, high adaptability to pollutant concentration changes and low energy consumption. It can realize harmless treatment of the exhaust gases. Biological purification has been successfully applied to volatile organic gases treatment in Germany and Netherlands. However, it is just limited for the pollutants in low concentration and gases in simple composition because it can't be used to recover the pollutants.

Nowadays, scientists all over the world are making efforts to construct the new strains that can purify waste gases rapidly through microbial engineering, genetic engineering, and other high biotechnologies. If some breakthrough progresses can be made in this area, biological purification for gas treatment will be applied more widely and effectively.

12.6.1.1 Principle of biological purification

1) Biological purification of hydrogen sulfide

Microbial purification of hydrogen sulfide (H_2S) is based on direct oxidation effect of

Thiobacillus ferroxidans and indirect oxidation effect of *T. denitrificans* and *T. thioparus*.

2) Biological purification of organic substances

The toxic substances in waste gases can be converted to less harmful or harmless substances on the basis of the nutrients requirement for microbial metabolism. When the organic substances are degraded and utilized by microorganisms, some part of the decomposed products are assimilated to new cells, and the other part supplies energy for propagation, growth and life activities of microorganisms. Finally all the organic substances are converted to substances with low or no toxicity.

12.6.1.2 Equipments for biological purification

1) Biological filtration devices

Biological filtration devices are commonly used for stench degradation, and necessary conditions are that the pollutants in waste gas must be adsorbed by filtration materials and utilized by microorganism, and the products of biological conversion should not impede the main conversion process.

The filtration materials, such as compost, solid wastes and so on, are usually used as the medium for microbial growth. In order to stabilize the filtration blanket and prolong the contact time, the gas flow velocity is usually controlled at 1-10cm/s and the materials should be loosened and replenished at regular intervals. Biological filtration devices are divided into biofilm filter and bio-trickling filter.

Biofilm filter and soil bed are applicable for treatment of aromatic compounds such as toluene and styrene, and aliphatic compounds such as propane and isobutane.

Bio-trickling filter is suitable for bio-reactions that are sensitive to pH changes because the pH value of the circulation fluids in the bio-trickling filter is easy to monitor. It's mainly used for biological treatment of the waste gases containing easily degradable halide (CH_2Cl_2).

2) Biological absorption device

Biological adsorption device is made up of absorber and waste gas treatment reactor, and the general process is shown in Fig.12.5.

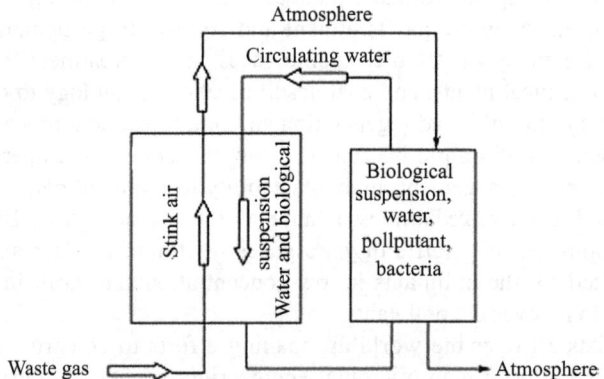

Fig.12.5 Biological absorption device

The waste gas is fed in from the bottom of the absorber and contacts with the absorbents in water in a backward flow, and consequently the pollutants will be absorbed by the water. The gas is emitted from the top of the absorber and enters the wastewater treatment bioreactor. The wastewater carrying pollutants flows out from the bottom of the absorber, enters the bioreactor, and is recovered through microbial reactions for recycling.

12.6.2 Membrane Separation Method

Membrane separation is a new method for atmospheric purification. Its basic principle is that when the mixed gas penetrates the stable membrane under the gradient pressure, the different components of the gaseous mixture can be separated according to their different permeation rate. Therefore, different gaseous pollutants can be separated by using the membrane in different structures.

In accordance with the types of membrane materials, the separation membrane can be divided into solid and liquid membrane. Liquid membrane technology is developed in recent decades and applicable for separation of SO_2, NO_x, CO_2, and H_2S in waste gas, but still not applied on an industrial scale.

At present the separation membrane used in practical industrial departments is the solid membrane. Various solid membranes can be classified as flat, tubular, hollow fiber, and spiral types according to the membrane shape, as homogeneous and composite membrane according to the membrane structure, as porous and nonporous membrane according to the pores of the membrane, and as inorganic and polymer membrane according to their materials. Porous membrane, such as sintered glass and porous celluloseacetate membrane, generally has the pore size between 0.5μm and 3μm. Nonporous membrane, like solid with ionic conductivity, silicone rubber, polycarbonate, homogeneous acetate fiber, also has pores but they are in small size.

Membrane separation has advantages of simple process, convenient control, high operation flexibility, room temperature operation, and low energy consumption (no phase transition energy). This technology has been used in petrochemical industry, hydrogen recovery in hydrogen synthesis, natural gas purification, oxygen enrichment in the air, and removal and recovery of carbon dioxide. It's expected that membrane separation will be widely applied for purification and recovery of the atmospheric pollutants.

The common membrane filter in gaseous membrane separation devices is classified as the following two types.

12.6.2.1 Prissm gas separator

The hollow fiber polysulfone is used as raw material and coated on the surface with a layer of hardenable polydimethylsilane in the thickness of 5-10μm to form the separator. This separator, as shown in Fig.12.6, has strong permeability and selection ability.

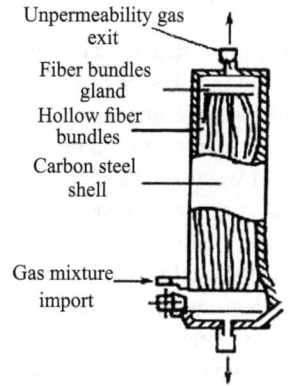

The structure of this device is basically similar to that of a heat exchanger, and it is composed mainly of shell, hollow fiber, and the tube-plate at the ends of the fiber. In practical use, after the raw gas enters the shell, the permeable fractions pass the hollow fiber into the center and then flow out, and the less-permeable fractions flow out from the outlet of the shell.

12.6.2.2 Spiral plate membrane separator

As shown in Fig.12.7, this spiral membrane separator

Fig.12.6 Structure of prissm gas separator

has a porous infiltration pipe and the membrane supports are wrapped outside the pipe. The high-pressure raw gases are fed into the "high-pressure duct" and the effused gases through the membrane enter the "penetration duct" and run out from the center (the separated component). The residual gases just move out from the flow channel outside the pipe. The membrane and the supports make up the membrane blade which is closed on the three sides, to isolate the raw gases and the permeable gases. The component is 200cm in diameter, 1 m in length, more than 8MPa in strength, and 6×10^4-$12 \times 10^4 m^3/h$ in material flow rate. The asymmetric membrane of acetate fiber is widely used for separation of hydrogen, acidic gas CO_2, H_2S and steam, carbohydrate, and oxygen.

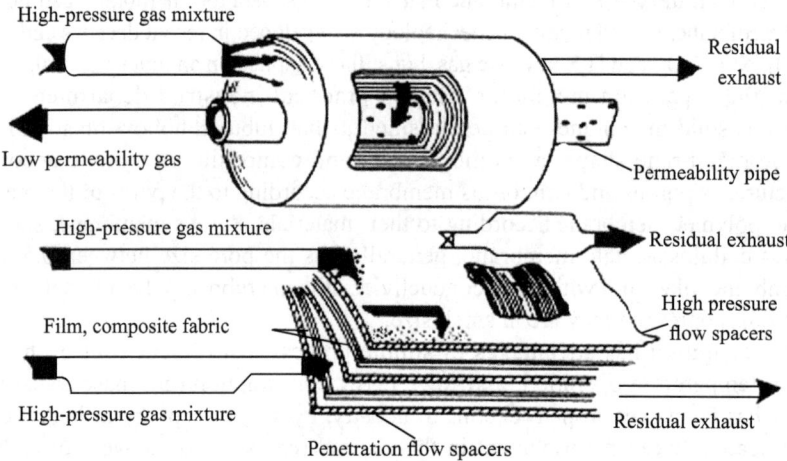

Fig.12.7 Membrane components of spiral gas separator made by separex

12.7 Biotreatment of Solid Wastes

It's learned that the amount of municipal garbage in China reached 135 million tons in 2004 and the amount of garbage is increasing at the rate of 4.8% annually along with acceleration of urbanization. The composition of municipal garbage is complicated, some parts are made up of glass, plastic, and metals and the other parts are composed of degradable solid organic substances such as paper, food waste, sewage, dry branches and fallen leaves, wastes produced by large-scaled farms, and so on. A great amount of garbage may contain or produce harmful substances during collection, transport and treatment. As a result, atmosphere, soil and water were polluted, affecting the municipal environmental quality and threatening human health, and it has become a scourge.

In recent years, a series of solid waste treatment technologies and devices adapted to domestic economic and technological development are used in succession in China. Some large and medium-sized cities have built solid wastes sanitary landfills, but most of them are still treating the garbage by means of stacking or simple landfill. Because the percolate collection system meeting the environmental protection requirements haven't been established and percolate can't be collected and purified, the water resource and the surrounding environment have been in serious pollution. To be worse, the fact that landfill gases are not collected and emitted has caused many explosion accidents. Scientists consider that solid waste treatment in China lags far behind the developed countries and the treatment technologies and devices can't meet the demands for social economic development at all.

Facing up to the fact that technologies and devices for domestic garbage collection, transportation, separation, landfill and incineration are so backward, the State Economic and Trade Commission has published "10th Five-year Plan of Environmental Industry Development" and increased investment in comprehensive treatment of paragenetic and associated minerals and the industrial three-waste. The composting technologies for municipal garbage have been developed. A series of highly mechanized composting plants, with matured pre-treatment, fermentation and post-treatment engineering and equipments have also been established in Wuxi, Hangzhou, Shanghai, etc., running in good condition, composting product quality, reliable operation, and environmental quality index have been in high level. Besides, a series of simple high-temperature composting systems with less mechanized but practical have also been established, and the mobile simple screening production lines are mainly used for stale garbage treatment, and they are playing important roles in treatments of the domestic garbage in large amount and the piled garbage.

12.7.1 Biotreatment of Solid Garbage

12.7.1.1 Landfill technology

Landfill technology has been the major way of domestic garbage treatment for a long time. It is an engineering process in which solid wastes are piled up in big pits or bottomlands, so as to restore the paleogeomorphology and maintain ecological equilibrium through scientific managements. Landfill method is able to treat the garbage in large amount, simple to operate, and with low investment. At present, 70% garbage is treated by landfill in both USA and England. However, if the technology is not operated properly, it will cause bad-looking appearance of the landfill and secondary pollution, like air pollution caused by odour. For example, abnormal smell will be produced, the breeding of flies and mosquitoes will take place and health conditions deteriorate, and the harmful wastes will affect metabolism of microbes in the landfill obviously along with harmful runoff occurrence or leakage into the groundwater to pollute the water resource gradually. In addition, the methane gas from fermented landfill garbage will easily cause explosions and other accidents. Now the traditional landfill technology has been modified. For landfill site selection, the bottom layer should be 4m higher than the groundwater level at least and there must be water-impermeable rock or clay layer. If there is no natural impermeable layer matrix, the water-impermeable materials (asphalt or plastic membrane) should be placed to avoid soil and water pollution by the percolate. The exhaust port should be set in the landfill so that methane can be emitted in time to avoid explosion and fire and the gas collection can be easily completed. Moreover, there must be monitoring systems for pollution situation of the groundwater, surface water, and the air in the environment. Sometimes the landfill needs some pretreatment for landfill things. The closed landfill constructed properly should be in better performance for waste treatment and methane produced during the process can be in commercial use.

In most Western countries, the landfill quantity decreases to reduce the demands for land, and this also enhances operation safety correspondingly. In the foreseeable future, landfill will continue to play important roles in solid waste management.

12.7.1.2 Composting

Composting is an important way to realize resource utilization and reduce the municipal wastes. Like the landfill technology, composting is also based on microbial metabolism. It's just microbial degradation that converts the organic wastes in the garbage to the stable humus. And these products reduce the volume of the raw wastes greatly and can be used as soil conditioner or fertilizer to return the environment safely. Composting is more suitable for treatment of household solid wastes with degradable organic substances in high contents. Composting is conducted at the bottom blanket covered by solid organic particles, in which the inherent microorganisms are growing and propagating. Successful composting requires for the optimal growth conditions for microorganisms. Because many operations demand for isolation and bio-heat is produced in microbial reactions, the internal heat of the compost is accumulated rapidly.

12.7.2 Waste Slag Bio-Leaching Technology

Bio-leaching mainly refers to microbial metallurgy or leaching technology, and is a new kind of engineering in the contemporary hydrometallurgical industry. The rare metals in commercial values can be recovered and extracted by means of microbial dissolution, i.e. bioleaching, when poor ores, waste ores, tailings, or fire-metallurgy slag are treated by bioleaching. Microbial mining is able to prevent loss of mineral resource, make the best use of the mineral reserves, and avoid or reduce solid waste pollution.

The metal bio-leaching reaction is involved in oxidation of mineral sulfur which can be conducted by bacteria, fungi, yeast, algae, and even protozoan. There are a large number of sulfur-contained substances in most of the minerals like iron sulfide, and the valuable metals can be dissociated through oxidation effects.

Chapter 13

Safety and Social Ethics of Contemporary Bioengineering Technologies

13.1 Ethical Problems with Human Cloning

The word "Clone" derives from "klon" in Greek, meaning branchlets or branches for propagation. Generally speaking, in English, clone is an asexual propagation line, that is to say, some progenies or the population composed of them are propagated from the same individual through asexual propagation and have genes completely identical to the parental individual. Its essential characteristic is perfect consistence among organism individuals in genetic composition. Some times we also call it embryonic cloning technique in life science and technology.

For plants, clone refers to direct regeneration of new individual plants from somatic cells, buds or branches. In such asexual reproduction, generation of new individuals needs neither fertilization process nor germ cells.

Preparation of plant clones is simple. Actually, the technique was utilized in propagation of good varieties long time ago. In today's planting industry, plenty of varieties are reproduced with cloning technique. Plant clones can be produced on a large scale for humans need in a factory, by utilizing plant cell and tissue culture technique, artificial induction and differentiation technique. Plant cloning technique is a common and effective approach for preservation and rapid propagation of rare plants and endangered plants, as well as preservation and rapid popularization of excellent traits of new crops or fruits, and will play a growing important role in the future production.

People used to think that all the somatic cells from higher animals have no totipotence, only eggs possess the capacity to develop into a new whole individual. Therefore, as for animals, so called asexual reproduction still needs involvement in eggs. And hence, animal cloning refers to artificial transfer of somatic cell nuclei into eggs without nuclei which then develop new individuals.

Nuclear transfer technique can be divided into two categories. One is less difficult but has high practicability, as the nuclei come from cells with no differentiation. With the technique it is easy to culture cloned animals, but it doesn't arouse enormous international response for its little difficulty. The other one is usually called somatic cloning technique for the nuclei coming from differentiated somatic cells, which is much more difficult in cloning, because somatic cells lose their totipotency after differentiation, i.e., plenty of genes are in rest state, and cannot express. A method called "serum starvation" should be adopted to conduct dormancy treatment of somatic cells, so as to restore their totipotency and make it possible to clone animals with somatic cells. Success of this technique is a miracle of life science.

Success in animal cloning will provide effective means for protection of endangered species and it will also play an indispensable role in preservation of species diversity and biological gene resources. For example, white-flag dolphins living in Yangtze River and pandas are very rare and endangered in China. White-flag dolphins have not been seen there in Yangtze River for many years, because human development destroys and interferes with their living environment. The only male dolphin raised artificially has already become aged but hasn't found its mate. If we don't take effective measures, the rare species, white-flag dolphin, will go into extinction in the near future. In such severe circumstances, animal cloning technique may be used to culture and preserve dolphins' somatic cells. With the eggs from the aquatic mammal animal cowfish genetically similar to dolphin as the receptor for nuclear transfer, allogeneic nuclear transfer may be performed transplantation, so that it is possible to culture and obtain the dolphin clone, preventing this human's friend and precious genes from dying out.

It is also useful to utilize the technology in rapid reproduction, popularization and application of fine varieties of animals. When some animals have been cultured to produce important protein drugs with genetic engineering technology, in order to apply them to production as fast as possible and amplify the production scale, it is necessary to reproduce the transgenic animals displaying excellent characters rapidly. Natural sexual reproduction must take much time because it will happen when such transgenic animals have their sex matured. At the same time, artificially introduced exogenous target genes will segregate according the genetic rule during meiosis or lose during passage. And so most reproduced progenies are not necessary for us. Now, utilization of somatic cloning will well solve those problems. As new individuals are produced through mitosis, they have the same genetic materials and production performances as their parents. In a word, the technology saves time, labour and material.

Somatic cell cloning technology is also used in culture of human organs, which promotes development and application of artificial culturing technology of human tissues and organs. The organs for transplantation have two big problems, as the organs are only derived from allogeneic bodies, the first is that we can collect organs only by means of personal contributions, and the number is limited because everyone has just a set of necessary organs and normally people cannot contribute their organs by maiming themselves, the second is that we must consider the allogenic immune rejection in transplantation, even though we have had the organs for transplantation. It is reported that it is little possible to successfully transplant organs and make them run well, except for corneal transplantation. If we can use patients' own somatic cells to culture and produce organs to be transplanted with somatic cloning technology, it is possible for us to solve those problems above and achieve our dreams. Of course, it will depend on development of ESC culturing technique and artificially induced committed differentiation.

13.1.1 Problems with Cloned Animals

Any cell or individual of higher organisms has two sets of independent genetic systems. One is the well-known main system, composed of complete genes in cell nuclei, which is in charge of most life activities of organism structures and cells. The other is cytoplasmic genetic system, composed of mitochondrial DNA and chloroplast DNA in cytoplasm. Plants have two cytoplasmic genetic systems, i.e., mitochondrial DNA and chloroplast DNA; while animals have only the former one. The cytoplasmic genetic system has close association with the nuclear one, but the genes of mitochondrial DNA and chloroplast DNA will not be found in chromosomes in cell nuclei. So functions of the cytoplasmic genetic system cannot be replaced by the nuclear one. As animal sperms cannot provide cytoplasm for progenies, mitochondrias of higher animals are all from maternal cytoplasm, in other words, all of mitochondrias come from maternal lines and have no relation with paternal ones.

After the clone sheep "Dolly" had been born for a few years, it was found that its telomere was shorter than normal ones, indicating its short life span. Dolly is a duplicate of a Finnish ewe, because its cell nuclei came from the Finnish ewe, and its cytoplasm from a Scottish black-face ewe. However, Dolly's cytoplasm was different from the Finnish ewe, and the following mitochondrial DNA analysis also demonstrated that Dolly's cytoplasm came from the Scottish black-face ewe. So Dolly was not a copy completely identical to the female ewe as the nuclear donor hereditarily. With mitochondria's crucial functional roles in life activities of cells and close relationship with many physical activities, Dolly will also have obvious difference from its maternal chain physically.

Nevertheless, the birth of Dolly showed it was possible to get individuals hereditarily identical to females using biotechniques. If the egg without nucleus as a receptor and the somatic cell both come from the same individual, the genes in the obtained duplicate's cytoplasm and nucleus will be identical to the maternal individual. Only on this occasion can we say the clone is actually a copy of female.

However, the maximal mitosis frequency of animal somatic cells is limited, and the age maker will be left in chromosomes each time. So the physiological age of this actual copy will not be identical to its actual age, that is, cloned animals will appear premature senility. And physical examination of Dolly seemed to confirm the conclusion. Then, it may have some distinction between organisms reproduced by cloning and the ones obtained by normal sexual reproduction.

13.1.2 Application Value and Possible Compact of Cloning Technique

The storm of "cloned sheep" once astonished the earth, hit the headlines over the world, and was even accused of speculation. As for HGP, its significance and compact outweighs that of "Dolly". Then, how much do people know about it? Or, what response do people have? HGP is well-known in America, and Time Weekly reports its progress in Column of Science News every year. And media propaganda plays an important role for the public to know about genes, understand gene science and adapt to social development. While, it is also found that the blanket coverage of some terms such as "gene times", "gene revolution", "gene medicine"and so on has puzzled us. Meanwhile, many discourses such as "it's not a dream to have a long life by dealing with genes", "genes will design your perfect offspring", and "genetic weapon may arise to make races become extinct", create a great sensation. The international society has come to agreement about the response to gene science and its following risks and challenges. Biotechnologies do bring new opportunities and challenges, and people had better take control of our own destiny, rationally facing up to these challenges from humans.

The birth of Dolly has shown that it is possible to copy humans with biotechnologies theoretically. If sexual reproduction is replaced or partially replaced by asexual one, our family, the essential organizational unit of human society, will be heavily impacted, and the basic life style of our society will be changed greatly. All of these consequentially will endanger and shake our foundation of laws, ethics and morality, directly and fundamentally, break the basic framework of society, causing social unrest and disorder and then leading to civilization regression. Therefore, in less than two months from Dolly's birth, many countries like Britain, America and China successively claimed that research funds supported by the government would not be used in any studies on human cloning. Nevertheless, someone in the United States still put forward the plan to clone human, and wanted to put it into effect as soon as possible. Many Americans asked the government to control the studies by legislation.

As previously analyzed, we cannot get clones hereditarily identical to the original individual with male somatic cells, even if the technique has been completely mature. As the mitochondria in cloned individual is different from the one in original one and mitochondria has vital importance in physiological metabolism, physiologically the cloned individuals are most different from the original ones. However, the somatic cells and the oocytes from females may be used to clone and get the individuals completely identical to them in genetic composition and physical characteristics, but the ones completely identical to their psychological state cannot be obtained. Formation of psychology state has something to do with both individual heredity and one's social experience and environment. And hence, for human beings with mind and social attributes as their essential features, it will not have any progressive significance to clone them.

We always desire to keep healthy and maintain vital energy by replacing injured or aging organs with healthy and vigorous new ones through organ transplantation. Every year, it is estimated that millions of people long for transplantation because of organ injury or failure. According to statistical estimation, thousands of people die form lack of organs to be transplanted every year, while the number of people who want to repair defect tissues far exceeds the data. Only in America, one million patients with Parkinson disease need to repair their nerve tissues. Therefore, plenty of corporations have positive prospects for the market of tissue and organ culturing technique. It is estimated that the expected output value of the human organ culturing industry will reach trillion dollars.

Somatic cloning technique has provided technical possibility for culture of human organs. However, when obtaining human tissues or organs in this way, we have to destroy cloned embryos even though one's own eggs and somatic cells are used to conduct therapeutic cloning or organ culture, which is still a great challenge to human beings and their dignity. And there are also serious obstacles to people's minds and ethics, because eggs have sacred meaning for humans, it is entirely possible for embryo obtained through somatic cloning to develop into a living human being.

So, are there other methods or possibility for culture of target organs or tissues which do not need to use eggs and embryos? Recently, scientists have had some discoveries and researches on main genes in charge of cell differentiation, as well as culture and researches on embryogenic potential of embryonic cells and adult stem cells. All these results bring us with new hopes for culturing human tissues or organs without eggs and embryos.

13.1.3 Social and Ethical Problems with Human Cloning

The reason of Dolly's sensational response over the world is that if higher mammals like sheep can be cloned, cloning human may have little problem in theory and technology. And for a moment, there are the following worldwide discussions: Whether we improve human's genetic characterristics by extracting cells from great scientists, politicians and entrepreneurs over the world? Whether we design and create all kinds of supermen possessing higher physical abilities and intelligence by combining cloning and gene recombination technique? Whether we shall create an army of slaves when the technique was stolen by the underworld? Will the world be dominated by cloned humans after sexual reproduction are replaced gradually? Really, there is an evil shadow, the cloned human, wandering around the world, which probably infringes on human dignity, wrecks human relations and has unexpected social harm. Then, what are the social and ethical problems brought about by the cloned human?

13.1.3.1 Ethical problems with human experiments

In some ethicists' eyes, such behaviours as cloning animals like sheep, monkeys, pigs and cows may arouse problems of animal ethics or rights. But there are opposite views that we can avoid ethical problems by well treating these laboratory animals, instead of treating them cruelly. As for use of human bodies for cloning experiments, it just means extracting a somatic cell from an adult, which does not seem immoral. Just like doing the blood test, we can get numerous cells in a blood test. In the process of cloning human, one somatic cell combine with one oocyte without nucleus, and this is a problem of embryo experiment for one person. Some ethicists abroad think that it actually relates to an ethical issue, because the rights of human embryos have been infringed on, especially when this research is not for the therapeutic purpose. However, the opposite view is that embryos, especially early embryos are not an individual at all, and so protection of some rights or respect of some dignity is meaningless. But when implanting the cloned egg or embryo into its maternal body to make it develop and grow, we will face a series of moral and ethical problems, namely, rights and dignity of experimental women and children. As you know, science can make one success after numerous failures; "Dolly" was also the only one obtained from 434 pairs of experimental cells after plenty of failures. If we do experiments with human body, we will inevitably clone a large number of abnormal human beings, including monstrosity and the disability (physical or mental or genetic disability). In addition, we cannot forecast what will happen to those clones' offspring, and meanwhile the abortion rate of mothers is quite high during experiments. All these will bring unhappiness to the current generation and their descendants. Can it be said that it is humane or moral? Can it constitute a legal obligation? Just in terms of protecting rights of women and children, we also should forbid or limit such experiments. When some huge crisis hides in the whole experiment process, the experiment should not be done, and this is the first hot spot.

13.1.3.2 Problem about human dignity

Assuming it is true that cloning human is successful and genetic recombination technique is also employed in the process, we can duplicate thousands of talents, such as Einstein, the king of football Bailey and movie star Marilyn Monroe. However, if we can copy genius, can't we copy a gang of mobsters who have violent and criminal genes? Can't we copy a string of slaves to accept commands to clone the "Maker" and attain the goal? In terms of sociology, these will cause social disorder and the problem of the "Genetic Class"; while in a sense of ethic, one problem will arise, and that is whether development of human cloning will make human beings themselves become tools, but not goals. Undoubtedly, it's a challenge to human dignity. All these worries seem to be science fictions, so as to make up a horrific and farcical scene. But it has been possibe to come into reality for science in the 21st century, and we should avoid abusing science and technology. The longest journey starts with a single step, so just do it now.

13.1.3.3 Problems about human rights of progenies

Assuming that we can absolutely forbid the cloning technique and genetic engineering copying harmful genes, and just the ones of talents can be chosen to recombine, we will live in a good and kind society, where people are clever and competent, handsome and beautiful, positive and socially active and live a long and healthy life. It will be a great victory of eugenics. But immediately, it brings about an ethical problem, that is, we have no rights to impose our value criteria and views about the good and the bad, the beautiful and the ugly on our descendants with genetic methods, instead of education. If the technique advanced 200 years earlier, people could have mastered perfect cloning technique and genetic recombination method in the Qing Dynasty, and then, our contemporary females would all have a pair of bound feet, even though the feet were duplicated with genetic engineering (including cloning technique), rather than by cruelly binding the feet of women. We have no freedom to choose our own fate and no rights to choose our value, moral ethic and aesthetic judgments. This is a problem about human rights of progenies, or an issue of our responsibilities for them, and also an ethical problem.

13.1.3.4 Damage Problems to diversity of human gene bank

The species of human being has a rich gene bank which constitutes individual diversity, leading to unity and stability, and forming intrinsic natural regulation and spontaneous balance between characteristics and capacities among individuals. And that is so-called ecological balance of natural genes. There are all kinds of persons in society, clever or foolish, strong or weak; competent for some jobs (for example, public security) in characters and ability, and others for another job (for example, accounting work), some one is good at scientific work for his/her prominence of the left brain, and another one good at artistic work for his/her prominence of the right brain. All these form a series of balances. Of course, human beings are social animals, and the decisive factors of our behaviours are not only genetic characters, but also growth developments, as well as social environments and educations. But for convenience, we have to employ single factor analysis, popularized in contemporary science, to study human behaviour. Though we just take genes into consideration, it doesn't deny environmental and educational influences. But the analysis to identical twins shows that genes play much more decisive roles in human behaviors than environmental factors. In a word, only in terms of genes, genetics has already showed us that there are differences between characters and capacities among human individuals. The differences and the inherent distribution can keep natural genetic ecological balance. However, appearance of human cloning makes it possible to replace sexual reproduction through cloning and genetic recombination, which will break the natural genetic ecological balance. We have to establish a national planning commission about cloning to decide which and how many human beings to be copied. As mandatory planned economy is used to replace market economy, it won't work if we substitute mandatory planed mechanism for natural selection, regulation and balance. At least, it doesn't work at the present stage of human development. It will make human beings excessively intervene natural process, breaking the inherent balance of ecological system. It is an issue of ecological ethic, namely, what ethical viewpoint and attitude we should have to face with nature. If sexual reproduction was substituted with asexual one, natural balance and our gene bank, together with human diversity and spontaneous adjustment will be destroyed. We should realize that human beings are not master of the nature; and we cannot dominate it as conquerors. And we must get along well with the nature, just as one of its parts. Meanwhile, the problem should be considered at an ethical level to know about and study relationship between the mankind and the nature. In fact, just like protection of our existing ecological environment, protection of our safety of genetic materials is a new ethical principle.

In a word, human cloning brings about lots of problems, some of which are traditional, such as human dignity and rights, family ethics, social responsibility, public moral and so on, and others belong to new ethical ones, like human rights of progenies, natural, ecological and environmental ethics, as well as human status in the nature. All these researches are conductive to ethics development.

Considering the problems above, we should keep study and application of human cloning under supervision and control of governments and related departments. And experiments of human cloning are forbidden without basic conditions. Only when we can make sure that animal cloning technique works well on human body, and laws and regulations can avoid its social harms, can we permit to conduct human cloning experiments. This is our basic attitudes towards human cloning. However, we haven't revealed the value viewpoint and conflicts are hidden in prohibition against human cloning.

Although Marxism opposes the counting happiness just according to psychological factors of pleasure and pain, and states that mostly maximum happiness just relates to distribution of welfare, Marxism ethical viewpoint is in accordance with utilitarianism in some aspects. One of basic principles of Marxism moral doctrines is working for the interests of the vast majority of people, even at the cost of sacrificing one's personal interests, which is regarded as a virtue. Therefore, we propose to forbidding cloning experiments from a Marxist viewpoint, considering that human cloning encounters various social ethical problems. But everything in the nature and society undergoes continuous development and change, including human ethics. Scientific spirit should adjust to our ethical viewpoint, and scientists themselves need to take social responsibilities and conform to social norm, and then the social ethic itself will develop with scientific and technological progress. It is a two-way coordinated development. We should carefully research and forecast development of genetic techniques, as well as how it can be in accordance with our social ethic.

It has its two sides for science and technology in society, positive or negative. And now, our scientists and public think more of its positive influence. Some ordinary scientists tend to support cloning experiments, thinking their goals are pursuit for truth and working for people's happiness. It is argued that they do it for submitting to conscience. Of course, it is also possible to abuse the scientific achievements, but science and progress can overcome such abuse at last, just as invention of fire will cause some disasters but fire brigade can extinguish it finally. So generally, they will cherish optimistic attitudes towards development of cloning technique. While our country is just at a stage of realizing industrialization by relying on science and technology to create achievements, and the public also have higher expectations for positive roles of science and technology. They tend to have scientific optimism and usually neglect its negative effect. For example, abuse of pesticides brings us a silent spring without twittering; and atomic energy has endangered our survival and environment pollution would destroy our earth. However, there is no sufficient awareness of those problems above. It is just necessary for us to explain and reveal the potential hazards resulting from human cloning, as we should declare nuclear weapons' harm to human beings.

13.2 Bioweapons and Biowarfare Agents

13.2.1 Concept of Bioweapons and Biowarfare Agents

The so-called bioweapons refer to biowarfare agents or bullets made from pathogenic microorganisms or biotoxins and its carriers, serving for military purpose or war requirements. Sometimes, it will also be used by terror groups and be viewed as origin of terror. The bioweapons were also called bacterial weapons at early stage, because pathogenic bacteria are used as warfare agents. And with development of science and technology, the biowarfare agents have gone beyond the scope of bacteria. Bioweapons were first used in the First World War, but its large-scale research and development started after establishment of immunology and microbiology in the 1930s In 1936, Japanese aggression army established Japanese Germ Factory in Harbin, and threw bacterial bullets in many regions of China from 1939 to 1942. And American army also used bioweapons in subsequent Korean War. At present, the internationally accepted biowarfare agents can fall into two groups, potential and standard. There are more than 23 kinds of pathogenic microorganisms and toxins, falling into 6 categories, which can be used as biowarfare agents which usually are distributed by aerosol form on large scale. Among present weapons for mass destruction, area effect of biological weapons is the biggest. World Health Organization estimates that when a strategic bomber utilizes different weapons to attack unprotected crowds, the destructive areas are as followed: 1 million t of nuclear weapons, $300km^2$; 15t of nervous chemical agents, $60km^2$; and 10t of bioweapons, 10 thousand km^2. During the Second World War, British Army once threw a bomb of *Anthrax bacillus* into Georgia Island, and caused it to be still a desert island now. It is the evil of bioweapon that brings about public extreme indignations all over the world. In 1972, the United Nations signed the international convention, which prohibited developing, producing or storing of bioweapons and called for destroy bacterial and toxin weapons. However, the minority of developed countries never give up preparation for biological wars, just in a much more secret way. As bioweapons are much easier to be made and smuggled than others of mass destruction, its threats to the whole human beings are enlarged after the cold war, rather than eliminated.

Bioweapons can cause diseases by utilizing biowarfare agents; there are 8 kinds of biowarfare agents armed, and 10 kinds stored presently in Americas Army. Across the world, partial wars and terrorism still exist, but all the people longing for peace insist on resisting any war or terrorism activity. Research and application of bioweapons or agents will bring us pains and disasters, so it is of vital importance to learn about the agents (or weapons) and their developing history, as well as what we can do to defend against them in the period of peace and construction.

Biowarfare agents and weapons are complementary to each other; the former generally refers to pathogenic microorganisms, toxins and other biological active materials to injury human and livestock and harm crops in military operations. It generally has characters as follows: ①most strong pathogenicity, even in small dose; ②easy mass production without limit of seasons; ③steady virulence during production, storage, transport and distribution; ④rapid reproduction and strong adaptability, easy to spread due to small size of pathogenic bacteria; ⑤"engineered pathogens" with stronger destruction capacity; ⑥effective self-protective measures for users, such as prophylactic vaccines. However, we have to recognize that more virulent strains are resistant to antibiotics or vaccines.

With rapid development of contemporary biotechnology, genetic weapons with destructive capacity at the molecular level now have been an important part of biological ones. Though the international conventions have forbidden using bioweapons, some countries are still studying and developing them secretly in order to meet their own needs. And some developed countries also worry about that others have owned some of them. On one hand, bioweapons have bigger destruction areas than any others; on the other hand, they would never be willing to fall behind others and positively research and develop those weapons with stronger destructive power on the contrary. Especially, the utilization of DNA recombination technique is either more beneficial to develop and produce more powerful weapons, or provides potential conditions for developing new diverse weapons. And, consequently, traditional bioweapons are developed into the new stage of genetic ones.

The so-called genetic weapons have an extremely important status in the bioweapons. Firstly, biowarfare agents with original performances have been further improved, because of exogenous harmful pathogenic genes can be transferred into some pathogenic bacteria via molecular vectors, i.e., plasmids, so as to increase their virulence; and original effective vaccines and antibiotics and diagnosis methods become invalid by changing the antigenic structure of pathogenic microorganisms. Secondly, new microorganisms with stronger pathogenicity or toxins and resistance to drugs were bred directionally by using genetic engineering technology. These biowarfare agents of engineered microorganisms have extremely high fatality rate, or can cause diseases difficult to treat. For example, the bioweapons made by transferring the snake venom gene from cobras to flu virus will cause the attacked victims to suffer from poison of snake venom, besides flu. According to the report titled as *Bioweapons and Humanity*, which is compiled by committed experts of Consolidating Medicine Association, the genetic weapons in the future will work in two ways below. Firstly, specific functions of bioweapons will be improved with genetic technology to upgrade their own destruction power; secondly, probably gene weapons will be developed to attack specific genes or races theoretically, which inevitably results in race extinction.

At present, at least scores of countries and numerous terrorists in the globe have mastered the manufacturing technique of bioweapons. For instance, harm of manufacturing smallpox virus agents is no less than Ebola virus, while smallpox vaccination has stopped in most countries all over the world and so human bodies have low immunity to the virus and are easy to infect. By estimation, if we distribute 20 pound of ultra-concentrated kreotoxin to Washington, half of its citizens will die. However, manufacturers of these bioweapons would not stay at the level. If genetic technique and bioweapons are combined, new bioweapons with much more destruction power will be produced. Why does some one produce these weapons? The only explanation is that this is natural response of those war lunatics. War is what they need, what they think and what they persue. Though the international conventions have forbidden manufacturing and using bioweapons, those war lunatics would not stop relevant researches on novel genetic weapons. Whoever loves peace and is opposed to war should clearly realize that those war maniacs are taking chances to do something non-humanistic regardless of disagreements from the public, as well as that contemporary biotechnology amazingly develops and it can also serve wars. At the same time, we must keep keen vigilance. Some war maniacs in the history once used human body as vector to research bacterial weapons and brought extreme disasters to people; and on the other hand, relevant international organizations should take measures to regulate and supervise genetic weapons as soon as possible, in order to avoid abusing high and new biotechnologies being for military purpose.

Biowarfare agents are the determinant factor for destruction power of bioweapons, and can be classified into 6 categories based on their morphological and pathological traits.

13.2.1.1 Bacterial agents

It is bacteria that were used to make biological war agents earliest in wars. They have the following features in the optimum conditions: ①rapid reproduction, their generation time is about 20 to 30 minutes and won't be over 1 hour; ②high adaptability, especially for those screened bacterial strains; ③strong virulence, those bacteria with strong virulence to livestock and human bodies such as pathogens of pestis, cholera, typhia, tuberculosis, syphilis, malaria and diphtheria are chosen; ④belonging to the heterotrophic type of bacteria, the most common bacteria used for bioweapons are *Bacillus anthracis, Yersinia pestis, Vibrio cholerae, Pasteurella tularensis, Brucellsis leucosis, Bacillus mallei, Bacillus tularense, Clostridium botulinum*, and etc.

13.2.1.2 Virus gents

Virus is a kind of organisms which has non-cellular morphology, specialize in obligatory parasitic life and can replicate themselves. It consists of nucleic acids and proteins, and widely distributes in the nature to infect human beings, animals, plants and microorganisms. Among all of pathogenic microorganisms, virus ranks the first place, and accounts for 75%. Its reproduction entirely relies upon host cells for its obligatory parasitism, that is, it cannot survive without its host. The earlier virus agents were arboviruses, and then developed in the form of aerosol. After being infected, most hosts manifest the symptoms of corresponding diseases slowly, but disease progression is out of control. Common viruses used as agents are the yellow fever virus, *Venezuelan equine encephalitis virus*, *Smallpox virus*, *Marburg virus* and so on, and other people also have tried to make use of Ebola virus culture as biowarfare agents.

13.2.1.3 Rickettsia agents

Rickettsia is procaryotic organisms whose trophic mode is identical to virus, whose size and physiological characters lie between bacteria and virus. There are about 10 kinds of rickettsias pathogenic to human beings, and the common ones include epidemic typhus rickettsia, Q-heat rickettsia and so on, among which Q-heat rickettsi has extremely strong infective capacity. For example, it is reported that the human body will be infected as long as one person inhales one rickettsia.

13.2.1.4 Chlamydia agents

Chlamydia is prokaryotes between bacteria and virus, which generally parasitize in human beings and animals, belonging to intracellular parasitism. Only a few kinds of them can cause diseases. The main agent of this category is Chlamydia ornithosis, which is also called psittacosis Chlamydia, it has low effective infective dose, but high infective capacity. Its caused diseases have rapid progression, and some patient with severe diseases may be fatal. Besides, the pathogen just stimulates weak immunity, and its hosts are still easily infected after being vaccinated. During the Second World War, Japanese militarists once used it to infect carrier pigeons, in order to cause the group infection of Soviet Union army and pigeons.

13.2.1.5 Fungus agents

Coccidioides *immitis* and *Histoplasma capsulatum* are fungus agents that cause general lesions via the respiratory tract. As about 80%-90% infectious crop diseases are caused by fungi, it is the main agents for destroying crops. Experiments done by American Army proved that about 70% of crops would be infected only with 3g *Magnaporthe grisea* per hectare rice field.

13.2.1.6 Toxin agents

Toxins are poisonous chemicals metabolized by animals, plants and microorganisms, much more poisonous agents can be obtained through biotechnology, so we also call them as bio-chemical agents. Protein toxin produced by bacteria has strong toxicity and is easy to realize mass production. The common toxins used as biowarfare agents are botulin, enterotoxin of *Staphylococus* and so on, which infect human bodies via the respiratory and digestive tract.

In spite of various classification modes, we should learn about and keep a vigilant eye upon every kind of biowarfare agent, together with their harm, and take some countermeasures. Any biowarfare agent will lead to a series of unsafe factors to human beings and farming husbandry, and even probably cause to local prevalence of some diseases.

Among the bacterial biowarfare agents which are used most frequently, pathogenic ones

Chapter 13   Safety and Social Ethics of Contemporary Bioengineering Technologies

take up big proportion. For instance, *Bacillus anthracis* can cause anthracnose, an acute infectious disease, and both humans and animals may have this disease. It was discovered in some countries and regions where farming husbandry developed earlier, including four ancient civilized countries, Egypt, India, Ancient China and Babylon. Its spore from pathogenic ones has great vitality in the air so as to maintain pathogenicity in external environment for over ten years, even several years in animal skins, because the spore has strong resistance to heat, dryness and chemical disinfectants. Anthracnose is a universal infectious disease, for example, cutaneous anthrax is the most common, pulmonary anthrax and intestinal anthrax can cause serious systemic poisoning, and the three kinds of anthrax can all cause acute hemorrhagic meningitis. The bacterium has two virulence factors, one is capsular substances, protecting from phage invading as well as lysozyme dissolution in phagocytes; the other is *Colletotrichum* toxin, composed of edema fraction (EF), lethal fraction (LF) and protective antigens (PA), the second one is a dominant lethal factor. By the way, a fatal recurrence phenomenon also exists. For example, patients' condition may suddenly worsen and they will die soon, although intestinal anthrax has been almost cured; and any emergency treatment does not work. But generally speaking, any species, livestock or human, will be faced with a disaster when suffering from anthracnose. Persons whose lungs are infected are bound to have anthracnose, and 90% of them will eventually die. Once, a report provided by Department of American Technology Assessment in 1993 showed that it would cause death of 3 million people if biological bulletin was dispersed using a light aircraft which was loaded with 100kg of *Bacillus anthracis*. Application of contemporary biotechnologies enables the agents to be much more destructive. A British magazine (in 1997) declared that some country (Russia) had developed new *Bacillus anthracis* with transgenic technology, which can resist against the vaccines and antibiotics. At the same time, lethal genes of the anthracis was inserted into *E.coli* DNA producing a new "recombinant virulence stain". The new strain had much more powerful infectivity for wild animals and livestock (especially cows and sheep), causing massive hemorrhage and death; by the way, it can also infect human beings. In Russia, there were once varieties of reports that antibiotic didn't work well against anthracnose, while American government thought they deliberately created it due to its high vigilance against Russia. If these drug-resistant pathogens were used to make biowarfare agents, we would be faced with major disasters, so it must be forbidden. As for natural drug-resistant or drug-dependent pathogens, they do exist and tend to spread. It is a research subject in medicine, and some countermeasures should be taken. According to reports, the proportion of penicillin-resistant bacteria has risen from 6% to 44% in a decade since 1980. American researchers believe that some new bacteria and viruses, which all the existing antibiotics cannot inhibit and show powerful drug-resistance, will emerge in the future and should be researched more. However, characters of artificial drug-resistant pathogens are entirely different from natural ones, especially research and development for the purpose of war should be alerted to take corresponding measures and follow international treaties strictly. As for those natural drug-resistant pathogens, their existence or prevalence will also endanger our life, but we can reduce or avoid the outcome by strengthening effective control over studies on them.

13.2.2   Dispersion Modes and Infection Routines of Biowarfare Agents

Power of bioweapons relies on dispersion device and modes to play, such as plane dissemination, bullet firing, pollution of water and food and insect transmission. Generally speaking, there are three dispersion modes of bioweapons: pollution of water source or food, dissemination of aerosol of biowarfare agents and dispersion of insect carriers. For the aerosol of biological war agents, their dispersion is universal. The so-called aerosol refers to a colloid dispersal system formed by liquid or solid particles which disperse and suspend in the air. Almost all the biowarfare agents are dispersed in this mode of aerosol, which has the following features: First, it has high efficiency and wide covering area, for example, the aerosol dispersed with a bomber or missile can pollute ten thousand square kilometers in area. Second, in this mode bioweapons can invade into human body directly through the respiratory tract, and an adult can be infected in 1min in the contaminated area. Third, the aerosol is not easy to awake through human sensory organs or common reconnaissance instruments. Forth, it can invade into some equipment works, naval vessels, vehicles and buildings without protection against chemical or bioweapons. Fifth, it is able to infect human through water, food and other things indirectly. However, only most of the particles in the size of less than 5μm can come into the pulmonary capillary bronchia and the alveoli to cause diseases. The American army stipulates that the aerosol is safe if it is dispersed after 2h in the day or 8h in the evening. The stability of aerosol has been improved with continuous advancement of protective agents and microencapsulation technology. The American imperialists and Japanese militarists once used insects to disperse biological weapons in some countries such as China, Korea and so on. No matter in which mode they are dispersed, the biowarfare agents mainly infect humans or livestock through the digestive tract, the respiratory tract or the skin. Understanding these can help us get some knowledge basis for prevention against pathogenic bacteria infection and enhancement of self-protection awareness.

13.2.3 Detection Techniques of Biowarfare Agents

13.2.3.1 Application of biosensors

Biosensors are one of important approaches to detect biological samples, and usually used to analyze samples' properties, chemical composition and concentration. Allowing for such advantages as high selectivity, high specificity, high measurement sensitivity and automatic operations, the technique is applied in many fields, such as medical science, environmental monitoring and food hygiene, as well as bio-chemical warfare agent detection in military. The biosensor technology can be used to detect pathogenic organisms and biotoxins, whose mode of action has high site-specificity. It is thought that immunosensors are the best choice for microbiological tests, and meanwhile specificity and sensitivity of nucleic acid hybridization is also suitable for making biosensors. With development of biotechnologies, multifunctional sensors, mini-sensors and high-sensitivity ones (surface photo voltage immunosensor, integrated optical biosensor, etc.) will certainly be developed and get breakthrough in detection of biowarfare agents.

13.2.3.2 Application of biological detectors

A novel high-precision bio-detector was researched and made by an institute of the American navy, by simulating the immune mechanism in the human body. Human body itself will automatically produce antibodies fighting against antigens (bacteria) when pathogens infect human body, and based on this mechanism many biosensors have been designed. For these biological detectors, an extremely thin silica gel cantilever looking like people's hairline is employed as the carrier and the antibodies are coated on its surface. When antigens meet with the antibodies, the response of bending towards the antigen will occur and can be detected. Besides, the cantilever contacts with small magnetic balls coated with the antibody liquid in the magnetic field, and when it contacts with antigens it will also bend towards them. We can test the number of antigens by the change of bending degree and antigens.

Such bio-detector is expected to be on the market before long.

13.2.3.3 Application of Biochip technology

Biochip technology will become the most accurate tool to test bioweapons, apart from biosensors and bio-detectors. It was reported in an American magazine, Futurologist (in 1999), that biochip carrying encoding DNA could be used to determine some biowarfare agents such as virus and *Anthrax bacillus*. It has practical value in accurate diagnosis of the epidemic or other diseases. Affymetrik in America developed a genetic bar code on DNA chips, which made it possible to determine the potential incidence of some disease for human beings. In China, biochips have made important progress in research and development, which is hoped to achieve industrialization.

In a word, biosensors, bio-detectors and biochips are the most effective and precise approachs for detection of biowarfare agents and the caused diseases, and these have a promising future.

13.2.4 Countermeasures Against Bioweapons and Biowarfare Agents

People must maintain vigilance against biowarfare agents, no matter where they are from or how they are dispersed. Terrorists can use some virus or weapons of mass destruction to launch a surprise attack. For example, Japanese Aum Doomsday Cult once dispersed poisonous gas on the Tokyo subway in 1996. Similarly, war maniacs may use bombs or shells made from pathogens to serve warfare or terrorism incidents.

Rapid development of contemporary biotechnologies has its dualism. One is that contemporary biotechnologies applied to military will lead to two consequences, namely, more destructive weapons and more complete countermeasures; the other is that we can also apply contemporary biotechnologies for non-military purposes, such as industry, agriculture, medicine and environmental protection, and that is a welfare for all the peace-loving people.

In either the war or peace period, we should take some necessary countermeasures against biowarfare agents to protect not only human lives but also development of the peace situation, in a word, readiness is all.

13.2.4.1 Observe international treaties and forbid use of bioweapon

Clause 23 of Convention at Second Hague Peace Conference forbade using toxins and the weapons containing toxins in 1907. On June 17th, 1925, the protocol formulated at Geneva Conference firstly called for clearly every country to ban use of bacteria weapons, i.e. "poison gas or analogous poisons and bacteria weapons should be forbidden in the war". The American ex-president Nixon claimed in 1971 that the United States gave up using biological weapons, and on December 16th of the same year, the general assembly of the United Nations ratified Convention on Ban against Development, Production and Storage of Bioweapons or Toxin Weapons and Destruction of Them (abbreviated as Convention on Ban against Biological Weapons). And the convention was formally signed among others, by the United States, the United Kingdom and (former) Soviet Union in April of the next year. There were 150 countries and regions participating in its formulation and over 100 countries and regions had signed their names on the convention file. China had taken part in the convention in 1984 and claimed that having suffered grievously from bioweapon attacks in the past, China had so far never produced and would never produce or possess those weapons in the future. It was our country's stand and viewpoint in the whole world. And only if every country observes the convention, we can live a happy life.

13.2.4.2 Take some pertinent defensive measures

Anthrax bacillus can be dispersed through flies and cause diseases by blood-sucking worms' biting to invade in humans or livestock. Mosquitoes and flies are important media of many diseases, so it is an important work to wipe out them in both war period and peace period. In addition, other media such as rodents, lizards and birds may all be carriers with bacteria, and also be controlled strictly. Besides, some solutions of bleaching powder can kill *Anthrax bacillus*. It was reported that *Anthrax bacillus* (spore) treated with one bleaching agent were buried in sandy soil at the depth of 3-5ft (1ft equals to 30.5cm), then detected 10 years later, and it was found that its spores were still alive. This showed the strong survival ability of this bacterium. Finally, formation of highly poisonous substances by mixing bleaching powder with the nerve poison VX should be avoided.

13.2.4.3 Produce or store some preventive vaccines

An aerosol of a vaccine composed of four kinds of bacteria was once researched and made by former Soviet Union, and was sprayed over one city, which was called as "protecting all the citizens by absorbing vaccines together". In the initial phase of the Second World War, the United Kingdom once sprayed *Anthrax bacillus* vaccine in the islands to the northwest beach of Scotland. In the United States Ministry of Defense had researched and produced biological weapon vaccine to prevent fatal *Anthrax bacillus* infection and inoculated vaccines among over 2 million soldiers on service or reserve duty in the summer of 1998. In fact, application of the vaccine in the army was the safe bet for soldiers, because the vaccine was not living bacteria, so they would never bring potential danger to soldiers. And it showed that the United States was anxious about such destructive bioweapons appearing in other countries. So it is very effective to inoculate vaccines to prevent against the harm of biological war agents. Just as some specialists once concluded, epidemic with most high infectivity take only one month to spread over the whole world. But it takes at least two months to produce and disperse the vaccine, and it takes effect 2 weeks after inoculation. And it generally takes several years to research and develop a safe vaccine for people. So we should prepare vaccines earlier for prevention of bioweapons.

It is very potential to employ biotechnologies to improve the effect and prolong action time of vaccines. For example, subunit vaccines that are made of major antigens of pathogens can effectively avoid transmission of epidemic or destroy biowarfare agents. The gene of one subunit vaccine is transferred into another vaccine with genetic technology to improve its action; or, multivalent vaccines may be prepared to prevent multiple epidemics and reduce the frequency of inoculations. The United States has researched a vaccine called *Bacillus* Calmette-Guérin(BCG), which can be used to prevent against measles, diphtheria, tetanus, polio, and hepatitis. In 1999, American Congress provided 150 million dollars for the prevention of bio-chemical weapons, including drug pharmaceutical supply, disease surveillance and research of new generation vaccines.

13.2.4.4 Use drugs

Many different types of biological products such as interferon, immunoglobulin and so on are developed with genetic and protein engineering technology to treat the disease caused by biowarfare agents.

Injection of immunoglobulin to human body is regarded as passive immunity, and it usually can only provide the immune action of several weeks or months and is different from vaccines which can activate hosts' immune system to produce antibodies. In spite of difference from vaccines, immunoglobulin is worthy of prevention against biowarfare agents and has certain effect on biowarfare agents of virus and toxins.

Antibiotics make many contributions to human lives, do well in prevention and treatment of bacterial and rickettsial diseases, but can do nothing about virus and toxin diseases. It has been found in contemporary clinical medicine that some pathogens have powerful antibiotic resistance and tolerance, bringing about difficulties in use of antibiotics, especially for the synthetic drug-resistant pathogens. It is said that *Anthrax bacillus* once appeared in Russia, and all the antibiotics had no effect on them, which showed that the pathogen had powerful drug-resistance. So it not only causes many difficulties of clinical therapy, but also provides new subject and task to antibiotic researchers. And we should solve the problem at two levels, one is appropriately researching new antibiotics or multiple functional ones to cope with threat of any biowarfare agents; the other is strengthening the study on pathological mechanism of pathogenic microorganisms.

It is reported that American researchers have found a key gene called DAM gene in *Salmonella typhimurium*, which can control gene expression. But it will only work when the host is infected, while the expressed product can activate other genes. The DAM gene can exist not only in *Salmonella typhimurium*, but also in other pathogens such as *Vibrio comma, Plague bacillus, Shigella dysenteriae*. Of course, there may be other pathogens used as biowarfare agents containing the DAM gene. However, discovery of the new gene can possibly lead to birth of new type of vaccines and antibiotics.

13.2.4.5 Develop "universal gel" technology

It is possible for universal gel to become the universal decontaminant of chemical and biological weapons, or to be used for purification of industrial leakage. The gel researched and made by American Lintec Company is a mixture of two powerful oxidants, namely, H_2O_2 and compounds of three potassium salts which are made by DuPont Company. Mixed with ferric sulfate the gel can be used to spray in the polluted areas (peroxides can react with iron to produce highly active oxidants). The experiments have showed that the gel can rapidly convert major warfare agents to safe by-products, and wipe out highly resistant spore of *Anthrax bacillus* within minutes. And the gel will become dry to form powder after it is used for hours, so it is easy to clear up and may be safe to dispose. Besides, the concentration of peroxides in the gel just corresponds to that in collutory. Recently, American Ministry of Energy has used the universal gel to deal with the terrorism; it can purify all the facilities and can be used to cleanse spraying planes and insecticide dispersing devices. Meanwhile, it is also suitable to treat leakage of chemical pollutants, so it is worthy of the honour of universal gel. At the same time, it is cost-saving to produce the gel. Practical application has proved that 1L of gel which costs only 7 cents is enough to treat $1m^2$ of polluted areas.

Human beings have gone towards the 21st century. Looking back upon history and observing the present situation, we know that the world is not peaceful in the past or at present, partial wars still happen now and then. The ambitions of hegemonists to conquer the world have not yet died out, and they use the forbidden weapons publically, including biological ones; meanwhile, we should realize existence of bio-terroristic organizations exist, and they also possess the bioweapons. And no matter for what purpose, any use of biological weapons will bring unexpected disasters to people. China had suffered grievously from biological war aroused by American imperialists and Japanese militarists in the past, especially the latter once used their killing machine, the pathogens of *Plague, Anthrax*, and *Vibrio comma* (The biowarfare agents of former Army 731),to do bacterial experiments in ten thousands Chinese common fellows and prisoners-of-war from Allied Army. Japanese militarists once planned to launch a bacterial campaign against American army in Saipan Island in June, 1944, but later the plan was aborted. For our country, the attitude and stand towards bioweapons is clear-cut, that is, we have always insisted on forbidding and destroying, totally and thoroughly, all the bioweapons. We oppose to any country's researching, producing and storing bioweapons, or any country's dispersing them in any way. As one of the state members to the Convention, BWC, we have fulfilled our obligations earnestly and completely. With rapid development of contemporary biotechnologies, research achievements must never serve warfare. Because our desire is to master the technologies and to utilize them protect ecological environment and mankinds survival, and also for making contributions to the cause of peace through the balanced development.

At present, among all the worries about potential risks of biotechnologies, only bioweapons have turned into reality. Chemical wars have already struck us; biological ones are a threatening devil for mankind. Chemical toxins are nonliving, while bacteria, virus and other biological toxins have the capabilities of infection and reproduction. If they survive somewhere, they can multiply, not to mention the fact that bioweapons will have increasing harm as years pass. Some bio-toxins can cause disability and some are fatal. For example, Ebola virus can make 90% of victims who are infected with it die within about one week. And at the terminal stage, patients will have tics, shiver and shake sharply till death, with the contaminated blood around their bodies. Ebola patients are incurable and its transmission mode is also ambiguous for people. We cannot determine whether one is infected by human body by closely touching the infected blood or body or through the air around a person who's infected. Recently, Ebola infection has broken out in Zaire, which directly leads to separation of any region in this country till the virus naturally disappears.

Fortunately, terrorism initiatives involving in bioweapons have been rare so far. In September, 1984, about 750 people were sick after having meals in several restaurants in the Dalles town located in Oregon, USA. In 1986, Sheila•Anand•Ma confessed in a Federal interrogation that she once added *Typhi salmonella* into salad in 4 restaurants with a gang of fanatics after a conflict with the natives in Oregon. And those *Typhi salmonella* were cultured in a lab in their big pasture. It was reported according to the Office of Technology Assessment (OTA) U.S in 1992 that the minority used biological toxins as warfare agents. Maybe, the first practice of biological weapon use happened in the 14th century when an army embattled in Black Sea threw *Plague* bacteria within the wall. In the 21st century, the only one being convicted is that Japan launched the biological attack on Chinese with *plague* bacteria and other bacteria in the 1930s and the 1940s.

There are few events about use of bioweapons in the human history, which can be explained from multiple aspects. Their users may have no knowledge and don't know how to use pathogens as weapon; by the way, they were also afraid of being infected. As for some countries and terrorists, they would not like to use them due to their unpredictability. With time passing, bacteria or virus can produce new toxicity or lose original toxicity through mutation, which may turn against the thrower's strategic intention. When being dispersed in one environment, pathogens will bring threats to any person in this environment, which causes difficulties in conquering the area.

At present, potential biowarfare agents include *Anthrax bacillus* (causing fatal *Anthrax* disease, and vaccines or antibiotic can prevent against small doses of the *Anthrax bacillus*), *Botulinum toxin* (causing *Botulinum* poison, usually death, and antitoxin sometimes may prevent from disease deterioration), *Yesinia pestis* (causing lymph gland plague, i.e. black death called in the median century. Its mortality can reach 90%. The vaccine can stimulate immunity on this pathogen. If drug is administered in time, antibiotics may play effect on this disease.), *Ebla* virus easiest to infect through contact, high lethal rate. So far it has not been clear which treatment method has efficacy on it. And the effective countermeasures against bioweapons are dust mask, protective screen, purifying agents and vaccines, as well as antibiotics and reconnaissance systems.

Generally speaking, in the competition between bioweapons and defense system, the latter is relatively weak. However, as long as research on anti-biological weapons are strengthened by making use of a hearty dislike of biological weapons, the threats will reduce as soon as possible.

13.3 Safety of Genetically Modified food

Transgenic technology can transfer genes of any source into plants, which not only expand the sources of important agricultural genes, but also realizes the purpose of orientated improvement of plants. Just for this point, transgenic technology has promising potential in improvement of crop varieties.

Transgenic animals are obtained by introducing genes of one species into zygote of another one or developing embryos to get new organisms with some features of the two species. Many methods, such as microinjection, reverse transcription virus infection, embryonic stem cell and somatic cell cloning technique, can be used to obtain transgenic animals at present. Microinjection is the earliest and most universal technique which has the advantages of high transfer rate and short test period, and long transferred genes. Its shortcomings are complicated operation, expensive equipment, difficulty in popularization, uncontrolled copy number of target genes, and probable mutation of host DNA. As for reverse transcription virus infection, it can transfer single copy genes and is easy to analyze inserting sites; it also won't damage target genes or cause mutation. Of course, the technique has its own disadvantages, for example, the transferred genes are short, and its test periods are long; and it needs the mosaic route and may cause replication and expression of transferred virus genes.

13.3.1 Worries about Safety of Genetically Modified Food

With rapid development of biotechnologies and increasing appearance of genetically modified food (GMF), people gradually focus on their safety, namely, whether GMF would be a threat to human health or have negative consequences for our environment and the food chain. According to continual research and application of GMF, it has no clear distinction from common food. However, people still have query on its safety, for example, whether foreign genes are safe, or whether the structures of genes are stable and they have mutation to harm human health, whether they would produce new harmful genetic characters or unhealthy compositions after transferred into plants, whether they would lead to change of contents and absorption of nutrients and other functional components, whether they would cause resistance of human flora to antibiotics through plasmid transmission by means of GMF if we use the genes to antagonize clinical antibiotics, and whether food allergy would increase. All of these are proposed for further research and observation.

Currently, common evaluation methods and systems already cannot be completely applied in the safety evaluation of GMF. So, all the countries in the world, especially international organizations, are actively discussing and establishing evaluation methods and systems of safety of GMF, and promoting international coordination and mutual accreditation, in order to preserve human health and environmental safety while making progress of biological technologies. Recently, safety of GMF has become a subject of public focus, and now some debates on its common doubts from the public to media home and abroad are to be discussed.

Firstly, we will discuss the conception of food safety. In 1991, Organization for Economic Cooperation and Development, in short for OECD, gave us a definition that the food was safe if we could reasonably make sure it was not harmful to food consumers. As for evaluation on safety of GMF, it has an internationally substantial equivalence principle to adopt and employ universally. The principle stresses that the purpose for evaluating GMF safety is relatively safety compared with similarly non-transgenic food, rather than absolute safety. We should emphasize on "a single case study", that is, we should not generalize about the safety. If we make the case analysis with substantial equivalence principle, it's easy to have a conclusion that all the GMF products that have been approved to conduct commercialized production are safe. For example, the first approved genetically modified food tomato is safe as food, compared with non- transgenic one. The transferred foreign genes may produce antisense mRNA which can inhibit activity of the ethylene synthetase genes, and have no detectable gene product. As food, the transgenic tomato is as safe as the non-transgenic one.

Another example is insect-resistant plants. It is always questioned whether insect-resistant plants would be poisonous to people if they can kill insects. So we need to explain its mechanism of killing insects. The insect-resistant plants was cultured, and they generally have genes of insecticidal protein from *Bacillus thuringiensis* (*Bt*), the bacterium, commonly found in the nature, contains a large group of insect-killing genes which have been widely applied as pesticides for over 60 years, and researchers home and abroad have made thorough study on their mechanism of killing insects. The Insecticidal Crystal Proteins (ICPs) of *Bt* genes can exist in the parasporal crystal, which dissolves to produce protoxin in alkaline conditions after being eaten by insects, due to action of enzyme system in the midgut, the protoxin releases active toxin which plays toxic effects by binding to specific receptor in the gut of insects. However, the gastric juice of mammal animals is strong acidic and there is no receptor of Bt toxin in their stomachs, so the Bt proteins will be degraded completely under the action of gastric juice after coming into mammal's stomach. It has been repeatedly proved for years that the ICPs are safe to mammals, birds and fish, as well as some non-target insects. And it has no reports that people and livestock have been poisoned by insect-resistant plants in America where plants insect-resistant transgenic corn and cotton are planted in a large scale. Moreover, the technology nowadays can make transferred genes expressed in certain tissues and under particular conditions. For example, while breeding rice resistant to stem borer, we can make rice plants produce insect-killing proteins only in stem or only when they are eaten, rather than in the seed. So we should be convinced that we need not worry about the food which is made from plants with ICPs genes.

Someone worries about that foreign genes would change people's genes and pass them on to offspring after consuming transgenic food. Those people who have such a worry do not know facts, i.e., we eat billions of genes from a morsel of food. People have never worried about that the genes from animals, plants and microorganisms may affect people inheritance even if human society has survived and evolved for millions of years. And another fact we should know is that plenty of genes are homologous among genomes of animals, plants and microorganisms, which is found in genome sequencing projects on a large scale in recent years. In other words, animals, plants and microorganisms we often mention generally refer to the organisms whole, but many genes are not yet characteristic for animals or plants at the gene level.

13.3.1.1 Effect of Transgenic crops on eco-environment

The genetic characters of plants are determined by their inherent DNA. If we extract the gene from some plants or cells and transfer it into another plant cell, the new plant will possess new character of the foreign gene, such as antivirus, pest-resistance, and herbicide resistance. And the new plant variety produced through such gene transfer is considered to be a transgenic plant.

Any gene may be transferred into plants with transgenic technology, which not only expands resources of important genes with agronomic characters, but also realizes oriented improvement of plants. It is the feature that makes transgenic technology have high potential in improvement of crop varieties and universal application. At present, we have got transgenic plants with resistance to diseases, drought, virus and herbicides, because these characters are determined by single gene. Once a plant has some resistances, high yield will increase naturally.

The genetic factor of transgenic plants with resistance to herbicides can be transferred to wild plants via pollination and seed migration, so the wild plants can have the same resistance with transgenic ones appear. And some of them may become super weeds that possess resistance to common herbicides. Besides, such phenomenon that transgenic plants become weed themselves or invade into new ecological areas to destroy ecological balance has drawn public's attention. To remove these weeds, more powerful and higher concentration of herbicides have to be used; and this must cause soil compaction, soil quality deterioration and drug residue, so as to destroy our ecological environment.

13.3.1.2 Transmission of marker genes may lead to antibiotic resistance

The marker genes used in genetic engineering are generally antibiotic resistance genes. People worry about that foods contain marker genes can influence microorganisms in the intestinal tract or cause resistance to antibiotics. Naturally, impact of ICPs has also been a focus for people. If we apply the marker genes with antibiotic resistance in transgenic crops, they may be transferred into microorganisms, or retain and express in the human intestinal tract or ruminant stomach. So it is suggested that we should not use the marker genes with resistance to some important clinical antibiotics, such as Ampicillin, Vancomycin and so on. Besides, the microorganisms with maker genes must have no pathogenicity. In 1993, the symposium titled as "Marker Genes in Transgenic Plants and Health" was held by WHO to discuss possibility of antibiotic resistance gene transfer, the conclusion was that it had no evidence to prove gene transfer from plants to intestinal microflora. The major reason is that it is a complex process for antibiotic resistance gene transfer, including gene transfer, expression and its effect on antibiotic efficacy. In addition, the antibiotic resistance genes can be expressed only under control of appropriate bacterial promoter, so the genes controlled by plant promoter cannot be expressed in microorganisms. In 1996, "Conference on Experts" Consultation on Biological Technologies and Food Safety" and the WHO symposium were jointly held by FAO and WHO. At these conferences experts thought the exogenous genes in transgenic plants had little possibility to transfer to intestinal microorganisms, but the possibility still existed. FAO and WHO also held experts' consultation conference to discuss about under which circumstance the antibiotic resistant genes mustn't be used in transgenic plant food.

13.3.1.3 GMF may cause food allergy

Genetic modification may increase the level of some natural toxin in one plant unintentionally while the function of one target gene is developed. Some natural plant toxin genes, such as genes of potatoes solanine, cassava cyanide, legumes protease inhibitor and so on, may be activated to improve expression of such toxins to harm consumers. Potential allergens in GMF cause human immune system to produce antigen-specific IgE reaction. The technological consultation meeting on food allergy held by FAO in 1995 gave a conclusion that the common food that can produce anaphylactic reactions through IgE was fish, peanuts, soybeans, milk, eggs, crustaceans, wheat and nuts. Common anaphylactic food accounts for approximately 90% of food anaphylactic reactions. FDA and EPA require biotech corporations to report whether their GMF contains problematic proteins, but there is still worry about unknown allergens which may slip out from the registration system.

There have been examples that donors' anaphylactic characters are transferred into plant receptors, for instance, in 1996, Pioneer Hi Bred International Corporation once introduced genes from Brazil nuts into soybeans to improve the protein content of animal feeds, which made some consumers become allergic to Brazil nuts to produce anaphylactic reactions. In September, 2000, the Cry9C insect-killing protein gene from StarLink Bt transgenic corn were detected in cornmeal pancake sold by Kraft in American food stores, and all of such food products was withdrawn from store shelves at once. The Cry9C insect-killing protein may be allergen for its heat resistance and indigestibility. And the corn with *Bt* gene above was once approved to be used in feeds, instead of food, and Aventis Corporation had stopped to sell its seeds.

Safe evaluation on GMF allergy is to determine whether it has potential allergy by evaluating whether they contain the same proteins as allergens. If the resource of transferred genes has no allergy history, we can determine whether it is possible to have potential allergy via analysis of amino acid sequences and physical and chemical analysis and tests.

13.3.1.4　Others

The other controversy on GMF still includes their toxicity and human ethics.

Potential toxins such as drugs and pesticides are expressed via plant engineering, which may be harmful to other non-target living things. For example, toxin of *Bt* may be a threat for useful insects when it goes into insects' food chains. And we still cannot ensure whether it is poisonous or safe to people, though there is yet no report about GMF's threat to human body. But in fact, as a new thing with excessive intervention to the nature and less technical maturation, transgenic technology may have potential negative effect and should draw the public's attention. We should also face the state that modification of organisms via transgenic technology may bring about problems of something or others, while we are rejoiced over accomplishment of evolution in labs which need thousands of years in the nature. After all, the present genome in the earth is product of long period evolution; and we cannot confirm whether it will have harmful influence or not if randomly and artificially inserting, replacing or changing some genes. Only through repeated tests and trials can we make the final decision.

Some host genes may be destroyed due to the insert sites of foreign genes; the inserted genes and their products may induce expression of silent genes.

Products of insertion may react with some enzymes in some metabolic pathway to disturb it, and make some metabolic product to accumulate or lose in hosts. So we cannot ignore any change of cell components in new transgenic plants. Some debates on GMF safety include not only allergy, toxicity, carcinogenicity, nutrition, quality change and environment, but also some issues aroused by tags or other questions. Potential danger and other controversial issues are discussed in the following text.

Foreign genes may cause unpredictable changes in food's nutritional value, and for the changes both decrease and increase are possible. Besides, currently there are few data on effect of changes of nutrients in edible plants and animals on their nutritional value, interaction between nutrients, bioavailability of nutrients, nutritional potential and metabolism.

China is a multi-nationality country and people of various nationalities have their own unique dietary habits and limitations. So we have some worries about GMF: certain religious bodies forbid animal's genes being transferred into their common edible animals (e.g. transferring pig's genes into sheep), which violates some people's diet precepts; and vegetarians may be confused about plant food with animal genes. It is also hard to accept that human genes are transferred into animals or organisms containing human genes are used as animal feeds.

13.3.2 Evaluation of Safety of Genetically Modified Food

Some people think that common food is safe, while GMF is not safe for it's modified artificially. In fact, it is a serious misunderstanding. Firstly, genetic modification does not start with genes transfer technology. It is a conceptual mistake for media to call genetically transgenic organisms (GMO), rather than traditional crop varieties as GMO. As for crops as our major food, they have had been modified natural genetically for ten thousands of years, and contain thousands of modified genes, so there is little "natural" food in human diet. Secondly, the natural food is not certainly safe. Most plants in the nature will produce toxins in different quantities, forming a defensive mechanism to protect themselves against animals or insects. Mankinds' improvement of animals and plants with a history of thousands of years has greatly reduced toxins in food, but it is well known that food has still different negative impacts on human bodies, because many diseases have something to do with food. Besides, the safety we called is just relative, and it can be safe for some people but harmful for others. For example, prawns and crabs and other seafood will produce allergic reaction, though they are delicious for most people.

There are even people who think that we should make sure the long-term effect of GMF before approval of commercialization. But we all know that our food nowadays is safe though it has never got toxicity test and long-term safety evaluation before it becomes food. As for other food (high fat and high sugar) harmful for human health due to long-term eating, nobody thinks it should be forbidden. There is still no reports about unsafety of GMF, while transgenic crops have been planted for ten years and consumers still have been up to billions. The long-term effect of GMF safety can be seen from the above-mentioned facts. The relatively scientific attitudes towards GMF safety is following definition of "safety" and insisting on the principle of "substantial equivalence".

According to this principle, if the protein produced by an introduced gene is confirmed to be safe, or compared with the original crop the transgenic one has no special changes in major nutrients (fat, protein and carbohydrates), morphology, and production of nutritional factors, toxic substances and allergic proteins, the GMF is the same as the original crop in safety. That is to say, the principle of "substantial equivalence" states that GMF is similar to non-GM one in the aspect of impact on human beings.

Scientifically, for the food which has no short-term toxicity and safe problems, if we have a suspicion on its potential danger, we must take time to test its long-term toxicity and safety, besides, such long-term monitoring and tracking usually needs 10-20 years. So it is important and reasonable to strengthen management of GMF, before we get certain conclusion about their safety. The international safety standards must be stipulated and the principle of consumers' right to know should be implemented, i.e. we should tag on the pack of GMF and make consumers themselves to decide whether to buy. According to the general international evaluation methods of biological safety and the harm of GMF to human beings, plants and animals, microorganisms and eco-environment, agricultural transgenic organisms are divided into 4 levels: safety level Ⅰ (without danger), safety level Ⅱ (low danger), safety level Ⅲ (medium danger) and safety level Ⅳ (high danger). It can be seen that we have taken actions about GMF safety, and safety of GMF will be guaranteed entirely as long as we stick to the principle of "prevention in advance".

In terms of scientific development, any major academic achievement has its dual nature. With development of science and technology, traditional production life styles and ethic viewpoints of human society formed before have changed respectively, which is a fact. We need not be afraid of research and application of transgenic technology; of course, there has been never any major scientific achievement being stopped due to people's worries. However, compared with other scientific technologies, transgenic technology is much more important. So it is too late to regret if all the worries come true. But we don't have to worry too much about transgenic technology; otherwise it will interfere with scientific development. So making scientific exploration and experiments valiantly and spreading and applying research achievements carefully are the right attitudes towards transgenic technology we should hold.

13.4 Research and Application of Human Genes and Its Impact

13.4.1 Background of Human Genome Project

HGP is the third largest scientific project following Manhattan Project and Apollo Moon-Landing Project. In fact, it is just a product of the residual problem of Manhattan Project, even though viewed as "Moon-Landing Project" in life science.

In 1945, it was success of Manhattan Project that made the United States throw atomic bombs in Hiroshima and Nagasaki of Japan, causing death of hundred thousands of civilians and survivors received a high dose of radiation. We know radiation can destroy structure of DNA and then lead to gene mutation.

To research effect of radiation on human beings, the U.S. Congress asked Atomic Energy Commission (AEC), later known as America's Department of Energy(DOE), to start the study on effect of radiation on human gene mutation lasting for decades. In theory, gene mutation frequency of persons who have received radiation is three times higher than the control group. However, they hadn't determined gene mutation in 12,000 victims due to the technology at that time. Because it needed analysis of plenty of DNA sequences, which was impossible according to DNA analysis technology at that time. As a result, the researchers hadn't found any difference of DNA structure between those victims and the control group, though they had shown mutation characters.

In May 1985, American DOE held a meeting in Santa Cruz, California at which they firstly proposed the protocol to determine human complete nucleic acid sequence, the draft of later HGP, which was an audacity of imagination exactly. The meeting was in an uproar in May 1986, when Smith, the director of HGP, claimed the project in Cold Spring Harbor Meetings. Senior biologists, including three Nobel Peace Prize winners (Walk Gilbert, Paul Berg, and James Watson) firmly supported it, but it was objected by many biologists, especially young ones. The reasons for objection included not only the project itself, but also its potential effect on contemporary biological studies and its management ways. Their worries were various, and the most important was the problem on ethic, law and society.

13.4.2 Related ethical, legal and social problems with HGP

One important reason for considerable stir caused by drawing HGP draft is that it may have far-reaching impact on ethical, legal and other social problems, besides the great scientific value itself. Though the other two projects in the scientific history, Manhattan Project and Apollo Moon-landing Project, also have something with social value issues, they only cause problems in application. However, the HGP itself involves in many social value problems. So the Project contains a subproject at first to study its ethical, legal and social issues (ELSI). American DOE and

National Institutes of Health provide 3%-5% budgets of HGP every year to study ELSI about use of genetic information. ①Fairness of genetic information usage. Should it be used by insurance companies, employers, courts, schools, asylums, the law enforcement departments and armies? How should we use it? ②Privacy and secrecy. Who should possess and control such genetic information? ③Mental influence and harm. How dose one's gene defect impact on others' viewpoint to him/her in the society? ④Genetic assay (before birth, carrier and pre-symptom test) and census (infant, premarital and professional). Do parents have rights to let their kids examine potential diseases when becoming adults? Are examinations and explanations from healthcare teams trusted? Should they be tested without treatments? ⑤Reproduction, including informed consent procedure, genetic information use and reproductive rights. Do medical personnel inform parents of risks and limitations of genetic engineering appropriately? Which fetal genetic test method is reliable and useful? ⑥Gene therapy used for treatment and prevention of genetic defect. What's the normal gene? What's defect or disability? What is the dominant factor for defect or disability? Is disability disease? Should they be prevented and treated? Dose seeking for treatment of it depreciate those who suffer from disability? ⑦Increase in genes, including those controlling some traits (e.g. body height) which parents want to give to their kids by utilizing gene therapy method, but have nothing with disease treatment and prevention. What are the issues of safety and ethics? How do they affect diversity of gene bank? ⑧ Fairness in application of genetic technology. Who has the rights to use these expensive technologies? Who is to pay for it? ⑨Clinical issues, including medical service suppliers, parents, and common public educations, as well as the standard of quality control in the test and its implementation. How to make clear, reliable and useful evaluation on genetic test? ⑩Product commercialization, including utilization of property rights, data and information, and who possesses the genes and other DNA fragments. ⑪Concepts about human beings and issues of philosophy, including free will and genetic determinism, concepts about diseases and health. Can all the unique behaviours be decided completely by genes? Can persons control their behaviours? Which genetic distinction can be accepted?

ELSI project emphasizes on different focuses in different periods due to the rapid developments of HGP process. All the issues on ethics and society are novel and complex, bound up with our interests. Research on all the issues is significant for healthy development of HGP, legislation of public rules on medicine, and understanding of the project. ELSI research is an important and essential part of the HGP.

Accomplishment of the human genome draft and further study will provide details of human bodies for us and make us understand ourselves. In the draft most genes seem to be standardized replicas, i.e., every individual among human beings shares the same genes which account for about 99.9% of the total human genes and their uniqueness is dominated by the rest 0.1% of the human genes. Scientists now can already identify various gene defect types which possibly causes diseases. One of the ideals of HGP is that the detailed knowledge of the human genome can make us prevent or reduce potential sufferings, make diagnoses become more accurate, and make treatment become more effective and thorough. However, on the other hand, people concern that those who have genetic defects will suffer from unfair treatment in employment and social insurance. For example, insurance companies may banish those people to the cold palace, or get a cold shoulder from their employers. So many people think that those including employers, insurance companies and other non-relevant people should not know their genetic test results. Even people want nobody to know the results, including their spouse and relatives, because it is just their own business. So can insurance companies, entrepreneurs and commercial partners know their genetic makeup? Can their spouse and relatives know it? How about themselves, or related government departments?

Human genetic information is closely related with human body health, so we should first know our gene constitution so as to take measures to exclude unhealthy factors. For example, some person are born with cancer genes, but they cannot present the symptom at once, or the disease incidence is 50% at one's age of 30, while 70% at the age of 50 and 90% at the age of 70. If the cancer can be treated by simple surgical treatment at early period, then it is very useful to know gene constitution. So generally speaking, we should know about our own gene constitution.

However, on the other hand, it will have no use to know our own gene constitution if there is no effective method for treating the cancer, rather it will increase our mental or material burden. So in such a case we should not make our relatives know our gene constitution.

Many people think that the spouse has rights to know partner's gene constitution, if they want only one child. After all, a healthy and clever baby is their wish. They can avoid giving birth to babies with fatal genes or prevent and treat them before their birth or in the early embryonic stages. Babies with genetic diseases will also bring sorrow to parents, not only the babies themselves. However, many people think that the couple should know each other's genetic constitution, some people think that the normal partner will discriminate their defected spouse, even decide to divoice, leading to broken families, if they know their spouse's genetic defect. Still, there is a viewpoint that it's inhumane to forbid defected babie's birth, after all it's also a life. So it's better not know others' genetic constitution.

In terms of economic interests, insurance companies, employers or commercial partners want to know whether one's genes are abnormal, because it will result in diseases possibly or death early in some time. If business partners don't know cooperator's genetic defects, it likely results in loss of business when the loss is bigger even includes their shareholders, consumers or clients. So it is reasonable for them to know the truth. However, it is not very convincing. Because all the economic decisions have their risk, genes, certainly, should be seen as one of enterprise decision and analysis risks. If all the commercialized activities aim at knowing one's gene for interests, no one can have his or her genetic privacy. Or, if one loses his healthcare insurance, job or commercial partners, successive social issues will be more complex than unknown. Especially, genes defect is not viewed as disease in the early period, and it possibly has something with surroundings. So it's not wise to make commercial decisions just based on gene constitution.

Doctors and hygienic administration institutions may need patients' genetic information. It can be helpful to keep one's health, for example, they can provide guidance for patients' life and effective treatment of diseases. So it is generally thought that doctors should understand patients' gene constitution. But doctors should respect patients' autonomies and keep secrets before they know their own genetic information.

It is believed for hygienic administration institutions that they should know citizens' genetic information; because this can help them know their health conditions or diseases and make public hygienic policies to increase social welfare. However, the administration institutions should follow the autonomic and voluntary principle, rather than force them to provide their own genetic information, after all, it's not their responsibilities. In addition, people should collect those messages in an anonymous way in order to protect personal information from leakage, and then reduce potential hurt or social discrimination.

13.4.3 Social Effect of Human Genome Science

Social effect of scientific events should be analyzed from three aspects. The first is prediction in advance and analysis after events. Industrialized atomic bombs were tested without adequate prediction and analysis on social or ethical outcomes. The American War Department firstly tried out the atomic bomb in the world, and even forbade over three scientists to discuss the event, it is an event without forecast of social effect. However, environmental pollution from industrialization and inequalities in wealth has been repeatedly analyzed and introspection. Some environmentalists and postmodernists have been discussing them till now. The danger brought about from the atomic bomb is the thing that concerns all contemporary politicians and people around the world. The Roslin Institute, located in Edinburgh of Britain, hadn't estimated its social effect during cloning Dolly, while the international organization of HGP has established an ethical committee, because the HGP contains a subproject called as EISL to debate its ethical, legal and social implications. In addition, the research fund is up to about 30 million dollars, which can show much stress on prediction of the social implication and value of the scientific project and research in advance.

The second aspect for analysis of social effect of scientific events is analysis of short-term and long-term social effects. Social effect of scientific events presents as a causal chain, or rather causal network. The called analysis of short range social effects is direct social effect of scientific events and research. For example, how interpretation of the human genome takes revolutionary effect on research exploration in biology, so as to discover life mysteries; how the scientific event makes healthcare affairs change much, such as raising its efficiency and making the focus of medical service move toward disease prevention; how it increases patents in biological research and medical device a lot, as well as promotes establishment of many unprecedented medical companies; how it brings business opportunities to entrepreneurs and employment to workers and intellectuals; how it changes agriculture fundamentally to create much more nutritional food; how it arouses fight for genetic resources, then leads to contradiction, competition and conflict between developed countries and undeveloped ones; how it shocks social insurance affairs and individual privacy, all the outcomes are relatively direct and well-defined to be used for forecast of social effect. As for the far-reaching social effect, it is the indirect outcomes of one scientific event, and requires us to imagine the outcomes of something far from the causal chain. Analysis of it is of little definiteness and much necessity, which can make persons imagine something incredible. For example, how genetic engineering changes human evolutionary orientation and humanity; and the cloning technology may perish families and make social structure change greatly. Of course, human intelligence is limited, the farther from causal chains, the less definite and the more diverse. It will bring many problems if we regard the analysis as the inevitable results of social development, or think accomplishment as the only goal, even organize parties to fight for it. But social science must be involved in analysis and studies on every possible situation in order to put forward every hypothesis, which is called the analysis of far-reaching social effect.

Certainly, the last aspect of result analysis of scientific events is their positive and negative effect. Scientific optimists see only the positive effect, while pessimists see only the negative one. A wise scientist is just anxious, neither an optimist nor a pessimist, but both of them.

References

1. Cao JW, Ma HW. 2004. Microbial Engineering. Science Press, Beijing.
2. Cao W. 2000. Trends in the Research and Development of Bioproducts. Letters in Biotechnology, Beijing.
3. Cen PL. 2000. Industrial Microbiology. Chemical Industry Press, Beijing.
4. Cen PL. 2004. Introduction to Bioengineering. Chemical Industry Press, Beijing.
5. Chen HL. 2003. Environmental Biotechnology. Chemical Industry Press, Beijing.
6. Chen SF. 2003. Contemporary Microbial Genetics. Chemical Industry Press, Beijing.
7. Deng PJ. 2003. Transgenic Food Safety and Nutritional Quality Evaluation and Validation. People's Health Publishing House, Beijing.
8. Feng BS, Wang QY, Hu YX. 2000. Principles and Practice of Animal Cell Engineering. Science Press, Beijing.
9. Guo BY. 2000. Genetic Engineering Pharmacy. Second Military Medical University Press, Shanghai.
10. Guo Y. 1996. Enzyme Engineering. China Light Industry Press, Beijing.
11. Han YR. 2002. Molecular Cell Biology. Science Press, Beijing.
12. He H. 2003. Analysis of Biological Pharmaceutical. Chemical Industry Press, Beijing.
13. He ZX, Jing GZ, Xu ZL, *et al*. 1999. Introduction to Contemporary Biotechnology. Beijing Normal University Press, Beijing.
14. He XX. 2003. Principles of Biotechnology. Chemical Industry Press, Beijing.
15. Huang SQ. 2001. Introduction to Contemporary Life Sciences. Higher Education Press, Beijing.
16. Jia SR. 2003. Bioreaction Engineering Principles. Science Press, Beijing.
17. Jiao RS. 1994. Introduction to Bioengineering. Chemical Industry Press, Beijing.
18. Li JH. 2002. Biological Engineering. China Medical Science and Technology Press, Beijing.
19. Liu GF. 2001. Introduction to Contemporary Life Sciences. Science Press, Beijing.
20. Liu H. 2004. Contemporary Food Microbiology. China Light Industry Press, Beijing.
21. Luo GM. 2002. Enzyme Engineering. Chemical Industry Press, Beijing.
22. Luo LX, Pan L. 2002. Cell Engineering. South China University of Technology Press, Shanghai.
23. Ma F, Feng YJ, Nan Q. 2003. Environmental Biotechnology. Chemical Industry Press, Beijing.
24. Mei LH. 2001. Biochemical Production Technology. Science Press, Beijing.
25. Peng ZY. 1999. Food Biotechnology. China Light Industry Press, Beijing.
26. Qi XJ. 2004. Contemporary Biopharmaceutical Technology. Chemical Industry Press, Beijing.
27. Pu EQ. 1999. Air Pollution Treatment Projects. Higher Education Press, Beijing.
28. Song SY, Lou SL. 2001. Introduction to Biotechnology. Science Press, Beijing.
29. Sun SH. 1993. Nucleic acid Vaccine. Second Military Medical University Press, Shanghai.
30. Wang MY, Zhou KF, Xu L. 2000. Molecular Biology and Traditional Chinese Medicine Research. Shanghai University of Traditional Chinese Medicine Press, Shanghai.

31. Wang YF. 2001. Water Pollution Control Technology. Chemical Industry Press, Beijing.
32. Wei SZ. 1996. Biochemistry. China Light Industry Press, Beijing.
33. Wu NH. 1998. Principles of Gene Engineering. Science Press, Beijing.
34. Wu QY. 2002. Essentials of Life Science. Higher Education Press, Beijing.
35. Wu WT. 1993. Biopharmaceutical Technology. China Medical Science and Technology Press, Beijing.
36. Xi DL, SunYS, Liu XY. 1999. Environmental Monitoring. Higher Education Press, Beijing.
37. Xiao DG. 2003. Principles of Microbial Engineering. China Light Industry Press, Beijing.
38. Xiong ZG. 2000. Principles of Fermentation Process. China Medical Science and Technology Press, Beijing.
39. Xiong ZG. 1999. Biotechnological Pharmaceutics. Higher Education Press, Beijing.
40. Yan GQ. 2003. Introduction to Life Science and Technology. Science Press, Beijing.
41. Zhai LJ, Gu H, Hu P, *et al.* 1998. Introduction to Contemporary Biotechnology. Higher Education Press, Beijing.
42. Zhai ZH, Wang XZ, Ding MX. 2000. Cell Biology. Higher Education Press, Beijing.
43. Zhang SZ. 1984. Enzyme preparation Industry. Science Press, Beijing.
44. Zhang WJ. 1999. Introduction to Life Science. Higher Education Press, Beijing.
45. Zhang ZP. 2003. Microbial Pharmaceuticals. Chemical Industry Press, Beijing.
46. Zhang ZL. 2004. Advances in Contemporary Life Sciences. Science Press, Beijing.
47. Zhou DQ. 1997. Microbiology Tutorial. Higher Education Press, Beijing.
48. Zhu YX, Li Y. 1997. Contemporary Molecular Biology. Higher Education Press, Beijing.
49. Colin R. 2002. Basic Biotechnology (Photocopy Edition). Science Press, Beijing.
50. Zhong JJ. 2002. Advances in Applied Biotechnology. East China University of Science and Technology Press, Shanghai.